FINITE ELEMENT METHOD FOR ELECTROMAGNETICS

IEEE/OUP SERIES ON ELECTROMAGNETIC WAVE THEORY

The IEEE/OUP Series on Electromagnetic Wave Theory consists of new titles as well as reprintings and revisions of recognized classics that maintain long-term archival significance in electromagnetic waves and applications.

BOOKS IN THE IEEE/OUP SERIES ON ELECTROMAGNETIC WAVE THEORY

Christopoulos, C., *The Transmission-Line Modeling Methods: TLM*
Clemmow, P. C., *The Plane Wave Spectrum Representation of Electromagnetic Fields*
Collin, R. E., *Field Theory of Guided Waves*, Second Edition
Dudley, D. G., *Mathematical Foundations for Electromagnetic Theory*
Elliot, R. S., *Electromagnetics: History, Theory, and Applications*
Felsen, L. B., and Marcuvitz, N., *Radiation and Scattering of Waves*
Harrington, R. F., *Field Computation by Moment Methods*
Jones, D. S., *Methods in Electromagnetic Wave Propagation*, Second Edition
Lindell, I. V., *Methods for Electromagnetic Field Analysis*
Peterson et al., *Computational Methods for Electromagnetics*
Tai, C. T., *Generalized Vector and Dyadic Analysis: Applied Mathematics in Field Theory*
Tai, C. T., *Dyadic Green Functions in Electromagnetic Theory*, Second Edition
Van Bladel, J., *Singular Electromagnetic Fields and Sources*
Wait, J., *Electromagnetic Waves in Stratified Media*

FINITE ELEMENT METHOD FOR ELECTROMAGNETICS

Antennas, Microwave Circuits, and Scattering Applications

IEEE/OUP Series on
Electromagnetic Wave Theory

John L. Volakis
University of Michigan

Arindam Chatterjee
Hewlett-Packard

Leo C. Kempel
Mission Research Corp.

IEEE Antennas & Propagation Society, *Sponsor*

The Institute of Electrical
and Electronics Engineers, Inc.,
New York

Oxford University Press
Oxford, Tokyo,
Melbourne

This book and other books may be purchased at a discount
from the publisher when ordered in bulk quantities. Contact:

IEEE Press Marketing
Attn: Special Sales
445 Hoes Lane, P.O. Box 1331
Piscataway, NJ 08855-1331
Fax: (732) 981-9334

For more information about IEEE PRESS products,
visit the IEEE Home Page: http://www.ieee.org/

Printed in the United States of America

10 9 8 7 6 5 4 3 2 1

ISBN 0-7803-3425-6
IEEE Order Number: PC5698

OUP ISBN 0 19 850479 9

Library of Congress Cataloging-in-Publication Data

Volakis, John Leonidas, 1956–
Finite element method for electromagnetics : with applications to antennas, microwave circuits, and scattering / John L. Volakis, Arindam Chatterjee, Leo C. Kempel.
p. cm.
Includes bibliographical references and index.
ISBN 0-7803-3425-6 (alk. paper)
1. Electromagnetic fields—Mathematical models. 2. Finite element method. 3. Antennas (Electronics) 4. Microwave circuits. 5. Electrons—Scattering. I. Chatterjee, A. (Arindam) II. Kempel, Leo C.
TK7867.2.V65 1998 97-48768
530.14′1—dc21 CIP

To our families

IEEE Press
445 Hoes Lane, P.O. Box 1331
Piscataway, NJ 08855-1331

IEEE Antennas & Propagation Society, *Sponsor*
AP-S Liaison to IEEE Press, Robert Mailloux

Cover design: William T. Donnelly, *WT Design*

Oxford University Press
Walton Street, Oxford OX2 6DP

Oxford New York
Athens Auckland Bangkok Bombay
Calcutta Cape Town Dar es Salaam Delhi
Florence Hong Kong Istanbul Karachi
Kuala Lumpur Madras Madrid Melbourne
Mexico City Nairobi Paris Singapore
Taipei Tokyo Toronto
and associated companies in
Berlin Ibadan

Contents

CHAPTER 6 THREE-DIMENSIONAL PROBLEMS: RADIATION AND SCATTERING 183

CHAPTER 7 THREE-DIMENSIONAL FE-BI METHOD 227

Preface

The finite element method (FEM) and its hybrid versions (finite element-boundary integral, finite element-absorbing boundary condition, finite element-mode matching, etc.) is one of the most successful frequency domain computational methods for electromagnetic simulations. It combines geometrical adaptability and material generality for modeling arbitrary geometries and materials of any composition. The latter is particularly important in electromagnetics since nearly most applications dealing with antennas, microwave circuits, scatterers, motor and generator modeling, etc. require the simulation of nonmetallic/composite materials. Also, the hybridization of the finite element method with integral equation techniques leads to fully rigorous approaches which combine the best aspects of volume and surface formulation techniques.

Because of its unique features, the finite element method is becoming the workhorse for electromagnetic modeling and simulations. Many research and development codes are now available from universities and industry, and these have demonstrated the utility and capability of the method. Also, a number of commercial finite element analysis packages are currently available. Typically, these packages do not yet incorporate the more rigorous hybrid versions of the FEM. However, they are rapidly evolving to more sophisticated and capable packages which incorporate new technologies in geometrical modeling, simulation engines, and solvers.

With the increasing importance of electromagnetics simulation packages using the FEM, this book should serve as a valuable text for students, practicing engineers, and researchers in electromagnetics. The original goal of writing the book was to serve as a text for beginning graduate students interested in the application of the finite element method and its hybrid versions to electromagnetics. However, the authors also recognized a need to report (in a coherent manner) the many recent advances in applying the method(s) to traditional and new problems in electromagnetics. The result is a book that can serve both beginning students and more advanced practitioners. The first half of the book has already been used in the classroom as part of a course on numerical electromagnetics at the University of Michigan. The second half of the book covers primarily work on three-dimensional (3D) developments and applications which have primarily appeared in the literature over the past 5 years.

The book assumes that the reader is a first-year graduate student who has likely taken one advanced course in electromagnetics beyond the standard undergraduate courses. For practicing engineers, it is assumed that the reader is familiar with concepts of electromagnetic radiation and has an understanding of Maxwell's equations and their implications. No previous experience in numerical methods is necessary, but such experience will, of course, help the reader in understanding the procedure of casting analytical equations into discrete systems for numerical solution.

For classroom use, it is expected that the first four chapters will be thoroughly covered with the exception of Chapter 2, which describes a variety of basis/expansion functions. At the introductory stage, only the initial sections of Chapter 2 need be covered. The reader may then return to Chapter 2 as needed. Chapters 3 and 4 [one-dimensional (1D) and two-dimensional (2D) formulations and applications] are written in a step-by-step process with the assumption that this is the first exposure of the reader to numerical methods and the finite element method in general. Chapters 5 through 7 introduce the finite element method and its hybrid versions for 3D simulations with applications to microwave circuits, scattering, and conformal antennas. These chapters are written at a more advanced level and cover the latest applications and successes of the method in electromagnetics. Chapter 5 (closed-domain 3D applications) is a straightfoward extension of the two-dimensional development in Chapter 4 and can be part of a quarter or semester course which includes Chapters 1–5. FEM implementations with absorbing boundary conditions and the finite element-boundary integral method for 3D applications are described in Chapters 6 and 7, respectively. These are realistic practical simulations and should be of particular interest to practicing engineers and researchers in the field. Their 2D counterparts are described in Chapter 4 at a significant level of detail along with explicit formulas for developing computer codes.

Chapter 8 describes some recent developments on the implementation of the boundary integral methods for mesh truncation in conjunction with fast integral methods. Fast integral methods have shown dramatic reductions in CPU and memory. They are currently the subject of research and will impact the utility and development of the finite element-boundary integral method.

Finally, Chapter 9 presents an overview of storage techniques for sparse systems, iterative solvers, preconditioning, parallelization, and a variety of details pertinent to the development of finite element codes. These items were not mixed with the earlier chapters which discuss the mathematics and applications for the FEM. Thus, the reader can refer to Chapter 9 at different stages, and as needed, when developing finite element codes.

J. L. Volakis
A. Chatterjee
L. C. Kempel
June 1997
Ann Arbor, MI

Acknowledgments

Interest in the finite element method (FEM) at the Radiation Laboratory of the University of Michigan began in 1987 by the first author and his graduate students. The motivation was to model large domains without restrictions in geometry and material composition. At that time, two graduate students, Timothy J. Peters and Kasra Barkeshli, had completed successful implementations of boundary integral solutions using k-space methods. This $O(N \log N)$ approach paved the way for a fully $O(N)$ finite element-boundary integral algorithm which combined the rigor of the boundary integral for mesh truncation and the generality of the FEM for volume/domain modeling. The first of these hybrid implementations was developed by Dr. Jian-Ming Jin, a graduate assistant of John Volakis, resulting in a highly successful finite element-boundary integral computer program. Versions of this code are still in use by government, industrial, and academic researchers in the United States. Another graduate student of Professor Volakis, Dr. Jeffery D. Collins, furthered this work to a body-of-revolution with an integral mesh enclosure. His later students—among them the co-authors, Dr. Chatterjee and Dr. Kempel, and Dr. Daniel C. Ross, Dr. Jian Gong, and Dr. Tayfun Özdemir—made significant contributions toward the understanding of 3D problems in antennas, scattering, and microwave circuits.

The authors are indebted to the entire research group of Professor Volakis for graciously helping in the preparation of the manuscript. We would particularly like to mention Hristos Anastassia, Lars Andersen, Youssry Botros, Arik Brown, and Tayfun Özdemir for proofreading and in providing data and figures. The authors are grateful to Dr. Sunil Bindiganavale for co-authoring the section on fast integral methods in Chapter 8. He also helped in proofreading various sections of the manuscript. The acknowledgments would be incomplete without mentioning Mr. Richard Carnes, whose expertise in LATEX made the typesetting of the book a much easier task. Ruby Sowards typed some sections of the book, and Patti Wolfe helped in preparing several figures. The authors are thankful to each of them.

A lot of encouragement was received from several people throughout the project. The authors would like to particularly acknowledge Professor Donald G. Dudley, Series Editor of IEEE Press, who was instrumental in publishing this book with the Press and was supportive throughout the preparation of the manuscript. The comments and constructive criticisms of the early chapters and the final manuscript by Professor Andreas Cangellaris, Professor Jin-Fa Lee, Dr. Daniel T. McGrath, and the anonymous reviewers are very much appreciated. The authors are also thankful to the entire IEEE Press staff (John E. Griffin, Linda C. Matarazzo, Christy Coleman, and Savoula Amanatidis) for their help.

On the personal front, our acknowledgments would not be complete without mentioning the support of our families. John Volakis would like to express gratitude to his wife, Maria, and children, Leonithas and Alexandro, for their patience, sacrifice, and understanding during the preparation of the manuscript. Arindam Chatterjee would like to express deep appreciation for the constant encouragement he received from his parents. He is also grateful to his wife for her insightful criticisms. Leo Kempel would like to express his appreciation for the support and patience provided by his wife, Cathy.

John L. Volakis
Arindam Chatterjee
Leo C. Kempel
August 1997
Ann Arbor, MI

1

Fundamental Concepts

The material presented in this book is generally considered to be at the level of a graduate student or a practising engineer. Thus, the reader is assumed to have been exposed to electromagnetics either through a suitable graduate course or practical experience. In this chapter, fundamental electromagnetic concepts and theorems are presented along with the notation used throughout this text so that a common base is available to all readers.

Many good texts on general electromagnetics principles and techniques are currently in print. Some are considered classical treatises such as [1]–[2] while others are more recent vintage such as [3]–[4]. Although this chapter presents the minimal introductory material necessary for the study of the finite element method as applied to electromagnetics, the interested reader is encouraged to consult these references for a more complete treatment of electromagnetics.

Upon assumption of a harmonic field, the phasor or time-harmonic form of Maxwell's equations are presented along with complex material definitions which permit the incorporation of loss mechanisms. The natural boundary conditions are derived followed by fictitious, though useful, approximate resistive and impedance conditions. Electrostatic and magnetostatic formulations are discussed for use in later examples of the finite element method. Several useful electromagnetic concepts are presented including the Poynting, uniqueness, superposition, and duality theorems. Time-harmonic Maxwell's equations will be covered first.

1.1 TIME-HARMONIC MAXWELL'S EQUATIONS

Maxwell's equations were originally written as a set of coupled, time-dependent integral equations. However, of primary interest in this text is the study of harmonically varying fields (i.e., frequency domain) with an angular frequency of

$\omega = 2\pi f$ rad/sec since the finite element method for electromagnetics utilizes time-harmonic fields. The interested reader is referred to one of the excellent general electromagnetics texts cited in this chapter's introduction for a discussion of the time-dependent form of Maxwell's equations. For our purposes, we begin with the time-harmonic form of Maxwell's equations.

The time-harmonic electric field is related to the time-dependent electric field (assuming that $j = \sqrt{-1}$) by

$$\begin{aligned}\mathcal{E}(x, y, z; t) &= \mathrm{Re}[\mathbf{E}(x, y, z)e^{j\omega t}] \\ &= \hat{x}E_{x0}\cos(\omega t + \phi_x) + \hat{y}E_{y0}\cos(\omega t + \phi_y) + \hat{z}E_{z0}\cos(\omega t + \phi_z)\end{aligned} \tag{1.1}$$

where the complex vector

$$\mathbf{E}(x, y, z) = \hat{x}E_{x0}e^{j\phi_x} + \hat{y}E_{y0}e^{j\phi_y} + \hat{z}E_{z0}e^{j\phi_z} \tag{1.2}$$

is referred to as the field phasor, and similar representations can be employed for the other field quantities. Introducing these into the time-dependent Maxwell's equations [5], we obtain a simplified set

$$\nabla \times \mathbf{H} = \mathbf{J} + j\omega\epsilon\mathbf{E} \tag{1.3}$$

$$\nabla \times \mathbf{E} = -\mathbf{M} - j\omega\mu\mathbf{H} \tag{1.4}$$

$$\nabla \cdot (\mu\mathbf{H}) = \rho_m \tag{1.5}$$

$$\nabla \cdot (\epsilon\mathbf{E}) = \rho \tag{1.6}$$

where the corresponding vector field and current phasors are

$\mathbf{E}$ = electric field intensity in volts/meter (V/m)

$\mathbf{H}$ = magnetic field intensity in amperes/meter (A/m)

$\mathbf{J}$ = electric current density in amperes/meter^2 (A/m^2)

$\mathbf{M}$ = magnetic current density in volts/meter^2 (V/m^2)

and the two scalar charge phasors are

ρ = electric charge density in coulombs/meter^3 (C/m^3)

ρ_m = magnetic charge density in webers/meter^3 (Wb/m^3)

Both the magnetic current density (**M**) and the magnetic charge density (ρ_m) are fictitious quantities introduced for convenience.

Implied in these time-harmonic equations are constitutive relations for an isotropic medium

$$\mathbf{D} = \epsilon\mathbf{E} = \epsilon_0\epsilon_r\mathbf{E} \tag{1.7}$$

$$\mathbf{B} = \mu\mathbf{H} = \mu_0\mu_r\mathbf{H} \tag{1.8}$$

$$\mathbf{J} = \sigma\mathbf{E} \tag{1.9}$$

$$\mathbf{M} = \sigma_m\mathbf{H} \tag{1.10}$$

where two additional phasors

$\mathbf{D}$ = electric flux density in coulombs/meter^2 (C/m^2)

$\mathbf{B}$ = magnetic field density in webers/meter^2 (Wb/m^2)

are related to **E** and **H**, respectively. These constitutive equations are an important link between the original time-dependent form of Maxwell's equations and the time-harmonic form used in the finite element method. Similarly, the phasor forms of the continuity equations [5] are given by

$$\nabla \cdot \mathbf{J} + j\omega\rho = 0 \tag{1.11}$$

$$\nabla \cdot \mathbf{M} + j\omega\rho_m = 0 \tag{1.12}$$

In (1.3)–(1.12), the material constants are given by

ϵ_0 = free space permittivity = 8.854×10^{-12} farads/meter (F/m)

μ_0 = free space permeability = $4\pi \times 10^{-7}$ henrys/meter (H/m)

ϵ_r = medium's relative permittivity constant

μ_r = medium's relative permeability constant

σ = electric current conductivity in mhos/m ($\frac{1}{\Omega}$/m)

σ_m = magnetic current conductivity in ohms/m (Ω/m)

The first two of these (ϵ_0 and μ_0) are fundamental constants while the others describe the specific material. For example, ϵ_r is a measure of the material's electric storage capacity while σ is a measure of the material's ability to conduct electric currents or alternatively as an Ohmic loss mechanism. The relative permeability μ_r and magnetic conductivity σ_m are the magnetic field analogues to ϵ_r and σ, respectively. For the purposes of the finite element method, all four of these material quantities may vary spatially (inhomogeneous) and spectrally (dispersive).

The current densities **J** and **M** appearing in (1.3) and (1.4) do not include the presence of impressed sources. In general, **J** and **M** can be written as a sum of impressed (or excitation) and induced (or conduction) currents as

$$\mathbf{J} = \mathbf{J}_i + \mathbf{J}_c = \mathbf{J}_i + \sigma\mathbf{E} \tag{1.13}$$

$$\mathbf{M} = \mathbf{M}_i + \mathbf{M}_c = \mathbf{M}_i + \sigma_m\mathbf{H} \tag{1.14}$$

where the subscript "i" denotes impressed currents while the subscript "c" refers to conduction currents. When these are substituted into (1.3) and (1.4), the familiar form of Maxwell's equations are obtained

$$\nabla \times \mathbf{H} = \mathbf{J}_i + j\omega\epsilon_0\dot{\epsilon}_r\mathbf{E} \tag{1.15}$$

$$\nabla \times \mathbf{E} = -\mathbf{M}_i - j\omega\mu_0\dot{\mu}_r\mathbf{H} \tag{1.16}$$

where

$$\dot{\epsilon}_r = \epsilon_r - j\frac{\sigma}{\omega\epsilon_0} = \epsilon' - j\epsilon'' = \epsilon'(1 - j\tan\delta) \tag{1.17}$$

and

$$\dot{\mu}_r = \mu_r - j\frac{\sigma_m}{\omega\mu_0} = \mu' - j\mu'' = \mu'(1 - j\tan\delta_m) \tag{1.18}$$

represent equivalent relative complex permittivity and permeability constants. For notational convenience, the dot over the relative constitutive parameters will be omitted with the understanding that these still represent all possible material losses.

Any one of the representations given in (1.17) and (1.18) are likely to be found in the literature with the quantities

$$\tan\delta = \frac{\epsilon''}{\epsilon'} \tag{1.19}$$

$$\tan\delta_m = \frac{\mu''}{\mu'} \tag{1.20}$$

referred to as the material's electric and magnetic loss tangents, respectively.

To summarize, Maxwell's equations in phasor form for isotropic media are

$$\nabla\times\mathbf{H} = \mathbf{J}_i + j\omega\epsilon\mathbf{E} \tag{1.21}$$

$$\nabla\times\mathbf{E} = -\mathbf{M}_i - j\omega\mu\mathbf{H} \tag{1.22}$$

$$\nabla\cdot(\mu\mathbf{H}) = -(\nabla\cdot\mathbf{M}_i)/j\omega \tag{1.23}$$

$$\nabla\cdot(\epsilon\mathbf{E}) = -(\nabla\cdot\mathbf{J}_i)/j\omega \tag{1.24}$$

where the phasor form of (1.11) and (1.12) was employed to rewrite (1.5) and (1.6) as given above.

The corresponding integral representations of (1.21)–(1.24) are

$$\oint_C \mathbf{H}\cdot d\mathbf{l} = \iint_S (\mathbf{J}_i + j\omega\epsilon\mathbf{E})\cdot d\mathbf{S} \tag{1.25}$$

$$\oint_C \mathbf{E}\cdot d\mathbf{l} = -\iint_S (\mathbf{M}_i + j\omega\mu\mathbf{H})\cdot d\mathbf{S} \tag{1.26}$$

$$\oiint_{S_c} \mu\mathbf{H}\cdot d\mathbf{S} = -\iiint_V \frac{\nabla\cdot\mathbf{M}_i}{j\omega}\,dV = \iiint_V \rho_m\,dV \tag{1.27}$$

$$\oiint_{S_c} \epsilon\mathbf{E}\cdot d\mathbf{S} = -\iiint_V \frac{\nabla\cdot\mathbf{J}_i}{j\omega}\,dV = \iiint_V \rho\,dV \tag{1.28}$$

where C is the contour bounding the open surface S illustrated in Fig. 1.1 and $d\mathbf{S} = \hat{n}\,dS$. The circle through the single integral indicates integration over a closed contour, whereas the same symbol through the surface integral denotes integration over the closed surface S_c which encloses the corresponding volume V. The surface S associated with the integrals (1.25) and (1.26) is completely unrelated to S_c which encloses the volume V.

Expressions (1.21) and (1.22) imply six scalar equations for the solution of the six components associated with **E** and **H**. Thus, for time-harmonic fields, (1.21) and

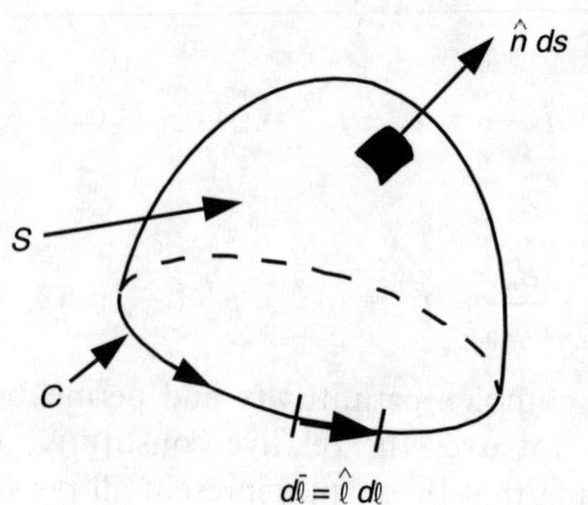

Figure 1.1 Illustration of the differential element ds and the contour C.

(1.22) or (1.25) and (1.26) are sufficient for a solution of the electric and magnetic fields. The divergence conditions (1.23) and (1.24), or their integral counterparts, (1.27) and (1.28), are superfluous. In fact, these two equations follow directly from the first two upon taking their divergence and observing that $\nabla \cdot (\nabla \times \mathbf{A}) = 0$ [6] for any vector $\mathbf{A}$. Equations (1.21)–(1.28) can be easily modified for anisotropic material as well. This requires that $\epsilon\mathbf{E}$ and $\mu\mathbf{H}$ be replaced by $\overline{\overline{\epsilon}} \cdot \mathbf{E}$ and $\overline{\overline{\mu}} \cdot \mathbf{H}$, respectively, where $\overline{\overline{\epsilon}}$ and $\overline{\overline{\mu}}$ represent 3×3 tensors [7].

In this text, open scattering and radiation problems will be considered. Consequently, any valid and unique solution of the electric and/or magnetic fields must also satisfy the Sommerfeld radiation condition, which describes the field behavior at infinity

$$\lim_{r \to \infty} r \left[\nabla \times \begin{Bmatrix} \mathbf{E} \\ \mathbf{H} \end{Bmatrix} + jk_0 \hat{r} \times \begin{Bmatrix} \mathbf{E} \\ \mathbf{H} \end{Bmatrix} \right] = 0 \tag{1.29}$$

where k_0 is the free-space wavenumber ($k_0 = 2\pi/\lambda_0 = \omega\sqrt{\epsilon_0\mu_0}$) and λ_0 is the corresponding free space wavelength. This simply states that the field is outgoing and of the form $e^{-jk_0 r}/r$ as $r \to \infty$.

1.2 WAVE EQUATION

Ampère-Maxwell's Law (1.21) and Faraday's Law (1.22) are independent first-order vector equations, and as noted earlier, they lead to a unique solution subject to the specified boundary conditions. They may be combined together to yield a single second-order vector equation in terms of $\mathbf{E}$ or $\mathbf{H}$ known as the wave equation. The finite element method is used to numerically approximate the solution of the wave equation.

Specifically by taking the curl of (1.21) or (1.22) and making use of the other, the following vector wave equations are obtained:

$$\begin{aligned} \nabla \times \left\{ \frac{\nabla \times \mathbf{E}}{\mu_r} \right\} - k_0^2 \epsilon_r \mathbf{E} &= -jk_0 Z_0 \mathbf{J} - \nabla \times \left\{ \frac{\mathbf{M}}{\mu_r} \right\} \\ \nabla \times \left\{ \frac{\nabla \times \mathbf{H}}{\epsilon_r} \right\} - k_0^2 \mu_r \mathbf{H} &= -jk_0 Y_0 \mathbf{M} + \nabla \times \left\{ \frac{\mathbf{J}}{\epsilon_r} \right\} \end{aligned} \tag{1.30}$$

where the upper set of equations are for solution of the electric field while the lower set is for solution of the magnetic field. In (1.30), ϵ_r denotes the relative permittivity of the media and μ_r indicates the relative permeability of the media. For free space, these two quantities are both unity. Also, $Z_0 = 1/Y_0 = \sqrt{\mu_0/\epsilon_0}$ is the free space wave impedance. In materials other than free space, the wave impedance and wavenumber are given by $Z = 1/Y = \sqrt{\mu/\epsilon}$ and $k = \omega\sqrt{\epsilon\mu}$, respectively.

Utilizing the vector identity

$$\nabla \times (\phi \mathbf{A}) = \nabla\phi \times \mathbf{A} + \phi \nabla \times \mathbf{A} \tag{1.31}$$

(1.30) can be written in a more convenient form

$$\left(\frac{1}{\mu_r}\right) \nabla \times \nabla \times \mathbf{E} - k_0^2 \epsilon_r \mathbf{E} + \left[\nabla \left(\frac{1}{\mu_r}\right) \times \nabla \times \mathbf{E} \right] = -jk_0 Z_0 \mathbf{J} - \nabla \times \left\{ \frac{\mathbf{M}}{\mu_r} \right\} \tag{1.32}$$

for the electric field and

$$\left(\frac{1}{\epsilon_r}\right)\nabla \times \nabla \times \mathbf{H} - k_0^2\mu_r\mathbf{H} + \left[\nabla\left(\frac{1}{\epsilon_r}\right) \times \nabla \times \mathbf{H}\right] = -jk_0 Y_0\mathbf{M} + \nabla \times \left\{\frac{\mathbf{J}}{\epsilon_r}\right\} \tag{1.33}$$

for the magnetic field. The significance of this form of the wave equation is that for homogeneous materials, the terms within the bracket are zero. Most implementations of the finite element method assume a homogeneous material within each finite element and hence this bracketed term can be set to zero for those cases.

Another important version of (1.30) for homogeneous media is obtained by utilizing the vector identity

$$\nabla \times \nabla \times \mathbf{E} = \nabla\nabla \cdot \mathbf{E} - \nabla^2\mathbf{E} \tag{1.34}$$

in (1.32) to get

$$-\nabla(\nabla \cdot \mathbf{E}) + \nabla^2\mathbf{E} + k_0^2\epsilon_r\mu_r\mathbf{E} = jk_0 Z_0\mu_r\mathbf{J} + \nabla \times \mathbf{M} \tag{1.35}$$

In a source-free region, (1.35) simplifies to

$$\nabla^2\mathbf{E} + k^2\mathbf{E} = 0 \tag{1.36}$$

These equations represent three vector field components each of which satisfies the Helmholtz or scalar wave equation

$$\nabla^2\psi + k^2\psi = 0 \tag{1.37}$$

where ψ denotes E_x, E_y, E_z. Similar partial differential equations can be formed for the magnetic field from (1.33).

1.3 ELECTROSTATICS AND MAGNETOSTATICS

Although for the majority of this text we are concerned with dynamic electromagnetics, some examples of static electromagnetics are included to illustrate basic finite element principles. Therefore, we present the basic equations of electrostatics and magnetostatics, namely Poisson's equation and the potential relations.

In this section, we present a basic review of electrostatic and magnetostatic expressions sufficient for this text. It is not a comprehensive review of either electro- or magnetostatics, and the interested reader is encouraged to study [8] or [9] for further information. We begin with electrostatics.

1.3.1 Electrostatics

The fundamental equations of electrostatics are forms of Faraday's equation (1.4) and Gauss' Law (1.6), namely

$$\begin{aligned} \nabla \times \mathbf{E} &= 0 \\ \nabla \cdot (\epsilon\mathbf{E}) &= \rho_v \end{aligned} \tag{1.38}$$

The electric field can be expressed in terms of a gauge condition involving a scalar quantity, $\phi(\mathbf{r})$

$$\mathbf{E} = -\nabla\phi(\mathbf{r}) \tag{1.39}$$

With this field expression (1.39), (1.38) reduces to Poisson's equation (here given in terms of general scalar fields and sources since Poisson's equation is also used for magnetostatics)

$$\nabla \cdot [\epsilon\nabla\phi(\mathbf{r})] = f(\mathbf{r}) \tag{1.40}$$

For electrostatics, the scalar quantity is a voltage ($\phi(\mathbf{r}) = V(\mathbf{r})$) and the sources are volume charges ($f(\mathbf{r}) = -\rho_v(\mathbf{r})$), hence (1.40) reduces to

$$\nabla \cdot [\epsilon\nabla V(\mathbf{r})] = -\rho_v(\mathbf{r}) \tag{1.41}$$

Solution of (1.41), subject to the appropriate boundary conditions (Dirichlet, Neumann, and/or impedance conditions), is equivalent to solving the original Maxwell's equations.

Solution of (1.41) is often accomplished either in closed form or numerically. Closed form solutions are available for only a limited number of boundary conditions [9]. Hence, it is usually more practical to employ a numerical method such as the finite element or boundary element methods.

The potential attributed to a volume charge is given by the integral relation

$$V^i(\mathbf{r}) = \iiint_V \frac{\rho_v(\mathbf{r}')}{\epsilon} G^{3D}(\mathbf{r}, \mathbf{r}')\, dV' \tag{1.42}$$

where the three-dimensional static Green's function is given by

$$G^{3D}(\mathbf{r}, \mathbf{r}') = \frac{1}{4\pi|\mathbf{r} - \mathbf{r}'|} = \frac{1}{4\pi R} \tag{1.43}$$

and volume charges are denoted by ρ_v. This representation is the solution of (1.41) in unbounded space and a pictorial relationship of the primed and unprimed parameters is shown in Fig. 1.2.

For two-dimensional situations (e.g., the sources of excitation run from $z = -\infty$ to $z = \infty$ and are invariant with respect to z), the following potential integral is appropriate

$$V^i(\mathbf{r}) = \iint_S \frac{\rho_s(\mathbf{r}')}{\epsilon} G^{2D}(\mathbf{r}, \mathbf{r}')\, dS' \tag{1.44}$$

where ρ_s denotes surface charges and G^{2D} is the two-dimensional static Green's function

$$G^{2D}(\mathbf{r}, \mathbf{r}') = \frac{1}{2\pi} \ln\left(\frac{1}{|\mathbf{r} - \mathbf{r}'|}\right) \tag{1.45}$$

These integral relations are used to determine the potential, V^i, at some point in space due to an impressed charge distribution. They are used to derive integral equations for the total potential due to an impressed source subject to boundary conditions on surfaces within the domain. Such an integral equation (for surface problems) is given by

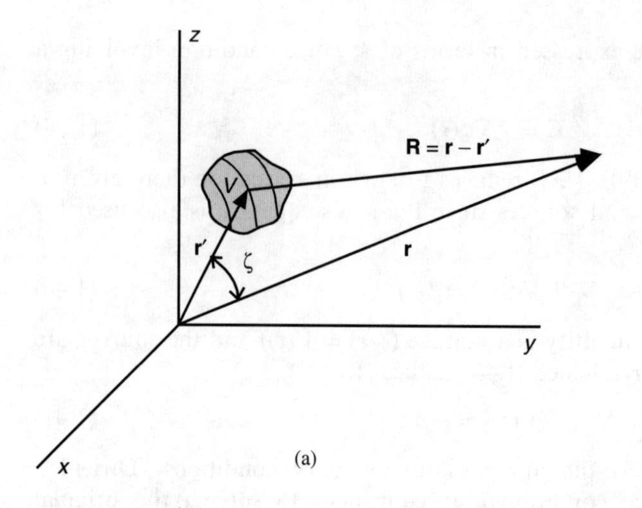

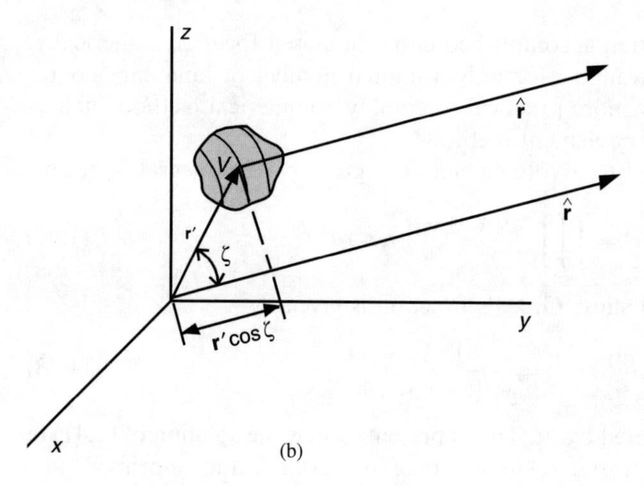

Figure 1.2 Illustration of the geometrical parameters associated with field representations; (a) near zone setup; (b) far zone setup.

$$V(\mathbf{r}) = V^i(\mathbf{r}) - \iint_S \left[G(\mathbf{r}, \mathbf{r}') \frac{\partial V(\mathbf{r}')}{\partial n'} - \frac{\partial G(\mathbf{r}, \mathbf{r}')}{\partial n'} V(\mathbf{r}') \right] dS' \tag{1.46}$$

where $\partial V(\mathbf{r})/\partial n = \hat{n} \cdot \nabla V(\mathbf{r})$, $\partial G(\mathbf{r}, \mathbf{r}')/\partial n' = \hat{n}' \cdot \nabla' G(\mathbf{r}, \mathbf{r}') = -\hat{n}' \cdot \nabla G(\mathbf{r}, \mathbf{r}')$ and $\hat{n}'$ denotes the outward normal to the integration surface S. Note that $\hat{n}' = n(\mathbf{r}')$ implies that the unit normal is a function of the integration variables, where $\hat{n} = \hat{n}(\mathbf{r})$ is a function of the observation variables.

For perfect electric conductors, (1.46) can be rewritten in terms of unknown surface charges. Specifically, by making use of (1.39), and relation $\rho_s/\epsilon = \hat{n} \cdot \mathbf{E} = -\hat{n} \cdot \nabla V(\mathbf{r})$, we obtain the usual expression

$$V(\mathbf{r}) = V^i(\mathbf{r}) + \iint_S \left[G(\mathbf{r}, \mathbf{r}') \frac{\rho_s(\mathbf{r}')}{\epsilon} \right] dS' \tag{1.47}$$

where G represents either (1.43) or (1.45), as appropriate.

1.3.2 Magnetostatics

The solution of Maxwell's equations for a stationary magnetic field is similar to the procedure given above for electrostatics. In this case, Ampère's and Gauss' Magnetic Laws are

$$\begin{aligned} \nabla \times \mathbf{H} &= \mathbf{J} \\ \nabla \cdot \mathbf{B} &= 0 \end{aligned} \tag{1.48}$$

where the static field density is assumed to be related to the magnetic field intensity by the expression

$$\mathbf{B} = \mu \mathbf{H} \tag{1.49}$$

Note that in (1.48), we have not assumed a fictitious magnetic charge density. Rather, the fundamental sources of static magnetic fields are currents, **J**.

We can define a magnetic vector potential in terms of these currents

$$\mathbf{A}(\mathbf{r}) = \mu \iiint_V \mathbf{J}(\mathbf{r}') G(\mathbf{r}, \mathbf{r}')\, dV' \tag{1.50}$$

where the Green's function is the same three-dimensional function used for electrostatics (1.43) or the two-dimensional function (1.45). For the latter case, the integral must be reduced to a two-dimensional one over the domain of **J**. With the introduction of this vector potential, solution of (1.48) with (1.49) yields the expression

$$\mathbf{B} = \nabla \times \mathbf{A} = -\mu \iiint_V \mathbf{J}(\mathbf{r}') \times \nabla G(\mathbf{r}, \mathbf{r}')\, dV' \tag{1.51}$$

The derivation of (1.51) can be found in most introductory electromagnetic texts [2, 3, 5], and similar expressions are possible for two-dimensional currents where the integration is taken over a surface rather than a volume.

Integral equations may be formed for magnetostatics in a similar manner to electrostatics and the interested reader is referred to [9].

1.4 SURFACE EQUIVALENCE

Surface equivalent currents are very useful in the formulation and execution of a numerical solution of Maxwell's equations. Their introduction can be readily justified in the context of the surface equivalence principle, e.g., *two sources that produce the same field within a region are said to be equivalent within that region.*

The surface equivalence principle states that the field exterior (or interior) to a given (possibly fictitious) surface may be exactly represented by equivalent currents placed on that surface and allowed to radiate into the region external (or internal) to that surface. For the exterior case, these equivalent currents are given in terms of the total exterior $(\mathbf{E}, \mathbf{H})$ fields while the interior fields are assumed to be zero (this is Love's equivalence principle). The appropriate currents for representing the fields exterior to the surface are given by

$$\begin{aligned} \hat{n} \times \mathbf{H} &= \mathbf{J} \\ \mathbf{E} \times \hat{n} &= \mathbf{M} \end{aligned} \tag{1.52}$$

For the interior fields, the negative of (1.52) are used. The radiated fields due to these equivalent currents are given by the integral expressions

$$\begin{aligned} \mathbf{E}(\mathbf{r}) = &-\oiint_{S_c} \nabla \times \overline{\overline{G}}(R) \cdot \hat{n}' \times \mathbf{E}(\mathbf{r}')\, dS' \\ &+ jk_0 Z_0 \oiint_{S_c} \overline{\overline{G}}(R) \cdot \hat{n}' \times \mathbf{H}(\mathbf{r}')\, dS' \end{aligned} \tag{1.53}$$

$$\begin{aligned} \mathbf{H}(\mathbf{r}) = &-\oiint_{S_c} \nabla \times \overline{\overline{G}}(R) \cdot \hat{n}' \times \mathbf{H}(\mathbf{r}')\, dS' \\ &- jk_0 Y_0 \oiint_{S_c} \overline{\overline{G}}(R) \cdot \hat{n}' \times \mathbf{E}(\mathbf{r}')\, dS' \end{aligned} \tag{1.54}$$

where $R = |\mathbf{r} - \mathbf{r}'|$, $\mathbf{r}$ and $\mathbf{r}'$ denote the observation and integration point, respectively, and $\hat{n}'$ is the outward directed unit normal at the point $\mathbf{r}'$. The closed surface on which the equivalence theorem is applied is denoted by S_c. These geometrical quantities are illustrated in Figs. 1.2 and 1.3. In (1.53) and (1.54), a dyadic Green's function is required which at least satisfies the radiation condition (1.29). When $\mathbf{J}$ and $\mathbf{M}$ are radiating in free space, the dyadic Green's function is given in closed form by

$$\overline{\overline{G}}_0 = -\left(\overline{\overline{I}} + \frac{\nabla\nabla}{k_0^2}\right) G_0(R) \tag{1.55}$$

where $\overline{\overline{I}} = \hat{x}\hat{x} + \hat{y}\hat{y} + \hat{z}\hat{z}$ is the unit dyad and the corresponding scalar Green's function is given by

$$G_0(R) = G_0(\mathbf{r}, \mathbf{r}') = \frac{e^{-jk_0 R}}{4\pi R} \tag{1.56}$$

Also,

$$\nabla \times \overline{\overline{G}}_0(\mathbf{r}, \mathbf{r}') = -\nabla \times [\overline{\overline{I}} G_0(\mathbf{r}, \mathbf{r}')] = -\nabla G_0(\mathbf{r}, \mathbf{r}') \times \overline{\overline{I}} \tag{1.57}$$

implying $\nabla \times \overline{\overline{G}}_0(\mathbf{r}, \mathbf{r}') \cdot \mathbf{M}(\mathbf{r}') = -\nabla G_0(\mathbf{r}, \mathbf{r}') \times \mathbf{M}(\mathbf{r}')$.

When (1.55)–(1.57) are introduced into (1.53) and (1.54), and after the use of common vector and dyadic identities, we obtain the representations

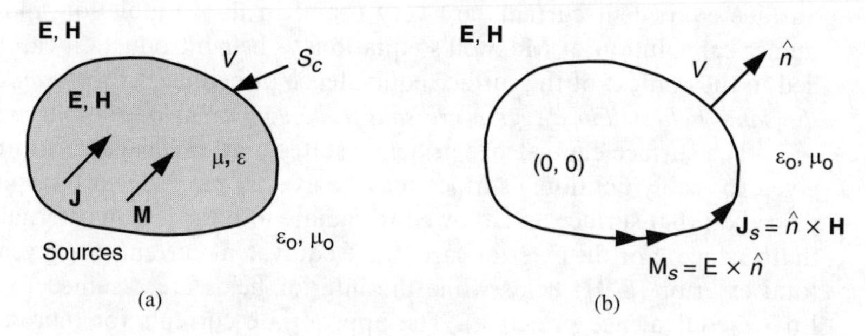

Figure 1.3 Illustration of the application of the surface equivalence principle.

$$\mathbf{E}(\mathbf{r}) = \oiint_{S_c} \Big[\mathbf{M}(\mathbf{r}') \times \nabla G_0(\mathbf{r},\mathbf{r}') - jk_0 Z_0 \mathbf{J}(\mathbf{r}')\, G_0(\mathbf{r},\mathbf{r}') - j\frac{Z_0}{k_0}\, \mathbf{J}(\mathbf{r}') \cdot \nabla\nabla G_0(\mathbf{r},\mathbf{r}')\Big]\, dS' \tag{1.58}$$

$$\mathbf{H}(\mathbf{r}) = \oiint_{S_c} \Big[-\mathbf{J}(\mathbf{r}') \times \nabla G_0(\mathbf{r},\mathbf{r}') - j\frac{k_0}{Z_0}\, \mathbf{M}(\mathbf{r}')\, G_0(\mathbf{r},\mathbf{r}') - j\frac{1}{k_0 Z_0}\, \mathbf{M}(\mathbf{r}') \cdot \nabla\nabla G_0(\mathbf{r},\mathbf{r}')\Big]\, dS' \tag{1.59}$$

More explicit expressions for **E** and **H** can be obtained by introducing the identities

$$\nabla\, G_0(\mathbf{r},\mathbf{r}') = \frac{d\,G(R)}{dR}\,\nabla R = -\left(jk_0 + \frac{1}{R}\right) G_0(R)\,\hat{R}$$

$$\nabla\nabla\, G_0(\mathbf{r},\mathbf{r}') = \hat{R}\hat{R}\left[\frac{1}{R^2} + \left(jk_0 + \frac{1}{R}\right)^2\right] G_0(\mathbf{r},\mathbf{r}') - (\bar{\bar{I}} - \hat{R}\hat{R})\left(jk_0 + \frac{1}{R}\right)\frac{G_0(\mathbf{r},\mathbf{r}')}{R}$$

in which $\hat{R} = (\mathbf{r} - \mathbf{r}')/|\mathbf{r} - \mathbf{r}'|$, as depicted in Fig. 1.2. Specifically, (1.58) becomes

$$\begin{aligned}\mathbf{E} = &-jk_0 \oiint_{S_c} [\mathbf{M}(\mathbf{r}') \times \hat{R}]\left(1 + \frac{1}{jk_0 R}\right) G_0(\mathbf{r},\mathbf{r}')\, dS' \\ &- jk_0 Z_0 \oiint_{S_c} \left\{\left[1 - \frac{j}{k_0 R} - \frac{1}{(k_0 R)^2}\right] \mathbf{J}(\mathbf{r}') \right. \\ &\left. - \left[1 - \frac{3j}{k_0 R} - \frac{3}{(k_0 R)^2}\right][\mathbf{J}(\mathbf{r}') \cdot \hat{R}]\hat{R}\right\} G_0(\mathbf{r},\mathbf{r}')\, dS'\end{aligned} \tag{1.60}$$

and a corresponding expression for **H** can be obtained by invoking duality ($\mathbf{E} \to \mathbf{H}$, $\mathbf{H} \to -\mathbf{E}$, $\mathbf{J} \to \mathbf{M}$, $\mathbf{M} \to -\mathbf{J}$, $\mu \to \epsilon$, $\epsilon \to \mu$, $Z_0 \to Y_0$, $Y_0 \to Z_0$).

We can rewrite (1.58) in a less singular form by noting the identities, $\nabla G = -\nabla' G$,

$$\nabla' \cdot [\mathbf{J}(\mathbf{r}') G_0(\mathbf{r},\mathbf{r}')] = G_0(\mathbf{r},\mathbf{r}') \nabla' \cdot \mathbf{J}(\mathbf{r}') + \mathbf{J}(\mathbf{r}') \cdot \nabla' G_0(\mathbf{r},\mathbf{r}') \tag{1.61}$$

$$\begin{aligned}\nabla[\nabla \cdot \mathbf{J}(\mathbf{r}') G_0(\mathbf{r},\mathbf{r}')] &= \nabla \mathbf{J}(\mathbf{r}') \cdot \nabla G_0(\mathbf{r},\mathbf{r}') + \mathbf{J}(\mathbf{r}') \cdot \nabla\nabla G_0(\mathbf{r},\mathbf{r}') \\ &= \mathbf{J}(\mathbf{r}') \cdot \nabla\nabla G_0(\mathbf{r},\mathbf{r}')\end{aligned} \tag{1.62}$$

to deduce that

$$\begin{aligned}\oiint_{S_c} \mathbf{J}(\mathbf{r}') \cdot \nabla\nabla G_0(\mathbf{r},\mathbf{r}')\, dS' &= -\nabla \oiint_{S_c} \mathbf{J}(\mathbf{r}') \cdot \nabla' G_0(\mathbf{r},\mathbf{r}')\, dS' \\ &= +\oiint_{S_c} \nabla' \cdot \mathbf{J}(\mathbf{r}') \nabla G_0(\mathbf{r},\mathbf{r}')\, dS'\end{aligned}$$

Introducing these into (1.58) yields the expression

$$\mathbf{E} = \oiint_{S_c} \left\{ \mathbf{M}(\mathbf{r}') \times \nabla G_0(\mathbf{r}, \mathbf{r}') - jk_0 Z_0 \mathbf{J}(\mathbf{r}') G_0(\mathbf{r}, \mathbf{r}') - j\frac{Z_0}{k_0} [\nabla' \cdot \mathbf{J}(\mathbf{r}')] \nabla G_0(\mathbf{r}, \mathbf{r}') \right\} dS'$$

which is most commonly used for integral equation numerical solutions and is also valid for open surfaces since the normal components of **J** to the perimeter edges of the surface vanish. The corresponding **H** field expression is again obtained by duality.

For far zone computations ($r \to \infty$), the Green's function (1.56) can be simplified as

$$G_0(\mathbf{r}, \mathbf{r}') = \frac{e^{-jk_0 r}}{4\pi r} e^{jk_0(\mathbf{r}' \cdot \hat{r})} \tag{1.63}$$

Using this in (1.58) and (1.59), carrying out the vector derivative operations, and retaining only the terms that decay[1] as $\mathcal{O}(1/r)$, we get

$$\mathbf{E}(\mathbf{r}) \approx jk_0 \frac{e^{-jk_0 r}}{4\pi r} \oiint_{S_c} [+\hat{r} \times \mathbf{M}(\mathbf{r}') + Z_0 \hat{r} \times (\hat{r} \times \mathbf{J}(\mathbf{r}'))] e^{jk_0(\mathbf{r}' \cdot \hat{r})} dS' \tag{1.64}$$

$$\mathbf{H}(\mathbf{r}) \approx jk_0 \frac{e^{-jk_0 r}}{4\pi r} \oiint_{S_c} [-\hat{r} \times \mathbf{J}(\mathbf{r}') + Y_0 \hat{r} \times (\hat{r} \times \mathbf{M}(\mathbf{r}'))] e^{jk_0(\mathbf{r}' \cdot \hat{r})} dS' \tag{1.65}$$

These are referred to as the far zone field expressions and are typically used for the evaluation of antenna radiated fields or for the calculation of the radar cross section (RCS) of a target. An acceptable criterion for using (1.64) and (1.65) is

$$r \geq \frac{2D^2}{\lambda_0} \tag{1.66}$$

where D is the largest antenna or target dimension. In this case, the phase error in the intervening approximations is maintained at less than $\frac{\pi}{8}$. A typical setup for computing the radiation from volume sources at points in the near and far zones is depicted in Fig. 1.2. Figure 1.4 also shows the spherical angles commonly used in electromagnetics and to be used throughout this text.

Note that use of (1.55) requires both equivalent currents (1.52). Alternatively, when **J** and **M** radiate in the presence of certain bodies such as an infinite metallic plane, a cylinder or a half-plane, a dyadic Green's function can be chosen to satisfy the boundary conditions appropriate for that surface, hence avoiding the need for currents to be placed on that surface. For example, the Dirichlet condition

$$\hat{n} \times \overline{\overline{G}}_1 = 0 \qquad \text{on S} \tag{1.67}$$

can be imposed for relating the electric or magnetic fields in the exterior region to the magnetic or electric currents, respectively. The radiated field expressions, (1.53) and (1.54), now simplify to

[1]The notation $\mathcal{O}(1/r)$ is read as "order of $1/r$."

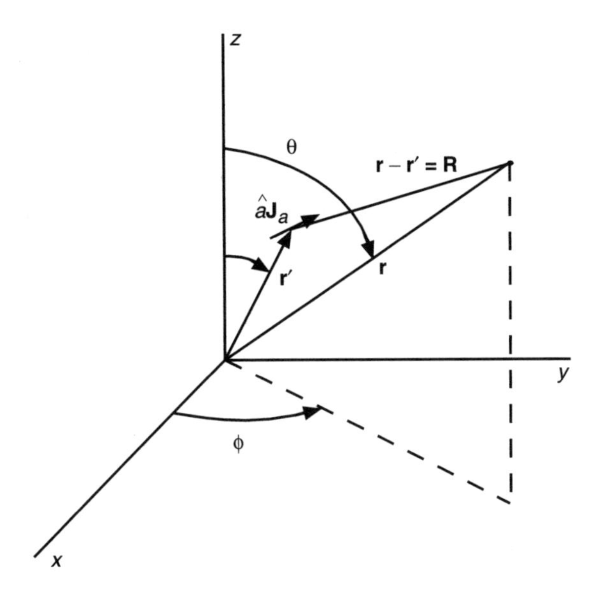

Figure 1.4 Illustration of an infinitesimal source $\mathbf{J} = \hat{a}J_a\, d\ell$ and the associated coordinates and angles.

$$\mathbf{E}(\mathbf{r}) = -\oint\!\!\oint_{S_c} \nabla \times \overline{\overline{G}}_1(R) \cdot \hat{n}' \times \mathbf{E}(\mathbf{r}')\, dS' \tag{1.68}$$

$$\mathbf{H}(\mathbf{r}) = -\oint\!\!\oint_{S_c} \nabla \times \overline{\overline{G}}_1(R) \cdot \hat{n}' \times \mathbf{H}(\mathbf{r}')\, dS' \tag{1.69}$$

implying that only a single current is required for the representation of each field quantity. One use of this Green's function is in the calculation of the scattering by a perfect magnetic conductor (a fictitious material) through the use of equivalent electric currents.

If the Neumann boundary condition is imposed

$$\hat{n} \times \nabla \times \overline{\overline{G}}_2 = 0 \qquad \text{on S} \tag{1.70}$$

the resulting field integral expressions are again given in terms of only one current

$$\mathbf{E}(\mathbf{r}) = +jk_0 Z_0 \oint\!\!\oint_{S_c} \overline{\overline{G}}_2(R) \cdot \hat{n}' \times \mathbf{H}(\mathbf{r}')\, dS' \tag{1.71}$$

$$\mathbf{H}(\mathbf{r}) = -jk_0 Y_0 \oint\!\!\oint_{S_c} \overline{\overline{G}}_2(R) \cdot \hat{n}' \times \mathbf{E}(\mathbf{r}')\, dS' \tag{1.72}$$

This Green's function ($\overline{\overline{G}}_2$) is useful for calculating the scattering by a perfect electric conductor using electric currents. Also, this Green's function will be used to relate the electric field quantities of an interior finite element formulation to the magnetic field of the bounding surface.

The above dyadic Green's functions, $\overline{\overline{G}}_1$ and $\overline{\overline{G}}_2$, are commonly termed the first and the second kind dyadic Green's functions, respectively. A good discussion of dyadic Green's functions used in electromagnetics is given in [10].

The above expressions are for three-dimensional fields. In the case of two-dimensional fields (e.g., one dimension, such as z, is invariant), similar expressions are used. These are scalar and typically written in terms of TM_z and TE_z polarizations. For TM_z ($H_z = E_x = E_y = 0$), the electric field on or outside a contour C due to fields on that surface is given by

$$E_z = \oint_C E_z(\mathbf{r}')\left[\frac{\partial}{\partial n'} G^{2D}(\mathbf{r}, \mathbf{r}')\right] dl' - jk_0 Z_0 \oint_C H_t(\mathbf{r}')[G^{2D}(\mathbf{r}, \mathbf{r}')]\, dl' \tag{1.73}$$

since $\partial E_z/\partial n = +jk_0 Z_0 H_t$. The corresponding expression for TE_z ($E_z = H_x = H_y = 0$) polarization is

$$H_z = \oint_C H_z(\mathbf{r}')\left[\frac{\partial}{\partial n'} G^{2D}(\mathbf{r}, \mathbf{r}')\right] dl' + jk_0 Y_0 \oint_C E_t(\mathbf{r}')[G^{2D}(\mathbf{r}, \mathbf{r}')]\, dl' \tag{1.74}$$

In (1.73) and (1.74), the subscript "t" denotes the tangential component of the field along the unit vector $\hat{t} = \hat{\ell}$, where $\hat{n} \times \hat{t} = \hat{z}$. Also, the two-dimensional Green's function is given by

$$G^{2D}(\mathbf{r}, \mathbf{r}') = -\frac{j}{4} H_0^{(2)}(k_0|\mathbf{r} - \mathbf{r}'|) \tag{1.75}$$

In (1.75), $H_0^{(2)}(\cdot)$ denotes the zeroth-order Hankel function of the second kind. We observe that (1.73) and (1.74) can be reduced from (1.46) by replacing the potential V with E_z or H_z and making use of the equivalent current relations (1.52) and Maxwell's equations to relate $\mathbf{E}$ and $\mathbf{H}$. For the special cases of the TM_z and TE_z polarizations, the first two of Maxwell's equations imply

$$\mathbf{H} = -\frac{jY_0}{k_0} \hat{z} \times \nabla E_z, \qquad \mathbf{E} = \frac{jZ_0}{k_0} \hat{z} \times \nabla H_z \tag{1.76}$$

All of the expressions presented in this section are given for currents radiating in free space. Fields within a homogeneous media can be determined by replacing k_0 with k and Z_0 with Z in all of these expressions. Throughout this text, k will denote the wavenumber in the homogeneous media ($k = \sqrt{\epsilon_r \mu_r}\, k_0$) and Z is the intrinsic impedance of that material ($Z = \sqrt{\mu/\epsilon}$).

1.5 NATURAL BOUNDARY CONDITIONS

Maxwell's equations cannot be solved without the specification of the required boundary conditions at material interfaces. The pertinent boundary conditions can be derived directly from the integral form of Maxwell's equations. Specifically, (1.25) is applied to the contour illustrated in Fig. 1.5(a) with S being the area enclosed by C. Assuming $\Delta\ell$ is small, $\Delta h \to 0$ and $\Delta\ell \gg \Delta h$, (1.25) gives

$$(\mathbf{H}_1 - \mathbf{H}_2) \cdot \hat{t} = [\mathbf{J}_i \cdot (\hat{n}_1 \times \hat{t})]\, \Delta h \tag{1.77}$$

and in deriving this we set

$$\lim_{\Delta h \to 0} \tfrac{1}{2} \Delta h[\epsilon_1 \mathbf{E}_1 + \epsilon_2 \mathbf{E}_2] \cdot \hat{n}_1 = 0$$

which is valid provided $\epsilon\mathbf{E}$ is finite at the interface. When (1.26) is applied to the same contour in Fig. 1.5(a) we find that

$$(\mathbf{E}_1 - \mathbf{E}_2) \cdot \hat{t} = -[\mathbf{M}_i \cdot (\hat{n}_1 \times \hat{t})]\, \Delta h \tag{1.78}$$

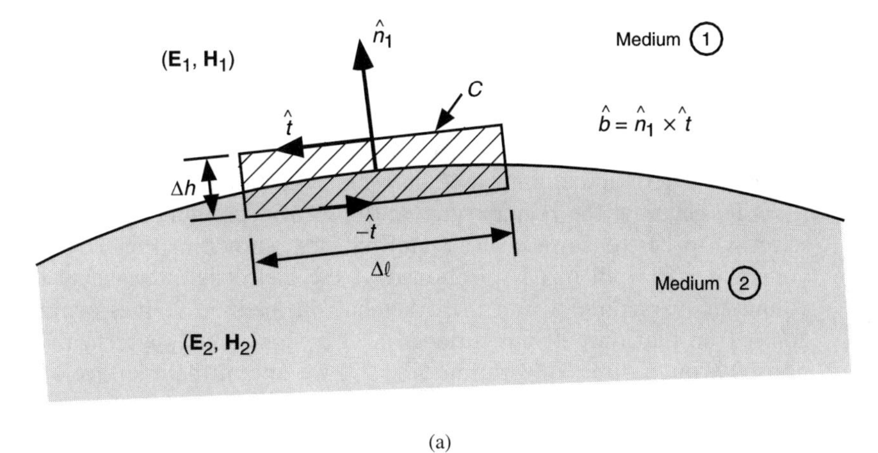

(a)

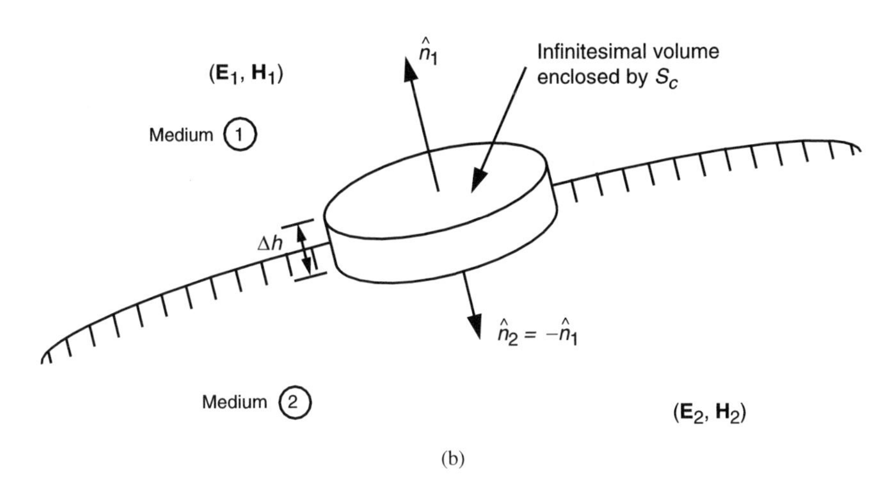

(b)

Figure 1.5 Geometries for deriving the boundary conditions (a) for tangential components, and (b) for normal components.

where we have again set

$$\lim_{\Delta h \to 0} \tfrac{1}{2} \Delta h [\mu_1 \mathbf{H}_1 + \mu_2 \mathbf{H}_2] = 0$$

implying that $\mu \mathbf{H}$ is finite at the dielectric interface.

The conditions (1.77) and (1.78) can be rewritten in vector form and more compactly by introducing the definitions

$$\mathbf{J}_{is} = \mathbf{J}_i \Delta h \tag{1.79}$$

$$\mathbf{M}_{is} = \mathbf{M}_i \Delta h \tag{1.80}$$

giving the conditions

$$\hat{n}_1 \times (\mathbf{H}_1 - \mathbf{H}_2) = \mathbf{J}_{is} \tag{1.81}$$

$$\hat{n}_1 \times (\mathbf{E}_1 - \mathbf{E}_2) = -\mathbf{M}_{is} \tag{1.82}$$

The quantities $\mathbf{J}_{is}$ and $\mathbf{M}_{is}$ are referred to as the impressed electric and magnetic surface current densities in A/m and V/m, respectively, at the interface. Note that if $\mathbf{E}_2$ and $\mathbf{H}_2$ are zero, these conditions are identical to (1.52) except that in this case $\mathbf{J}_{is}$ and $\mathbf{M}_{is}$ refer to actual impressed currents rather than equivalent currents.

To generate the boundary conditions corresponding to (1.27) and (1.28), we select S_c to be the surface of a small pill box, shown in Fig. 1.5(b), enclosing the volume V. The pill box is positioned at the dielectric interface so that half of its volume is in medium 1 and the other half in medium 2. It is again assumed that $\Delta h \to 0$ so that only its flat surfaces need be considered in performing the integrations. Through direct integration of (1.27) we obtain the interface conditions

$$\hat{n}_1 \cdot (\mu_1 \mathbf{H}_1 - \mu_2 \mathbf{H}_2) = \rho_{ms} \tag{1.83}$$

$$\hat{n}_1 \cdot (\epsilon_1 \mathbf{E}_1 - \epsilon_2 \mathbf{E}_2) = \rho_s \tag{1.84}$$

where ρ_s denotes the unbounded electric surface charge density in C/m^2 at the interface and ρ_{ms} is the corresponding fictitious surface magnetic charge density in Wb/m^2.

The boundary/interface conditions (1.81)–(1.84), although derived for time-harmonic fields, are applicable for instantaneous fields as well. In the time-harmonic case, only (1.81) and (1.82) are required in conjunction with (1.23) and (1.24) for a unique solution of the fields.

If we ignore the fictitious magnetic currents and charges appearing in (1.81)–(1.84), the boundary conditions are

$$\hat{n}_1 \times (\mathbf{H}_1 - \mathbf{H}_2) = \mathbf{J}_{is} \tag{1.85}$$

$$\hat{n}_1 \times (\mathbf{E}_1 - \mathbf{E}_2) = 0 \tag{1.86}$$

$$\hat{n}_1 \cdot (\mu_1 \mathbf{H}_1 - \mu_2 \mathbf{H}_2) = 0 \tag{1.87}$$

$$\hat{n}_1 \cdot (\epsilon_1 \mathbf{E}_1 - \epsilon_2 \mathbf{E}_2) = \rho_s \tag{1.88}$$

The first two of these state that the tangential electric fields are continuous across the interface whereas the tangential magnetic fields are discontinuous at the same location by an amount equal to the impressed electric current. Unless a source (i.e., free charge) is actually placed at the interface, $\mathbf{J}_{is}$ is also zero and in that case, the tangential magnetic fields will be continuous across the media as well.

When medium 2 is a perfect electric conductor then $\mathbf{E}_2 = \mathbf{H}_2 = 0$. In addition, M_{is} and ρ_{ms} vanish and (1.81)–(1.84) reduce to

$$\hat{n}_1 \times \mathbf{H}_1 = \mathbf{J}_{is} \tag{1.89}$$

$$\hat{n}_1 \times \mathbf{E}_1 = 0 \tag{1.90}$$

$$\hat{n}_1 \cdot (\mu_1 \mathbf{H}_1) = 0 \tag{1.91}$$

$$\hat{n}_1 \cdot (\epsilon_1 \mathbf{E}_1) = \rho_s \tag{1.92}$$

The first two of these now imply that the tangential electric field vanishes on the surface of the perfect electric conductor whereas the tangential magnetic field is equal to the impressed electric surface current on the conductor.

1.6 APPROXIMATE BOUNDARY CONDITIONS

In the previous section, the boundary conditions which must be imposed at the interface of different dielectrics were presented. Sometimes, it is difficult to utilize these conditions since excessive computational cost is required or the resulting formulation is numerically unstable such as the case of a thin dielectric sheet. In many cases, much simpler approximate boundary conditions that account for the presence of an inhomogeneous medium, coated metallic surface, or a thin dielectric layer can be employed to simulate the actual surface. Below we discuss two types of such approximate conditions: impedance boundary and sheet transition conditions. The interested reader is directed to [11] for a general treatment of approximate conditions.

1.6.1 Impedance Boundary Conditions

The most widely applied approximate conditions are referred to as the Standard Impedance Boundary Conditions (SIBC) or Leontovich Boundary conditions. It is derived by considering a plane wave impinging upon a material half-space. Consider a material-air interface which corresponds to the $y = 0$ plane. The SIBC takes the form

$$E_z = -\eta Z_0 H_x, \qquad E_x = \eta Z_0 H_z \tag{1.93}$$

where the free space impedance is given by $Z_0 = \sqrt{\mu_0/\epsilon_0}$ and the normalized material characteristic impedance is η. An important concept to understand is that these conditions are applied slightly above the interface (assuming a plane wave originating in the upper half-space) at $y = 0^+$. Combining these two conditions, the vector form of the SIBC is given by

$$\hat{n} \times (\hat{n} \times \mathbf{E}) = -\eta Z_0 \hat{n} \times \mathbf{H} \tag{1.94}$$

where the outward directed unit normal, $\hat{n}$, is shown in Fig. 1.6. This form of the SIBC is not restricted to a particular interface [as is the case with (1.93)] and is commonly applied to convex surfaces such as a sphere, cylinder, etc.

All of the quantities used in (1.94) are familiar and well defined except for the normalized impedance, η. One means of deriving this quantity is to demand that the reflected field attributed to (1.94) is identical to that due to the natural boundary conditions. Then,

$$\eta = \sqrt{\frac{\mu_r}{\epsilon_r}} \tag{1.95}$$

This is exact for an infinite planar interface while it is approximate for a curved boundary provided that

$$|\mathrm{Im}\,(\sqrt{\epsilon_r \mu_r})| k_0 r_i \gg 1 \tag{1.96}$$

where $\mathrm{Im}\,(\cdot)$ denotes the imaginary part of the complex argument and the principle radii of curvature, r_i, is associated with the surface at a point. This condition assures that the material is sufficiently lossy so that the fields which penetrate into the material does not re-emerge at some other point.

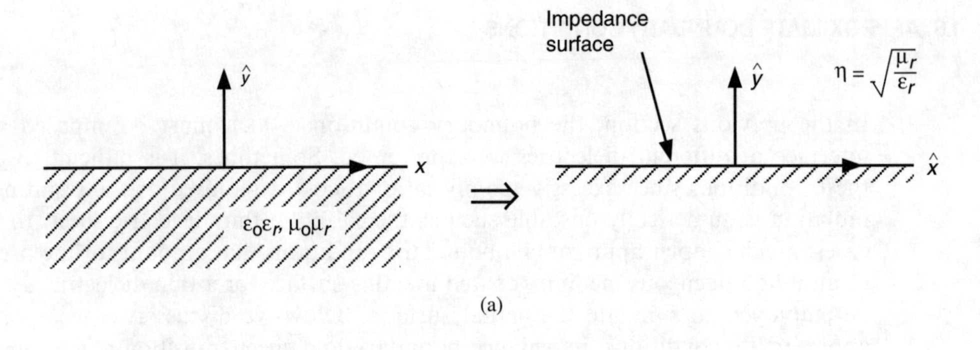

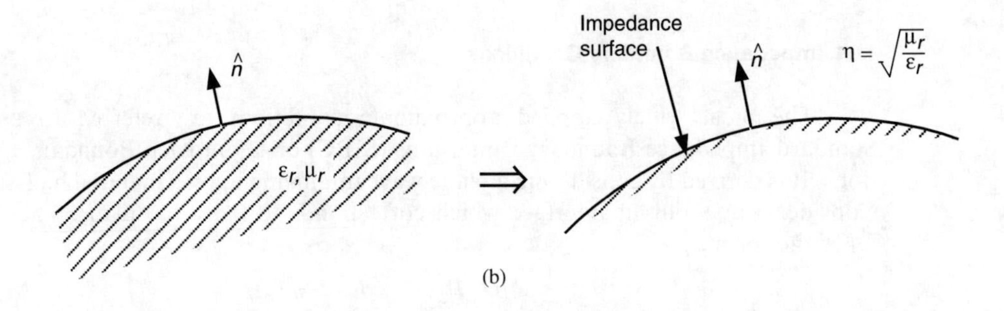

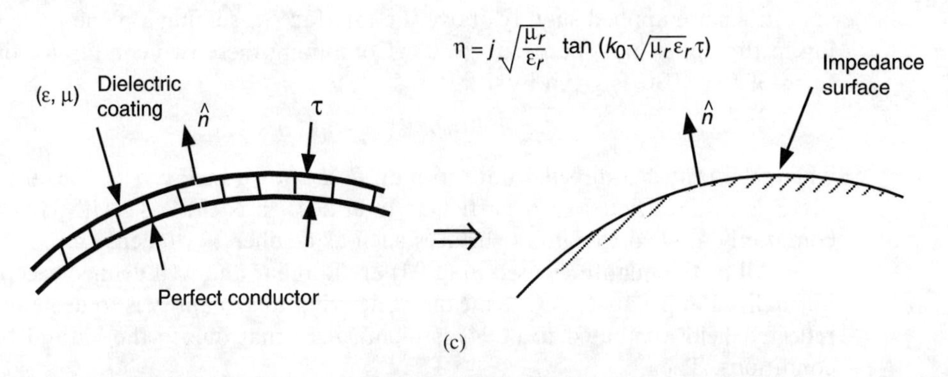

Figure 1.6 Simulation of dielectric boundaries and coatings with SIBCs.

For a coated conductor, the choice of η typically is found by considering a shorted transmission line model with length corresponding to the coating thickness, τ:

$$\eta = j\sqrt{\frac{\mu_r}{\epsilon_r}}\tan(k_0\sqrt{\epsilon_r\mu_r}\,\tau) \tag{1.97}$$

However, this condition is derived at normal incidence and deteriorates at oblique angles with increasing inaccuracy for thicker coatings.

The SIBCs can be applied for modeling surfaces whose material properties vary slowly in the transverse plane. For a planar interface, the coating can have a varying composition in the normal dimension, and Rytov [12] found the following impedance

$$\eta = \sqrt{\frac{\mu_r}{\epsilon_r}} \left\{ 1 + \frac{1}{j2k_0 N} \frac{\partial}{\partial y} \ln (Z_0 N) + \mathcal{O}(N^{-2}) \right\} \tag{1.98}$$

is useful where $N = \sqrt{\mu_r \epsilon_r}$ is the index of refraction and the normal derivative is applied at the surface.

More accurate approximate conditions can be developed by incorporating higher order derivatives in their constructions. These are referred to as *Generalized Impedance Boundary Conditions* (*GIBCs*), and these are discussed in [11].

1.6.2 Sheet Transition Conditions

A thin dielectric layer may be replaced with an equivalent sheet model to simplify the analysis. Consider a thin dielectric layer with thickness τ, as shown in Fig. 1.7. This layer has conductivity σ and will support a volume current density

$$\mathbf{J} = \sigma \mathbf{E} \tag{1.99}$$

where $\mathbf{E}$ is the electric field within the layer. For $\tau \ll \lambda$, this volume current may be replaced with an equivalent sheet current (having units of A/m)

$$\mathbf{J}_s = \tau \mathbf{J} \tag{1.100}$$

and from (1.99)

$$\mathbf{E} = \frac{\mathbf{J}_s}{\tau\sigma} = Z_0 R_e \mathbf{J}_s \tag{1.101}$$

This condition is referred to as a *resistive* sheet transition condition which only supports a single surface current, $\mathbf{J}_s$. The parameter R_e is the normalized *resistivity* of the sheet and is measured in Ohms per square.

The electric field was assumed to be tangential to the sheet in the above derivation. A more general expression for the resistive sheet condition is given by

$$\hat{n} \times (\hat{n} \times \mathbf{E}) = -Z_0 R_e \mathbf{J}_s \tag{1.102}$$

Furthermore, it is desirable to utilize fields just outside or inside the sheet surface. Since the tangential electric field is continuous across the sheet

$$\begin{aligned} \hat{n} \times [\hat{n} \times (\mathbf{E}^+ + \mathbf{E}^-)] &= -2Z_0 R_e \mathbf{J}_s \\ \hat{n} \times (\mathbf{E}^+ - \mathbf{E}^-) &= 0 \end{aligned} \tag{1.103}$$

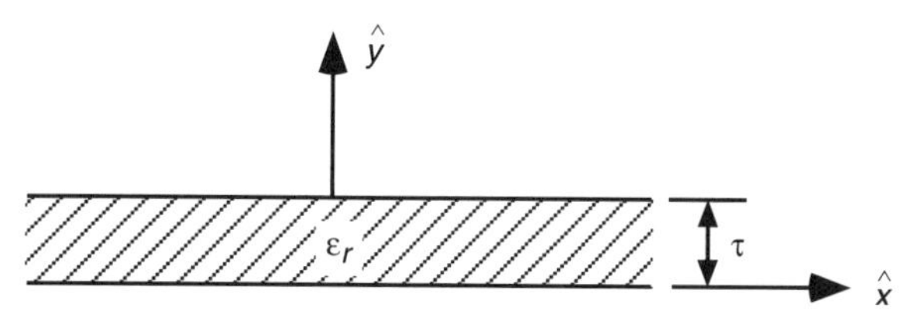

Figure 1.7 Simulation of thin dielectric layer with a sheet condition.

The superscripts $\pm$ denote the fields just above and below the sheet, and the introduction of the second equation in (1.103) is necessary to maintain equivalency with (1.102). The natural boundary conditions may be used to rewrite (1.103) as

$$\begin{aligned} \hat{n} \times [\hat{n} \times (\mathbf{E}^{+} + \mathbf{E}^{-})] &= -2Z_0 R_e \hat{n} \times (\mathbf{H}^{+} - \mathbf{H}^{-}) \\ \hat{n} \times (\mathbf{E}^{+} - \mathbf{E}^{-}) &= 0 \end{aligned} \tag{1.104}$$

As long as the loss in the layer is sufficient to assure that no multiple field penetrations will occur, these resistive transition conditions may be used for curved layers.

The dual to the resistive sheet condition is the *conductive* sheet condition which supports only a surface *magnetic* current. It is given by

$$\begin{aligned} \hat{n} \times [\hat{n} \times (\mathbf{H}^{+} + \mathbf{H}^{-})] &= 2Y_0 R_m \hat{n} \times (\mathbf{E}^{+} - \mathbf{E}^{-}) \\ \hat{n} \times (\mathbf{H}^{+} - \mathbf{H}^{-}) &= 0 \end{aligned} \tag{1.105}$$

The normalized *conductivity* of this sheet is denoted by R_m with units Mhos per square. This condition is required for the simulation of materials which have nontrivial permeability. Also, a special combination of coincident resistive and conductive sheets with respective resistivity and conductivity

$$R_e = \frac{\eta}{2}, \qquad R_m = \frac{1}{2\eta} \tag{1.106}$$

yields the same result as an impedance sheet with impedance η and (1.106) implies that $4R_e R_m = 1$. This set of sheets are useful in simplifying the analysis of a planar impedance sheet since coplanar resistive and conductive sheets are uncoupled.

The resistivity of a dielectric layer can be determined by considering the equivalent polarization current

$$\mathbf{J} = jk_0 Y_0 (\epsilon_r - 1)\mathbf{E} \tag{1.107}$$

and (1.100). It follows that the tangential components of the field are given by

$$E_t = Z_0 R_e J_{st} \tag{1.108}$$

with

$$R_e = \frac{-j}{k_0 \tau (\epsilon_r - 1)} \tag{1.109}$$

A dual conductive sheet is given by

$$R_m = \frac{-j}{k_0 \tau (\mu_r - 1)} \tag{1.110}$$

More accurate representations can be formed using *Generalized Sheet Transition Conditions* (*GSTCs*) [11] which incorporate higher order derivatives in their construction.

1.7 POYNTING'S THEOREM

The quantity ("*" indicates complex conjugation)

$$\mathbf{S} = \tfrac{1}{2}\mathbf{E} \times \mathbf{H}^{*} \tag{1.111}$$

is known as the complex Poynting vector and has units of Watts/m^2. It represents the complex power density of the wave, and it is therefore important to understand the source and nature of this power. To do so, we refer to (1.21) and (1.22), where by dotting each equation with $\mathbf{E}$ and $\mathbf{H}^*$, we have

$$\mathbf{E}\cdot\nabla\times\mathbf{H}^* = \mathbf{J}_i^*\cdot\mathbf{E} - j\omega\epsilon^*\mathbf{E}^*\cdot\mathbf{E} = \mathbf{J}_i^*\cdot\mathbf{E} - j\omega\epsilon^*|\mathbf{E}|^2 \tag{1.112}$$

$$\mathbf{H}^*\cdot\nabla\times\mathbf{E} = -\mathbf{M}_i\cdot\mathbf{H}^* - j\omega\mu\mathbf{H}\cdot\mathbf{H}^* = -\mathbf{M}^i\cdot\mathbf{H}^* - j\omega\mu|\mathbf{H}|^2 \tag{1.113}$$

From the vector identity [6]

$$\nabla\cdot(\mathbf{E}\times\mathbf{H}^*) = \mathbf{H}^*\cdot\nabla\times\mathbf{E} - \mathbf{E}\cdot\nabla\times\mathbf{H}^* \tag{1.114}$$

we then obtain

$$\nabla\cdot(\mathbf{E}\times\mathbf{H}^*) = j\omega\epsilon^*|\mathbf{E}|^2 - j\omega\mu|\mathbf{H}|^2 - \mathbf{J}_i^*\cdot\mathbf{E} - \mathbf{M}_i\cdot\mathbf{H}^* \tag{1.115}$$

which is an identity valid everywhere in space. Integrating both sides of this over a volume V containing all sources, and invoking the divergence theorem yields

$$\frac{1}{2}\oiint_{S_c}(\mathbf{E}\times\mathbf{H}^*)\cdot d\mathbf{s} = \frac{1}{2}\iiint_V [j\omega\epsilon^*|\mathbf{E}|^2 - j\omega\mu|\mathbf{H}|^2 - \mathbf{J}_i^*\cdot\mathbf{E} - \mathbf{M}_i\cdot\mathbf{H}^*]\,dv \tag{1.116}$$

which is commonly referred to as Poynting's theorem. Since S_c is closed, based on energy conservation one deduces that the right hand side of (1.116) must represent the sum of the power stored or radiated, i.e., escaping, out of the volume V. Each term of the volume integral of (1.116) is associated with a specific type of power but before proceeding with their identification, it is instructive that ϵ^* be first replaced by $\epsilon_0\epsilon_r + j\frac{\sigma}{w}$. Equation (1.116) can then be rewritten as

$$\frac{1}{2}\,\mathrm{Re}\oiint_{S_c}(\mathbf{E}\times\mathbf{H}^*)\cdot d\mathbf{s} = P_{ei} + P_{mi} - P_d \tag{1.117}$$

$$\frac{1}{2}\,\mathrm{Im}\oiint_{S_c}(\mathbf{E}\times\mathbf{H}^*)\cdot d\mathbf{s} = 2\omega[W_e - W_m] - \frac{1}{2}\,\mathrm{Im}\iiint_V [\mathbf{J}_i^*\cdot\mathbf{E} + \mathbf{M}_i\cdot\mathbf{H}^*]\,dv \tag{1.118}$$

where

$$P_{ei} = -\frac{1}{2}\iiint_V \mathrm{Re}(\mathbf{J}_i^*\cdot\mathbf{E})\,dv = \text{averaging outgoing power due to the impressed current } \mathbf{J} \tag{1.119}$$

$$P_{mi} = -\frac{1}{2}\iiint_V \mathrm{Re}(\mathbf{M}_i\cdot\mathbf{H}^*)\,dv = \text{average outgoing power due to the impressed current } \mathbf{M}_i \tag{1.120}$$

$$P_d = \frac{1}{2}\iiint_V \sigma|\mathbf{E}|^2\,dv = \text{average power dissipated in } V \tag{1.121}$$

$$W_e = \frac{1}{4}\iiint_V \epsilon_0\epsilon_r|\mathbf{E}|^2\,dv = \text{average electric energy stored in } V \tag{1.122}$$

$$W_m = \frac{1}{4}\iiint_V \mu_0\mu_r|\mathbf{H}|^2\,dv = \text{average magnetic energy stored in } V \tag{1.123}$$

The time-averaged power delivered to the electromagnetic field outside V is clearly the sum of P_{ei} and P_{mi}, whereas P_d is that dissipated in V due to conductor losses. Thus, we may consider

$$P_{\rm av} = \frac{1}{2}\,\mathrm{Re}\oiint_{S_c} (\mathbf{E} \times \mathbf{H}^*) \cdot d\mathbf{s} \tag{1.124}$$

to be the average or radiated power outside V if σ is zero in V. Expression (1.118) gives the reactive power, i.e., that which is stored within V and is not allowed to escape outside the boundary of S_c.

1.8 UNIQUENESS THEOREM

Whenever one pursues a solution to a set of equations it is important to know a priori whether this solution is unique and if not, what are the required conditions for a unique solution. This is important because depending on the application, different analytical or numerical methods will likely be used for the solution of Maxwell's equations. Given that Maxwell's equations (subject to the appropriate boundary conditions) yield a unique solution, one is then comforted to know that any convenient method of analysis will yield the correct solution to the problem.

The most common form of the uniqueness theorem is: *In a region V completely occupied with dissipative media, a harmonic field* $(\mathbf{E}, \mathbf{H})$ *is uniquely determined by the impressed currents in that region plus the tangential components of the electric or magnetic fields on the closed surface S_c bounding V.* This theorem may be proved by assuming for the moment that two solutions exist, denoted by $(\mathbf{E}_1, \mathbf{H}_1)$ and $(\mathbf{E}_2, \mathbf{H}_2)$. Both fields must, of course, satisfy Maxwell's equations (1.21) and (1.22) with the same impressed currents $(\mathbf{J}_i, \mathbf{M}_i)$. We have

$$\begin{aligned} \nabla \times \mathbf{H}_1 &= \mathbf{J}_i + j\omega\epsilon\mathbf{E}_1, & \nabla \times \mathbf{H}_2 &= \mathbf{J}_i + j\omega\epsilon\mathbf{E}_2 \\ \nabla \times \mathbf{E}_1 &= -\mathbf{M}_i - j\omega\mu\mathbf{H}_1, & \nabla \times \mathbf{E}_2 &= -\mathbf{M}_i - j\omega\mu\mathbf{H}_2 \end{aligned} \tag{1.125}$$

and when these are subtracted we obtain

$$\nabla \times \mathbf{H}' = j\omega\epsilon\mathbf{E}' \tag{1.126}$$

$$\nabla \times \mathbf{E}' = -j\omega\mu\mathbf{H}' \tag{1.127}$$

where $\mathbf{E}' = \mathbf{E}_1 - \mathbf{E}_2$ and $\mathbf{H}' = \mathbf{H}_1 - \mathbf{H}_2$. To prove the theorem it is then necessary to show that $(\mathbf{E}', \mathbf{H}')$ are zero or equivalently, if no sources are enclosed by a volume V, the fields in that volume are zero for a given set of tangential electric and magnetic fields on S_c.

As a corollary to the uniqueness theorem, it can be shown that *if a harmonic field has a zero tangential electric or magnetic field on a surface enclosing a source-free region V occupied by dissipative media, the field vanishes everywhere within V.*

The usual proof of the uniqueness theorem can be found in many electromagnetics texts (see, for example, [5]).

1.9 SUPERPOSITION THEOREM

The superposition theorem states that for a linear medium, the total field intensity due to two or more sources is equal to the sum of the field intensities attributed to each individual source radiating independent of the others. In particular, let us consider two electric sources $\mathbf{J}_1$ and $\mathbf{J}_2$. On the basis of the superposition theorem, to find the total field caused by the simultaneous presence of both sources, we can consider the field due to each individual source in isolation. The fields $(\mathbf{E}_1, \mathbf{H}_1)$ due to $\mathbf{J}_1$ satisfy the equations

$$\nabla \times \mathbf{H}_1 = \mathbf{J}_1 + j\omega\epsilon\mathbf{E}_1 \tag{1.128}$$

$$\nabla \times \mathbf{E}_1 = -j\omega\mu\mathbf{H}_1 \tag{1.129}$$

and the fields corresponding to $\mathbf{J}_2$ satisfy

$$\nabla \times \mathbf{H}_2 = \mathbf{J}_2 + j\omega\epsilon\mathbf{E}_2 \tag{1.130}$$

$$\nabla \times \mathbf{E}_2 = -j\omega\mu\mathbf{H}_2 \tag{1.131}$$

By adding these two sets of equations, it is clear that the total field due to both sources combined is given by

$$\mathbf{E} = \mathbf{E}_1 + \mathbf{E}_2, \qquad \mathbf{H} = \mathbf{H}_1 + \mathbf{H}_2 \tag{1.132}$$

where $(\mathbf{E}_1, \mathbf{H}_1)$ and $(\mathbf{E}_2, \mathbf{H}_2)$ are obtained by solving separately (1.128)–(1.129) and (1.130)–(1.131), respectively.

1.10 DUALITY THEOREM

The duality theorem relates to the interchangeability of the electric and magnetic fields, currents, charges, or material properties. We observe from (1.3) and (1.4) that the first can be obtained from the second via the interchanges

$$\begin{aligned} \mathbf{M} &\rightarrow -\mathbf{J} \\ \mathbf{E} &\rightarrow \mathbf{H} \\ \mathbf{H} &\rightarrow -\mathbf{E} \\ \mu &\rightarrow \epsilon \end{aligned} \tag{1.133}$$

Similarly, (1.4) can be obtained from (1.3) via the interchanges

$$\begin{aligned} \mathbf{J} &\rightarrow \mathbf{M} \\ \mathbf{E} &\rightarrow \mathbf{H} \\ \mathbf{H} &\rightarrow -\mathbf{E} \\ \epsilon &\rightarrow \mu \end{aligned} \tag{1.134}$$

The duality theorem can reduce formulation and computational effort when one is able to invoke it for a particular application.

1.11 NUMERICAL TECHNIQUES

For numerical solutions, all governing equations can be written in operator form as

$$\mathcal{L}u - f = 0 \tag{1.135}$$

subject to appropriate boundary or transition conditions

$$B(u) = 0 \tag{1.136}$$

within the domain (Ω) and on its boundary ($S_c = \partial\Omega$). In these, the operator $\mathcal{L}$ is based on one of the following: an integral representation of the fields such as (1.52)–(1.53), on the vector wave equation (1.30), or the Helmholtz equation (1.36) for scalar fields. It is understood that u must be replaced by a vector field $\mathbf{u}$ when dealing with the vector wave equations (1.30) or (1.35). The forcing function f is a known excitation function while u or $\mathbf{u}$ is the unknown quantity. Throughout this text, u or $\mathbf{u}$ will denote a field or current density.

Unfortunately, very few analytical solutions for (1.135) are available in electromagnetics. One such solution, the fields due to a magnetic dipole in the presence of an infinite metallic plane or cylinder, will be used in Chapter 7 to form the appropriate dyadic Green's function for those geometries. However, most useful electromagnetic scattering and radiation problems cannot be solved using analytical methods. Rather, an approximate numerical solution is sought which in some way closely resembles the exact solution. Two methods of formulating such an approximate solution are: the Ritz method and the method of weighted residuals.

1.11.1 The Ritz Method

The Ritz or Rayleigh–Ritz method[2] [13, pp. 74–78], [14, pp. 13–63] seeks a stationary point of a variational functional. For operators which are *self-adjoint* and *positive-definite* (see later subsection for definitions), the stationary point of the following functional

$$F(\tilde{u}) = \tfrac{1}{2}\langle \mathcal{L}\tilde{u}, \tilde{u}\rangle - \langle f, \tilde{u}\rangle \tag{1.137}$$

is an approximate solution of (1.135). In (1.137), the *inner product* over the domain Ω (volume, surface, or contour) of the two functions is defined as

$$\langle a, b\rangle = \int_\Omega ab\, d\Omega \tag{1.138}$$

or for vector functions

$$\langle \mathbf{a}, \mathbf{b}\rangle = \int_\Omega \mathbf{a}\cdot\mathbf{b}\, d\Omega \tag{1.39}$$

The choice of this inner product extends the validity of the variational expressions to vectorial fields. When the operator $\mathcal{L}\tilde{u}$ and f in (1.137) are chosen as

[2]The method was originally introduced by Rayleigh in 1877 and was extended by Ritz in 1909.

$$\mathcal{L}\mathbf{u} = \nabla \times \left\{ \frac{\nabla \times \mathbf{u}}{\mu_r} \right\} - k_0^2 \epsilon_r \mathbf{u} \tag{1.140}$$

$$\mathbf{f} = -jk_0 Z_0 \mathbf{J} - \nabla \times \left(\frac{\mathbf{M}}{\mu_r} \right) \tag{1.141}$$

it can be shown that setting the first variation of $F(\mathbf{u})$ to zero is equivalent to satisfying the vector wave equation (1.30) over the computational domain Ω. Similarly, when

$$\mathcal{L}u = \nabla^2 u + k_0^2 u \tag{1.142}$$

setting the first variation of $F(u)$ to zero is equivalent to satisfying the inhomogeneous Helmholtz wave equation

$$\nabla^2 u + k_0^2 u = f \tag{1.143}$$

To show that setting the first variation of $F(u)$ to zero is equivalent to satisfying the Helmholtz equation (1.143), we begin by rewriting the functional (we assume a two-dimensional domain)

$$\begin{aligned} F(u) &= \tfrac{1}{2} \langle \nabla^2 u + k_0^2 u, u \rangle - \langle f, u \rangle \\ &= \frac{1}{2} \int\!\!\int_\Omega [\nabla^2 u + k_0^2 u] u \, d\Omega - \int\!\!\int_\Omega f u \, d\Omega \end{aligned} \tag{1.144}$$

as

$$F(u) = \frac{1}{2} \int\!\!\int_\Omega (-\nabla u \cdot \nabla u + k_0^2 u^2) \, d\Omega + \frac{1}{2} \oint\!\!\oint_C u \frac{\partial u}{\partial n} \, dl - \int\!\!\int_\Omega f u \, d\Omega \tag{1.145}$$

in which $\nabla u \cdot \hat{n} = \partial u / \partial n$, C denotes the contour enclosing the region Ω (see Fig. 1.8) and $\hat{n}$ is the unit normal vector to C. Note that in deriving (1.145) we used the identity

$$\psi(\nabla \cdot \nabla u) = -\nabla u \cdot \nabla \psi + \nabla \cdot (\psi \nabla u) \tag{1.146}$$

and the divergence theorem

$$\int\!\!\int_\Omega \nabla \cdot (\nabla u) \, d\Omega = \oint_C \nabla u \cdot \hat{n} \, ds \tag{1.147}$$

Next we proceed to evaluate the first variation of $F(u)$ given by

$$\delta F = F(u + \Delta u) - F(u) \tag{1.148}$$

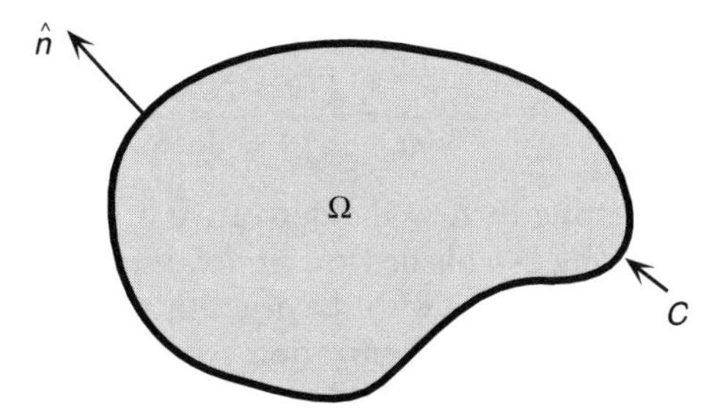

Figure 1.8 Illustration of the region Ω and the enclosing contour C.

where $\Delta \to 0$ is a scalar quantity. The evaluation of δF involves the quantities

$$(u + \Delta u)^2 = u^2 + 2(\Delta u)u + (\Delta u)^2 \tag{1.149}$$

$$[\nabla u + \nabla(\Delta u)] \cdot [\nabla u + \nabla(\Delta u)] = \nabla u \cdot \nabla u + 2\nabla(\Delta u) \cdot \nabla u + \nabla(\Delta u) \cdot \nabla(\Delta u) \tag{1.150}$$

$$(u + \Delta u) \frac{\partial}{\partial n}(u + \Delta u) = u\frac{\partial u}{\partial n} + u\frac{\partial(\Delta u)}{\partial n} + \Delta u\frac{\partial u}{\partial n} + \Delta u\frac{\partial(\Delta u)}{\partial n} \tag{1.151}$$

These can be simplified by neglecting the last term of each expansion which is of order Δ^2. Doing so yields the approximation

$$\begin{aligned} F(u + \Delta u) \approx\ & \frac{1}{2}\iint_\Omega [-\nabla u \cdot \nabla u + k_0^2 u^2]\, d\Omega - \iint_\Omega fu\, d\Omega \\ & + \frac{1}{2}\oint u\frac{\partial u}{\partial n}\, dl \\ & + \Delta\iint_\Omega [-u\nabla u \cdot \nabla u + k_0^2 u^2]\, d\Omega \\ & + \frac{\Delta}{2}\oint_C \left[u\frac{\partial u}{\partial n} + u\frac{\partial u}{\partial n}\right] dl - \Delta\iint_\Omega fu\, d\Omega \end{aligned} \tag{1.152}$$

When this expression is compared with that for $F(u)$ in (1.145), we have

$$\begin{aligned} F(u + \Delta u) \approx F(u) + \Delta & \iint_\Omega u[\nabla \cdot \nabla u + k_0^2 u - f]\, d\Omega \\ & - \Delta\oint_C u\frac{\partial u}{\partial n}\, dl + \frac{\Delta}{2}\oint_C \left[u\frac{\partial u}{\partial n} + u\frac{\partial u}{\partial n}\right] dl \end{aligned} \tag{1.153}$$

where we also used the divergence theorem (1.147) and the identity (1.146) to obtain the second and third terms. Clearly, the last two terms in (1.153) cancel each other leading to

$$\begin{aligned} \delta F &= F(u + \Delta u) - F(u) \\ &= \Delta\iint_\Omega u[\nabla^2 u + k_0^2 u - f]\, d\Omega \end{aligned} \tag{1.154}$$

Thus, setting $\delta F = 0$ implies that $\nabla^2 u + k_0^2 u = f$ provided Δu is nonzero. That is, from (1.154) we conclude that the extremization of F obtained by setting

$$\delta F = 0 \quad \text{or} \quad \frac{\partial F}{\partial u} = 0$$

is equivalent to enforcing the Helmholtz wave equation over the domain Ω. In practice, the condition $\delta F = 0$ is enforced by setting

$$\frac{\partial F}{\partial u} = \left.\frac{F(u + \Delta u) - F(u)}{\Delta u}\right|_{\Delta \to 0} = \left.\frac{\delta F}{\Delta u}\right|_{\Delta \to 0} = 0 \tag{1.155}$$

i.e., by setting to zero the derivative of the functional with respect to u.

Having established the equivalence between $\delta F = 0$ and the Helmholtz equation we can proceed with the discretization of $F(u)$ and δF to obtain a discrete system of equations. The discretization begins with the trial function, $\tilde{u}$, expanded in terms of N basis functions

$$\tilde{u} = \sum_{j=1}^{N} u_j w_j = \{u\}^T \{w\} \tag{1.156}$$

where w_j are the basis functions and u_j are the unknown expansion coefficients. In (1.156), column data vectors are denoted with $\{\cdot\}$ while row data vectors involve a transposition $\{\cdot\}^T$. Substituting (1.156) into (1.137), the functional becomes

$$F(\tilde{u}) = \tfrac{1}{2}\{u\}^T \left[\iint_\Omega \{w\}\mathcal{L}\{w\}^T \, d\Omega \right] \{u\} - \{u\}^T \iint_\Omega \{w\} f \, d\Omega \tag{1.157}$$

where we used the innerproduct definition

$$\langle \{u\}, \{v\} \rangle = \{u\}^T \{v\}$$

for discrete data vectors. This functional is extremized by allowing all partial derivatives with respect to the coefficients, $\{u\}$, to vanish

$$\frac{\partial}{\partial u_i} F(\tilde{u}) = \frac{1}{2} \left[\iint_\Omega w_i \mathcal{L}\{w\}^T \, d\Omega \right] \{u\} + \tfrac{1}{2}\{u\}^T \iint_\Omega \{w\} \mathcal{L} w_i \, d\Omega - \iint_\Omega w_i f \, d\Omega = 0 \tag{1.158}$$

A single equation is obtained by differentiating with respect to each u_i. For $i = 1, 2, \ldots, N$ we obtain N equations which can be written as a matrix system

$$[A]\{u\} = \{b\} \tag{1.159}$$

The elements of the matrix $[A]$ and excitation vector $\{b\}$ are given by

$$\begin{aligned} A_{ij} &= \iint_\Omega w_i \mathcal{L} w_j \, d\Omega \\ b_i &= \iint_\Omega w_i f \, d\Omega \end{aligned} \tag{1.160}$$

A word of caution: electromagnetics differs from other branches of engineering in that no physical significance can be attached to the stationary point of the functional (1.137). In mechanical systems, for example, minimizing this functional represents minimization of the total potential energy of the system. However, since electromagnetics involves complex quantities, such a statement may not be asserted.

1.11.2 Functionals for Anisotropic Media

In three dimensions with anisotropic media [15, 16] the appropriate operator is of the form

$$\mathcal{L}(\mathbf{u}) = \begin{cases} \nabla \times (\overline{\overline{\mu}}_r^{-1} \cdot \nabla \times \mathbf{u}) - k_0^2 \overline{\overline{\epsilon}}_r \cdot \mathbf{u}, & \mathbf{u} = \mathbf{E} \\ \nabla \times (\overline{\overline{\epsilon}}_r^{-1} \cdot \nabla \times \mathbf{u}) - k_0^2 \overline{\overline{\mu}}_r \cdot \mathbf{u}, & \mathbf{u} = \mathbf{H} \end{cases} \tag{1.161}$$

with the corresponding source function given by

$$\mathbf{f} = \begin{cases} -j\omega\mu_0 \mathbf{J} - \nabla \times (\overline{\overline{\mu}}_r^{-1} \cdot \mathbf{M}), & \mathbf{u} = \mathbf{E} \\ -j\omega\epsilon_0 \mathbf{M} + \nabla \times (\overline{\overline{\epsilon}}_r^{-1} \cdot \mathbf{J}), & \mathbf{u} = \mathbf{H} \end{cases} \tag{1.162}$$

The associated functional to be extremized is then of the form (for $\mathbf{u} = \mathbf{H}$)

$$F(\mathbf{H}) = \frac{1}{2}\int_{\Omega} [\nabla \times \mathbf{H} \cdot \overline{\overline{\epsilon}}_r^{-1} \cdot \nabla \times \mathbf{H} - k_0^2 \mathbf{H} \cdot \overline{\overline{\mu}}_r \cdot \mathbf{H}]\, dV - \int_{\Omega} \mathbf{H} \cdot \mathbf{f}\, dV \tag{1.163}$$

where

$$\overline{\overline{\epsilon}} = \begin{pmatrix} \epsilon_{xx} & \epsilon_{xy} & \epsilon_{xz} \\ \epsilon_{yx} & \epsilon_{yy} & \epsilon_{yz} \\ \epsilon_{zx} & \epsilon_{zy} & \epsilon_{zz} \end{pmatrix}, \qquad \overline{\overline{\mu}}_r = \begin{pmatrix} \mu_{xx} & \mu_{xy} & \mu_{xz} \\ \mu_{yx} & \mu_{yy} & \mu_{yz} \\ \mu_{zx} & \mu_{zy} & \mu_{zz} \end{pmatrix} \tag{1.164}$$

are the permittivity and permeability tensors of the media and Ω represents a volume. In general, for arbitrary anisotropy, this functional will lead to an asymmetric (non-Hermitian) system. One way to obtain a symmetric system is to use the functional

$$F(\mathbf{u}) = \langle \mathcal{L}\mathbf{u}, \mathbf{u}_a \rangle - \langle \mathbf{u}, \mathbf{f}_a \rangle - \langle \mathbf{u}_a, \mathbf{f} \rangle \tag{1.165}$$

where $\mathbf{u}_a$ and $\mathbf{f}_a$ satisfy the partial differential equation

$$\mathcal{L}_a \mathbf{u}_a = f_a \tag{1.166}$$

in which $\mathcal{L}_a$ is the adjoint operator to $\mathcal{L}$. That is,

$$\langle \mathcal{L}\mathbf{u}, \mathbf{v} \rangle = \langle \mathbf{v}, \mathcal{L}_a \mathbf{u} \rangle \tag{1.167}$$

1.11.3 Method of Weighted Residuals

The method of weighted residuals [17], [18] begins with the residual

$$\mathcal{R} = \mathcal{L}\tilde{u} - f \tag{1.168}$$

and seeks a solution for $\tilde{u} = u$ by satisfying the condition $\mathcal{R} = 0$ within the domain Ω. In general, such a solution cannot be achieved at all points in Ω. Instead, it is more practical to find a solution which satisfies the residual condition in some average or weighted sense over N subdomains of Ω, viz.

$$\int_{\Omega} \left[t_i \mathcal{L}\{w\}^T \{u\} - t_i f \right] d\Omega = 0, \qquad i = 1, 2, 3, \ldots, N \tag{1.169}$$

In general any testing function, t_i (also referred to as trial or weighting functions), may be used; however, since these functions modify the enforcement of the boundary conditions throughout the domain, the choice of testing functions affects the quality of the solution (1.169). One popular testing procedure is called *collocation* or *point matching*. In this, the testing or weighting function is a Dirac delta function, $t_i = \delta(x - x_i)$, which implies enforcement of the boundary conditions only at discrete points (e.g., x_i, $i = 1, 2, 3, \ldots, N$). Another popular choice is termed *Galerkin's procedure*. When employing the Galerkin's testing procedure, the testing function is identical to the expansion function used in (1.156), e.g., $t_i = w_i$ and the weighted residual equation is given by

$$\left[\int_{\Omega} w_i \mathcal{L} w_j \, d\Omega \right] \{u\} = \int_{\Omega} w_i f \, d\Omega \tag{1.170}$$

which is identical to the Ritz procedure given above. Thus, Galerkin's method leads to the same linear system (1.159) as the Ritz method.

As a generalization, when $F(u)$ is chosen as

$$F(u) = \frac{1}{2}\langle \mathcal{L}u, u^*\rangle - \langle f, u^*\rangle \tag{1.171}$$

where the "*" indicates complex conjugation, the extremization of $F(u)$ leads to a linear system that is identical to that obtained from the weighted residual method with $t_i = w_i^*$.

1.11.4 Vector and Matrix Norms in Linear Space

A norm is a real valued function that provides a measure of the size or "length" of a multicomponent mathematical quantity such as a vector or a matrix. It is used in numerical analysis to provide a measure of how well a given vector approximates the exact solution. For matrices, norms provide a single value to quantify the "size" of the matrix $[A]$. They are often used in evaluating the numerical system's condition, which in turn affects the stability of the solution. That is, of interest is how a small change in the excitation or right hand side of the matrix system (1.159) affects the solution data vector.

1.11.4.1 Vector Norms

EUCLIDEAN NORM. The most popular form for a given discrete data vector $\{u\} = \{u_1, u_2, u_3, \ldots, u_N\}^T$ is the *Euclidean* norm. It is defined by[3]

$$\|\mathbf{u}\|_2 = \sqrt{\sum_{i=1}^{N}(u_i)^2} = \langle\{u\}, \{u\}\rangle = \{u\}^T\{u\} \tag{1.172}$$

for real valued $\{u\}$ and by

$$\|\mathbf{u}\|_2 = \sqrt{\sum_{i=1}^{N}|u_i|^2} = \langle\{u\}, \{u\}\rangle = \{u\}^T\{u^*\} \tag{1.173}$$

for complex vector $\{u\}$. Here $|u_i|$ implies the absolute value of the quantity. Throughout the book, the notation $\|u\|$ will imply the Euclidean norm of a vector or data column unless otherwise noted.

INFINITY NORM. The infinity norm of a data vector is defined by

$$\|\mathbf{u}\|_\infty = \text{max of } |u_i| \qquad 1 \le i \le N \tag{1.174}$$

This norm is also referred to as the *uniform vector* norm or *maximum magnitude* norm.

HÖLDER NORM. The *Hölder* or *p-norm* is a generalization of the Euclidean norm and is defined as

$$\|\mathbf{u}_p\| = \left\{\sum_{i=1}^{N}|u_i|^p\right\}^{1/p} \tag{1.175}$$

where $|u_i|^p$ denotes the pth power of the quantity $|u_i|$.

[3]Here $\mathbf{u}$ is a simpler notation for $\{u\}$.

1.11.4.2 Matrix Norms

FROBENIUS NORM. The *Frobenius* matrix norm is a generalization of the Euclidean vector norm. For a square matrix [A], it is given by

$$\|\mathcal{A}\|_F = \sqrt{\sum_{i=1}^{N}\sum_{j=1}^{N} |A_{ij}|^2} \tag{1.176}$$

where A_{ij} denotes the (i, j) entry of the [A] matrix.[4] Note that (1.176) can be generalized to nonsquare matrices and to the *p*-matrix norms.

MATRIX INFINITY NORM. The *infinity* or *uniform* matrix norm is defined by

$$\|\mathcal{A}\|_\infty = \max_{1\leq j\leq N} \text{ of } \left(\sum_{i=1}^{N} |A_{ij}|\right) \tag{1.177}$$

This specific norm is also referred to as the *row-sum* norm. Similarly the *column-sum* matrix norm is given by

$$\|\mathcal{A}\|_\infty = \max_{1\leq i\leq N} \text{ of } \left(\sum_{j=1}^{N} |A_{ij}|\right) \tag{1.178}$$

The infinity norm is referred to as the *natural* norm of [A]. It can be shown that

$$\|\mathcal{A}\|_\infty = \max_{\|\mathbf{u}\|=1}(\|[A]\{u\}\|) = \max_{1\leq i\leq N} \text{ of } \left(\sum_{j=1}^{N} |A_{ij}|\right) \tag{1.179}$$

For Hermitian matrices (1.178) is identical to (1.177).

MATRIX CONDITION NUMBER. The condition of a matrix is related to the natural norm of [A] as

$$\text{Cond}(\mathcal{A}) = \|\mathcal{A}\|\|\mathcal{A}^{-1}\| \geq \frac{|\lambda_{\max}|}{|\lambda_{\min}|} \tag{1.180}$$

where $\|\mathcal{A}^{-1}\|$ refers to the natural norm of the inverse of [A] (the Frobenius norm is not a natural norm). Here $\lambda_{\max}$ and $\lambda_{\min}$ denote the maximum and minimum eigenvalues of the matrix, respectively. Since $|\lambda_{\max}|$ is a lower bound for the natural norm of [A], the ratio $|\lambda_{\max}|/|\lambda_{\min}|$ gives a conservative estimate for the condition of [A].

As an example of the importance of Cond($\mathcal{A}$), let us assume that due to truncation or arithmetic errors, [A] is instead approximated by $[A_\Delta]$. A computer with *t* decimal digits of accuracy gives

$$\frac{\|\mathcal{A} - \mathcal{A}_\Delta\|}{\|\mathcal{A}\|} = 10^{-t}$$

[4]Here we use the notation $\|\mathcal{A}\| = \|[A]\|$, i.e., the calligraphic capital letter implies a matrix. Such a notation will be used later in Chapter 9.

which is a measure of the normalized error in approximating $[A]$. If the condition of the matrix is $c = \text{Cond}(\mathcal{A})$, the corresponding (normalized) error in the computed solution $\{u\}$ will then be [19, 20]

$$\frac{\|\mathbf{u} - \mathbf{u}_\Delta\|}{\|\mathbf{u}\|} = 10^{-s}$$

where

$$s \geq t - \log_{10} c \tag{1.181}$$

That is, s is always smaller than t, implying a larger error for the final solution vector. More specifically, an error in the seventh decimal place for the norm of $[A]$ (i.e., $t = 7$) translates to an error in the third decimal place for (the norm of) the solution vector when the matrix condition number is 10^4. Alternatively, if $\text{Cond}(\mathcal{A}) = 1$, the solution vector error is of the same order (i.e., to the same decimal place) as the matrix error itself.

1.11.5 Some Matrix Definitions

In Table 1.1 we give the mathematical and descriptive definitions of matrices often encountered in numerical analysis.

Additional definitions can be found in several numerical mathematics books (e.g., see [19] and [20]). As can be understood, the operators which generated the matrices in Table 1.1 also carry the same definition. That is, an operator is referred to as Hermitian or self-adjoint if the resulting matrix is also Hermitian.

TABLE 1.1 Definitions of Matrices Often Encountered in Numerical Analysis

Mathematical Statement	Descriptive Statement
$[A]^T = [A]$	Symmetric
$[A] = [A^*]^T$	Hermitian (self-adjoint)
$[A] = -[A^*]^T$	Skew Hermitian
$[A]^{-1} = [A^*]^T$	Unitary
$[A^T][A] = [I]$ $[I]$ = identity matrix	Orthogonal
$[A^*]^T[A] = [A][A^*]^T$	Normal
$A_{ij} > 0$	Positive-definite
$A_{ij} \geq 0$	Nonnegative or positive semidefinite
$A_{ij} = 0$ for $i \neq j$	Diagonal
$A_{ij} = 0$ for $i > j$	Upper triangular
$A_{ij} = 0$ for $i \geq j$	Strictly upper triangular
$A_{ij} > \sum_{\substack{j=1 \\ i \neq j}}^{N} \lvert A_{ij} \rvert$ for all i	Diagonally dominant
$\{u\}^T[A]\{u\} > 0, \{u\} \neq 0$	Positive-definite
$\{u\}^T[A]\{u\} \geq 0, \{u\} \neq 0$	Nonnegative
$(\{u\}^T[A]\{u\})(\{v\}^T[A]\{v\}) < 0$ $\{u\} \neq 0, \{v\} \neq 0$	Indefinite

TABLE 1.2 Some Common Relationships Between Operators and Eigenvalues

Operator Type	Eigenvalue (λ) Properties
Hermitian	Real eigenvalues and >0
Unitary	Eigenvalues on unit circle
Skew Hermitian	Eigenvalues on imaginary axis
Positive semidefinite	Eigenvalues ≥ 0
Positive-definite	Eigenvalues >0
Indefinite	Some eigenvalues are >0 and some are <0

Given the nature of the matrix or operator, we can immediately make a statement about the eigenvalues of that operator. Some of the most common relationships between operators and eigenvalues are given in Table 1.2.

1.11.6 Comparison of Solution Methods and Their Convergence

The Rayleigh–Ritz and Galerkin's methods are standard solution approaches for solving differential equations arising in practical engineering problems. Both methods project a continuous space onto a finite separable Hilbert space.[5] The mathematical problem is then rephrased to seek a discrete solution set whose entries are the coefficients of the expansion. The premise of each method is summarized in Table 1.3. The third entry in the table is referred to as the *Least Squares Method* and is generally more robust than the other two, as will be noted later.

It was shown in Section 1.11.3 that Galerkin's method is equivalent to the Rayleigh–Ritz approach when the first variation is set to zero. At the heart of this statement is the assumption that the operator $\mathcal{L}$ (or the resultant discrete matrix $\mathcal{A}$) is positive-definite. Unfortunately, in most practical problems in electromagnetics, particularly as k_0 becomes larger, the operators $\mathcal{L}u = \nabla u^2 - k_0^2 u$ and $\mathcal{L}\mathbf{u} = \nabla \times (\mu_r^{-1} \nabla \times \mathbf{u}) - k_0^2 \epsilon_r \mathbf{u}$ do not guarantee positive-definiteness. That is, if the operator is not positive-definite, the Rayleigh–Ritz method fails to ensure minimization of the functional since a global stationary point may not exist. However, the application of Galerkin's method to yield a discrete system does not require that the operator is positive-definite or even symmetric.[6] The resultant solution is simply a stationary point which is not guaranteed to be a minimum.

TABLE 1.3 Mathematical Statement of Solution Methods

Method	Mathematical Statement (subject to boundary conditions)
Rayleigh–Ritz	minimize $\frac{1}{2}\langle \mathcal{L}u, u\rangle - \langle f, u\rangle$
Galerkin	solve $\langle \mathcal{L}u, u_j\rangle - \langle f, u_j\rangle = 0$
Least Squares	minimize $Q(u) = \langle \mathcal{L}u - f, \mathcal{L}u - f\rangle$

[5] Hilbert space refers to a linear space where a given interproduct has been defined and which is complete with respect to this interproduct.

[6] Because of the complex ϵ_r and μ_r, in electromagnetics, the operators may be symmetric but not Hermitian (i.e., self-adjoint). From Section 1.11.5, Hermitian operators have positive and real eigenvalues and are therefore positive.

In some cases, the problem statement is that of minimizing a functional and consequently the Rayleigh–Ritz procedure is the natural method of choice. Examples include problems associated with system energy minimizations and resonance computations. However, in solving the Helmholtz or vector wave equations (subject to given boundary conditions), the minimization of the functional must simultaneously imply a solution of those equations. For those cases, Galerkin's method is the appropriate approach for constructing the linear system. However, introduction of an appropriate variational functional can simplify the problem statement when dealing with boundary conditions other than Neumann-type (also referred to as natural conditions). As seen from the derivation given in Section 1.11.1, the presence of the boundary integral term provides a direct means for imposing the Dirichlet, impedance, or other type of boundary conditions. In the case of Galerkin's method, the boundary terms are introduced after application of the divergence theorem. Finally, we note that in some cases a variational (minimization) statement of the boundary value problem may not be possible. In those cases, Galerkin's method is the only approach for constructing a linear system of equations.

When dealing with operators that are not positive-definite, apart from the breakdown of the functional minimization process, most iterative linear system solvers also break down. Specifically, convergence of the approximate solution $\tilde{u}$ to the exact solution u cannot be proven [21], [22] without invoking positive-definiteness. When in doubt, a positive-definite operator (and a corresponding positive-definite matrix) can be generated by instead solving the differential equation

$$\mathcal{L}^a \mathcal{L} u = \mathcal{L}^a f \tag{1.182}$$

where $\mathcal{L}^a$ satisfies

$$\langle \mathcal{L} u, v \rangle = \langle u, \mathcal{L}^a v \rangle \tag{1.183}$$

and is referred to as the *adjoint* operator. Clearly, (1.182) is obtained by multiplying the right and left hand sides of $\mathcal{L} u = f$ by $\mathcal{L}^a$. The new operator $\mathcal{P}$, where

$$\mathcal{P} u = g \tag{1.184}$$

($g = \mathcal{L}^a f$) is now positive-definite and self-adjoint. The corresponding matrix system resulting from (1.184) is of the form

$$[A^*]^T [A]\{u\} = [A^*]^T \{f\} \tag{1.185}$$

or

$$[B]\{u\} = \{g\}$$

where $[B] = [A^*]^T [A]$. It is thus seen that the desired property of positive-definiteness comes at the price of squaring the matrix condition number. As is well known, large matrix conditions lead to less accurate solutions and slower convergence when an iterative solver is used.

It should be remarked that minimization of the functional for the *Least Squares Method* is equivalent to solving the differential equation (1.182). Consequently, the Least Squares Method leads to positive-definite systems at the expense of squaring the matrix condition. Also, the Least Squares Method minimizes the square of the norm (as $\tilde{u} \to u$), viz.

$$\lim_{N\to\infty} \|\mathcal{L}\tilde{u} - f\|^2 \to 0 \tag{1.186}$$

whereas Galerkin's method minimizes the norm

$$\lim_{N\to\infty} \|\mathcal{L}\tilde{u} - f\| \to 0 \tag{1.187}$$

Again, it is important to note that nothing can be said about convergence unless the operator is positive-definite.

1.11.7 Field Formulation Issues

The finite element method can assume various forms depending on the desired field quantity. Many applications prefer either a total or secondary electric field formulation. Other applications desire a result in terms of either the total or secondary magnetic field. Some applications can utilize a potential formulation. Thus, even though Maxwell's equations relate these various quantities, an accurate field computation often demands a particular formulation. The advantages and disadvantages of each of these formulations are discussed below.

The total electric field formulation is a very popular choice. This is because enforcement of the boundary conditions associated with perfect electric conductors (pec) is particularly easy. Since the tangential electric fields on a pec surface must vanish, the edges of the mesh associated with those surfaces are a priori set to zero. Three methods are commonly used in practice to enforce this condition. The first is accomplished by forcing a null field condition to zero out all entries of the matrix associated with that edge (except for the self-term which is set to unity), and by also setting the excitation entry to zero. Thus, as the unknown fields are solved, the edges lying on pec surfaces are forced to zero. The second method involves a preprocessing step where the edges associated with a pec surface are removed from the list of unknowns. Thus, the number of edges is greater than the number of unknowns and matrix entries for these pec edges are never computed. This approach has the advantage of reducing the order of the matrix and therefore reducing memory and compute cycle demands. The third method, useful when an iterative matrix solver is employed, involves forcing the unknowns associated with the pec edges to zero during each iteration.

Thus, use of a total electric field formulation and edge-based elements (see Chapter 2 for a discussion on edge-elements) reduces the order of the matrix and computational burden. However, this is not the only valid formulation and in certain circumstances, a scattered field or a magnetic field formulation may be preferred. Scattered (or secondary) field formulations are used to simplify the use of absorbing boundary conditions (see Chapters 4 and 6). However, they also have an added advantage when a boundary integral is used for mesh closure. Experience has shown that phase errors in the computed interior field tend to increase within the mesh at locations distant from boundaries on which boundary conditions are imposed. This is due to unavoidable numerical inaccuracies that increase as the effect of the boundary conditions propagate throughout the mesh. That is, previous errors in the adjoining field are incorporated and magnified as the field is evaluated at a more distant field point. Since boundary conditions always are enforced with total fields, the total field formulation enforces such conditions only on the boundaries of the mesh and pec surfaces (for E-field formulations). For very large computational domains, significant distance can lie between a field point within the interior of the mesh and the mesh boundary. Hence, the potential for error propagation throughout the mesh. A scattered field formulation enforces the boundary conditions on the mesh boundary, pec surfaces, and dissimilar material interfaces. Therefore, the distance between a boundary condition and any interior field point is reduced and accordingly the phase error throughout the mesh may be also reduced. The scattered field formulation has the disadvantage of higher matrix order (i.e., more unknowns and equations) and explicit enforcement of the boundary conditions associated with pec surfaces and material discontinuities.

Magnetic field formulations are also possible and can be obtained by applying duality to the corresponding electric field equations. A total magnetic field formulation has the same advantages for a fictitious perfect magnetic conductor (pmc) as the total electric field has for pec surfaces. A scattered magnetic field formulation also can reduce phase error propagation in the same manner as the scattered electric field formulation.

Magnetic field formulations are preferred for applications where the desired result is the magnetic field within the computational domain. This is due to the fact that although Maxwell's equations relate the electric and magnetic fields, in practice one quantity cannot be accurately obtained from the other by numerical differentiation. This is due to the inherent error occurring when continuous derivatives are replaced with discrete differences. Rather, a computationally expensive integral expression is necessary for accurate field differentiation provided a suitable Green's function is available. Hence, an accurate solution demands a formulation consistent with the desired result.

Finally, some finite element practitioners utilize a potential formulation which employs the scalar or vector potentials as the unknown quantities (see Chapter 5). The use of this approach is related to the hybrid finite element-boundary integral method where the singularity of the integral equation associated with the boundary can be reduced with a potential formulation. However, if this reduction is present, we note that a numerical differentiation operation may be required to obtain the desired field quantity and this operation may lead to inaccuracy.

REFERENCES

[1] J. A. Stratton. *Electromagnetic Theory*. McGraw-Hill, New York, 1941.

[2] R. E. Collin. *Field Theory of Guided Waves*. IEEE Press, New York, 1991.

[3] C. A. Balanis. *Advanced Engineering Electromagnetics*. McGraw-Hill, New York, 1989.

[4] D. S. Jones. *Methods in Electromagnetic Wave Propagation*. IEEE Press, New York, 1994.

[5] R. F. Harrington. *Time-Harmonic Electromagnetic Fields*. McGraw-Hill, New York, 1961.

[6] C. T. Tai. *Generalized Vector and Dyadic Analysis*. IEEE Press, New York, 1992.

[7] J. A. Kong. *Theory of Electromagnetic Waves*. Wiley InterScience, New York, 1975.

[8] J. D. Jackson. *Classical Electrodynamics*. Wiley InterScience, New York, 1975.

[9] J. van Bladel. *Electromagnetic Fields*. Hemisphere Publishing Corp., New York, 1985.

[10] C. T. Tai. *Dyadic Green's Functions in Electromagnetic Theory*. IEEE Press, New York, 1994.

[11] T. B. A Senior and J. L. Volakis. *Approximate Boundary Conditions in Electromagnetics*. IEE Press, London, 1995.

[12] S. M. Rytov. Computation of the skin effect by the perturbation method. *J. Exp. Theor. Phys.*, 10:180–189, 1940. Translation by V. Kerdemelidis and K. M. Mitzner, Northrop Navair, Hawthorne, CA 90250.

[13] S. G. Mikhlin. *Variational Methods in Mathematical Physics*. Macmillan, New York, 1964.

[14] J. N. Reddy. *An Introduction to the Finite Element Method*. McGraw-Hill, New York, 1984.

[15] A. Konrad. Vector variational formulation of electromagnetic fields in anisotropic media. *IEEE Trans. Microwave Theory Tech.*, MTT-24:553–559, September 1976.

[16] C. H. Chen and C. D. Lieu. The variational principle for non-self-adjoint electromagnetic problems. *IEEE Trans. Microwave Theory Tech.*, MTT-28:878-886, August 1980.

[17] R. F. Harrington. *Field Computation by Moment Methods*. Macmillan, New York, 1968.

[18] J. J. H. Wang. *Generalized Moment Methods in Electromagnetics*. John Wiley & Sons, New York, 1991.

[19] R. L. Burden and J. D. Faires. *Numerical Analysis*. PWS Pub. Co., Boston, fifth edition, 1993.

[20] G. H. Golub and C. F. Van Loan. *Matrix Computations*. Johns Hopkins Univ. Press, Baltimore, MD, 1983.

[21] G. Strang and G. J. Fix. *An Analysis of the Finite Element Method*. Prentice Hall, Inc., Englewood Cliffs, NJ, 1973.

[22] D. G. Dudley. *Mathematical Foundations for Electromagnetic Theory*. IEEE Press, New York, 1994.

2

Shape Functions for Scalar and Vector Finite Elements

2.1 INTRODUCTION

The finite element method is used for modeling a wide class of problems by breaking up the computational domain into elements of simple shapes. Suitable interpolation polynomials (commonly referred to as *shape* or *basis* functions) are used to approximate the unknown function within each element. Once the shape functions are chosen, it is possible to program the computer to solve complicated geometries by solely specifying the basis functions. The element choice, however, needs human intervention and intelligence to ensure a reliable solution of the problem at hand. As will be shown later in this chapter, the development of a special class of elements which mimic the character of electric/magnetic fields has proved to be the key in obtaining robust solutions to three-dimensional problems in electromagnetics.

In this chapter, we will discuss the derivation of *node-based* and *edge-based* shape functions for one-, two- and three-dimensional finite elements. Node-based shape functions have been used extensively in civil and mechanical engineering applications as well as in scalar electromagnetic field problems. However, a full three-dimensional vector formulation brings out numerous deficiencies in these traditional element shape functions [1], [2]. *Edge-based* vector basis functions with unknowns associated with element edges havc thus been derived to overcome the problems related to nodal basis and these are now extensively used for solving three-dimensional electromagnetics problems. We will also describe the hierarchical nature of the edge-based functions and their possible applicability in p-based refinement techniques.

2.2 FEATURES OF FINITE ELEMENT SHAPE FUNCTIONS

The polynomials used to interpolate finite element solutions on specific element shapes have some distinct features over the wide variety of basis functions used in other partial differential equation (PDE) or integral equation (IE) techniques.

2.2.1 Spatial Locality

Finite element shape functions have compact support within each element, i.e., their scope of influence is limited only to the immediate neighboring elements. This feature plays a pivotal role in the viability of finite elements over integral equation (IE) methods. The limited scope of influence for the basis functions is a distinguishing property of PDE techniques and leads to very sparse matrices in finite elements, whereas IE techniques give rise to full, dense matrices resulting in poor scalability as problem size increases.

2.2.2 Approximation Order

The order of the approximation depends on the completeness of the polynomials making up the finite element basis functions. Moreover, the form of the polynomial function must remain unchanged under a linear transformation from one Cartesian coordinate system to another. This requirement is satisfied if the polynomials are complete to a specific order such as

$$u(x, y) = c_1 + c_2x + c_3y + c_4x^2 + c_5xy + c_6y^2 \tag{2.1}$$

or when the extra terms are symmetric with respect to one another, as in the following incomplete third-order polynomial

$$u(x, y) = c_1 + c_2x + c_3y + c_4x^2 + c_5xy + c_6y^2 + c_7x^2y + c_8xy^2 \tag{2.2}$$

Such approximation functions have the characteristic that, for fixed x or y, they are always complete polynomials in the other variable. In general, we seek expansion polynomials that will yield the highest order of approximation for a minimum number of unknowns associated with that element shape. The two examples shown above apply to two dimensions, but their extension to three-dimensional elements is straightforward. Typically, the higher the order of the approximating polynomial, the lower the error in the final solution if element size remains constant. As usual, there is a tradc-off here between the desired accuracy and the degrees of freedom required to solve the problem.

2.2.3 Continuity

The order of the differential equation to be solved determines the order of shape function to be employed. Functions with continuous derivatives up to the nth order are said to be C^n continuous. For elliptic PDEs of order $2k$ ($k = 1, 2$), the continuity requirement is C^{k-1} for Galerkin methods. In most electromagnetic problems, functions which exhibit C^0 continuity (i.e., function continuity) are used since the discontinuous first derivatives are integrable. However, it is difficult to impose continuity of order 1 and higher since the determination of suitable shape functions is very complicated. For example, ninth-order polynomials are required to obtain C^1 continuity for tetrahedral elements. In electromagnetics, Wong and Cendes [3] used C^1 node-based triangles to avoid the problem of spurious modes in the determination of cavity resonances. All shape functions derived in the following sections impose function continuity or C^0 continuity (not derivative continuity) between elements.

2.3 NODE-BASED ELEMENTS

In node-based finite elements, the form of the sought function in the element is controlled by the function values at its nodes. The approximating function can then be expressed as a linear combination of basis functions weighted by the nodal coefficients. If the function values u_i^e at the nodes are taken as nodal variables, then the approximating function for a two-dimensional element e with p nodes has the form

$$u^e(x, y) = \sum_{i=1}^{p} u_i^e N_i^e(x, y) \tag{2.3}$$

Since the expression (2.3) must be valid for any nodal variable u_i^e, the basis function $N_i^e(x, y)$ must be unity at node i and zero for all remaining nodes within the element.

Shape functions can be derived either by inspection (*Serendipity* family) or through simple products of appropriate polynomials (*Lagrange* family). It is easier and more systematic to construct higher order bases in the Lagrange family while progression to higher orders is difficult in the Serendipity family. However, Lagrange shape functions have undesirable interior nodes and more unknowns than Serendipity shape functions of the same order.

2.3.1 One-Dimensional Basis Functions

One-dimensional finite elements are employed for solving problems where the discretization domain involves a curve or a contour around a two-dimensional structure; for example, the bounding curve of the cross section of an infinite cylinder. These basis functions can also be used in conjunction with higher-dimensional finite elements when the modeled structure can be decomposed into a single dimension without loss of accuracy. It is most convenient to derive one-dimensional basis functions in terms of Lagrange polynomials [4]. Let us consider a straight line with endpoints x_1 and x_2. The basis functions for element e are then defined as

$$N_1^e(x) = \frac{x_2^e - x}{x_2^e - x_1^e}$$

$$N_2^e(x) = \frac{x - x_1^e}{x_2^e - x_1^e}$$

The basis functions have unit magnitude at one node and vanish at all others with linear variation between the nodes. Higher order basis functions can be constructed by inserting nodes between the endpoints of the finite element. If x_i^e, $i = 1, 2, \ldots, n$, are the nodes of the one-dimensional element, and we are interested in finding the basis function for the kth node x_k^e, then the corresponding Lagrange polynomial describing this basis function is given by [4]

$$N_k^e(x) = \frac{(x - x_1^e)(x - x_2^e)\ldots(x - x_{k-1}^e)(x - x_{k+1}^e)\ldots(x - x_n^e)}{(x_k^e - x_1^e)(x_k^e - x_2^e)\ldots(x_k^e - x_{k-1}^e)(x_k^e - x_{k+1}^e)\ldots(x_k^e - x_n^e)} \tag{2.4}$$

The basis function defined above is of $(n-1)$th order and correspondingly passes through zero $n-1$ times.

2.3.2 Two-Dimensional Basis Functions

Two-dimensional finite elements have found widespread use in modeling structures whose third dimension is significantly larger or smaller than the cross section, thus ensuring little variation in the unknown parameters in this third direction. Two-dimensional finite elements have also been used to obtain reliable estimates of three-dimensional problems since the computational cost for obtaining two-dimensional solutions is vastly less expensive than for three dimensions.

2.3.2.1 Rectangular and Quadrilateral Elements. The simple shape of the rectangular element permits its shape functions to be written down merely by inspection. On examining the element shape given in Fig. 2.1, the shape functions can be cast in the form

$$N_1^e = \frac{1}{A^e}\left(x_c^e + \frac{h_x^e}{2} - x\right)\left(y_c^e + \frac{h_y^e}{2} - y\right)$$

$$N_2^e = \frac{1}{A^e}\left(-x_c^e + \frac{h_x^e}{2} + x\right)\left(y_c^e + \frac{h_y^e}{2} - y\right)$$

$$N_3^e = \frac{1}{A^e}\left(-x_c^e + \frac{h_x^e}{2} + x\right)\left(-y_c^e + \frac{h_y^e}{2} + y\right)$$

$$N_4^e = \frac{1}{A^e}\left(x_c^e + \frac{h_x^e}{2} - x\right)\left(-y_c^e + \frac{h_y^e}{2} + y\right)$$

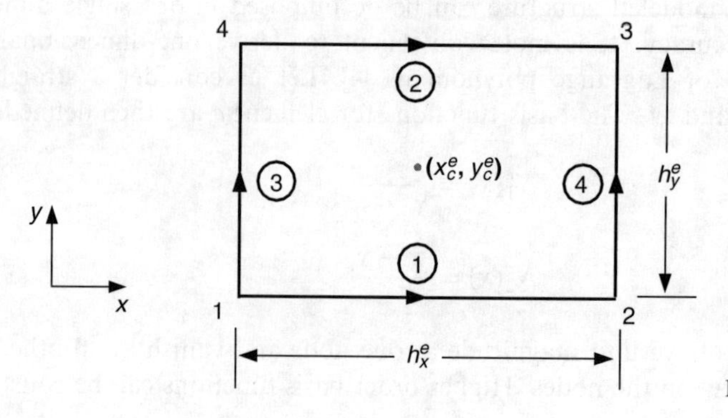

Figure 2.1 Rectangular element.

where x_c^e and y_c^e denote the coordinates of the midpoints of the element edges, h_x^e and h_y^e represent the edge lengths and A^e denotes the area of the element. Each basis function N_i^e has unit magnitude at the ith node, vanishes at the remaining three nodes, and varies linearly. Hence it can be used in (2.3) to represent u^e. Higher order rectangular elements include the eight node (three equispaced nodes per edge) and the twelve node (four equispaced nodes per edge) quadrilateral element discussed in [4]. However, these elements can model only regular geometries and decline in accuracy with excessive shape distortion. Thus, often they are not very useful in practice.

Irregular geometries can be modeled by using quadrilateral elements which can also be viewed as distorted rectangles. To construct basis functions for a quadrilateral element, we need to use a transformation that maps a quadrilateral element in the xy plane to a square element in the $\xi\eta$ plane (Fig. 2.2). Such a transformation can be found by satisfying the following relation at the four nodes of the quadrilateral element

$$x = a + b\xi + c\eta + d\xi\eta \qquad y = a' + b'\xi + c'\eta + d'\xi\eta \tag{2.5}$$

The unknown coefficients—a, b, c, d and a', b', c', d'—are solved by mapping the four corners of the quadrilateral in the xy plane to the corner points of the unit square in the $\xi\eta$ plane. The eight equations thus obtained are

$$\begin{aligned} x_1 &= a - b - c + d, & y_1 &= a' - b' - c' + d' \\ x_2 &= a + b - c - d, & y_2 &= a' + b' - c' - d' \\ x_3 &= a + b + c + d, & y_3 &= a' + b' + c' + d' \\ x_4 &= a - b + c - d, & y_4 &= a' - b' + c' - d' \end{aligned} \tag{2.6}$$

On solving for the unknown coefficients in the above equation, the basis functions can be cast in the following form

$$N_i^e = \tfrac{1}{4}(1 + \xi_0)(1 + \eta_0), \qquad i = 1, \ldots, 4 \tag{2.7}$$

where $\xi_0 = \xi\xi_i$ and $\eta_0 = \eta\eta_i$ and

$$x = \sum_{i=1}^{4} N_i^e(\xi, \eta)x_i^e, \qquad y = \sum_{i=1}^{4} N_i^e(\xi, \eta)y_i^e \tag{2.8}$$

The variables (ξ_i, η_i) denote the coordinates of the ith node in the (ξ, η) coordinate system. The linear quadrilateral is also known as an *isoparametric* element since the shape functions defining the geometry and the nodal values are the same.

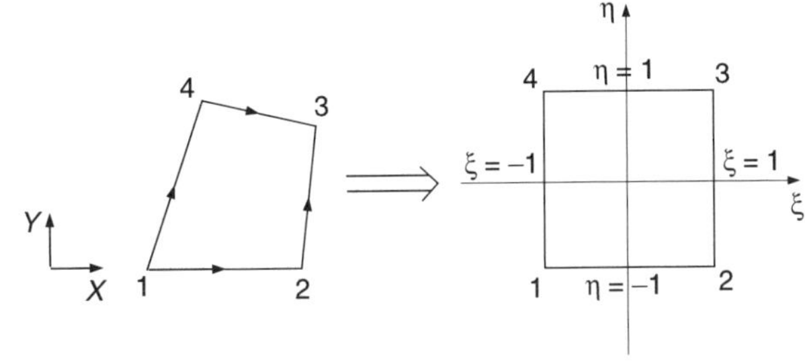

Figure 2.2 Transformation of a quadrilateral element in the xy plane to a unit square in the $\xi\eta$ plane.

Higher order quadrilateral elements include the eight-node element (four corner nodes and four midside nodes) and the twelve-node element (four equispaced nodes per edge). The basis functions for such elements can be found in [5].

Due to the irregular shape of the quadrilateral element, it is not easy to integrate the basis functions in the xy plane. To facilitate and generalize the integration process, the concept of the *Jacobian* is introduced. The Jacobian matrix, or more specifically its determinant, transforms the infinitesimal area or volume element from one coordinate system to another. If we consider the above example, we are essentially transforming between the global (x, y) coordinates and the local (ξ, η) coordinates. By the chain-rule of partial differentiation, we can express the ξ derivative of N_i^e as

$$\frac{\partial N_i^e}{\partial \xi} = \frac{\partial N_i^e}{\partial x}\frac{\partial x}{\partial \xi} + \frac{\partial N_i^e}{\partial y}\frac{\partial y}{\partial \xi} \tag{2.9}$$

Similarly, taking the η derivative and combining the expressions, we can write the result in matrix form as

$$\begin{Bmatrix} \partial N_i^e/\partial \xi \\ \partial N_i^e/\partial \eta \end{Bmatrix} = \begin{bmatrix} \partial x/\partial \xi & \partial y/\partial \xi \\ \partial x/\partial \eta & \partial y/\partial \eta \end{bmatrix} \begin{Bmatrix} \partial N_i^e/\partial x \\ \partial N_i^e/\partial y \end{Bmatrix} = [J] \begin{Bmatrix} \partial N_i^e/\partial x \\ \partial N_i^e/\partial y \end{Bmatrix}$$

Since (x, y) are known explicitly in terms of the local coordinates (ξ, η), the Jacobian matrix can be found explicitly in terms of local coordinates. Care must be taken in the choice of the local coordinate system such that the Jacobian matrix is nonsingular. To find the derivatives with respect to x and y, we merely need to invert the Jacobian matrix to yield

$$\begin{Bmatrix} \partial N_i^e/\partial x \\ \partial N_i^e/\partial y \end{Bmatrix} = [J]^{-1} \begin{Bmatrix} \partial N_i^e/\partial \xi \\ \partial N_i^e/\partial \eta \end{Bmatrix} \tag{2.10}$$

The technique can easily be generalized for n-dimensional transformations if necessary. The infinitesimal area element dA can now be written as

$$dx\,dy = \det[J]\,d\xi\,d\eta \tag{2.11}$$

Thus the integration of the irregular quadrilateral element is simplified considerably by performing it over the local coordinate system instead of the global Cartesian coordinate system.

2.3.2.2 Triangular Elements. Triangular elements are popular because they can model arbitrary geometries. We will determine the shape functions of triangular elements by using Lagrange interpolation polynomials. In their final expression, the shape functions will be expresed in terms of the so-called *area coordinates*. Let us consider a point P within a triangular element (Fig. 2.3) located at (x, y), where (x_i^e, y_i^e) denote the coordinates of the ith triangle node. The area of the smaller triangle formed by points P, 2, and 3 is given by

$$\Delta_1 = \frac{1}{2}\begin{vmatrix} 1 & x & y \\ 1 & x_2^e & y_2^e \\ 1 & x_3^e & y_3^e \end{vmatrix} \tag{2.12}$$

The area coordinate L_i^e is then given by

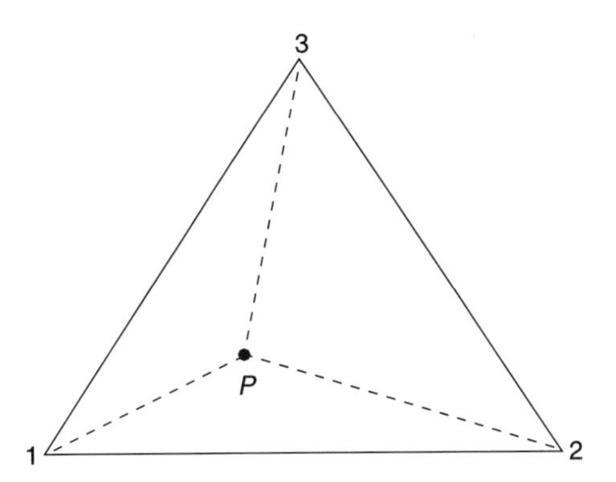

Figure 2.3 Triangular element.

$$L_1^e = \frac{\Delta_1}{\Delta} = \frac{\text{Area } P23}{\text{Area } 123} \tag{2.13}$$

where Δ is the area of the whole triangle and can be found from (2.12) by replacing x and y in the first row with x_1 and y_1. Similarly, the two remaining area coordinates L_2 and L_3 are given by

$$L_2^e = \frac{\Delta_2}{\Delta} = \frac{\text{Area } P31}{\text{Area } 123}$$

$$L_3^e = \frac{\Delta_3}{\Delta} = \frac{\text{Area } P12}{\text{Area } 123}$$

The values for x and y inside the triangular element reduce to

$$x = \sum_{i=1}^{3} L_i^e x_i^e, \quad y = \sum_{i=1}^{3} L_i^e y_i^e, \quad 1 = \sum_{i=1}^{3} L_i \tag{2.14}$$

where the latter condition $\sum_{i=1}^{e} L_i = 1$ is a result of the area identity $\Delta_1 + \Delta_2 + \Delta_3 = \Delta$. Alternatively, L_1^e, L_2^e, and L_3^e can be obtained in terms of x, y, and the vertex coordinates by solving the system of equations (2.14).

The coordinate L_1^e is zero on the edge opposite to vertex 1 and unity at vertex 1. Its variation along the height of the triangle is displayed in Fig. 2.4. The remaining two area coordinates associated with the other two vertices behave similarly, vanishing on the edge opposite to the corresponding vertex and having unit magnitude at the vertex it belongs to. This feature combined with spatial locality and C^0 continuity qualifies the area coordinates as suitable basis functions N_i^e for a triangle when the interpolation order is linear. That is,

$$N_i^e = L_i^e \tag{2.15}$$

Higher order basis functions for triangles can be derived using the procedure given in [4] and [6]. In general, the shape function N_i^e for node i, labeled as (I, J, K), is given by

$$N_i^e = P_I^n(L_1^e) P_J^n(L_2^e) P_K^n(L_3^e), \qquad I + J + K = n \tag{2.16}$$

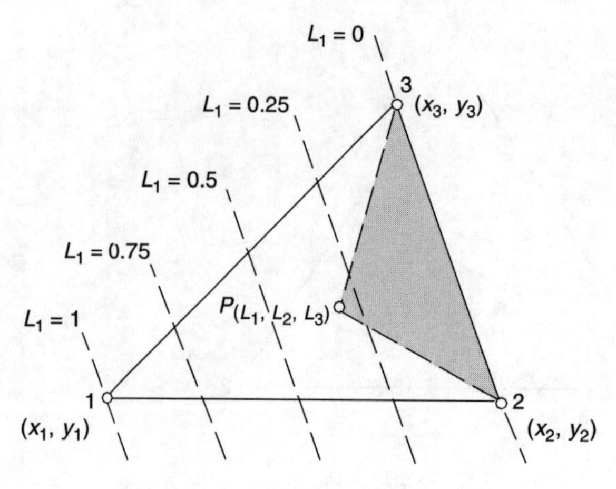

Figure 2.4 Area coordinates of a triangle.

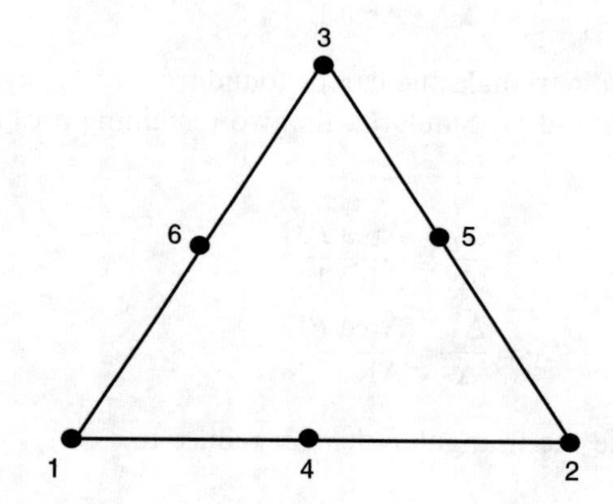

Figure 2.5 Six-noded triangular element supporting quadratic bases.

where L_i^e, $i = 1, \ldots, e$ are area cordinates defined previously and $P_I^n(L_1^e)$ is the polynomial

$$P_I^n(L_1^e) = \frac{1}{I!}\prod_{s=0}^{I-1}(nL_1^e - s), \qquad I > 0$$
$$P_0^n = 1 \tag{2.17}$$

Similar definitions apply for $P_J^n(L_2^e)$ and $P_K^n(L_3^e)$.

Using the formulae given above, the basis functions for a quadratic triangle (see Fig. 2.5) can be conveniently defined as

$$\begin{aligned} &N_i^e = L_i^e(2L_i^e - 1), \quad i = 1, 2, 3 && \text{CORNER NODES} \\ &N_4^e = 4L_1^eL_2^e, \quad N_5^e = 4L_2^eL_3^e, \quad N_6^e = 4L_3^eL_1^e, && \text{MIDSIDE NODES} \end{aligned} \tag{2.18}$$

Besides simplifying the treatment of basis functions for a triangle, simplex coordinates greatly facilitate the integration of an arbitrary function over a triangular region. A very useful integration formula in terms of the area coordinates—L_1, L_2, and L_3—over a triangular domain is given by

$$\int_\Delta (L_1^e)^a(L_2^e)^b(L_3^e)^c \, dx\, dy = \frac{a!b!c!}{(a+b+c+2)!}2\Delta \tag{2.19}$$

where a, b, and c are integers and Δ is the area of the triangular region.

2.3.3 Three-Dimensional Basis Functions

Shape functions for three-dimensional elements can be described in a precisely analogous way to their two-dimensional counterpart. However, the simple rules for inter-element continuity given previously must be modified. The nodal field values should now interpolate to give continuous fields across the face of each element.

2.3.3.1 Rectangular Bricks. The simplest polynomial approximation to a rectangular brick element is the trilinear function

$$u^e(x, y, z) = a^e + b^e x + c^e y + d^e z + e^e xy + f^e yz + g^e zx + h^e xyz \tag{2.20}$$

whose eight parameters are uniquely determined by matching $u^e(x, y, z)$ to the field values u_i^e at the eight corners of the brick. This results in eight equations that are solved to determine the coefficients $a^e, b^e, \ldots, h^e$. The final expression is in the form

$$u^e(x, y, z) = \sum_{i=1}^{8} u_i^e N_i^e(x, y, z) \tag{2.21}$$

However, this approach of formulating (2.21) is cumbersome and can be easily avoided on writing down the required basis functions by mere inspection. Since the basis function N_i^e must be unity at node i and zero at the remaining nodes, the eight interpolation functions can be written down as

$$N_1^e = \frac{1}{V^e}\left(x_c^e + \frac{h_x^e}{2} - x\right)\left(y_c^e + \frac{h_y^e}{2} - y\right)\left(z_c^e + \frac{h_z^e}{2} - z\right)$$

$$N_2^e = \frac{1}{V^e}\left(-x_c^e + \frac{h_x^e}{2} + x\right)\left(y_c^e + \frac{h_y^e}{2} - y\right)\left(z_c^e + \frac{h_z^e}{2} - z\right)$$

$$N_3^e = \frac{1}{V^e}\left(-x_c^e + \frac{h_x^e}{2} + x\right)\left(-y_c^e + \frac{h_y^e}{2} + y\right)\left(z_c^e + \frac{h_z^e}{2} - z\right)$$

$$N_4^e = \frac{1}{V^e}\left(x_c^e + \frac{h_x^e}{2} - x\right)\left(-y_c^e + \frac{h_y^e}{2} + y\right)\left(z_c^e + \frac{h_z^e}{2} - z\right)$$

$$N_5^e = \frac{1}{V^e}\left(x_c^e + \frac{h_x^e}{2} - x\right)\left(y_c^e + \frac{h_y^e}{2} - y\right)\left(-z_c^e + \frac{h_z^e}{2} + z\right)$$

$$N_6^e = \frac{1}{V^e}\left(-x_c^e + \frac{h_x^e}{2} + x\right)\left(y_c^e + \frac{h_y^e}{2} - y\right)\left(-z_c^e + \frac{h_z^e}{2} + z\right)$$

$$N_7^e = \frac{1}{V^e}\left(-x_c^e + \frac{h_x^e}{2} + x\right)\left(-y_c^e + \frac{h_y^e}{2} + y\right)\left(-z_c^e + \frac{h_z^e}{2} + z\right)$$

$$N_8^e = \frac{1}{V^e}\left(x_c^e + \frac{h_x^e}{2} - x\right)\left(-y_c^e + \frac{h_y^e}{2} + y\right)\left(-z_c^e + \frac{h_z^e}{2} + z\right)$$

where x_c^e, y_c^e, and z_c^e denote the coordinates of the center of the element, h_x^e, h_y^e, and h_z^e represent the edge lengths of the element and V^e is the element volume.

Brick elements of the Serendipity family are derived in [4]. To obtain higher order basis functions, progressively larger number of nodes are uniformly placed on the element edges. Bricks with nth-order interpolation functions (with $n+1$ per edge) require $8+12(n-1)$ degrees of freedom. Higher order bricks are rarely used in electromagnetics applications since the regular shape requirement and decline of accuracy with excessive shape distortion place severe limitations on the generality of the geometry to be modeled.

Shape functions for hexahedral elements or distorted bricks can be derived by mapping the element in the xyz coordinate system onto a standard cube in a new $\xi\eta\zeta$ coordinate system. We proceed along lines similar to the derivation of bases for the quadrilateral element in Section 2.3.2. We express the Cartesian coordinates (x, y, z) in terms of (ξ, η, ζ) as follows:

$$\begin{aligned} x &= a_1 + b_1\xi + c_1\eta + d_1\zeta + e_1\xi\eta + f_1\eta\zeta + g_1\zeta\xi + h_1\xi\eta\zeta \\ y &= a_2 + b_2\xi + c_2\eta + d_2\zeta + e_2\xi\eta + f_2\eta\zeta + g_2\zeta\xi + h_2\xi\eta\zeta \\ z &= a_3 + b_3\xi + c_3\eta + d_3\zeta + e_3\xi\eta + f_3\eta\zeta + g_3\zeta\xi + h_3\xi\eta\zeta \end{aligned} \tag{2.22}$$

The unknown coefficients—$a_i, b_i, c_i, d_i, i = 1, 2, 3$—can be obtained by a one-to-one mapping of the corners of the hexahedral element to the corner points of the unit cube. The desired transformation thus yields the basis functions N_i^e of the hexahedral finite element

$$N_i^e = \tfrac{1}{8}(1+\xi_i\xi)(1+\eta_i\eta)(1+\zeta_i\zeta) \tag{2.23}$$

with (ξ_i, η_i, ζ_i) denoting the coordinates of the ith node in the $\xi\eta\zeta$ coordinate system. As before, the relationship between the (x, y, z) and (ξ, η, ζ) coordinates is given by

$$x = \sum_{i=1}^{8} N_i^e(\xi, \eta, \zeta)x_i^e; \quad y = \sum_{i=1}^{8} N_i^e(\xi, \eta, \zeta)y_i^e; \quad z = \sum_{i=1}^{8} N_i^e(\xi, \eta, \zeta)z_i^e \tag{2.24}$$

As in the case of the two-dimensional quadrilateral element, the Jacobian comes to our aid for calculating the volume integral over the arbitrary hexahedral domain. The Jacobian matrix $[J]$ is of order three and is given by

$$\begin{Bmatrix} \partial N_i^e/\partial\xi \\ \partial N_i^e/\partial\eta \\ \partial N_i^e/\partial\zeta \end{Bmatrix} = \begin{bmatrix} \partial x/\partial\xi & \partial y/\partial\xi & \partial z/\partial\xi \\ \partial x/\partial\eta & \partial y/\partial\eta & \partial z/\partial\eta \\ \partial x/\partial\zeta & \partial y/\partial\zeta & \partial z/\partial\zeta \end{bmatrix} \begin{Bmatrix} \partial N_i^e/\partial x \\ \partial N_i^e/\partial y \\ \partial N_i^e/\partial z \end{Bmatrix} = [J] \begin{Bmatrix} \partial N_i^e/\partial x \\ \partial N_i^e/\partial y \\ \partial N_i^e/\partial z \end{Bmatrix}$$

The volume element transformation from the global xyz to the local $\xi\eta\zeta$ coordinate system is expressed as

$$dx\,dy\,dz = \det[\boldsymbol{J}]\,d\xi\,d\eta\,d\zeta \tag{2.25}$$

2.3.3.2 Tetrahedral Elements. The three-dimensional analogue of a two-dimensional triangle is a tetrahedron (four-faced element). Once again, we can introduce special coordinates, called *volume coordinates* or *simplex coordinates*, to simplify the derivation of shape functions. If P is a point within the eth tetrahedron shown in Fig. 2.6, the four volume coordinates are given by

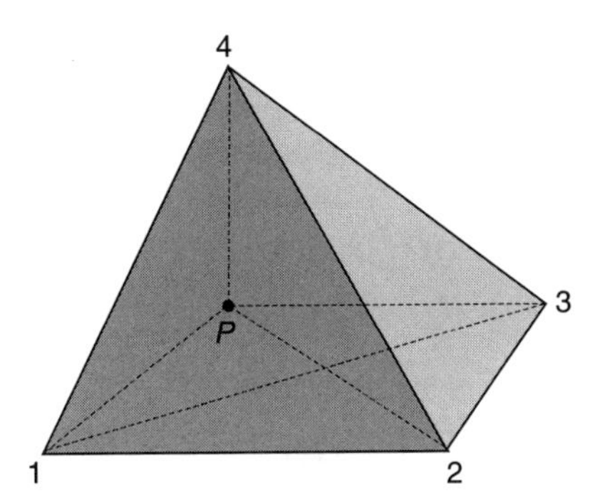

Figure 2.6 Tetrahedral element.

$$L_1^e = \frac{\text{Volume } P234}{\text{Volume } 1234}$$
$$L_2^e = \frac{\text{Volume } P341}{\text{Volume } 1234}$$
$$L_3^e = \frac{\text{Volume } P412}{\text{Volume } 1234}$$
$$L_4^e = \frac{\text{Volume } P123}{\text{Volume } 1234} \tag{2.26}$$

and any position within the element is specified by

$$x = \sum_{i=1}^{4} L_i^e x_i; \quad y = \sum_{i=1}^{4} L_i^e y_i; \quad z = \sum_{i=1}^{4} L_i^e z_i$$

with (x_i, y_i, z_i) being the coordinates of node i. As for the two-dimensional case (triangular elements), the basis functions N_i^e are equal to the volume coordinates, i.e.,

$$N_i^e = L_i^e, \qquad i = 1, \ldots, 4 \tag{2.27}$$

Quadratic shape functions for a tetrahedron necessitates the use of ten node points: the four corner nodes and the remaining six at the midpoints of the edges. The shape functions for the quadratic tetrahedron are given by

$$\begin{aligned} N_i^e &= L_i^e(2L_i^e - 1), \quad i = 1, \ldots, 4 && \text{CORNER NODES} \\ N_i^e &= 4L_j^e L_k^e, \quad i = 5, \ldots, 10, && \text{MIDSIDE NODES} \\ & j \text{ and } k \text{ are endpoints of each edge} \end{aligned} \tag{2.28}$$

Similarly to triangular elements, volume coordinates greatly simplify integration over tetrahedral elements. A useful formula for integrating over the volume of a tetrahedron is

$$\int_{\text{Volume}} (L_1^e)^a (L_2^e)^b (L_3^e)^c (L_4^e)^d \, dx\, dy\, dz = \frac{a!b!c!d!}{(a+b+c+d+3)!} 6V \tag{2.29}$$

where a, b, c, and d are integers and V is the volume of the tetrahedron.

2.3.3.3 Triangular Prism Elements. Other three-dimensional elements that have simple shapes include the triangular prism and isoparametric elements. To ensure that a small number of elements can model a relatively complex region, distorted prisms can be used in conjunction with rectangular bricks. The shape functions for the first-order prism element (shown in Fig. 2.7) is given by

$$\begin{aligned} N_i^e &= \tfrac{1}{2}L_1^e(1+\zeta_0), \qquad i = 1,4 \\ N_i^e &= \tfrac{1}{2}L_2^e(1+\zeta_0), \qquad i = 2,5 \\ N_i^e &= \tfrac{1}{2}L_3^e(1+\zeta_0), \qquad i = 3,6 \end{aligned} \tag{2.30}$$

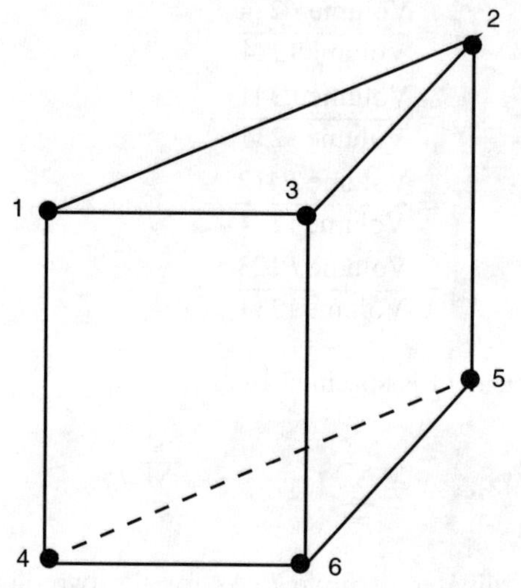

Figure 2.7 Linear triangular prism element.

where $\zeta_0 = 2(z - z_c)/height$ varies linearly from -1 to $+1$ over the height of the prism and is zero at the midpoint z_c of the vertical edge (joining nodes 1–4, 2–5, or 3–6 in Fig. 2.7). Here L_i^e refer to the area coordinates of the triangle that forms the cross section of the prism (i.e., the triangle formed by nodes 123 or 456).

The quadratic node-based triangular prism has 15 nodes—one each at the corners and at the midpoints of each of the nine edges. Shape functions for the quadratic and cubic triangular prisms can be found in [4]. For a more efficient discretization using the fewest unknowns, prisms and bricks can be combined. This is easily done because prisms and bricks can be readily connected by sharing the same nodes and edges at their boundaries.

2.4 EDGE-BASED ELEMENTS

In electromagnetics, we encounter serious problems when node-based elements are employed to represent vector electric or magnetic fields. First, spurious modes are observed when modeling cavity problems using node-based elements [7]. Nodal basis functions impose continuity in all three spatial components whereas edge bases guarantee continuity only along the tangential component. This feature mimics the behavior of field components along discontinuous material boundaries, and its importance in resolving the problem of spurious solutions will be discussed in detail in Chapter 5. Second, nodal bases require special care for enforcing boundary conditions at material interfaces, conducting surfaces, and geometry corners [8]. The first limitation can also jeopardize the near-field results of a scattering problem, the far-field typically escapes contamination since spurious modes do not radiate.

Edge-based finite elements, whose degrees of freedom are associated with the edges and the faces of the finite element mesh, have been shown to be free of the above shortcomings. The details of how edge bases avoid the pitfalls of nodal basis functions will be discussed in Chapter 5. Edge basis functions were described by Whitney [9] over 35 years ago and have been revived by Bossavit and Verite [1], Nedelec [10], and Hano [11] in the recent past. It was Nedelec's landmark paper [10] that laid down the guidelines for constructing finite element basis functions that span his curl conforming space with degrees of freedom associated with the edges, faces, and elements of a finite element mesh. Mur and de Hoop [12]; van Welij [13]; Barton and Cendes [14]; Jin and Volakis [15], [16]; and Lee et al. [17] among several others have extended their applicability to various two- and three-dimensional shapes and even constructed higher order elements for a more accurate approximation of the field values. More recently, Monk [18] provided error estimates and convergence proofs for edge bases. The derivations leading up to the convergence proof is beyond the scope of this book but the final result is stated below. If we denote by $H^s(\Omega)$ the standard Sobolev space of functions of order s in Ω and by $\|\cdot\|_s$ the norm on this space, we can then define the space

$$H(curl; \Omega) = \{\boldsymbol{u} \in (L^2(\Omega))^3 | \nabla \times \boldsymbol{u} \in (L^2(\Omega))^3\}$$

and its corresponding norm

$$\|\boldsymbol{u}\|_{H^c} = (\|\boldsymbol{u}\|_0^2 + \|\nabla \times \boldsymbol{u}\|_0^2)^{1/2}$$

If $\mathbf{E}$ is the exact solution of Maxwell's equation in Ω and $\mathbf{E}^h$ is the finite-dimensional approximation to it (in essence, the edge basis functions), then assuming that $\mathbf{E} \in (H^{k+1}(\Omega))^3$, the following result holds

$$\|\mathbf{E} - \mathbf{E}^h\|_{H^c} \leq Ch^k \|\mathbf{E}\|_{k+1} \tag{2.31}$$

provided ω is not an interior eigenvalue and h is sufficiently small. In (2.31), h is the discretization parameter, k is the order of the basis function and C is a constant independent of h. Thus higher order bases lead to lower errors in the solution when the sampling size is sufficiently small. The convergence is also optimal in h.

2.4.1 Two-Dimensional Basis Functions

2.4.1.1 Rectangular Elements. We consider the rectangular element first since its vector basis function is usually the easiest to formulate. For the element shown in Fig. 2.1, we can find its edge-based finite element basis function merely by inspection. If the edges are numbered according to Table 2.1 the vector basis functions can be written as

TABLE 2.1 Edge Numbering for Rectangular Element

Edge No.	i_1	i_2
1	1	2
2	4	3
3	1	4
4	2	3

$$\boldsymbol{W}_1^e = \frac{1}{h_y^e}\left(-y + y_c^e + \frac{h_y^e}{2}\right)\hat{\mathbf{x}}$$
$$\boldsymbol{W}_2^e = \frac{1}{h_y^e}\left(y - y_c^e + \frac{h_y^e}{2}\right)\hat{\mathbf{x}}$$
$$\boldsymbol{W}_3^e = \frac{1}{h_x^e}\left(-x + x_c^e + \frac{h_x^e}{2}\right)\hat{\mathbf{y}}$$
$$\boldsymbol{W}_4^e = \frac{1}{h_x^e}\left(x - x_c^e + \frac{h_x^e}{2}\right)\hat{\mathbf{y}}$$

where $\hat{\mathbf{x}}$, $\hat{\mathbf{y}}$, and $\hat{\mathbf{z}}$ are the unit vectors in the Cartesian coordinate system. The above basis functions have unity value along one edge and zero over all others, i.e.,

$$\boldsymbol{W}_i^e \cdot \hat{e}_j = \delta_{ij}$$

where δ_{ij} is the Kronecker delta and $\hat{e}_j$ is the unit vector along the jth edge.

A graphical illustration of the $\boldsymbol{W}_1^e$ vector basis function is given in Fig. 2.8. In this figure, the largest value of $|\boldsymbol{W}_1^e|$ corresponds to the largest vector length.

Using the above $\boldsymbol{W}_i^e$ the electric field within the finite element can be represented as

$$\mathbf{E}^e = \sum_{i=1} E_i^e \boldsymbol{W}_i^e \tag{2.32}$$

where now E_i^e denotes the average tangential field along the ith edge. The basis functions $\boldsymbol{W}_i^e$ guarantee tangential continuity across inter-element boundaries since they have a tangential component only along the ith edge and none along the other edges. They are also divergenceless within the element and possess a constant nonzero curl. It should be noted that by taking the cross-product of $\hat{\mathbf{z}}$ with $\boldsymbol{W}_i^e$, we obtain

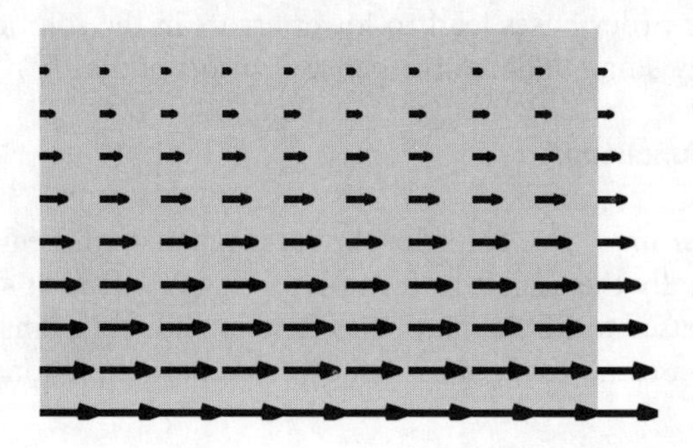

Figure 2.8 Illustration of $\boldsymbol{W}_1^e$ for rectangular element.

basis functions which possess normal continuity across element boundaries, have zero curl and non-zero divergence. The latter are ideal for representing surface current densities and are known as rooftop basis functions in electromagnetics. They have found extensive use in the solution of integral equations [19] and hybrid finite element–boundary integral implementations.

Edge-based vector bases for quadrilateral elements can be derived by carrying out the transformation detailed in the derivation of nodal basis for quadrilaterals in the previous section and then taking the gradient of the resulting expression for each edge. These bases have two shortcomings. First, the integrals associated with edge-based quadrilateral elements do not lend themselves to easy evaluation. Second, they may not be divergence free. However, their ability to model complicated shapes with a lesser number of unknowns than tetrahedra and the inherent property of enforcing tangential continuity across elements makes them attractive for use in two-dimensional vector formulations.

2.4.1.2 Triangular Elements. We consider again the triangular element depicted in Fig. 2.3. Since the edges of an arbitrary triangular element are not parallel to the x- or y-axis, it is not easy to guess the form of the vector basis function by inspection. Therefore, the vector basis for a triangular element will be expressed in terms of its area coordinates, L_1^e, L_2^e, and L_3^e. These are the *Whitney* elements. If the local edge numbers are defined according to Table 2.2, then edge bases for a triangular element are defined as

$$\boldsymbol{W}_k^e = \boldsymbol{N}_{ij}^e = \ell_{ij}(L_i^e \nabla L_j^e - L_j^e \nabla L_i^e), \qquad i, j = 1, 2, 3 \tag{2.33}$$

where $\boldsymbol{W}_k^e$ denotes the basis function for the kth edge of the eth element and $\ell_{ij} = \ell_k$ is the length of the edge formed by nodes i and j of the triangle. The vector field inside the triangular element can, therefore, be expanded as

$$\mathbf{E}^e = \sum_{k=1}^{e} E_k^e \boldsymbol{W}_k^e \tag{2.34}$$

where E_k^e denotes the tangential field along the kth edge. It can be easily shown that the edge-based functions defined in (2.33) have the following properties within the element

$$\nabla \cdot \boldsymbol{N}_{ij}^e = 0$$
$$\nabla \times \boldsymbol{N}_{ij}^e = 2\ell_{ij} \nabla L_i^e \times \nabla L_j^e$$

If $\hat{\mathbf{e}}_1$ is the unit vector pointing from node 1 to node 2 in Fig. 2.3, then $\hat{\mathbf{e}}_1 \cdot \nabla L_1 = -1/\ell_1$ and $\hat{\mathbf{e}}_1 \cdot \nabla L_2 = 1/\ell_1$. Since L_1 is a linear function that varies

TABLE 2.2 Edge Numbering for Triangular Element

Edge No.	Node i_1	Node i_2
1	1	2
2	2	3
3	3	1

from unity at node 1 and zero at node 2 and L_2^e is unity at node 2 and zero at node 1, we have

$$\hat{\mathbf{e}}_1 \cdot \boldsymbol{N}_{12}^e = L_1^e + L_2^e = 1 \tag{2.35}$$

along the entire length of edge 1. This implies that $\boldsymbol{N}_{12}^e$ has a constant tangential component along edge 1. Moreover, since L_1^e vanishes along edge 2, ∇L_1^e is normal to edge 2, L_2^e vanishes along edge 3, and ∇L_2^e is normal to edge 3, $\boldsymbol{N}_{12}^e$ has no tangential component along these edges. Similar observations apply to $\mathbf{N}_{23}^e$ and $\mathbf{N}_{31}^e$. Thus, tangential continuity is preserved across inter-element boundaries but normal continuity is not. Fig. 2.9 shows the actual variation of the basis function for the edge of a right triangle that is opposite to the node associated with the right angle. A different method of constructing edge bases for triangular elements is given in [20], [21].

Higher order vector basis functions involve adding a node at the midpoint of each edge and including the contribution of *facet* elements to the approximating function. Unknowns in the triangular element are assigned as shown in Fig. 2.10 [17].

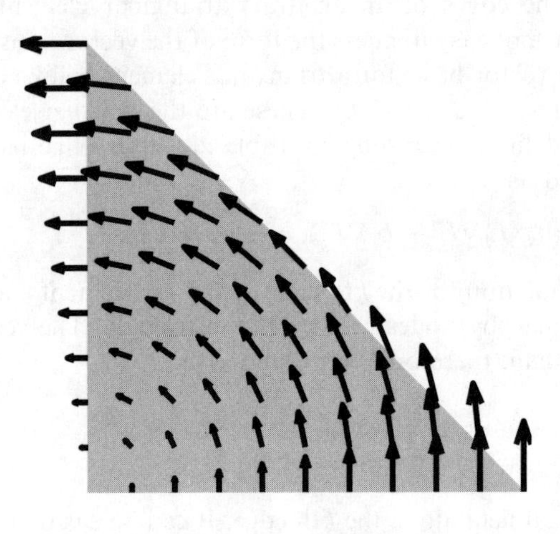

Figure 2.9 Variation of edge basis function for the edge opposite to the right angle.

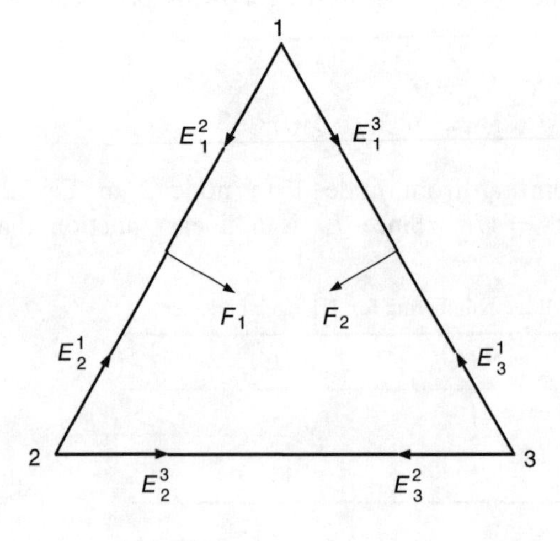

Figure 2.10 Triangular edge element with two unknowns each per edge and per face.

The tangential projection of the vector field along edge $\{i, j\}$ is determined by two unknowns E_i^j and E_j^i and two facet unknowns—F_1 and F_2—are provided to allow a quadratic approximation of the normal component along two of the three edges. Only two facet unknowns are required to make the range space of the curl operator complete to first order. Therefore, there are eight degrees of freedom for each triangular element. Since the edge variables provide common unknowns across element boundaries, tangential continuity of the field over the boundary is assured. However, an obvious disadvantage of these elements is that the two-facet variables cannot be symmetrically assigned. This disadvantage can be avoided by employing third-order edge bases [22]. The higher order approximation to the vector field within the element is given by

$$\mathbf{E}^e = \sum_{i=1}^{3} \sum_{j=1}^{3} E_i^{j^e} L_i^e \nabla L_j^e + F_1^e L_i^e (L_j^e \nabla L_k^e - L_k^e \nabla L_j^e) + F_2^e L_j^e (L_k^e \nabla L_i^e - L_i^e \nabla L_k^e) \qquad i \neq j \tag{2.36}$$

where we have arbitrarily chosen the facet variables to lie on edges 1 and 2. These variables are local unknowns associated with each separate triangular element and are included to provide a linear approximation for $\nabla_t \times \mathbf{E}_t$, where the subscript t denotes the tangential components of the operator. This property turns out to be very important in the selection of the order of the basis function to be used in the modeling process. The basis described by (2.36) can be classified as belonging to the $H^1(curl)$ space. The $H^k(curl)$ space consists of those vectors whose inner products are square integrable and whose curl consists of complete polynomials of order k. The basis given in (2.33) thus belongs to the $H^0(curl)$ space, since its curl is merely constant within the finite element. The basis generated by excluding the facial contribution would result in six unknowns—two per edge—but the order of the approximation would still be $H^0(curl)$ and does not add to the accuracy of modeling the **H**-field while doubling the unknown count. It should also be noted that the form of the facet bases in (2.36) are different in the original paper [17]. This is due to recent analysis [22] that shows that the Nedelec constraints [10] are met by (2.36), resulting in smaller dispersion error and better conditioned matrices.

2.4.2 Three-Dimensional Basis Functions

Edge-based elements have facilitated to a great degree the finite element analysis of three-dimensional structures in electromagnetics. Linear nodal bases with their problem of spurious modes and difficulty in maintaining only tangential continuity across material interfaces are not as convenient for electromagnetic field simulations in three dimensions. On the other hand, the introduction of edge-based shape functions provides a robust way of treating general three-dimensional problems having material inhomogeneities and structural irregularities like sharp edges and corners.

In the following section, we will consider first the simple rectangular bricks and will proceed to present edge-based shape functions for more complicated finite elements such as tetrahedrals and curvilinear hexahedrals. The chapter is concluded with a brief discussion on hierarchical edge elements.

2.4.2.1 Rectangular Bricks and Hexahedrals. As in the two-dimensional case, we derive the edge-based shape function for a rectangular brick (see Fig. 2.11) by simple inspection. Since a constant tangential field component must be assigned to each edge of the element, we can express the shape function along each edge of the element as [15]

$$\boldsymbol{W}_1^e = \frac{1}{h_y^e h_z^e}\left(-y + y_c^e + \frac{h_y^e}{2}\right)\left(-z + z_c^e + \frac{h_z^e}{2}\right)\hat{\mathbf{x}}$$

$$\boldsymbol{W}_2^e = \frac{1}{h_y^e h_z^e}\left(y - y_c^e + \frac{h_y^e}{2}\right)\left(-z + z_c^e + \frac{h_z^e}{2}\right)\hat{\mathbf{x}}$$

$$\boldsymbol{W}_3^e = \frac{1}{h_y^e h_z^e}\left(-y + y_c^e + \frac{h_y^e}{2}\right)\left(z - z_c^e + \frac{h_z^e}{2}\right)\hat{\mathbf{x}}$$

$$\boldsymbol{W}_4^e = \frac{1}{h_y^e h_z^e}\left(y - y_c^e + \frac{h_y^e}{2}\right)\left(z - z_c^e + \frac{h_z^e}{2}\right)\hat{\mathbf{x}}$$

$$\boldsymbol{W}_5^e = \frac{1}{h_x^e h_z^e}\left(-z + z_c^e + \frac{h_z^e}{2}\right)\left(-x + x_c^e + \frac{h_x^e}{2}\right)\hat{\mathbf{y}}$$

$$\boldsymbol{W}_6^e = \frac{1}{h_x^e h_z^e}\left(z - z_c^e + \frac{h_z^e}{2}\right)\left(-x + x_c^e + \frac{h_x^e}{2}\right)\hat{\mathbf{y}}$$

$$\boldsymbol{W}_7^e = \frac{1}{h_x^e h_z^e}\left(-z + z_c^e + \frac{h_z^e}{2}\right)\left(x - x_c^e + \frac{h_x^e}{2}\right)\hat{\mathbf{y}}$$

$$\boldsymbol{W}_8^e = \frac{1}{h_x^e h_z^e}\left(z - z_c^e + \frac{h_z^e}{2}\right)\left(x - x_c^e + \frac{h_x^e}{2}\right)\hat{\mathbf{y}}$$

$$\boldsymbol{W}_9^e = \frac{1}{h_x^e h_y^e}\left(-x + x_c^e + \frac{h_x^e}{2}\right)\left(-y + y_c^e + \frac{h_y^e}{2}\right)\hat{\mathbf{z}}$$

$$\boldsymbol{W}_{10}^e = \frac{1}{h_x^e h_y^e}\left(x - x_c^e + \frac{h_x^e}{2}\right)\left(-y + y_c^e + \frac{h_y^e}{2}\right)\hat{\mathbf{z}}$$

$$\boldsymbol{W}_{11}^e = \frac{1}{h_x^e h_y^e}\left(-x + x_c^e + \frac{h_x^e}{2}\right)\left(y - y_c^e + \frac{h_y^e}{2}\right)\hat{\mathbf{z}}$$

$$\boldsymbol{W}_{12}^e = \frac{1}{h_x^e h_y^e}\left(x - x_c^e + \frac{h_x^e}{2}\right)\left(y - y_c^e + \frac{h_y^e}{2}\right)\hat{\mathbf{z}}$$

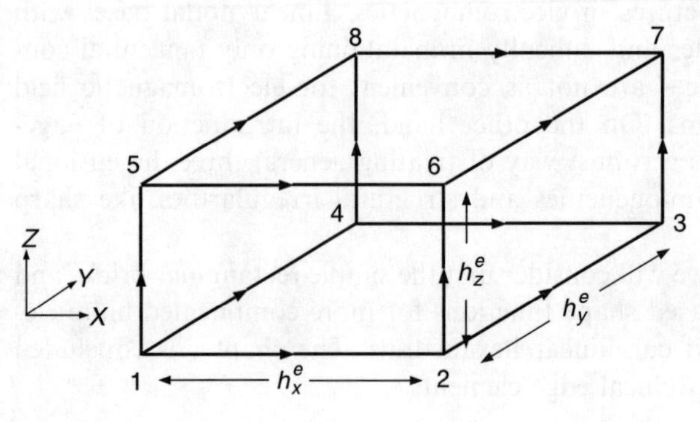

Figure 2.11 Rectangular brick element. The numbers denote the local node numbering scheme.

where h_x^e, h_y^e, h_z^e denote the edge lengths in the x, y, and z directions, respectively, and the center coordinates of the brick are given by (x_c^e, y_c^e, z_c^e). If the local edge numbers are defined as in Table 2.3, the vector field within the element can be expressed as

$$\mathbf{E}^e = \sum_{k=1}^{12} E_k^e \boldsymbol{W}_k^e \tag{2.37}$$

where E_k^e represents the value of the electric field along the kth edge of the eth element. The vector bases $\mathbf{W}_k^e$ defined for the rectangular brick element have zero divergence and a nonzero curl. Furthermore, the expansion (2.37) guarantees tangential continuity of the electric field across the surfaces of the elements.

A rectangular brick element has limitations in the sense that it is unable to model irregular geometries. For this reason, the analog of the two-dimensional quadrilateral (the hexahedral element) is more attractive for modeling practical three-dimensional problems. As in the case of the quadrilateral element in two dimensions, a hexahedral element in Cartesian coordinates can be seen as the image of a unit cube under a trilinear mapping to the $\xi\eta\zeta$ coordinate system (see Fig. 2.12).

Let us consider those faces for which $\xi =$ constant. Therefore, $\nabla\xi$ must then possess only a normal component on that face. Since ξ varies linearly along the edges that are parallel to the ξ-axis, the vector function $\nabla\xi$ has nonzero tangential components only along those edges that are parallel to the ξ-axis. Using the node-based

TABLE 2.3 Edge Definition for Rectangular Brick

Edge No.	Node i_1	Node i_2
1	1	2
2	4	3
3	5	6
4	8	7
5	1	4
6	5	8
7	2	3
8	6	7
9	1	5
10	2	6
11	4	8
12	3	7

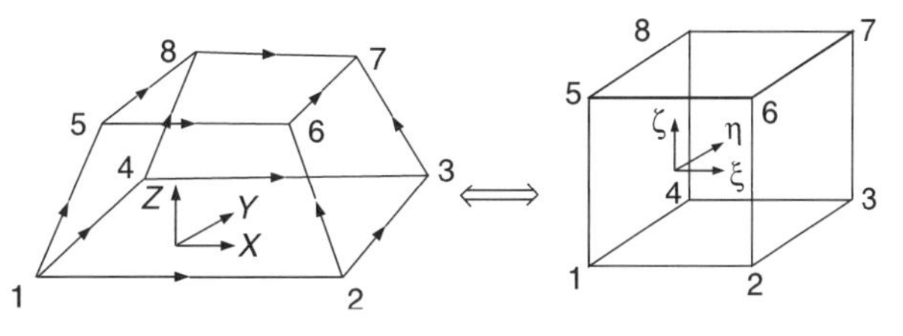

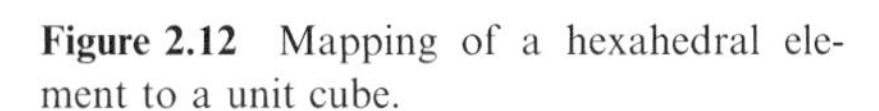
Figure 2.12 Mapping of a hexahedral element to a unit cube.

expression for the shape function in a hexahedral element given in (2.23), we may write the corresponding edge bases as

$$W_k^e = \frac{h_k^e}{8}(1 + \eta_k \eta)(1 + \zeta_k \zeta)\,\nabla\xi \qquad \text{edges} \parallel \text{to } \xi\text{-axis} \tag{2.38}$$

$$W_k^e = \frac{h_k^e}{8}(1 + \xi_k \xi)(1 + \zeta_k \zeta)\,\nabla\eta \qquad \text{edges} \parallel \text{to } \eta\text{-axis} \tag{2.39}$$

$$W_k^e = \frac{h_k^e}{8}(1 + \xi_k \xi)(1 + \eta_k \eta)\,\nabla\zeta \qquad \text{edges} \parallel \text{to } \zeta\text{-axis} \tag{2.40}$$

where (ξ_k, η_k, ζ_k) denote the coordinates at the kth edge and h_k^e is the length of the kth edge belonging to the eth element.

The vector bases derived above possess all the desired continuity properties of edge elements and generally result in about half the number of unknowns generated by tetrahedral gridding. The difficulty in generating a finite element mesh of an arbitrary structure using hexahedra can be a serious limitation. In practice, often, a combination of hexahedra and tetrahedra in a finite element mesh is used and continuity is imposed across the different element interfaces to solve the problem. However, this may result in ill-conditioned matrices.

2.4.2.2 Tetrahedral Elements. Tetrahedra are, by far, the most popular element shapes to be employed for three-dimensional applications. This is because the tetrahedral element is the simplest tessellation shape capable of modeling arbitrary three-dimensional geometries and is also well suited for automatic mesh generation. The derivation of shape functions for these elements follow the same pattern as that for triangular vector basis functions. If we consider the tetrahedron shown in Fig. 2.6 and define the edge numbers according to Table 2.4, we have

$$W_k^e = N_{ij}^e = \ell_{ij}(L_i^e \nabla L_j^e - L_j^e \nabla L_i^e), \qquad i, j = 1, \ldots, 4 \tag{2.41}$$

where again $\ell_{ij} = \ell_k$ denotes the length of the edge between nodes i and j, which in turn define the kth edge. The vector field within the element can then be expanded as

$$\mathbf{E} = \sum_{k=1}^{6} E_k^e W_k^e \tag{2.42}$$

TABLE 2.4 Edge Definition for Tetrahedron

Edge No.	Node i_1	Node i_2
1	1	2
2	1	3
3	1	4
4	2	3
5	4	2
6	3	4

where the coefficients E_k^e represent the average value of the field along the kth edge of the eth element.

An elegant explanation of the physical character of the edge-based interpolation function is given by Bossavit [23]. Let us consider edge number 1 connecting nodes 1 and 2 in Fig. 2.6. Since ∇L_2^e is orthogonal to facet {134} and ∇L_1^e is orthogonal to facet {234}, the field turns around the axis 3–4 and is normal to planes containing nodes 3 and 4. The field thus has only tangential continuity across element faces. Edge elements can also be described as Whitney elements of degree one and can be broadly classified as belonging to the $H^0(curl)$ space.

Whitney elements of the second degree are called *facet* elements because they are constant over the face of the tetrahedron. The vector function for the facet element can be written as

$$N_{ijk}^e = 2(L_i^e \nabla L_j^e \times \nabla L_k^e + L_j^e \nabla L_k^e \times \nabla L_i^e + L_k \nabla L_i^e \times \nabla L_j^e), \qquad i, j, k = 1, \ldots, 4 \tag{2.43}$$

As explained in [23], we now have a central field (as if emanating from node 4 in Fig. 2.6) on each of the two tetrahedra that share the face $\{1, 2, 3\}$. The field can be imagined as coming from the 'source' 4, growing, crossing the facet, and vanishing into the 'well' $4'$, the fourth vertex of the other tetrahedron. Thus, this field has normal continuity and the flux across the facet forms the degree of freedom for the element.

Alternative expressions for linear basis inside a tetrahedron have been derived in [14]. They are given by

$$\boldsymbol{W}_{7-i}^e = \begin{cases} \mathbf{f}_{7-i} + \mathbf{g}_{7-i} \times \mathbf{r}, & \mathbf{r} \text{ in the tetrahedron} \\ 0, & \text{otherwise} \end{cases} \tag{2.44}$$

with

$$\mathbf{f}_{7-i} = \frac{h_{7-i}}{6V_e} \mathbf{r}_{i_1} \times \mathbf{r}_{i_2} \tag{2.45}$$

$$\mathbf{g}_{7-i} = \frac{h_i h_{7-i} \hat{\mathbf{e}}_i}{6V_e} \tag{2.46}$$

in which $i = 1, 2, \ldots, 6$, V_e is the volume of the tetrahedral element, $\hat{\mathbf{e}}_i = (\mathbf{r}_{i_2} - \mathbf{r}_{i_1})/h_i$ is the unit vector of the ith edge, and $h_i = |\mathbf{r}_{i_2} - \mathbf{r}_{i_1}|$ is the length of the ith edge with $\mathbf{r}_{i_1}$ and $\mathbf{r}_{i_2}$ denoting the position vector of the i_1 and i_2 nodes. It can be shown that (2.41) is identically equal to (2.44) when simplified. Therefore,

$$\mathbf{g}_{7-i} = h_{7-i}(\nabla L_{i_1}^e \times \nabla L_{i_2}^e), \qquad i = 1, \ldots, 6$$

where i_1 and i_2 are given in Table 2.4. The basis functions given in (2.44) have zero divergence and constant curl ($\nabla \times \boldsymbol{W}_i^e = 2\mathbf{g}_i^e$). The form of the basis functions given in (2.44) is similar to the zeroth-order edge elements postulated by Nedelec [10].

The order of the polynomial approximation for the first-order edge element given in (2.41) or (2.44) can be taken as 0.5. This is because the value of the basis function is constant, i.e., $O(1)$, along the edge it supports and is linear everywhere else within the element. Mur and de Hoop [12] presented edge elements which are consistently linear, yielding a linear approximation of the field both inside each tetrahedron and along its edges and faces. However, the curl of the basis is still

$O(1)$. Since this requires two unknowns per edge, there are twelve degrees of freedom per element. The basis functions in [12] are derived by first defining the outwardly directed vectorial areas of the faces as

$$\boldsymbol{A}_i = \boldsymbol{r}_j \times \boldsymbol{r}_k + \boldsymbol{r}_k \times \boldsymbol{r}_l + \boldsymbol{r}_l \times \boldsymbol{r}_j \tag{2.47}$$

where $\boldsymbol{r}_i, i = 1, \ldots, 4$ denote the position vectors of the vertices of the tetrahedron and i, j, k, l are cyclic. Then the edge-based vectorial expansion function is defined by

$$\boldsymbol{N}_{ij}^e(\boldsymbol{r}) = -\frac{\phi_i(\boldsymbol{r})\boldsymbol{A}_j}{3V}, \quad i, j = 1, \ldots, 4, \qquad i \neq j \tag{2.48}$$

where V is the volume of the tetrahedron and $\phi(\boldsymbol{r})$ is a linear scalar function of position given by

$$\phi_i(\boldsymbol{r}) = \frac{1}{4} - \frac{(\boldsymbol{r} - \boldsymbol{r}_b) \cdot \boldsymbol{A}_i}{3V}$$

in which $\boldsymbol{r}_b$ is the position vector of the centroid of the tetrahedron. We observe that $\phi_i(\boldsymbol{r})$ equals unity when $\boldsymbol{r} = \boldsymbol{r}_i$ and zero for the remaining vertices of the tetrahedral element. In that sense, they are very similar to the simplex or volume coordinates mentioned earlier. They also satisfy the following equalities:

$$\boldsymbol{r} = \sum_{i=1}^{4} \phi_i(\boldsymbol{r})\, \boldsymbol{r}_i, \qquad \sum_{i=1}^{4} \phi_i(\boldsymbol{r}) = 1$$

The edge basis function $\boldsymbol{N}_{ij}^e$ is a linear vector function of position inside the tetrahedral element, and its tangential component vanishes on all edges of the element except the one joining vertices i and j. $\boldsymbol{N}_{ij}^e$ varies linearly along the edge formed by nodes i and j such that $\boldsymbol{N}_{ij}^e \cdot \boldsymbol{r}_j = 0$ while

$$\boldsymbol{N}_{ij}^e \cdot (\boldsymbol{r}_i - \boldsymbol{r}_j) = 1$$

These basis functions have nonzero values of divergence and curl.

An inspection of the expressions for the vectorial areas $\mathbf{A}_i$ reveals that the form is identical to that obtained by taking the gradient of one of the simplex or volume coordinates mentioned earlier. In other words, the three components of the vector $\boldsymbol{A}_1$ have the same functional dependence as that obtained by ∇L_1^e

$$\nabla L_1^e = \frac{1}{6V}\left[\hat{\mathbf{x}} \det \begin{vmatrix} y_2 & 1 & z_2 \\ y_3 & 1 & z_3 \\ y_4 & 1 & z_4 \end{vmatrix} - \hat{\mathbf{y}} \det \begin{vmatrix} x_2 & 1 & z_2 \\ x_3 & 1 & z_3 \\ x_4 & 1 & z_4 \end{vmatrix} + \hat{\mathbf{z}} \det \begin{vmatrix} x_2 & 1 & y_2 \\ x_3 & 1 & y_3 \\ x_4 & 1 & y_4 \end{vmatrix}\right]$$

where L_1 is the volume coordinate for a tetrahedron defined in (2.26), "det" indicates the value of the determinant of the matrix and (x_i, y_i, z_i) denote the coordinates of the ith vertex. This is only to be expected since the gradient of the shape function is normal to its corresponding edge in two dimensions and normal to its corresponding face in three dimensions. The basis functions with consistently linear interpolation in the tetrahedron can thus be rewritten in a more convenient notation as

$$\boldsymbol{W}_k^e = \boldsymbol{N}_{ij}^e = \ell_{ij} L_i^e \nabla L_j^e, \qquad i, j = 1, \ldots, 4, \qquad i \neq j \tag{2.49}$$

where the normalization factor ℓ_{ij} was introduced. Note how the first order edge basis is similar in form to the zeroth-order edge basis in (2.41).

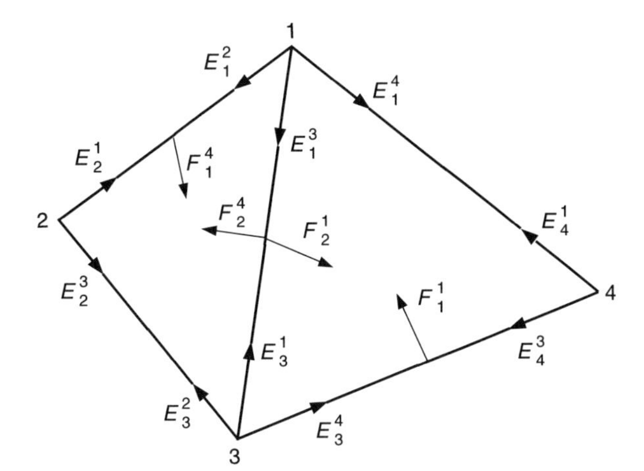

Figure 2.13 Tetrahedral element.

Still higher order basis functions are sometimes necessary for rapidly varying fields. The second-order edge basis ($O(r^{1.5})$) for a tetrahedral element was first presented by Lee, Sun, and Cendes [24]. We need 20 degrees of freedom to achieve a quadratic approximation of the vector field inside a tetrahedron (see Fig. 2.13). Accordingly, the field within a tetrahedron can be written as

$$\mathbf{E}^e = \sum_{i=1}^{4}\sum_{j=1}^{4} E_i^j L_i^e \nabla L_j^e + \sum_{\substack{i=1\\ i\neq j}}^{4} F_1^i L_i^e (L_j^e \nabla L_k^e - L_k^e \nabla L_j^e) + F_2^i L_j^e (L_k^e \nabla L_i^e - L_i^e \nabla L_k^e) \tag{2.50}$$

where i, j, k form cyclic indices. The facet variables F_1^i and F_2^i are common unknowns for two tetrahedra that share the same face. Even higher order edge-based elements complete up to polynomial order two can be constructed. Each tetrahedral element now has 30 unknowns—three along each edge and three on each face.

2.4.2.3 Triangular Prism Elements. The primary attraction of triangular prisms lies in the fact that they yield fewer unknowns than tetrahedrals while retaining the ability to mesh arbitrary geometries unlike hexahedrals. Moreover, it is sometimes possible to extrude the volume mesh out of an existing surface mesh using triangular prisms. This feature, however, may not always lead to good quality elements, especially when the geometry is non-planar with sharp corners. Finally, it is not easy to construct edge basis functions for such elements. Özdemir and Volakis [25] proposed edge-based shape functions for right-angled and distorted triangular prisms. The vector basis functions derived in [25] are a combination of edge basis over the triangular cross section and a linear variation over the height of the prism. A sketch of the basis function over the triangular and quadrilateral faces is presented in Fig. 2.14. One of the shortcomings of these bases is the lack of tangential continuity across element faces when the prisms are distorted, i.e., the vertical arms are not at right angles to the plane of the triangular faces.

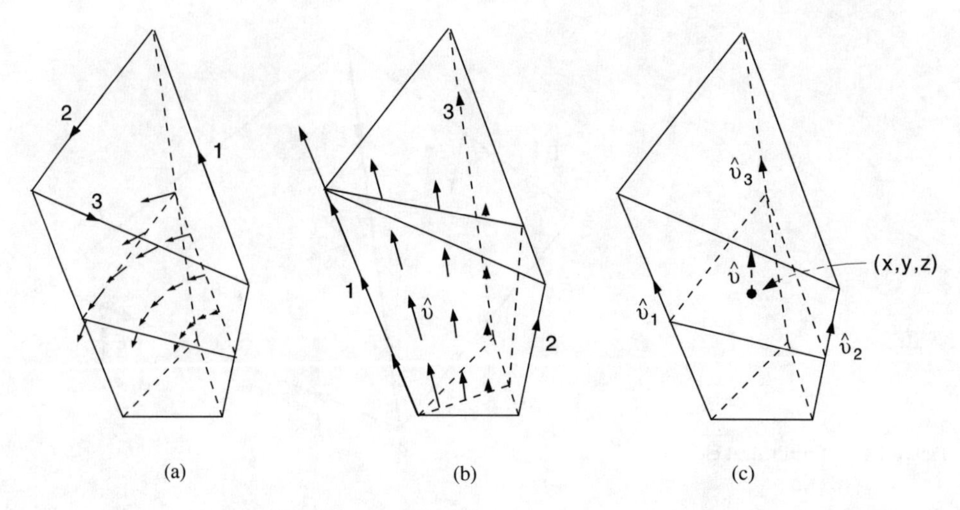

Figure 2.14 Sketch of edge basis function over triangular and quadrilateral faces of prism element. [*Courtesy of T. Özdemir.*]

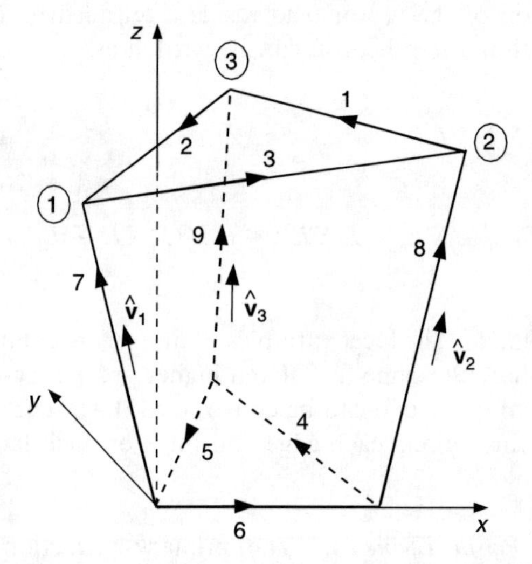

Figure 2.15 Distorted triangular prism.

The geometry of an arbitrary triangular prism is shown in Fig. 2.15. The edge basis functions for the top triangle edges are given by

$$\boldsymbol{W}_k^e = \boldsymbol{N}_{ij}^e = b_{ij}(L_i^e \nabla L_j^e - L_j^e \nabla L_i^e)\, s, \qquad i, j = 1, 2, 3; \qquad k = 1, 2, 3 \tag{2.51}$$

and those for the bottom edges are

$$\boldsymbol{W}_k^e = \boldsymbol{N}_{ij}^e = b_{ij}(L_i^e \nabla L_j^e - L_j^e \nabla L_i^e)(1 - s), \qquad i, j = 4, 5, 6; \qquad k = 4, 5, 6 \tag{2.52}$$

and the vertical edges are

$$\boldsymbol{W}_k^e = \hat{\mathbf{v}}_i(\zeta, \eta)\, L_i^e(\zeta, \eta), \qquad i = 1, 2, 3; \qquad k = 7, 8, 9 \tag{2.53}$$

In the above equations, L_i^e are the node-based shape functions (area coordinates of the triangle) defined earlier and s is a normalized parameter which is zero at the bottom face and unity at the top face of the prism. It should be noted that the basis functions for the top and the bottom edges are exactly similar to that of a triangular basis scaled by a dimensionless parameter. For the vertical edges, the vector $\hat{\mathbf{v}}$ is a linear weighting of the unit vectors $\hat{\mathbf{v}}_1$, $\hat{\mathbf{v}}_2$, $\hat{\mathbf{v}}_3$ associated with the vertical arms and is defined as

$$\boldsymbol{v}(\zeta, \eta) = \sum_{i=1}^{3} L_i^e(\zeta, \eta)\hat{\mathbf{v}}_i$$
$$\hat{\mathbf{v}}(\zeta, \eta) = \frac{\boldsymbol{v}(\zeta, \eta)}{|\boldsymbol{v}(\zeta, \eta)|} \tag{2.54}$$

This particular choice of $\hat{\mathbf{v}}$ minimizes tangential discontinuity across inter-element faces.

It is worthwhile to note that the edge basis functions for the triangular prism stated above are not Nedelec-type [10] elements since the first-order bases are not of the form

$$\boldsymbol{W}_k^e = \boldsymbol{\alpha}_k + \boldsymbol{\beta}_k \times \boldsymbol{r}, \qquad k = 1, \ldots, \text{ number of edges} \tag{2.55}$$

where $\boldsymbol{\alpha}_k$, $\boldsymbol{\beta}_k$ are constants and $\boldsymbol{r}$ is the position vector inside the finite element. The edge bases for the top and bottom faces fall into the Nedelec form but the bases for the vertical forms do not, hence the loss of tangential continuity across element faces.

2.4.2.4 Curvilinear Elements. Wang and Ida proposed a systematic method for the construction of curvilinear elements in [26]. The vector shape function is expressed in the following form:

$$\boldsymbol{W}_k^e(\boldsymbol{r}) = \phi_k(\xi, \eta, \zeta)\,\boldsymbol{v}_k(\boldsymbol{r}), \qquad k = 1, \ldots, M \tag{2.56}$$

where $\phi_k(\xi, \eta, \zeta)$ are completely defined in the local coordinate system, $\boldsymbol{v}_k$ contains the edge and facet information and M denotes the number of degrees of freedom in the element. These basis functions differ from the bases described earlier in the chapter in that they are constructed in the local coordinate system. Since the direction vectors are defined in global coordinates, the bases are uniquely defined. The joint effect of ϕ_k and $\boldsymbol{v}_k$ ensures that $\boldsymbol{W}_k^e$ is unity at node k and zero elsewhere. These basis functions usually lead to a symmetric system of equations. Antilla and Alexopoulos also proposed a curved brick *superparametric* element with 8, 27, and 64 nodes in [27] for solving scattering problems. Superparametric elements attach unknowns to a lesser number of points than required to define the geometry. The advantage of curvilinear elements lies in the fact that they can model curved surfaces with more accuracy and lesser number of unknowns than rectilinear elements. Analytical surfaces and even complicated non-planar surface features can thus be modeled exactly at low computational cost. However, many mesh generation packages cannot construct curvilinear elements for arbitrary geometries.

2.4.2.5 Hierarchical Vector Elements. Finite elements are said to be hierarchical when the basis functions for an element are a subset of the basis functions for any element of higher order [4]. Hierarchical elements find use in a class of adaptive finite elements (called *p*-refinement) where the order of approximation is improved by refining the order of the polynomial basis functions instead of refining the mesh density. However, there is usually a trade-off when higher order basis functions extract a heavy price in terms of computer resources. A major problem with going to higher order bases is the increased density of the finite element matrix and the likely worsening of the matrix condition number. Moreover, some structures may have features for which a lower order approximation is sufficient to model the field variations. This is especially the case where field variations are either uniform or constant and a higher order of interpolation may actually degrade the solution. It turns out that the vector finite elements defined by Nedelec [10] and subsequently derived by Barton and Cendes [14] and Lee et al. [17] have a hierarchical structure. Vector elements complete up to polynomial order two are available, and basis functions of a given order are fully compatible with basis functions of lower or higher orders. Thus elements of different orders could be used in the same mesh. Specifically, lower order elements could be used in regions where field variation is slowly varying and higher order elements in regions where the field varies rapidly.

The implementation of hierarchical vector elements can be difficult, especially at the transition boundaries where elements of one order merge into the elements of higher or lower order. If several vector elements share an edge, the field tangent to the edge must be made identical in each of the tetrahedra. This is done by carefully matching the coefficients of the vector basis function corresponding to that edge. For tangential continuity across a face, the same equality must be enforced between the coefficients of all the edge and facet functions associated with the face. Table 2.5 given in [28] shows the basis functions for hierarchical vector finite elements. It should be mentioned that for the zeroth-order edge element, the described polynomial approximation to be of order 0.5 is somewhat of a misnomer. It should be taken to mean that the field variation along the edge is constant, i.e., $O(r^0)$, and the variation normal to the edge is $O(r^1)$. Averaging the orders, albeit a mathematically dubious procedure, yields the described polynomial order. On the plus side, the table offers a concise view of the hierarchical nature of these edge elements. Higher order basis functions are constructed by systematically adding the extra terms up to the desired order. It should be noted that the bases for the tetrahedron with six and 20 unknowns shown in Table 2.5 is identical to the $H^0(curl)$ and $H^1(curl)$ edge basis given in (2.41) and (2.49), respectively.

TABLE 2.5 Hierarchical Basis Functions for a Tetrahedral Element

Element Type	Polynomial Order	Unknowns per Element	Basis Function
Edge	0.5	6	$L_i \nabla L_j - L_j \nabla L_i$
Edge	1	12	$\nabla(L_i L_j)$
Face	1.5	20	$L_i(L_j \nabla L_k - L_k \nabla L_j)$
Face			$L_j(L_k \nabla L_i - L_i \nabla L_k)$
Edge	2	30	$\nabla[L_i L_j(L_i - L_j)]$
Face			$\nabla[L_i L_j L_k]$

REFERENCES

[1] A. Bossavit and J. C. Verite. A mixed FEM-BIEM method to solve 3D eddy current problems. *IEEE Trans. Magnetics*, 18:431–5, March 1982.

[2] J. P. Webb. Edge elements and what they can do for you. *IEEE Trans. Magnetics*, 29:1460–1465, 1993.

[3] S. H. Wong and Z. J. Cendes. Combined finite element-modal solution of three-dimensional eddy current problems. *IEEE Trans. Magnetics*, 24(6), November 1988.

[4] O. C. Zienkiewicz. *The Finite Element Method*. McGraw-Hill, New York, Third edition, 1979.

[5] K. Tuncer, D. Norrie, and F. Brezzi. *Finite Element Handbook*. McGraw-Hill, New York, 1987.

[6] J. M. Jin. *The Finite Element Method in Electromagnetics*. John Wiley & Sons, New York, 1993.

[7] Z. J. Cendes and P. Silvester. Numerical solution of dielectric loaded waveguides: I—Finite element analysis. *IEEE Trans. Microwave Theory Tech.*, 118:1124–1131, 1970.

[8] X. Yuan, D. R. Lynch, and K. Paulsen. Importance of normal field continuity in inhomogeneous scattering calculations. *IEEE Trans. Microwave Theory Tech.*, 39:638–642, April 1991.

[9] H. Whitney. *Geometric Integration Theory*. Princeton Univ. Press, NJ, 1957.

[10] J. C. Nedelec. Mixed finite elements in R^3. *Numer. Math.*, 35:315–41, 1980.

[11] M. Hano. Finite element analysis of dielectric-loaded waveguides. *IEEE Trans. Microwave Theory Tech.*, 32:1275–1279, October 1984.

[12] G. Mur and A. T. de Hoop. A finite element method for computing three-dimensional electromagnetic fields in inhomogeneous media. *IEEE Trans. Magnetics*, 21:2188–2191, November 1985.

[13] J. S. van Welij. Calculation of eddy currents in terms of H on hexahedra. *IEEE Trans. Magnetics*, 21:2239–2241, November 1985.

[14] M. L. Barton and Z. J. Cendes. New vector finite elements for three-dimensional magnetic field computation. *J. Appl. Phys.*, 61(8):3919–3921, April 1987.

[15] J. M. Jin and J. L. Volakis. Electromagnetic scattering by and transmission through a three-dimensional slot in a thick conducting plane. *IEEE Trans. Antennas Propagat.*, 39(4):543–550, April 1991.

[16] J. M. Jin and J. L. Volakis. Scattering and radiation from microstrip patch antennas and arrays residing in a cavity. *IEEE Trans. Antennas Propagat.*, 39:1598–1604, November 1991.

[17] J. F. Lee, D. K. Sun, and Z. J. Cendes. Full-wave analysis of dielectric waveguides using tangential vector finite elements. *IEEE Trans. Microwave Theory Tech.*, MTT-39(8):1262–1271, August 1991.

[18] P. Monk. A finite element method for approximating the time-harmonic Maxwell equations. *Numer. Math.*, 63:243–261, 1992.

[19] D. H. Schaubert, D. R. Wilton, and A. W. Glisson. A tetrahedral modeling method for electromagnetic scattering by arbitrarily shaped inhomogeneous dielectric bodies. *IEEE Trans. Antennas Propagat.*, pp. 77–85, January 1984.

[20] D. R. Tanner and A. F. Peterson. Vector expansion functions for the numerical solution of Maxwell's equations. *Microwave Opt. Tech. Lett.*, 2(2):331–334, 1989.

[21] R. D. Graglia, D. R. Wilton, and A. F. Peterson. Higher order interpolatory vector bases for computational electromagnetics. *IEEE Trans. Antennas Propagat.*, pp. 329–342, March 1997.

[22] J. S. Savage and A. F. Peterson. Higher-order vector finite elements for tetrahedral cells. *IEEE Trans. Microwave Theory Tech.*, 44(6):874–879, June 1996.

[23] A. Bossavit. Whitney forms: A class of finite elements for three-dimensional computations in electromagnetism. *IEEE Proceedings*, 135, pt. A(8), November 1988.

[24] J. F. Lee, D. K. Sun, and Z. J. Cendes. Tangential vector finite elements for electromagnetic field computation. *IEEE Trans. Magnetics*, 27(5):4032–4035, September 1991.

[25] T. Özdemir and J. L. Volakis. Triangular prisms for edge-based vector finite element antenna analysis. *IEEE Trans. Antennas Propagat.*, pp. 788–797, May 1997.

[26] J. S. Wang and N. Ida. Curvilinear and higher order 'edge' finite elements in electromagnetic field computation. *IEEE Trans. Magnetics*, 29(2):1491–1494, March 1993.

[27] G. E. Antilla and N. G. Alexopoulos. Scattering from complex three-dimensional geometries by a curvilinear hybrid finite element-integral equation approach. *J. Opt. Soc. Am. A*, 11(4):1445–1457, April 1994.

[28] J. P. Webb and B. Forghani. Hierarchal scalar and vector tetrahedra. *IEEE Trans. Magnetics*, 29(2):1495–1498, March 1993.

3

Overview of the Finite Element Method: One-Dimensional Examples

3.1 INTRODUCTION

The *finite element method* (FEM) belongs to the class of partial differential equation (PDE) methods. Its origin is frequently traced to Courant [1] who in the 1940s first discussed piecewise approximations in the appendix of his paper. In the 1950s, Argyris [2] began putting together the many mathematical ideas (*domain partitioning, assembly, boundary conditions,* etc.) that comprise the FEM for aircraft structural analysis. The introduction of FEM to the engineering community occurred in the 1960s, and some feel that the conferences on finite elements held in 1965, 1968, and 1970 at the Wright Patterson Air Force Base in Dayton, Ohio, U.S. played an important role in advancing the method. Finite element activity in electrical engineering also began in the late 1960s with the papers by Silvester [3] (see also the reprints volume [4] and Arlett, Bahrani and Zienkiewicz [5]) addressing applications to waveguide and cavity analysis. Later developments on absorbing boundary conditions, perfectly matched absorbers and hybridizations with boundary integral methods have led to the successful application of the FEM to open domain problems in scattering, microwaves circuits, and antennas. The method's main advantage is its capability to treat any type of geometry and material inhomogeneity without a need to alter the formulation or the computer code. That is, it provides geometrical fidelity and unrestricted material treatment. Moreover, the application of the FEM leads to sparse matrix systems which can be stored with low memory requirements when iterative solvers are employed for the solution of these systems. We typically state that the FEM systems have $O(N)$ storage requirements, implying that the memory needed for a solution of an FEM system is proportional to the number of unknowns N. For most cases these memory requirements may range from $10N$ to $40N$ depending on the type of problem considered and the employed basis or expansion functions approximating the field within the computation domain. This is in contrast to boundary integral solutions which lead to fully populated systems having $O(N^2)$ storage and $O(N^3)$ CPU requirements. However, it should be pointed out that the number of unknowns for boundary integral equations are generally much less than those of FEM for the same problem. Nevertheless, when dealing with nonmetallic structures, the FEM and its hybrid versions is the most attractive choice.

3.2 OVERVIEW OF THE FINITE ELEMENT METHOD

The geometrical adaptability and low memory requirements of the FEM have made it one of the most popular numerical methods in all branches of engineering. Its application to boundary value problems [6] involves the subdivision of the computational domain (region where the fields are to be determined) into smaller elements [7], [8]. For two-dimensional problems, these elements are typically triangles or quadrilaterals as discussed in Chapter 2 and illustrated in Fig. 3.1. Additional example meshes are given in Figs. 3.2 and 3.3 with the latter referring to a three-dimensional mesh around a sphere.

The subdivision of the domain into small elements is referred to as meshing or discretization of the geometry and is an important part of the FEM solution procedure. By keeping the elements small enough (typically less than 1/10 of a wavelength per side), the field interior to the element can be safely approximated by some linear or, if necessary, higher order expansion. The collection of these elements and their associated expansion or shape function is therefore capable of modeling arbitrary and rather complex fields in terms of unknown coefficients which may represent the field values at the nodes (node-based basis) or the average field values over the edges (edge-based basis).

In the context of the FEM, the equations for the unknown coefficients of the expansions are constructed by enforcing the wave equation in a *weighted* (average) sense over each element. A subsequent step involves the application of the boundary conditions leading to a matrix system of the form

$$[A]\{x\} = \{b\} \tag{3.1}$$

where $\{b\}$ is a column matrix and is determined on the basis of the boundary conditions or the forced excitation (current source, incident field, etc.). The matrix $[A]$ is square of size $N \times N$, very sparse and typically symmetric unless nonreciprocal material exists in the computational domain. Its nonzero entries provide the relationship among field or voltage of adjacent elements within the computational domain, and its specific form is a characteristic of the problem geometry and discretization of the domain. Once the system (3.1) has been constructed, its solution proceeds with the application of an iterative or direct solver. Iterative solvers are primarily used for large systems (i.e., large numbers of unknowns, N) since these solvers avoid explicit storage of the entire matrix. That is, only the nonzero entries need be stored by

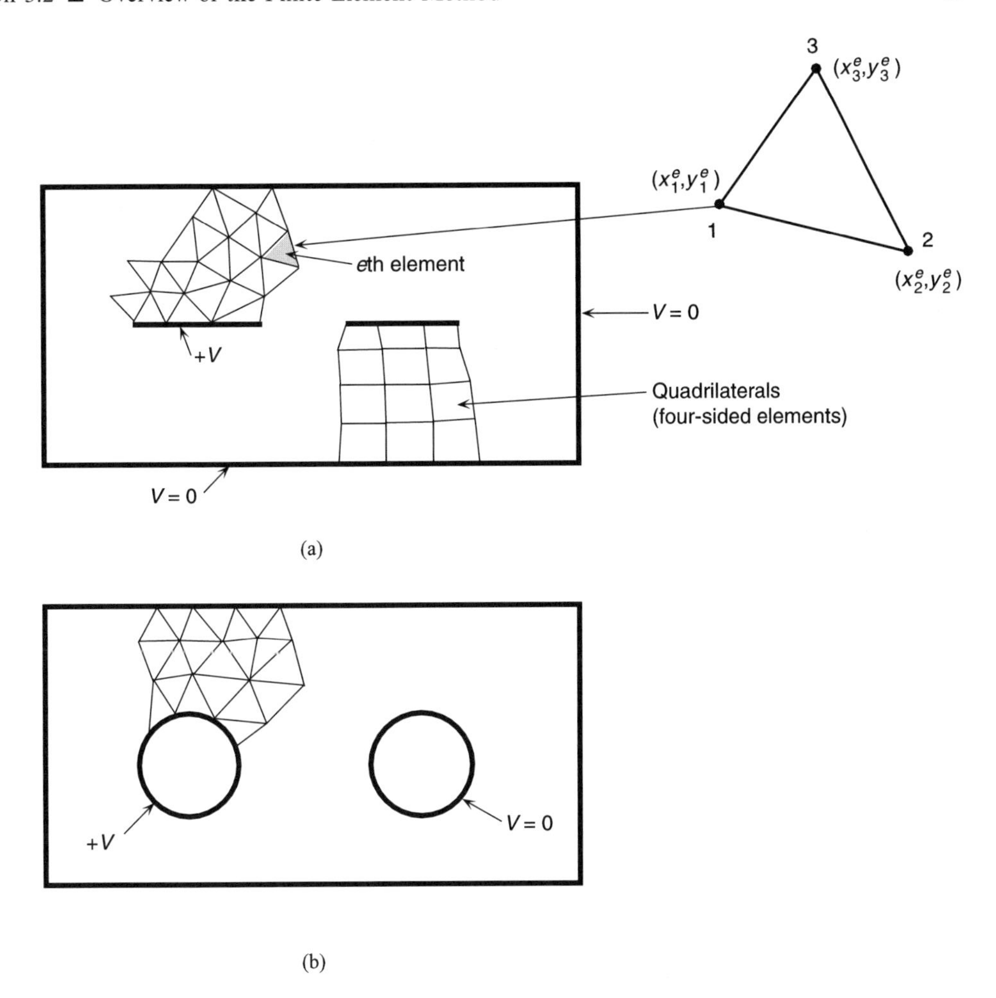

Figure 3.1 Example illustrations of finite element meshes: (a) shielded strip-conductor transmission line problem; (b) shielded circular conductor transmission line problem.

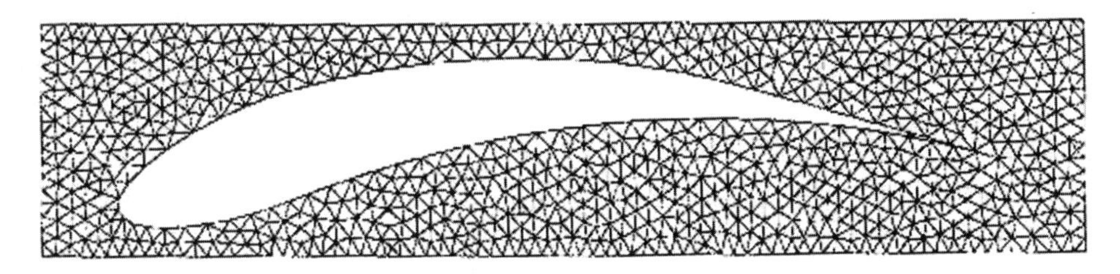

Figure 3.2 Finite element mesh around an airfoil for scattering computations. [*Courtesy of Daniel C. Ross.*]

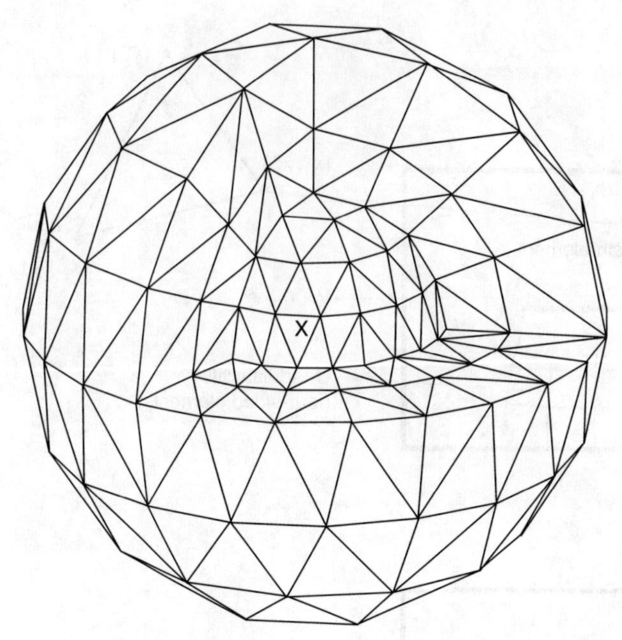

Figure 3.3 Structured tetrahedral mesh around a metallic sphere.

employing established storage schemes such as the compressed row and ITPACK formats discussed in Chapter 9. Direct solvers such as LU decomposition are still better suited for smaller size systems since they require storage of the entire matrix including its nonzero entries.

The steps involved in the generation and solution of an FEM system can be summarized as follows:

- Define the problem's computational domain
- Choose mesh truncation schemes (in the case of open domain problems)
- Choose discrete elements and shape functions
- Generate mesh (prepocessing)
- Enforce the wave equation over each element (or Laplace's/Poisson's equation for statics) to generate the element matrices
- Apply boundary conditions and assemble element matrices to form the overall sparse system (3.1)
- Ensure matrix symmetry (for domains with reciprocal materials)
- Choose solver and solve matrix system
- Postprocess field data to extract parameters of interest (such as eigenvalues, capacitance, impedance, insertion loss, scattering matrix, radar cross section and so on)

In this chapter we will present these steps for one-dimensional problems before discussing them for two-dimensional applications in Chapter 4. Although many of the two-dimensional problems are approximations of three-dimensional ones, they are nevertheless attractive because of their simplicity. Consequently, they can be used to illustrate the solution procedure without burdens due to geometrical and formulation complexity. Of course, one-dimensional problems provide the simplest way to illustrate the computational approach and below we begin with a brief illustration of the FEM for solving the classic Sturm-Liouville differential equation.

3.3 EXAMPLES OF ONE-DIMENSIONAL PROBLEMS IN ELECTROMAGNETICS

Consider the solution of the differential equation (one-dimensional Sturm-Liouville problem) [6]

$$-\frac{d}{dx}\left(p(x)\,\frac{dU}{dx}\right) + q(x)\,U(x) = f(x) \qquad 0 < x < x_a, \tag{3.2}$$

where $p(x)$, $q(x)$, and $f(x)$ are known functions and $U(x)$ is the unknown field or voltage quantity. Depending on the interpretation of $U(x)$, this equation can represent any of the following problems illustrated in Fig. 3.4.

Parallel Plate Capacitor

$U(x) = V(x)$: potential between plates
Boundary Conditions: $V(0) = 0$, $V(x_a) = V_a$,
$p(x) = -1$, $q(x) = 0$, $f(x) = -\rho/\epsilon$

$$\text{Differential equ.:} \qquad \frac{d^2}{dx^2}\,V(x) = -\frac{\rho}{\epsilon} \quad (\text{or } \nabla^2 V = -\rho/\epsilon) \tag{3.3}$$

Wave Between Parallel Plates

$U(x) = E_y(x)$: electric field between plates
Boundary Conditions: $E_y(0) = E_y(x_a) = 0$, $p(x) = -1/\mu_r$,
$q(x) = k_0^2\epsilon_r$, $f(x) =$ source function

$$\text{Differential equ.:} \qquad \frac{d}{dx}\left(\frac{1}{\mu_r}\,\frac{d}{dx}\,E_y\right) + k_0^2\epsilon_r E_y = f(x) \tag{3.4}$$

Reflection From a Coated Metallic Conductor

$$U(x) = \begin{cases} E_z(x): & \text{perpendicular polarization} \\ \quad \text{or} & \\ H_z(x): & \text{parallel polarization} \end{cases}$$

Boundary Conditions:

$$E_z(x=0) = 0, \qquad \left.\left(\frac{\partial E_z}{\partial x} + jk_0 E_z\right)\right|_{x=x_a} = 2jk_0 e^{jk_0 x_a}$$

or

(a) Infinite capacitor problem

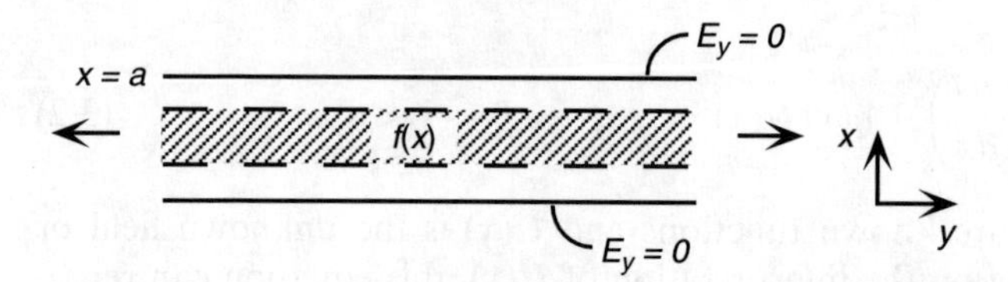

(b) Parallel plate

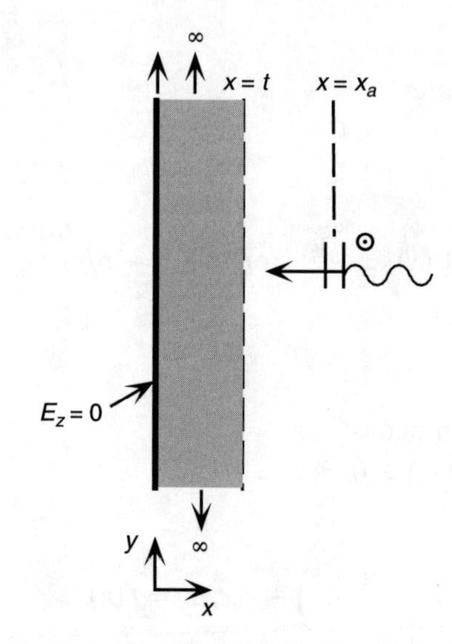

(c) Reflection from a coated metallic conductor

Figure 3.4 Problems represented by the one-dimensional differential equation. Here $x_a = a$ in (a) and (b).

$$\left.\frac{\partial H_z}{\partial x}\right|_{x=0} = 0, \qquad \left.\left(\frac{\partial H_z}{\partial x} + jk_0 H_z\right)\right|_{x=x_a} = 2jk_0 e^{jk_0 x_a}$$

$$\text{Diff. equ.:} \quad \frac{d}{dx}\left(\frac{1}{\mu_r}\frac{dE_z}{dx}\right) + k_0^2\epsilon_r E_z = 0 \quad \text{or} \quad \frac{d}{dx}\left(\frac{1}{\epsilon_r}\frac{dH_z}{dx}\right) + k_0^2\mu_r H_z = 0 \qquad (3.5)$$

In this latter case, the boundary condition at $x = x_a$ is a consequence of the fact that the reflected field must be of the form

$$\left\{\begin{matrix} E_z^r \\ H_z^r \end{matrix}\right\} = Re^{-jk_0 x} \qquad (3.6)$$

where R is the reflection coefficient of the coated ground plane and is not known until the FEM solution is completed. Thus, the total field

$$E_z = E_z^{\text{inc}} + E_z^{\text{r}}, \quad \text{or} \quad H_z = H_z^{\text{inc}} + H_z^{\text{r}} \tag{3.7}$$

satisfies the stated boundary condition. As indicated in Fig. 3.4, E_z^{inc}, and H_z^{inc} represent the z components of the plane wave incident upon the metal backed dielectric slab.

3.4 THE WEIGHTED RESIDUAL METHOD

We proceed now with the solution of (3.2) on the basis of the finite element method. As a first step we introduce the residual

$$R(x) = -\frac{d}{dx}\left(p(x)\,\frac{dU}{dx}\right) + q(x)\,U(x) - f(x) \tag{3.8}$$

which must of course be zero in accordance with the state problem. However, it is impractical to enforce $R(x) = 0$ at every point in the domain from $x = 0$ to $x = x_a = a$. Since $U(x)$ is not expected to vary substantially over a small distance, say Δx, we subdivide the domain into small segments (as shown in Fig. 3.5) and instead enforce the condition

$$\int_{\text{Domain of } W_m} W_m(x)\,R(x)\,dx = 0 \tag{3.9}$$

over each of the segments. Remarking that $W_m(x)$ is some weighting function to be defined later, (3.9) enforces the differential equation on an average sense over the mth segment. By changing the integration or testing interval (and weighting function W_m) from $m = 1$ up to $m = N$, we can construct N equations for the solution of the discretized potential or field values. Before proceeding to do so, we make the following observations about $W_m(x)$:

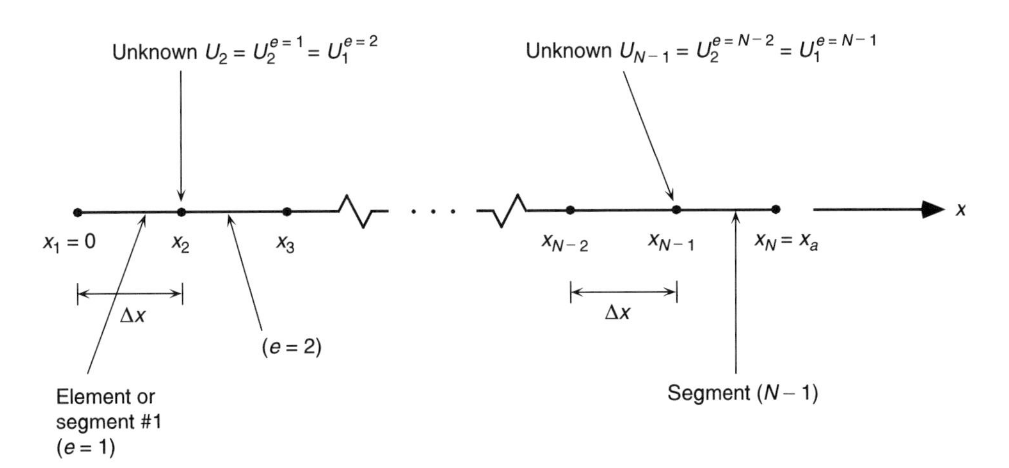

Figure 3.5 Tessellation of the line segment $0 < x < x_a$.

- If $W_m(x) = \delta(x - x_m)$ or $W_m(x) = \delta[x - (x_{m+1} + x_m)/2]$, the resulting weighted residual procedure is referred to as point matching and leads to a form of the finite difference method.
- If $W_m(x)$ is set equal to the basis functions used for the representation of $U(x)$, the procedure is referred to as Galerkin's method. This is the most popular testing/weighting method for casting the differential equation to a linear system.
- The choice of $W_m(x)$ is not completely arbitrary. For the mathematical steps in the FEM procedure to hold rigorously, $W_m(x)$ and its derivative must be at least square integrable over the domain. Specifically, for the problem at hand, it must satisfy the condition

$$\int_0^{x_a} \left\{ [W_m(x)]^2 + \left[\frac{d}{dx} W_m(x) \right]^2 \right\} dx < \infty$$

  In addition, $W_m(x)$ must satisfy conditions at the boundary nodes (endpoints at $x = 0$ and $x = x_a$ for the one-dimensional problem) which are compatible with the imposed boundary conditions. Certain smoothness conditions on $W_m(x)$ and $dW_m(x)/dx$ may also need to be imposed.

Before generating a linear system of equations from (3.9) subject to the boundary conditions, it is first necessary to cast it in a more suitable form by following the steps:

STEP 1 Take advantage of the weighting function $W_m(x)$ to reduce the order of the derivatives contained in $R(x)$. To do so we employ integration by parts, giving

$$-\int_0^{x_a} W_m(x) \frac{d}{dx}\left(p(x) \frac{dU}{dx} \right) dx = -\left[p(x)\, W_m(x) \frac{dU(x)}{dx} \right]_{x=0}^{x=x_a} + \int_0^{x_a} p(x) \frac{dW(x)}{dx} \frac{dU(x)}{dx}\, dx \tag{3.10}$$

The first right hand side (RHS) term can be evaluated by enforcing the known boundary conditions at the endpoints. Its effect on the overall system will be considered later.

STEP 2 Derive the *weak form* of the differential equation. The weak form of the differential equation is most appropriate for numerical solution and is obtained by substituting (3.10) into (3.9). We have

$$\int_0^{x_a} \left[p(x) \frac{dW_m}{x} \frac{dU}{dx} + q(x)\, W_m(x)\, U(x) - W_m(x) f(x) \right] dx - \left[p(x)\, W_m(x) \frac{dU}{dx} \right]_0^{x_a} = 0 \tag{3.11}$$

which holds provided (3.2) is valid. However, because of the integral, the *weak form* (3.11) enforces the differential equation on an average (and therefore weaker) sense. Equation (3.11) is often referred to as a *variational* statement of the problem. *What is remarkable about* (3.11) *is that it incorporates in a single mathematical statement the requirements imposed by the differential equation and the boundary conditions at the endpoints.* That is, upon substitution of the boundary conditions on $U(x)$ and

$dU(x)/dx$, (3.11) is not only an alternative statement of the differential equation (3.2), but also includes information about the boundary conditions which are essential for the uniqueness of the solution. This is at the heart of the FEM, and later in this chapter we will observe how the boundary conditions impact the discrete form of (3.11).

3.5 DISCRETIZATION OF THE "WEAK" DIFFERENTIAL EQUATION

The discretization of (3.11) to a linear set of equations is done by introducing an expansion for $U(x)$ and then making appropriate choices for the weighting functions $W_m(x)$. This is an essential step in all numerical solution procedures, and the previous chapter served to introduce the various classes of basis functions used in a discrete representation of the unknown function. We choose the linear representation

$$U(x) = \sum_{e=1}^{N_e} \sum_{i=1}^{2} U_i^e \, N_i^e(x) \tag{3.12}$$

where U_i^e are the unknown coefficients of the expansion and (see Fig. 3.6)

$$N_1^e(x) = \begin{cases} \dfrac{x_2^e - x}{x_2^e - x_1^e}, & x_1^e < x < x_2^e \\ 0 & \text{otherwise} \end{cases} \qquad N_2^e(x) = \begin{cases} \dfrac{x - x_1^e}{x_2^e - x_1^e}, & x_1^e < x < x_2^e \\ 0 & \text{otherwise} \end{cases} \tag{3.13}$$

are the shape functions discussed in Chapter 2. That is,

$$U(x) = \sum_{e=1}^{N_e} \left[U_1^e \frac{x_2^e - x}{x_2^e - x_1^e} + U_2^e \frac{x - x_1^e}{x_2^e - x_1^e} \right] \tag{3.14}$$

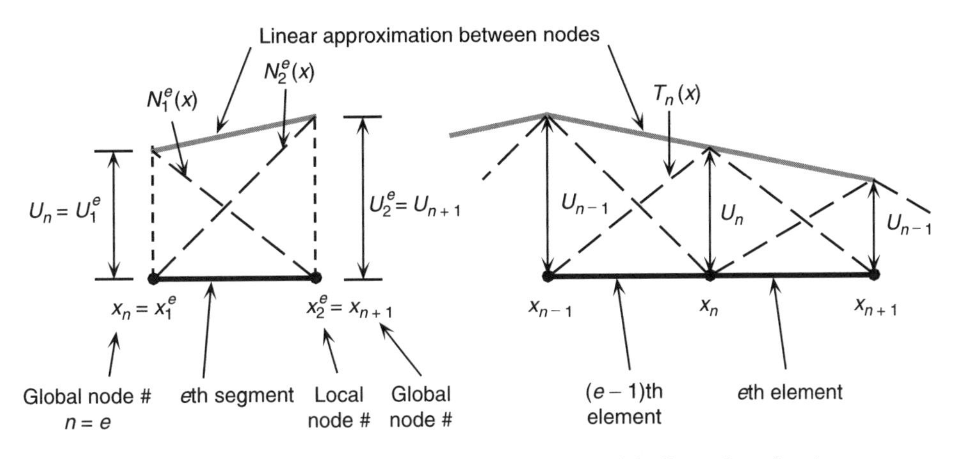

Figure 3.6 Illustration of the eth segment or element and the linear shape functions: (a) the nodal expansion functions; (b) overlay of the nodal expansion functions.

where U_1^e and U_2^e are the unknown values at each node. On the basis of this expansion, the field or potential over the domain $0 < x < x_a$ is determined by a linear combination of the field or potential values at each node. Although not necessary for this one-dimensional example, two types of node-numbering schemes are typically used to facilitate the programming and implementation of the finite element solution.

Local Node Numbers. These are assigned node numbers unique only within a single element. For the line segments in Fig. 3.5 each segment is formed by two nodes having the local node numbers 1 and 2, as illustrated in Fig. 3.6. That is, the notation x_1^e refers to the location of the local node 1 of the eth element. Similarly, U_1^e refers to the field or potential at node 1 of the eth element. Using this type of notation, we can develop formulations and equations for a single element which can then be incorporated into the overall solution by attaching the superscript e to all local or element variables. In this manner, the uniqueness of the equations is maintained when combined with those from the other elements. The elements (two-node segments for one-dimensional problems) in the computational domain are assumed to have a unique number from $e = 1$ to $e = N_e$.

Global Node Numbers. Each node of the discretized domain is also given a unique number from 1 to N, as shown for example in Fig. 3.5. The assignment of these global numbers is necessary since eventually all unknowns from each element must be collected (a process referred to as element assembly) into a matrix system such as that in (3.1). At that stage, it is necessary to maintain a single subscript or array dimension. This necessitates the correspondence between the notation x_i^e and x_n, where the first refers to the local numbering notation and hereon the single subscript will be reserved for the global numbering notation. As can be realized, since every node in Fig. 3.5 belongs to two elements, multiple local notations can refer to the same global node or field value. As an example (see Fig. 3.6), the node location x_n is identical to the locations implied by the notations x_1^e and x_2^{e-1}. Likewise, for the field values we can state that $U_e = U_1^e = U_2^{e-1}$ and so on.

When the expansion (3.14) is substituted into (3.11) we get

$$\sum_{e=1}^{N_e}\left\{\sum_{i=1}^{2} U_i^e \int_{x_1^e}^{x_2^e}\left[p(x)\,\frac{dW_m(x)}{dx}\,\frac{dN_i^e(x)}{dx} + q(x)\,W_m(x)\,N_i^e(x) - W_m(x)f(x)\right]dx\right\}$$
$$-\left[p(x)\,W_m(x)\left.\frac{dU}{dx}\right|_{x=x_a} - p(x)\,W_m(x)\left.\frac{dU}{dx}\right|_{x=0}\right] = 0 \qquad (3.15)$$

The latter terms in the brackets are due to contributions from the endpoints of the domain and their evaluation is subject to the specific boundary conditions. This equation now explicitly shows how the boundary conditions enter into the construction of the linear system. Hereon, we will refer to their contributions as [endpoints] since we have not yet specified the type of boundary condition to be imposed.

We are now ready to make different choices for the weighting function to generate a system of linear equations for the solution of $\{U_n\}$. As stated earlier, this step is also referred to as testing and Galerkin's method is usually employed in the finite element method. Specifically, we choose $W_m(x) = N_j^e(x)$ and for each of these testing or weighting functions a single linear equation is generated. From (3.15) we have

$$\left\{ \sum_{i=1}^{2} U_i^e \int_{x_1^e}^{x_2^e} \left[p(x) \frac{dN_i^e(x)}{dx} \frac{dN_j^e(x)}{dx} + q(x) N_i^e(x) N_j^e(x) - N_j^e(x) f(x) \right] dx \right\} + [\text{endpoints}] = 0 \tag{3.16a}$$

where

$$\begin{aligned}
[\text{endpoints}] &= \text{endpoints}_1 + \text{endpoints}_2 \\
&= \left[p(x) N_1^1(x) \left. \frac{dU}{dx} \right|_{x=0} - p(x) N_2^{N_e}(x) \left. \frac{dU}{dx} \right|_{x=x_a} \right] \\
&= \left[p(x) \left. \frac{dU}{dx} \right|_{x=0} - p(x) \left. \frac{dU}{dx} \right|_{x=x_a} \right]
\end{aligned} \tag{3.16b}$$

Since $N_j^e(x)$ is nonzero only over the eth element, the summation over the elements can be eliminated at this stage. In other words, integration is carried out only over the nonzero portion of the integrand. We can rewrite (3.16a) in matrix form as

$$[A_{ji}^e]\{U_i^e\} + [\text{endpoints}] = \{b_j^e\} \tag{3.17a}$$

which can be considered as the weighted discrete form of the differential equation. This matrix system provides a relationship only between the two nodes forming the eth element and is therefore a localized relationship among the node fields/potentials. The endpoint contributions appear only when $e = 1$ or $e = N_e$ and vanish when the Neumann boundary conditions $\left. \frac{dU}{dx} \right|_{x=0} = \left. \frac{dU}{dx} \right|_{x=x_a} = 0$ are imposed. Otherwise (3.17a) becomes

$$[A_{ij}^e]\{U_j^e\} = \{b_i^e\} \tag{3.17b}$$

This is commonly referred to as the *elemental matrix* system. Also, the matrix

$$[A_{ij}^e] = \begin{bmatrix} A_{11}^e & A_{12}^e \\ A_{21}^e & A_{22}^e \end{bmatrix} \tag{3.18}$$

is referred to as the element matrix and its entries are given by

$$A_{ij}^e = \int_{x_1^e}^{x_2^e} \left[p(x) \frac{dN_i^e(x)}{dx} \frac{dN_j^e(x)}{dx} + q(x) N_i^e(x) N_j^e(x) \right] dx \tag{3.19}$$

The excitation vector entries are specified by

$$b_i^e = \int_{x_1^e}^{x_2^e} N_i^e(x) f(x)\, dx \tag{3.20}$$

Since $N_i^e(x)$ are linear functions, the evaluation of A_{ij}^e can be carried out in closed form provided $p(x)$ and $q(x)$ are taken as constant over the integration of the eth element. Specifically, setting $p(x) \approx p^e$ and $q(x) \approx q^e$ for $x_1^e < x < x_2^e$, we find that

$$A_{11}^e = A_{22}^e = \frac{p^e}{|x_2^e - x_1^e|} + q^e \frac{|x_2^e - x_1^e|}{3} \tag{3.21}$$

$$A_{12}^e = A_{21}^e = -\frac{p^e}{|x_2^e - x_1^e|} + q^e \frac{|x_2^e - x_1^e|}{6} \tag{3.22}$$

Also, on approximating $f(x)$ by $f(x) \approx f^e$ over the eth element, we obtain

$$b_1^e = b_2^e = f^e \frac{|x_2^e - x_1^e|}{2} \tag{3.23}$$

The above testing procedure will result in $2N_e$ equations obtained by letting $e = 1, 2, \ldots, N_e$ in (3.17). Since only $N_e + 1$ unique unknowns exist, it is necessary to condense or combine the $2N_e$ equations down to $N_e + 1$. The additional set of equations is a result of testing at the same nth node from the left using the testing function $N_2^{e-1}(x)$ and from the right using the testing function $N_1^e(x)$. Their reduction to $N_e + 1$ equations is referred to as *assembly* of the element equations and is a standard step in all finite element solutions.

3.6 ASSEMBLY OF THE ELEMENT EQUATIONS

The essence of the assembly procedure is to take the average of the test equations from the left and right of the nth node. That is, we consider the weighted average,

$$\left[\int_{x_1^{e-1}}^{x_2^{e-1}} N_2^{e-1}(x) R(x)\, dx + \int_{x_1^e}^{x_2^e} N_1^e(x)\, R(x)\, dx\right] = 0$$

or

$$\int_{x_1^{e-1}=x_{m-1}}^{x_2^e = x_{m+1}} T_m(x)\, R(x)\, dx = 0, \qquad m = 2, 3, \ldots, N-1 \tag{3.24}$$

where $T_m(x)$ is illustrated in Fig. 3.6(b) and is given by

$$T_m(x) = \begin{cases} \dfrac{x - x_{m-1}}{\Delta x} & x_{m-1} < x < x_m \\ \dfrac{x_{m+1} - x}{\Delta x} & x_m < x < x_{m+1} \\ 0 & \text{otherwise} \end{cases} \tag{3.25}$$

From (3.24), the test equation at the mth node has the explicit form

$$\sum_{n=1}^{N} A_{mn} U_n = b_m, \qquad m = 2, 3, \ldots, N-1 \tag{3.26}$$

where

$$A_{mn} = \int_{x_{m-1}}^{x_{m+1}} \left[p(x) \frac{dT_m}{dx} \frac{dT_n}{dx} + q(x)\, T_m(x)\, T_n(x) \right] dx$$

$$= \begin{cases} A_{11}^{e=m} + A_{22}^{e=m-1}, & n = m \\ A_{12}^{e=m-1}, & n = m-1 \\ A_{21}^{e=m+1}, & n = m+1 \end{cases} \tag{3.27}$$

$$b_m = \int_{x_{m-1}}^{x_{m+1}} T_m(x) f(x)\, dx = b_2^{e=m-1} + b_1^{e=m} \tag{3.28}$$

We have temporarily excluded the cases of $m = 1$ and $m = N$ since testing at the boundary nodes 1 and N requires the inclusion of the endpoint contributions. These must be dealt with after the specification of the boundary conditions. Thus, (3.24) gives a set of $N - 2$ equations with the other two equations to be supplied later. When the Dirichlet boundary conditions $U(x = 0) = U(x = x_a) = 0$ are imposed at the endpoints of the domain, the implication is that U_1 (or U_1^1) $= U_N$ (or $U_2^{N_e}$) $= 0$ and thus the number of unknowns is reduced from N down to $N - 2$. Consequently, when $U(x)$ satisfies the Dirichlet boundary conditions on both ends of the domain, the system (3.24) is sufficient for the solution of the unknown vector $\{U_n\}$ provided the node fields or potentials are set to zero a priori.

In practice (i.e., in a computer implementation of the FEM), the construction of the linear system is done by manipulating the element matrix system (3.17). This procedure appears rather tedious on paper because of its repetitiveness but is most suitable for computer implementation. We will illustrate it below for the one-dimensional formulation and in connection with the three-element segment given in Fig. 3.7.

For convenience, let us assume that the Neumann boundary condition is satisfied at nodes 1 and 2, i.e., $\left.\frac{\partial U}{\partial x}\right|_{x=0} = \left.\frac{\partial U}{\partial x}\right|_{x=x_a} = 0$. Thus, the element matrix equations (3.17b) are applicable even when testing at elements 1 and 3. More specifically, for $e = 1$, (3.17b) becomes

$$\begin{bmatrix} A_{11}^1 & A_{12}^1 \\ A_{21}^1 & A_{22}^1 \end{bmatrix} \begin{Bmatrix} U_1^1 \\ U_2^1 \end{Bmatrix} = \begin{Bmatrix} b_1^1 \\ b_2^1 \end{Bmatrix} \tag{3.29}$$

Also, for $e = 2$ and $e = 3$, we have

$$\begin{bmatrix} A_{11}^2 & A_{12}^2 \\ A_{21}^2 & A_{22}^2 \end{bmatrix} \begin{Bmatrix} U_1^2 \\ U_2^2 \end{Bmatrix} = \begin{Bmatrix} b_1^2 \\ b_2^2 \end{Bmatrix} \tag{3.30}$$

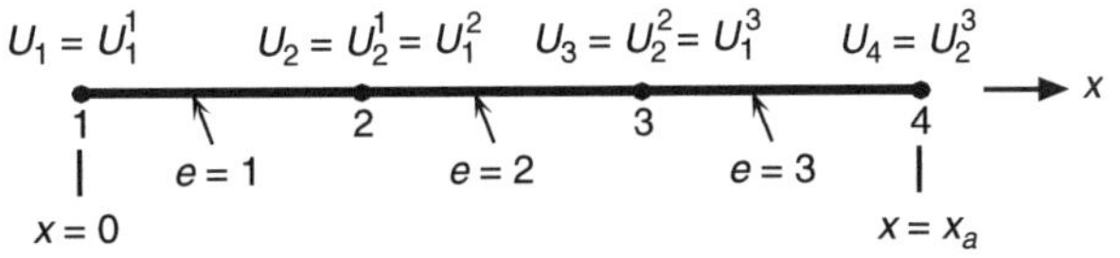

Figure 3.7 Three-element tessellation of a line segment.

and

$$\begin{bmatrix} A_{11}^3 & A_{12}^3 \\ A_{21}^3 & A_{22}^3 \end{bmatrix} \begin{Bmatrix} U_1^3 \\ U_2^3 \end{Bmatrix} = \begin{Bmatrix} b_1^3 \\ b_2^3 \end{Bmatrix} \tag{3.31}$$

These are six equations for the four unknown coefficients U_1, U_2, U_3, and U_4. To reduce (3.29)–(3.31) down to four equations, we simply add those which correspond to testing from the left and right of the node. This addition of equations amounts to simply performing the first sum in (3.15) and is not an arbitrary decision in the process. For nodes 1 and 4, there is only testing from one side and therefore the first equation of (3.29) and the second equation of (3.31) are left unchanged. For node 2, we add the second equation of (3.29) and the first equation of (3.30). Likewise for node 3, we add the second equation of (3.30) with the first equation of (3.31). The resulting (i.e., *assembled*) system of four equations is

$$\begin{bmatrix} A_{11}^1 & A_{12}^1 & 0 & 0 \\ A_{21}^1 & A_{22}^1 + A_{11}^2 & A_{12}^2 & 0 \\ 0 & A_{21}^2 & A_{22}^2 + A_{11}^3 & A_{12}^3 \\ 0 & 0 & A_{21}^3 & A_{22}^3 \end{bmatrix} \begin{Bmatrix} U_1 \\ U_2 \\ U_3 \\ U_4 \end{Bmatrix} = \begin{Bmatrix} b_1^1 \\ b_2^1 + b_1^2 \\ b_2^2 + b_1^3 \\ b_2^3 \end{Bmatrix} \tag{3.32}$$

where we employed the global node notation for the $\{U\}$ vector. In placing this into the compact notation

$$[A]\{U\} = \{b\} \tag{3.33}$$

we note that $[A]$ is a tridiagonal matrix regardless of the number of elements used for the tessellation of the line segment $0 < x < x_a$. That is, a maximum of three nonzero entries appear in each row of $[A]$ and except for the top and bottom row, the diagonal entry and its adjacent elements are the only nonzero entries. We can state that the bandwidth of the matrix is three regardless of the number of elements. Simply put, as the number of nodes/elements increases, the matrix takes the generic sparse form

$$[A] = \begin{bmatrix} x & x & 0 & 0 & \dots & & & & & & & \\ x & x & x & 0 & 0 & \dots & & & & & & \\ 0 & x & x & x & 0 & 0 & \dots & & & & & \\ 0 & 0 & x & x & x & 0 & 0 & \dots & & & & \\ 0 & 0 & 0 & x & x & x & 0 & 0 & \dots & & & \\ & \dots & 0 & 0 & x & x & x & 0 & 0 & \dots & & \\ & & & \ddots & \ddots & \ddots & & & & & & \\ & & & & & \cdot & \cdot & \cdot & 0 & 0 & x & x \end{bmatrix} \tag{3.34}$$

where the "x" symbols imply a nonzero entry and the " . . . " denote a continuation of the zero entries. In applied mechanics $[A]$ is referred to as the *stiffness* matrix because of the similarity of (3.33) to the equation $Kx = f$ for the deflection x of a linear spring with stiffness K under an applied force f. In electromagnetics, $[A]$ can be interpreted in several ways depending on the physical quantity represented by $U(x)$. For example, if $U(x) =$ voltage or electric field, then $[A]$ can represent an admittance matrix with $\{b\}$ being the electric current excitation. Alternatively, if $U =$ electric current or magnetic field, then $[A]$ may be compared to an impedance matrix.

Because of the sparsity of $[A]$, only its nonzero entries are stored when solving the matrix system (3.33). Also, global numbering as used in (3.26)–(3.28) must be employed in defining the assembled matrix system. Clearly, the matrix system in (3.32) is identical to that in (3.26)–(3.28) except for the first and last row of (3.32) which correspond to the omitted $m = 1$ and $m = N$ equations in (3.26). In practice, the assembly of $[A]$ is done by employing (3.27) and (3.28) directly or by implementing a double loop and keeping track of the correspondence between the local and global numbering. A possible double loop which generates the nonzero entries of $[A]$ is

```
C  Initialize the [A] matrix to zero
         DO 10 m = 1, N
         DO 10 n = 1, N
10       A(m, n) = 0.0
C  Loop through all elements and construct [A]
         DO 20 e = 1, N − 1
C  Compute element matrix [A^e]
         DO 30 i = 1, 2
         DO 30 j = 1, 2
30    Compute AE(i, j) from equations (3.21) and (3.22)
C  Assemble [A^e] into global matrix
         A(e, e) = A(e, e) + AE(1, 1)
         A(e + 1, e + 1) = A(e + 1, e + 1) + AE(2, 2)
         A(e, e + 1) = AE(1, 2)
20       A(e + 1, e) = AE(2, 1)
```

3.7 ENFORCEMENT OF BOUNDARY CONDITIONS

So far we have postponed a comprehensive discussion on the imposition of boundary conditions. As is the case with any differential equation, a unique solution can be obtained only after the specification of boundary conditions which *constrain* the values of the field/potential at the boundaries (endpoints for the one-dimensional case) of the domain. These boundary conditions (also referred to as *boundary constraints*) come in various forms. In their most typical form they provide a specification of the field at the end nodes 1 and N or a specification on the value of the normal derivative of the field. However, the boundary condition may simply provide a relationship between the normal derivative and the field at the boundary node(s). The derivative of $U(x)$ at the boundary nodes should not be approximated by using the expansion (3.12). This representation is an interpolation function between the boundary nodes, but the fields/potentials and their derivatives at the boundary nodes must be independently specified for a unique solution of the differential equation. As a reference, we remark that if the spatial variable x in (3.2) was replaced by the time variable t, the boundary conditions would become the initial conditions of the temporal response $U(t)$.

The various boundary conditions to be encountered in the solution of differential equations are associated with a well-established nomenclature. Below is a list of the boundary conditions typically imposed at the boundary nodes. The type of boundary condition to be used will depend on the physics of the problem, as discussed in Section 3.2.

3.7.1 Neumann Boundary Conditions (Homogeneous)

For one-dimensional problems, this boundary condition states that

$$\frac{\partial U}{\partial x} = 0 \tag{3.35}$$

at the left or right endpoint of the domain. In two- and three-dimensional problems, it can be stated as

$$\hat{\mathbf{n}} \cdot \nabla U = \frac{\partial U}{\partial n} = 0 \qquad \text{on } S \text{ or } C \tag{3.36}$$

where $\hat{\mathbf{n}}$ denotes the outgoing unit normal vector of the domain boundary, as illustrated in Fig. 3.8. In acoustics this is referred to as the hard boundary condition, and in electromagnetics the magnetic field obeys this condition on metallic boundaries.

The Neumann boundary condition is the easiest to be numerically enforced in FEM solutions. In this case, the [endpoint] contributions (3.16b) vanish and the elemental equations (3.17b) lead to the global system (3.32) without any special considerations.

3.7.2 Dirichlet Boundary Conditions (Homogeneous)

The Dirichlet boundary condition specifies a vanishing field or potential at the endpoints of the computational domain, i.e.,

$$U(x) = 0 \text{ at endpoints} \tag{3.37}$$

In acoustics this is referred to as the soft boundary condition. For two- and three-dimensional electromagnetic problems, the Dirichlet boundary condition is satisfied by the tangential electric fields on all metallic surfaces.

When the Dirichlet boundary condition must be satisfied in connection with the elemental equations (3.17a), the [endpoints] are not zero and must be considered in the assembly of the final system. Since the [endpoints] contribute only to those

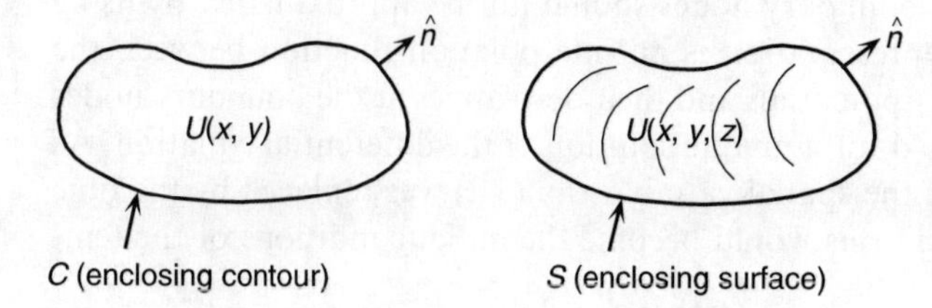

Figure 3.8 Illustration of the enclosures for two- and three-dimensional domains.

element equations resulting from testing at nodes 1 and N, the global system (3.32) for the three-element tessellation (see Fig. 3.7) takes the form

$$\begin{bmatrix} A_{11}^1 & A_{12}^1 & 0 & 0 \\ A_{21}^1 & A_{22}^1 + A_{11}^2 & A_{12}^2 & 0 \\ 0 & A_{21}^2 & A_{22}^2 + A_{11}^3 & A_{12}^3 \\ 0 & 0 & A_{21}^3 & A_{22}^3 \end{bmatrix} \begin{Bmatrix} U_1 \\ U_2 \\ U_3 \\ U_4 \end{Bmatrix} + \begin{Bmatrix} \text{endpoint}_1 \\ 0 \\ 0 \\ \text{endpoint}_2 \end{Bmatrix} = \begin{Bmatrix} b_1^1 \\ b_2^1 + b_1^2 \\ b_2^2 + b_1^3 \\ b_2^3 \end{Bmatrix} \tag{3.38}$$

However, $U_1 = U_4 = 0$, as dictated by (3.37), and thus $N - 2 = 2$ unknowns (U_2 and U_3) remain to be determined from the solution of (3.38). This allows us to discard the first and last equations of (3.38) provided we set $U_1 = U_4 = 0$ wherever they occur. In doing so, we have

$$\begin{bmatrix} A_{22}^1 + A_{11}^2 & A_{12}^2 \\ A_{21}^2 & A_{22}^2 + A_{11}^3 \end{bmatrix} \begin{Bmatrix} U_2 \\ U_3 \end{Bmatrix} = \begin{Bmatrix} b_2^1 + b_1^2 \\ b_2^2 + b_1^3 \end{Bmatrix} \tag{3.39}$$

which can be inverted to yield

$$\begin{Bmatrix} U_2 \\ U_3 \end{Bmatrix} = \frac{1}{(A_{22}^1 + A_{11}^2)(A_{22}^2 + A_{11}^3) - A_{21}^2 A_{12}^2} \begin{bmatrix} A_{22}^2 + A_{11}^3 & -A_{12}^2 \\ -A_{21}^2 & A_{22}^1 + A_{11}^2 \end{bmatrix} \begin{Bmatrix} b_2^1 + b_1^2 \\ b_2^2 + b_1^3 \end{Bmatrix}$$

Thus, even though the [endpoint] contributions are not computable, we can still solve for the node fields or potentials. We remark that if (3.38) was a system of N equations, the reduced system (3.39) will consist of $N - 2$ equations after the enforcement of the Dirichlet boundary conditions.

3.7.3 Nonzero Boundary Constraints (Inhomogeneous)

We may also encounter situations when the field or potential at the end node is assigned a specified value. An example of this situation is the parallel plate capacitor problem where the upper plate has a potential equal to V_a. As a more general example, let us consider the situation where in reference to the three-element example in Fig. 3.7, we set

$$U_1 = Q_0, \qquad \left.\frac{\partial U}{\partial x}\right|_{x=x_a} = Q_a \tag{3.40}$$

which are typically referred to as inhomogeneous Dirichlet and Neumann boundary conditions, respectively.

Substituting these values into the system (3.38) gives

$$\begin{bmatrix} A_{11}^1 & A_{12}^1 & 0 & 0 \\ A_{21}^1 & A_{22}^1 + A_{11}^2 & A_{12}^2 & 0 \\ 0 & A_{21}^2 & A_{22}^2 + A_{11}^3 & A_{12}^3 \\ 0 & 0 & A_{21}^3 & A_{22}^3 \end{bmatrix} \begin{Bmatrix} Q_0 \\ U_2 \\ U_3 \\ U_4 \end{Bmatrix} + \begin{Bmatrix} \text{endpoint}_1 \\ 0 \\ 0 \\ \tilde{Q}_a \end{Bmatrix} = \begin{Bmatrix} b_1^1 \\ b_2^1 + b_1^2 \\ b_2^2 + b_1^3 \\ b_2^3 \end{Bmatrix}$$

where $\tilde{Q}_a = -Q_a p(x_a) N_2^3(x_a) = -Q_a p(x_a)$. Clearly, since U_1 is already specified, the first of the equations can be discarded. After doing so and rearranging, we obtain the system

$$\begin{bmatrix} A_{22}^1 + A_{11}^2 & A_{12}^2 & 0 \\ A_{21}^2 & A_{22}^2 + A_{11}^3 & A_{12}^3 \\ 0 & A_{21}^3 & A_{22}^3 \end{bmatrix} \begin{Bmatrix} U_2 \\ U_3 \\ U_4 \end{Bmatrix} = \begin{Bmatrix} b_2^1 + b_1^2 \\ b_2^2 + b_1^3 \\ b_2^3 \end{Bmatrix} - \begin{Bmatrix} A_{21}^1 Q_0 \\ 0 \\ \tilde{Q}_a \end{Bmatrix} \tag{3.41}$$

which can be solved for U_2, U_3 and U_4. We again point out that, when dealing with N nodes, (3.41) would be a system of $N-1$ equations and with the exception of the first and last equations, the rest will have three nonzero elements as illustrated by the matrix (3.34).

The procedure of reducing the system (3.38) to (3.39) or (3.41) is often referred to as *condensation* of boundary conditions. This reduction is typically performed during the assembly process by eliminating for example the rows which test at a boundary node assigned a specified value. Finally, we remark that the condensation process modifies the excitation column implying that the specification of a potential or a field value on a boundary is equivalent to a source excitation. Thus, the excitation of the domain can either be specified through a nonzero $f(x)$ or through the enforcement of nontrivial boundary constraints.

3.7.4 Impedance Boundary Conditions

The impedance boundary condition provides a relationship between the field and its normal derivative. Referring to Fig. 3.8, it is typically stated as [9]

$$\frac{\partial U}{\partial n} + \alpha U = 0 \qquad \text{on } S \text{ or } C \tag{3.42}$$

where α is a constant. This boundary condition has been found very useful in modeling the presence of thin dielectric coatings without a need to tessellate the region interior to the dielectric. In finite element simulations, (3.42) also plays the role of the radiation condition or a first order absorbing condition (to be discussed later). These boundary conditions will be discussed extensively in later chapters and are used for truncating the computational domain of open domain problems as in the case of scattering by an airfoil (see Fig. 3.2). They basically provide a statement on the field behavior at the boundary nodes. The need to mesh the region beyond the boundary enclosure is therefore eliminated provided the absorbing boundary condition gives the proper field behavior beyond the boundary enclosure.

A generalization of (3.42) is

$$\frac{\partial U}{\partial n} + \alpha U = \beta \qquad \text{on } S \text{ or } C \tag{3.43}$$

where α and β are constants and we can refer to this as the inhomogeneous boundary condition. The treatment of (3.43) is no different than that for (3.42). For the one-dimensional problem considered here, (3.43) reduces to

$$\begin{aligned} \frac{\partial U}{\partial x} + \alpha_0 U &= \beta_0, \qquad x = 0 \\ \frac{\partial U}{\partial x} + \alpha_a U &= \beta_a, \qquad x = x_a = a \end{aligned} \tag{3.44}$$

When these conditions are used in the finite element solution of the three-element segment example (see Fig. 3.7), the resulting system is again of the same form as (3.38). From (3.16b), we have

$$\begin{aligned} [\text{endpoint}]_1 &= p(0)(\beta_0 - \alpha_0 U_1) \\ [\text{endpoint}]_2 &= -p(x_a)(\beta_a - \alpha_a U_4) \end{aligned}$$

and when these results are incorporated into (3.38), after rearranging, we obtain the system

$$\begin{bmatrix} A_{11}^1 - \alpha_0 p(0) & A_{12}^1 & 0 & 0 \\ A_{21}^1 & A_{22}^1 + A_{11}^2 & A_{12}^2 & 0 \\ 0 & A_{21}^2 & A_{22}^2 + A_{11}^3 & A_{12}^3 \\ 0 & 0 & A_{21}^3 & A_{22}^3 + \alpha_a p(x_a) \end{bmatrix} \begin{Bmatrix} U_1 \\ U_2 \\ U_3 \\ U_4 \end{Bmatrix}$$

$$= \begin{Bmatrix} b_1^1 \\ b_2^1 + b_1^2 \\ b_2^2 + b_1^3 \\ b_2^3 \end{Bmatrix} + \begin{Bmatrix} -\beta_0 p(0) \\ 0 \\ 0 \\ +\beta_a p(x_a) \end{Bmatrix} \tag{3.45}$$

This system can now be solved for $\{U\}$ and we remark that the middle two equations are unchanged by the imposition of the impedance boundary conditions. It is clear that when N equations are involved, the imposition of the impedance boundary condition will only alter the first and last equations of the overall system.

3.8 EXAMPLES

To illustrate the application of the various steps in constructing and assembling an FEM system, let us consider two simple examples.

EXAMPLE 3.1

Solve numerically the differential equation

$$-\frac{d^2 E_y(x)}{dx^2} + \pi^2 E_y(x) = 2\pi^2 \sin \pi x, \qquad 0 < x < 1 \tag{3.46}$$

using ten segments and subject to the boundary conditions

$$E_y(0) = E_y(1) = 0 \tag{3.47}$$

Compare the numerical results with the exact solution [10, Chapter 11]

$$E_y(x) = \sin \pi x \tag{3.48}$$

Solution

This problem corresponds to a source in a parallel plate waveguide as illustrated in Fig. 3.4(b). By comparison to the general form of the differential equation in (3.2), we note that

$$p(x) = 1, \quad q(x) = \pi^2, \quad f(x) = 2\pi^2 \sin \pi x$$

Also, from Fig. 3.9 and the formulae (3.21) and (3.22), we find that

$$A_{11}^e = A_{22}^e = \frac{1}{\Delta x} + \pi^2 \frac{\Delta x}{3} = 10 + \frac{0.1\pi^2}{3} = 10.328\,987$$

and

$$A_{12}^e = A_{21}^e = -\frac{1}{\Delta x} + \pi^2 \frac{\Delta x}{6} = -9.835\,507$$

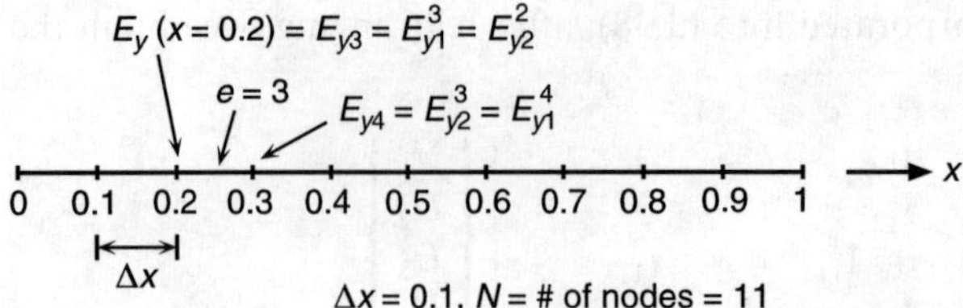

Figure 3.9 Tessellation of a line segment into ten equal length elements.

Thus, the elemental equations are given by

$$\begin{bmatrix} 10.328987 & -9.835507 \\ -9.835507 & 10.328987 \end{bmatrix} \begin{Bmatrix} E^e_{y1} \\ E^e_{y2} \end{Bmatrix} = \begin{Bmatrix} b^e_1 \\ b^e_2 \end{Bmatrix} \tag{3.49}$$

where from (3.20)

$$b^e_1 = 2\pi^2 \int_{x^e_1}^{x^e_2} \left(\frac{x^e_2 - x}{x^e_2 - x^e_1} \right) \sin(\pi x)\, dx, \qquad b^e_2 = 2\pi^2 \int_{x^e_1}^{x^e_2} \left(\frac{x - x^e_1}{x^e_2 - x^e_1} \right) \sin(\pi x)\, dx,$$

with $x^e_1 = (e-1)\,\Delta x$ and $x^e_2 = e\,\Delta x$ for $e = 1, 2, \ldots, 10$.

After assembling the elemental equations (3.49) subject to the Dirichlet boundary conditions, we get the 9×9 system

$$\begin{bmatrix} 20.6580 & -9.8355 & 0 & 0 & \ldots & & 0 & 0 & 0 \\ -9.8355 & 20.6580 & -9.8355 & 0 & 0 & \ldots & 0 & 0 & 0 \\ 0 & -9.8355 & 20.6580 & -9.8355 & 0 & 0 & 0 & 0 & 0 \\ & & \vdots & & & & & & \\ 0 & 0 & 0 & 0 & \ldots & 0 & -9.8355 & 20.6580 & -9.8355 \\ 0 & 0 & 0 & 0 & \ldots & 0 & 0 & -9.8355 & 20.6580 \end{bmatrix}$$

$$\cdot \begin{Bmatrix} E_{y2} \\ E_{y3} \\ E_{y4} \\ \vdots \\ E_{y9} \\ E_{y10} \end{Bmatrix} = \begin{Bmatrix} b_2 \\ b_3 \\ b_4 \\ \vdots \\ b_9 \\ b_{10} \end{Bmatrix}$$

with

$$\{b\}^T = \{0.605, 1.1507, 1.5838, 1.8619, 1.9577, 1.8619, 1.5838, 1.1507, 0.605\}$$

This system is identical in form to that in (3.26). The node fields E_{y1} and E_{y11} were not included since they are zero as dictated by the boundary conditions. Also, we remark that the symmetry of the system (i.e., $A_{ij} = A_{ji}$) is not unexpected. Electromagnetic problems with reciprocal permittivity and permeability tensors are inherently reciprocal, and this property is exhibited in the Hermitian form of the matrix.

Solving the above system via matrix inversion, for example, gives the following results:

Node #	E_y^{FEM}	E_y^{exact}	Error $= \lvert E_y^{\text{FEM}} - E_y^{\text{exact}} \rvert$
1	0	0	0
2	0.3103	0.3090	0.0013
3	0.5902	0.5878	0.0024
4	0.8123	0.8090	0.0033
5	0.9550	0.9511	0.0039
6	1.0041	1.0000	0.0041
7	0.9550	0.9511	0.0039
8	0.8123	0.8090	0.0033
9	0.5902	0.5878	0.0024
10	0.3103	0.3090	0.0013
11	0	0	0

As seen, the difference between the exact and numerical solution is in the third decimal place, indicating that the employed number of elements are sufficient for an accurate representation of the field distribution. To reduce the solution error, more elements can be used for the tessellation of the line segment. However, as $\Delta x \to 0$, the system condition $\|\mathscr{A}\|\|\mathscr{A}^{-1}\|$, where $\|\mathscr{A}\|$ is a natural norm of the matrix $[A]$. Eventually the error would not reduce with increasing N, unless the machine precision is also increased with the inclusion of additional decimal places in the calculation of the matrix entries and in carrying out the system solution. Numerical precision is particularly important for solving problems with many thousands of unknowns (the case with many practical problems).

Having the node fields, the field is found from (3.12) or (3.14). Specifically, since $x_1^e = 0.1(e-1)$ and $x_2^e = 0.1e$, we have

$$E_y(x) = 10\sum_{e=1}^{10}[E_{y1}^e(0.1e - x) + E_{y2}^e(x - 0.1e + 0.1)]\,P_{\Delta x}(x - 2e + 0.05)$$

$$= \sum_{n=1}^{9} E_{yn}\left|\frac{x - 0.1n}{0.1}\right| P_{2\Delta x}(x - 0.1n)$$

where

$$P_{\Delta x}(x) = \begin{cases} 1 & -\Delta x/2 < x < \Delta x/2 \\ 0 & \text{otherwise} \end{cases}$$

is a pulse function.

EXAMPLE 3.2

The field reflected by a metal-backed dielectric slab due to a plane wave excitation $E_z^{\text{inc}} = e^{jk_0x}$ is given by

$$E_z^{\text{r}} = Re^{-jk_0x}$$

where R is the unknown reflection coefficient to be determined.

The dielectric slab is of thickness t, as shown in Fig. 3.4(c). Consider the case of $t = 0.25\lambda_0$, $\epsilon_{\text{r}} = 4 - j\beta$, $\mu_{\text{r}} = 1$ and employ the finite element method to compute the reflection coefficient R as a function of the loss parameter β. Compare the result with the analytical reflection coefficient given by

$$R = -\frac{Z_0 - jZ\tan(k_0\sqrt{\mu_r\epsilon_r}t)}{Z_0 + jZ\tan(k_0\sqrt{\mu_r\epsilon_r}t)}$$

where

$$Z_0 = \sqrt{\frac{\mu_0}{\epsilon_0}} = 120\pi$$

and

$$Z = Z_0\sqrt{\frac{\mu_r}{\epsilon_r}}$$

are the wave impedances in free space and in the dielectric medium, respectively.

Solution

As discussed at the beginning of Section 3.2, the pertinent differential equation is

$$\frac{d^2 E_z}{dx^2} + k_0^2 \epsilon_r E_z = 0$$

subject to the boundary conditions

$$E_z(0) = 0; \quad \frac{\partial E_z}{\partial x} + jk_0 E_z = 2jk_0 e^{jk_0 x}, \quad x = x_a$$

We will set $x_a = t + \Delta t$, where Δt is the chosen element length, as illustrated in Fig. 3.10. The choice of Δt (the element size) should be somewhere between $\lambda_d/10$ and $\lambda_d/20$, where $\lambda_d = 2\pi/\mathrm{Re}(k_d)$ is the wavelength in the material. For our case, $k_d = (2\pi/\lambda_0)\sqrt{\mu_r \epsilon_r}$ and $\lambda_d = \lambda_0/\sqrt{\mathrm{Re}(\epsilon_r \mu_r)}$, and we will choose $\Delta t = 0.025\lambda_0$ to ensure that this criterion is satisfied. Since $t = 0.25\lambda_0$, this choice of Δt leads to $N_e = 11$ layers or elements, including the air-filled element occupying the region $t < x < t + \Delta t$. We also observe that for $\Delta t = 0.025\lambda_0$ the sampling rate is 20 elements per wavelength in the dielectric since $\mathrm{Re}(\epsilon_r) = 4$. From (3.21) and (3.22) the entries of the elemental equation are given by (with $p^e = 1$, $q^e = -k_0^2 \epsilon_{re}$)

$$A_{11}^e = A_{22}^e = \frac{1}{\Delta t} - \frac{k_0^2 \epsilon_{re}\,\Delta t}{3} = 40. - 0.32899\epsilon_{re}$$

$$A_{12}^e = A_{21}^e = -\frac{1}{\Delta t} - \frac{k_0^2 \epsilon_{re}\,\Delta t}{6} = -40. - 0.164493\epsilon_{re}$$

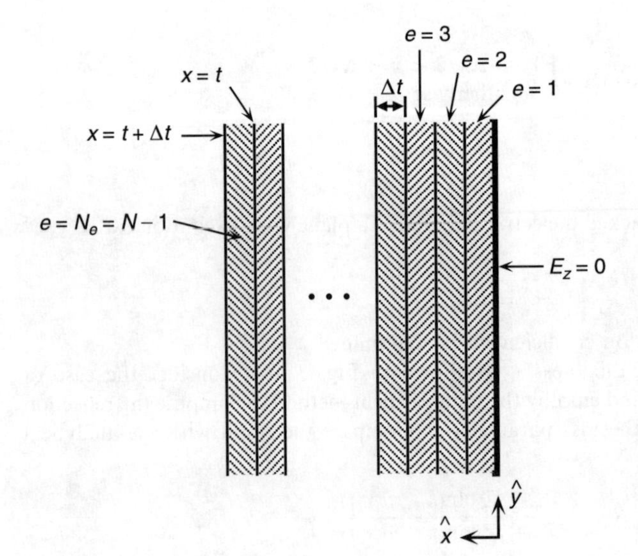

Figure 3.10 Tessellation of the dielectric slab using $N_e = N - 1$ layers (elements).

where ϵ_{re} denotes the relative permittivity of the eth element and we have suppressed the presence of the factor λ_0 since the free-space wavelength cancels out in the final result. After assembly, we will obtain 11 equations since $E_{z1} = E_z(x = 0) = 0$. Except for the last and first, all other equations of the assembled system will be of the form

$$[0, 0, \ldots, A_{(m-1)m}, A_{mm}, A_{m(m+1)}, 0, \ldots, 0]\{E_z\} = 0$$

where $\{E_z\}^T = \{E_{z2}, E_{z3}, \ldots, E_{z12}\}$ is the unknown node fields vector, $A_{mm} = A_{11}^m + A_{22}^{e=m-1}$, $A_{(m-1)m} = A_{21}^{m-1}$, and $A_{m(m+1)} = A_{12}^m$. The last equation of the system is obtained by recognizing that the boundary condition at $x = t + \Delta t$ is of the form (3.44) with $\alpha_a = jk_0$ and $\beta_a = 2jk_0 e^{jk_0 x_a}$. Thus,

$$[\text{endpoint}]_2 = +jk_0 E_{z12} - 2jk_0 e^{jk_0(t+\Delta t)}$$

and the tenth equation of the assembled system will be

$$A_{21}^{e=11} E_{z11} + (A_{22}^{e=11} + jk_0)E_{z12} = +2jk_0 e^{jk_0(t+\Delta t)}$$

That is, $b_{12} = 2jk_0 e^{jk_0(t+\Delta t)}$ is the only nonzero entry of the excitation column and is a result of the assumed plane wave incidence.

Upon solution of the assembled FEM system, the reflection coefficient is extracted from the node field values as

$$R^{\text{FEM}} = \frac{E_{z11} - E_z^{\text{inc}}(x = t + \Delta t)}{E_z^{\text{inc}}(x = t + \Delta t)}$$

A plot of R^{FEM} as a function of the loss parameter $\beta = \text{Im}(\epsilon_r)$ is given in Fig. 3.11 (courtesy of S. Legault).

It is seen that the computed values of R^{FEM} are in good agreement with the exact result even when the discrete layers in the dielectric are reduced from ten down to only $N_e - 1 = 2$ (corresponding to a sampling rate of eight elements per λ_d). In this case the decay of the field as it propagates in the dielectric is very rapid and the employed discretization of the dielectric slab is not sufficiently fine to pick up the large changes in the field values from one discrete slab (element) to the next. This is a good example of the fundamental assumptions made when discretizing the computational domain. For large β, the field decrease is slower and thus less samples are needed to approximate the field within the region while maintaining the same level of accuracy.

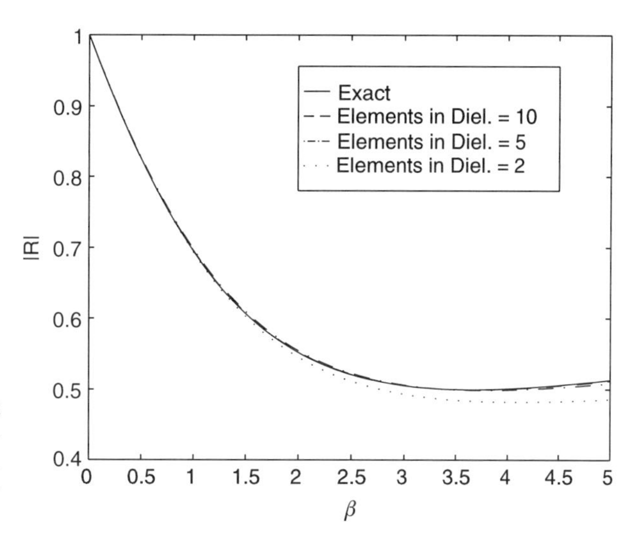

Figure 3.11 Plot of the reflection coefficient (at normal incidence) from a grounded dielectric slab of thickness $t = 0.25\lambda_0$ ($\epsilon_r = 4 - j\beta$, $\mu_r = 1$); comparison of the numerically computed R^{FEM} with the exact result R^{exact}.

EXAMPLE 3.3

Repeat Example 3.2 assuming that $\epsilon_r = \mu_r = 4 - j\beta$ and with all other parameters left unchanged.

Solution

By selecting $\mu_r = \epsilon_r$, the impedance of the wave inside and outside the dielectric is

$$Z_0 = \sqrt{\frac{\mu_0}{\epsilon_0}} = \sqrt{\frac{\mu_0 \mu_r}{\epsilon_0 \epsilon_r}} = 120\pi$$

and thus the wave does not exhibit any reflection at $x = t$ (i.e., at the dielectric interface). The interface is then referred to in the literature as *reflectionless*. Also, as it enters the dielectric, the wave is absorbed due to the nonzero imaginary components of the constitutive parameters. If β is chosen so that the wave has decayed to negligible levels by the time it exits into the air medium, the dielectric layer can be considered as "perfectly absorbing" since no reflected field is returned. Such layers have been proposed recently [11], and it has been shown that certain anisotropic layers can lead to perfectly matched interfaces for incidences away from normal. These types of layers are important in finite element simulations because they can be used for simulating a nonreflecting surface. The latter is essential in solving open domain problems, as is the case with scattering and antenna radiation problems considered later.

For the one-dimensional problem considered here, we will restrict ourselves to a simple isotropic dielectric layer having $\epsilon_r = \mu_r = 4 - j\beta$. The numerical results for a layer of thickness $t = 0.25\lambda_0$ are given in Fig. 3.12 (results are courtesy of S. Legault) as a function of β and for three different sampling rates ($N_e = 6, 11, 21$). It is again seen that as β increases, the numerically computed reflection coefficient begins to deviate from the analytical/exact result given by $R = -e^{-jk_0 t \epsilon_r}$ [12]. This is expected and is due to the rapid decay of the field within the absorber as β is increased. Thus, higher sampling is needed to better model the field values from one discrete layer to the other.

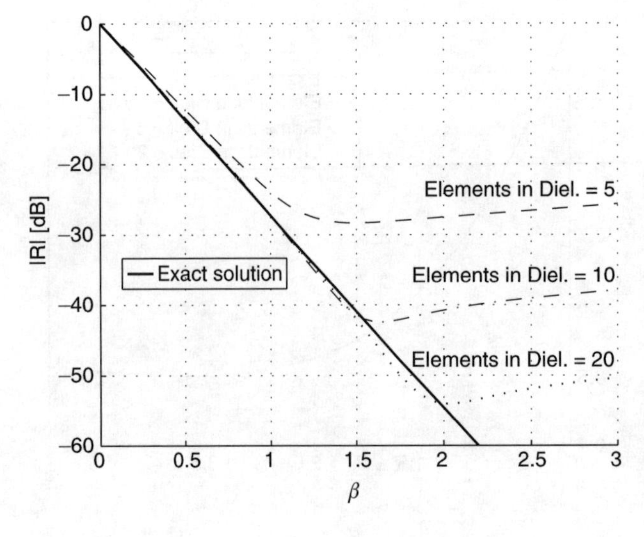

Figure 3.12 Plot of the reflection coefficient (at normal incidence) from a grounded dielectric slab of thickness $t = 0.25\lambda_0$ having $\epsilon_r = \mu_r = 4 - j\beta$.

APPENDIX 1: SAMPLE ONE-DIMENSIONAL MATLAB FEM ANALYSIS PROGRAM

```
%This MATLAB code can be used to reproduce the data in Fig. 3-11
%       Courtesy of LARS ANDERSEN
%t=thickness of slab
%k0=free space propagation constant
%xa=location of the left computational domain endpoint
%alphaa=alpha coeff. to be used in the boundary condition at xa; see (3.43)
%betaa=betaa coeff. to be used in the boundary condition at xa; see (3.43)
%p=p coefficient appearing in the differential equation (3.2)
%epr=relative permittivity of the slab
%beta=as defined in Fig. 3-11
%q=coefficient appearing in the differential equation (3.2)
%N=number of nodes (N-1=number of layers)
%R_FEM=computed reflection coefficient (to be plotted)

% Initialization

clear;
N=7;
t=0.25;
Dx=t/(N-2);
xa=t+Dx;
p=-1;
k0=2*pi;
f=0;
alphaa=j*k0;
betaa=2*j*k0*exp(j*k0*xa);
%values to generate plot in Fig. 3-11
Z=26;
%epsr=4-j*beta;
for z=1:Z
        beta=(z-1)/(Z-1)*5;
        epsr=4-j* beta;

% Initialize global matrix and vector

for m=1:N,
  b(m,1)=0;
  for n=1:N,
    A(m,n)=0;
  end
end

% Compute element matrices and assemble global matrix

for n=1:N-1,
  xe1=(n-1)*xa/(N-1);
  xe2=n*xa/(N-1);
  if n==N-1,
    eps=1;
```

```
    else
      eps=epsr;
    end
    q=eps*k0^2;
    Ae1(1,1)=p/abs(xe2-xe1)+q*abs(xe2-xe1)/3;
    Ae1(2,2)=Ae1(1,1);
    Ae1(1,2)=-p/abs(xe2-xe1)+q*abs(xe2-xe1)/6;
    Ae1(2,1)=Ae1(1,2);
    be1(1)=f*abs(xe2-xe1)/2;
    be1(2)=be1(1);
    A(n,n)=A(n,n)+Ae1(1,1);
    A(n,n+1)=A(n,n+1)+Ae1(1,2);
    A(n+1,n)=A(n+1,n)+Ae1(2,1);
    A(n+1,n+1)=A(n+1,n+1)+Ae1(2,2);
    b(n,1)=b(n,1)+be1(1);
    b(n+1,1)=b(n+1,1)+be1(2);
  end

  % Enforce boundary conditions

  A(1,1)=1;
  for n=2:N,
    A(n,1)=0;
    A(1,n)=0;
  end
  A(N,N)=A(N,N)+alphaa*p;
  b(N)=b(N)+betaa*p;

  % Solve the resulting NxN system

  x=inv(A)*b;

  % Compute reflection coefficient

  R_FEM=abs((x(N)-exp(j*k0*xa))/exp(j*k0*xa));
  zeta0=377;
  zeta1=sqrt(1/epsr)*zeta0;
  k1=sqrt(epsr)*k0;
  R_exact=abs((zeta1*tanh(j*k1*t)-zeta0)/(zeta1*tanh(j*k1*t)+zeta0));
  Rf(z,1)=beta;
  Rf(z,2)=R_FEM;
  Re(z,1)=beta;
  Re(z,2)=R_exact

  end

  %Plot Reflection Coefficient

  clf;
  plot(Rf(:,1),Re(:,2)'r');
  hold;
  plot(Re(:,1),Re(:,2));
```

```
xlabel('beta');
ylabel('|R|');

% End of program
```

APPENDIX 2: USEFUL INTEGRATION FORMULAE FOR ONE-DIMENSIONAL FEM ANALYSIS

$$I_1 = \left(\frac{1}{h_n}\right)^2 \int_{x_n}^{x_{n+1}} (x_{n+1} - x)(x - x_n)\, q(x)\, dx \approx \frac{h_n}{12} [q(x_n) + q(x_{n+1})] + 0(h_n^3)$$

$$I_2 = \left(\frac{1}{h_{n-1}}\right)^2 \int_{x_{n-1}}^{x_n} (x - x_{n-1})^2\, q(x)\, dx \approx \frac{h_{n-1}}{12} [3q(x_n) + q(x_{n-1})]$$

$$I_3 = \left(\frac{1}{h_n}\right)^2 \int_{x_n}^{x_{n+1}} (x_{n+1} - x)^2\, q(x)\, dx \approx \frac{h_n}{12} [3q(x_n) + q(x_{n+1})]$$

$$I_4 = \frac{1}{h_{n-1}} \int_{x_{n-1}}^{x_n} p(x)\, dx \approx \tfrac{1}{2}[p(x_n) + p(x_{n-1})]$$

$$I_5 = \frac{1}{h_{n-1}} \int_{x_{n-1}}^{x_n} (x - x_{n-1}) f(x) \approx \frac{h_{n-1}}{6} [2f(x_n) + f(x_{n-1})]$$

$$I_6 = \frac{1}{h_n} \int_{x_n}^{x_{n+1}} (x_{n+1} - x) f(x)\, dx \approx \frac{h_{n-1}}{6} [2f(x_n) + f(x_{n+1})]$$

In all cases, $h_n = |x_{n+1} - x_n|$ and $h_{n-1} = |x_n - x_{n-1}|$ are assumed to be small. These are derived by introducing a linear approximation for the functions $q(x)$, $f(x)$, and $p(x)$. For example, in deriving the approximation for I_4, $p(x)$ was approximated as

$$p(x) = p(x_{n-1}) \frac{x_n - x}{h_{n-1}} + p(x_n) \frac{x - x_n}{h_{n-1}}$$

REFERENCES

[1] R. Courant. Variational methods for a solution of problems of equilibrium and vibrations. *Bull. Amer. Math. Soc.*, 49:1–23, 1943.

[2] J. H. Argyris. Energy theorems and structural analysis. *Aircraft Engineering*, 26:347–356, 1954.

[3] P. Silvester. Finite element solution of homogeneous waveguide problems. *Alta Frequenza*, 38:313–317, 1969.

[4] P. Silvester and G. Pelosi, editors. *Finite Elements for Wave Electromagnetics: Methods and Techniques*. IEEE Press, New York, 1994.

[5] P. L. Alett, A. K. Bahrani, and O. C. Zienkiewicz. Application of finite elements to the solution of Helmholtz's equation, *Proc. IEE*, 115:1762–1766, 1968.

[6] I. Stakgold. *Boundary-Value Problems of Mathematical Physics*. Macmillan, New York, 1968. Volumes I and II.

[7] S. C. Charpa and R. P. Canale. *Numerical Methods for Engineers*. McGraw-Hill, New York, Second edition, 1988.

[8] R. L Burden and J. Faires. *Numerical Analysis*. PWS Pub. Co., Boston, Fifth edition, 1993.

[9] T. B. A. Senior and J. L. Volakis. *Approximate Boundary Conditions in Electromagnetics*. IEE Press, London, 1995.

[10] E. Kreyszig. *Advanced Engineering Mathematics*. John Wiley & Sons, New York, Fifth edition, 1983.

[11] D. M. Kingsland, J. Gong, J. L. Volakis, and J.-F. Lee. Performance of an anisotropic artificial absorber for truncating finite element meshes. *IEEE Trans. Antennas Propagat.*, 44:975–982, July 1996.

[12] S. R. Legault, T. B. A. Senior, and J. L. Volakis. Design of planar absorbing layers for domain truncation in FEM applications. *Electromagnetics*, 16(4):451–464, July–August 1996.

4

Two-Dimensional Applications

4.1 INTRODUCTION

Having discussed one-dimensional examples, we now proceed with the application of the FEM to two-dimensional problems. Although these represent simplifications of the real-world three-dimensional situations, they are much simpler to formulate and solve. Thus, they are appropriate for illustrating the FEM procedure and all aspects of the three-dimensional analysis can be discussed in the context of two dimensions. Specifically, matrix assembly, absorbing boundary conditions, and various hybrid formulations of the FEM can be discussed in two dimensions without loss of generality. This chapter is therefore important in understanding the machinery needed to carry out FEM analysis and in illustrating its capabilities.

Two-dimensional models can also be used to generate useful results for a number of practical problems in electromagnetics. For example, TE and TM mode analysis for straight waveguides of arbitrary cross section can be carried out using a two-dimensional formulation [1]–[2] or the capacitance of transmission lines, such as those in Fig. 3.1, can be found by solving Laplace's equation in two dimensions. Similarly, the inductance can be computed by carrying out a two-dimensional analysis of Maxwell's equations. Also, an understanding of three-dimensional problems can be obtained by performing an analysis of closely related two-dimensional problems. Since two-dimensional analysis is much simpler, it is often a quick way to obtain results for many practical open domain problems. The latter are among the most difficult to solve and include those of radiation and radar scattering. Scattering involves the computation of fields returned from a given structure due to plane wave excitation or, more generally, the impinging radar wave. Two-dimensional analysis has been extensively used in scattering and has provided engineers with an understanding of scattering phenomena in the absence of three-dimensional simulation which may be prohibitive due to their greater CPU and storage requirements.

In this chapter we cover many aspects of the finite element method (and hybrid versions) at sufficient detail to provide the reader with a comfortable level of understanding its implementation. Such an understanding is essential for a three-dimensional analysis where many of the steps must be discussed symbolically due to the size of the matrices even for very small problem sizes. We begin by first performing a reduction of Maxwell's equations to two-dimensional wave equations and proceed with the solution of the latter by following the same FEM steps discussed in the previous chapter. That is, we generate the *weak* form of the wave equation and carry out its discretization with the introduction of linear shape functions. Matrix assembly and boundary conditions are then discussed for determining the propagation constants in waveguides, and we give examples of this type of analysis. In proceeding with the solution of open domain problems, we first discuss absorbing boundary conditions and material absorbers for truncating the finite element mesh. The necessary modifications of the FEM system are presented and example calculations are given for scattering by cylindrical structures. The use of the boundary integral for truncating the FEM mesh is presented in Section 4.4, and an example application to scattering by two-dimensional recessed cavities is given. Finally, we discuss the implementation of two-dimensional solutions using edge-based basis functions since these are essential in three dimensions and are used extensively in subsequent chapters.

4.2 TWO-DIMENSIONAL WAVE EQUATIONS

Throughout this chapter we shall assume that the fields or potentials are either independent of the z coordinate or have a known z dependence as is the case of propagation in waveguides. That is, it is assumed that the geometry's cross section in any xy cut remain invariant for all z values. The appropriate assumption depends on the type of physical problem being considered as discussed below.

4.2.1 Transmission Lines

The characteristic impedance Z_c and phase velocity v_p of the transmission line can be computed by solving Laplace's equation

$$\nabla_t^2 V = \frac{\partial^2 V}{\partial x^2} + \frac{\partial^2 V}{\partial y^2} = 0 \tag{4.1}$$

where ∇_t^2 denotes the Laplacian in two dimensions. The characteristic impedance and capacitance of the transmission line is found by following these steps (see Fig. 4.1):

1. Carry out the FEM solution and find $V(x, y)$ in the absence of dielectrics.
2. Determine the charge per unit length of one of the conductors by carrying out the integration

$$Q_0 = \int_{\text{Contour}} \varepsilon \hat{n} \cdot \nabla V \, dl \tag{4.2}$$

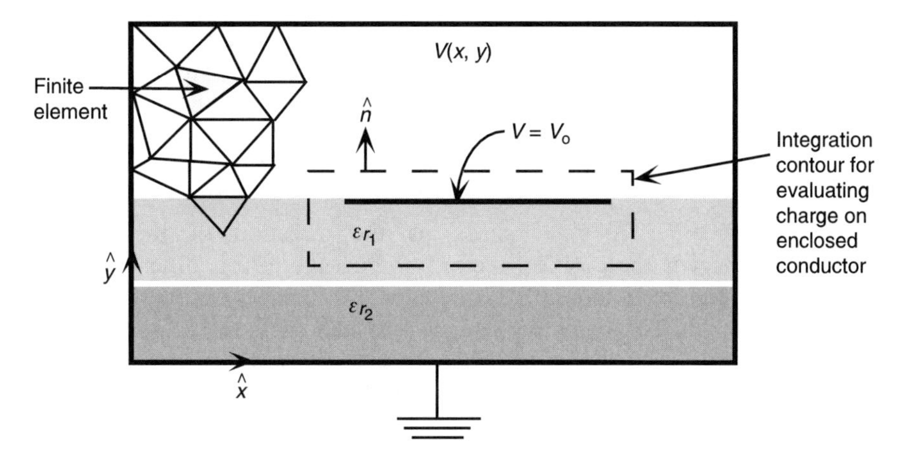

Figure 4.1 Shielded microstrip line.

where $\hat{n}$ denotes the outward directed unit normal and the contour is shown in Fig. 4.1.

3. Evaluate the capacitance per unit length of the free space filled transmission line from

$$C_0 = \frac{Q_0}{\Delta V} \tag{4.3}$$

where ΔV is the potential difference between the two conductors.

4. Repeat steps 1 to 3 for the original dielectrically filled transmission line to obtain the per unit length capacitance of the line $C = \frac{Q}{\Delta V}$.
5. Determine the line's characteristic impedance using

$$Z_c = \frac{1}{v_p C} = \frac{1}{v_0 \sqrt{C C_0}} \tag{4.4}$$

where v_0 denotes the wave velocity of the air-filled transmission line. We also note that $v_p = v_0 / \sqrt{\varepsilon_e}$ where ε_e is referred to as the effective dielectric constant of the substrate and, from (4.4)

$$\varepsilon_e = \frac{C}{C_0} \tag{4.5}$$

4.2.2 Two-Dimensional Scattering

In scattering analysis, the excitation is a plane wave of the form

$$\mathbf{H}^i = \hat{z} e^{jk_0(x \cos\phi_0 + y \sin\phi_0)} \tag{4.6}$$

for TE incidence (also referred to as H_z polarization) or

$$\mathbf{E}^i = \hat{z} e^{jk_0(x \cos\phi_0 + y \sin\phi_0)} \tag{4.7}$$

for TM incidence (also referred to as E_z polarization), as illustrated in Fig. 4.2. Both of these plane waves are impinging upon the scatterer at an angle ϕ_0. They may be alternatively written as

$$\begin{Bmatrix} \mathbf{E}^i \\ \mathbf{H}^i \end{Bmatrix} = \hat{z}e^{-jk_0\hat{k}^i\cdot\mathbf{r}}$$

where $\hat{k}^i = -(\hat{x}\cos\phi_0 + \hat{y}\sin\phi_0)$ is the direction of the incident wave and $\mathbf{r} = x\hat{x} + y\hat{y}$ is the position vector at which the field is observed. More generally, a magnetic current source $\hat{z}M_{iz}(x, y)$ for TE incidence or an electric current source $\hat{z}J_{iz}(x, y)$ for TM incidence can be assumed as the excitation. For these types of excitation, the fields scattered by the cylinder will also be z directed and thus for the TE case the vector wave equation becomes

$$\nabla_t \times \left[\frac{1}{\varepsilon_r}\nabla_t \times (\hat{z}H_z)\right] - \hat{z}k_0^2\mu_r H_z = -\hat{z}j\omega\varepsilon_0 M_{iz} \tag{4.8}$$

where we included the possibility of a magnetic source as the excitation ($k_0 = 2\pi/\lambda_0 = \omega\sqrt{\mu_0\varepsilon_0}$, $k = 2\pi/\lambda = \omega\sqrt{\mu\varepsilon}$). Using the identity[1]

$$\nabla_t \times \left[\frac{1}{\varepsilon_r}\nabla_t \times (\hat{z}H_z)\right] = \nabla_t \times \left[\frac{1}{\varepsilon_r}(-\hat{z} \times \nabla_t H_z)\right] = -\hat{z}\nabla_t \cdot \frac{1}{\varepsilon_r}\nabla_t H_z$$

the vector wave equation (4.8) is now reduced to the scalar wave equation

$$\nabla_t \cdot \left(\frac{1}{\varepsilon_r}\nabla_t H_z\right) + k_0^2\mu_r H_z = j\omega\varepsilon_0 M_{iz} \tag{4.9}$$

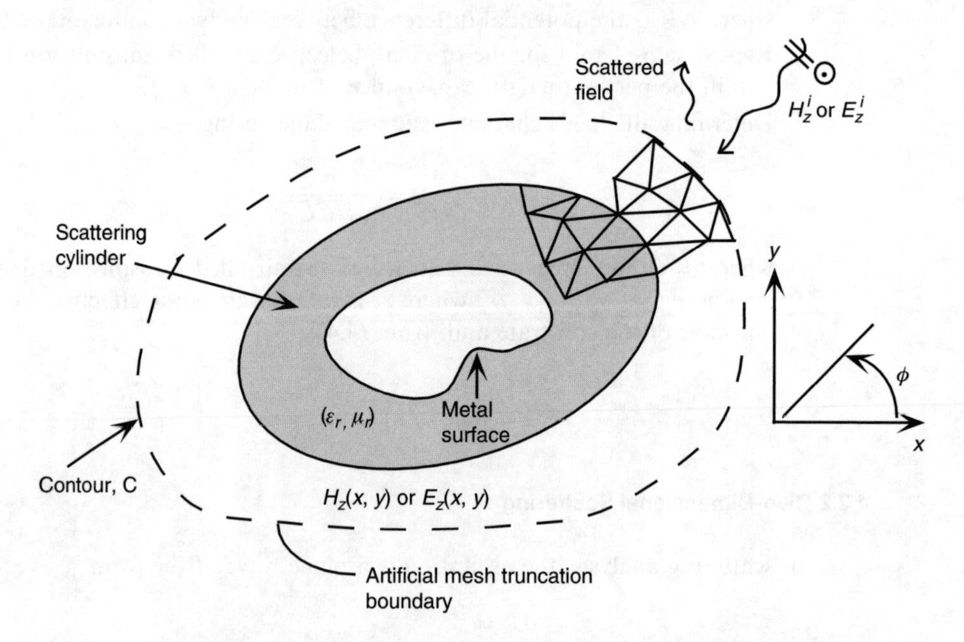

Figure 4.2 Two-dimensional scattering configuration.

[1]Derived from the vector identity [3] $\nabla \times (\mathbf{A} \times \mathbf{B}) = \mathbf{A}\nabla \cdot \mathbf{B} - \mathbf{B}\nabla \cdot \mathbf{A} + (\mathbf{B} \cdot \nabla)\mathbf{A} - (\mathbf{A} \cdot \nabla)\mathbf{B}$.

for inhomogeneous media. Here, H_z denotes the z component of the total magnetic field. In scattering, it is typically decomposed as

$$H_z = H_z^{\text{scat}} + H_z^i \tag{4.10}$$

where H_z^i is the component of the magnetic field generated by $\hat{z}M_{iz}$ in isolation or is simply given by (4.6). The component H_z^{scat} is referred to as the scattered magnetic field (z component only) and is the unknown quantity of interest. As will be discussed later, there are advantages in solving for H_z^{scat} directly and in this case the pertinent wave equation is

$$\nabla_t \cdot \left(\frac{1}{\varepsilon_r}\nabla_t H_z^{\text{scat}}\right) + k_0^2\mu_r H_z^{\text{scat}} = -\nabla_t \cdot \left(\frac{1}{\varepsilon_r}\nabla_t H_z^i\right) - k_0^2\mu_r H_z^i + j\omega\varepsilon_0 M_{iz} = f(x, y) \tag{4.11}$$

obtained by substituting (4.10) into (4.9). From duality, the corresponding wave equation for TM incidence is

$$\nabla_t \cdot \left(\frac{1}{\mu_r}\nabla_t E_z^{\text{scat}}\right) + k_0^2\varepsilon_r E_z^{\text{scat}} = -\nabla_t \cdot \left(\frac{1}{\mu_r}\nabla_t E_z^i\right) - k_0^2\varepsilon_r E_z^i + j\omega\mu_0 J_{iz} \tag{4.12}$$

where E_z^{scat} denotes the z component of the scattered electric field and E_z^i is the z component of the incident electric field generated by $\hat{z}J_{iz}$ in isolation or is simply given by (4.7).

4.2.3 Waveguide Propagation (Homogeneous Cross Section)

In waveguide propagation, the field is assumed to be of the form

$$\begin{Bmatrix} \mathbf{E} \\ \mathbf{H} \end{Bmatrix} = \begin{Bmatrix} \mathbf{U}_e(x, y) \\ \mathbf{U}_h(x, y) \end{Bmatrix} e^{-j\beta z} \tag{4.13}$$

where β is the propagation constant along z and $\mathbf{U}(x, y)$ is the field value over the waveguides cross section. For a waveguide whose nonmetallic region Ω is filled with a homogeneous material certain standard simplifications are afforded. In these waveguides, the configuration as well as the field behavior is independent of the third variable. This would be the z direction in a Cartesian coordinate system. Thus, in the absence of a source, the identity

$$\nabla \times \left(\frac{1}{\varepsilon_r}\nabla \times \mathbf{U}\right) = \frac{1}{\varepsilon_r}\nabla \times \nabla \times \mathbf{U} = -\frac{1}{\varepsilon_r}\nabla^2\mathbf{U} \tag{4.14}$$

holds. In (4.14), $\mathbf{U}$ denotes the electric or magnetic field vector.

The identity (4.14) can be used to rewrite the vector wave equations as (see Section 1.2)

$$\nabla^2\mathbf{H} + k^2\mathbf{H} = 0 \tag{4.15}$$

$$\nabla^2\mathbf{E} + k^2\mathbf{E} = 0 \tag{4.16}$$

with $k = k_0\sqrt{\mu_r\epsilon_r}$. Given that $\nabla^2\mathbf{H} = \hat{x}\nabla^2 H_x + \hat{y}\nabla^2 H_y + \hat{z}\nabla^2 H_z$, these equations permit a decoupling of the differential equation among the field components. Thus for TE modes ($E_z = 0$, $H_z \neq 0$), we only need to solve the scalar wave equation

$$\nabla_t^2 H_z + (k^2 - \beta^2) H_z = 0 \tag{4.17}$$

where ∇_t denotes the surface gradient and we set $\partial^2 H_z/\partial z^2 = -\beta^2 H_z$. This is also called the Helmholtz equation and is basically the scalar equivalent of the curl–curl vector formulation. We note that once H_z is found from (4.17), Maxwell's equations can be used to obtain the other field components using the expressions $E_y = (j\omega\mu/\gamma^2)(\partial H_z/\partial x)$, $E_x = -(j\omega\mu/\gamma^2)(\partial H_z/\partial y)$, $H_x = (-j\beta/\gamma^2)(\partial H_z/\partial x)$, $H_y = (-j\beta/\gamma^2)(\partial H_z/\partial y)$ (see also Section 1.4). The quantities

$$\gamma = \sqrt{k^2 - \beta^2}, \qquad Z^{TE} = \frac{\beta}{\omega\varepsilon} \tag{4.18}$$

denote the wavenumber and characteristic mode impedance, respectively. Alternatively, we could opt to solve the vector wave equation

$$\nabla_t \times \left(\frac{1}{\mu_r} \nabla_t \times \mathbf{E}_t\right) - (k_0^2 \varepsilon_r - \beta^2)\mathbf{E}_t = 0 \tag{4.19}$$

in which

$$\mathbf{E}_t = \hat{x} E_x + \hat{y} E_y \tag{4.20}$$

is the total transverse electric field in the guide. Again, (4.19) is valid only for cases where the field variation is independent of the third dimension.

Alternatively, for the TM modes ($H_z = 0, E_z \neq 0$), the appropriate scalar and vector wave equations are simply the duals of (4.17)–(4.19). However, the boundary conditions are of the Dirichlet type when solving for E_z and of the Neumann type when solving for H_z. As was shown in Chapter 3, the relation $\partial H_z/\partial n = 0$ serves as the boundary condition for the TE case.

4.2.4 Waveguide Propagation (Inhomogeneous Cross Section)

Let us again consider the waveguide in Fig. 4.3 consisting of a center conductor enclosed by an outer metallic shell. The region between the outer boundary of the inner conductor and inner boundary of the outer conductor is the pertinent computational domain. When the waveguide's material cross section is inhomogeneous, the fields can exhibit variation in the third dimension. Therefore, the normal component of the field is required for an accurate representation of waveguide propagation. Also, the resulting differential equation is not easily separable, which leads to a coupled differential equation in the tangential and normal components of the field.

To simplify the vector wave equation, we begin by rewriting the del operator using (4.13) as

$$\nabla = \nabla_t + \hat{z}\frac{\partial}{\partial z} = \nabla_t - j\beta\hat{z} = \hat{x}\frac{\partial}{\partial x} + \hat{y}\frac{\partial}{\partial y} - j\beta\hat{z} \tag{4.21}$$

implying that

$$\begin{aligned}\nabla \times \mathbf{E} &= (\nabla_t - j\beta\hat{z}) \times (\mathbf{E}_t + \hat{z}E_z) \\ &= \nabla_t \times \mathbf{E}_t + \nabla_t E_z \times \hat{z} - j\beta\hat{z} \times \mathbf{E}_t \\ &= \nabla_t \times \mathbf{E}_t + (\nabla_t E_z + j\beta\mathbf{E}_t) \times \hat{z}\end{aligned}$$

in which $\mathbf{E} = \mathbf{E}_t + \hat{z}E_z$ as before. Thus,

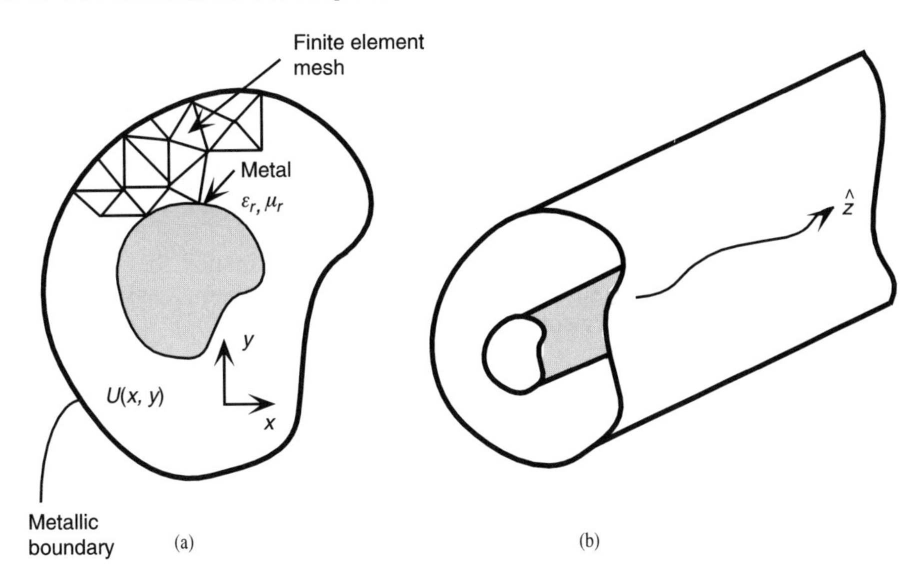

Figure 4.3 Waveguide configuration: (a) cross section of waveguide; (b) three-dimensional view.

$$
\begin{aligned}
\nabla \times \frac{1}{\mu_r} \nabla \times \mathbf{E} &= (\nabla_t - j\beta\hat{z}) \times \left(\frac{1}{\mu_r}\right) [\nabla_t \times \mathbf{E}_t + (\nabla_t E_z + j\beta\mathbf{E}_t) \times \hat{z}] \\
&= \nabla_t \times \frac{1}{\mu_r} \nabla_t \times \mathbf{E}_t - \hat{z} \times \frac{j\beta}{\mu_r} [(\nabla_t E_z + j\beta\mathbf{E}_t) \times \hat{z}] \\
&\quad + \nabla_t \times \left[\frac{1}{\mu_r} (\nabla_t E_z + j\beta\mathbf{E}_t) \times \hat{z}\right]
\end{aligned} \tag{4.22}
$$

Next we introduce (4.22) into the vector wave equation $\nabla \times (1/\mu_r)\nabla \times \mathbf{E} - k_0^2\varepsilon_r\mathbf{E} = 0$ and set each of the vector components to zero. This permits the decomposition of the original wave equation into a pair of differential equations [4, 5, 6]

$$
\nabla_t \times \frac{1}{\mu_r} \nabla_t \times \mathbf{E}_t - \frac{j\beta}{\mu_r} (\nabla_t E_z + j\beta\mathbf{E}_t) - k_0^2\varepsilon_r\mathbf{E}_t = 0 \tag{4.23}
$$

$$
\nabla_t \times \left[\frac{1}{\mu_r} (\nabla_t E_z + j\beta\mathbf{E}_t) \times \hat{z}\right] - k_0^2\varepsilon_r E_z\hat{z} = 0 \tag{4.24}
$$

for the transverse and z components of the wave equation. Clearly, (4.23) and (4.24) represent a pair of coupled differential equations which either needs to be decoupled or solved as is. Decoupling them using the divergence property yields a nonsymmetric generalized eigenvalue problem, the solution of which is numerically inefficient. However, using simple variable transformations [2],

$$
\begin{aligned}
\boldsymbol{e}_t &= j\beta\mathbf{E}_t \\
e_z &= E_z
\end{aligned} \tag{4.25}
$$

the coupled pair of differential equations (4.23) and (4.24) can be expressed as

$$\begin{aligned} \nabla_t \times \left(\frac{1}{\mu_r} \nabla_t \times \boldsymbol{e}_t\right) + \beta^2 \frac{1}{\mu_r} (\nabla_t e_z + \boldsymbol{e}_t) &= k_0^2 \varepsilon_r \boldsymbol{e}_t \\ \beta^2 \nabla_t \times \left[\frac{1}{\mu_r} (\nabla_t e_z + \boldsymbol{e}_t) \times \hat{z}\right] &= \beta^2 k_0^2 \varepsilon_r e_z \hat{z} \end{aligned} \tag{4.26}$$

The coupled pair of differential equations (4.26) can now be solved for β^2—the square of the propagation constant of the inhomogeneous waveguide—subject to the following boundary conditions:

$$\begin{aligned} \hat{n} \times \boldsymbol{e}_t &= 0 \\ e_z &= 0 \end{aligned} \tag{4.27}$$

on PEC surfaces and

$$\begin{aligned} (\nabla_t e_z + \boldsymbol{e}_t) \cdot \hat{n} &= 0 \\ \nabla_t \times \boldsymbol{e}_t &= 0 \end{aligned} \tag{4.28}$$

on PMC surfaces. The differential equation to be discretized is obtained by adding the two coupled differential equations and weighting them with the necessary weighting functions. This formulation leads to a symmetric generalized eigenvalue problem which is a direct result of the transformation (4.25).

4.3 DISCRETIZATION OF THE TWO-DIMENSIONAL WAVE EQUATION

From the above presentation, a general form of the two-dimensional wave equation is

$$\nabla \cdot [p(x, y) \nabla U(x, y)] + k_0^2 q(x, y) U(x, y) = f(x, y) \tag{4.29}$$

and it is understood that here $\nabla = \nabla_t = \hat{x}\frac{\partial}{\partial x} + \hat{y}\frac{\partial}{\partial y}$. This can be specialized to the problems of scattering and waveguide propagation by choosing $p(x, y)$ and $q(x, y)$ appropriately. When we consider TE (or H_z polarization) we must choose

$$U(x, y) = H_z(x, y), \quad p(x, y) = \frac{1}{\varepsilon_r}, \quad q(x, y) = \mu_r$$

while for TM or E_z polarization

$$U(x, y) = E_z(x, y), \quad p(x, y) = \frac{1}{\mu_r}, \quad q(x, y) = \varepsilon_r$$

The steps to be followed for the solution of (4.29) via the FEM parallel those given in Sections 3.4 to 3.7 for the solution of the corresponding one-dimensional (ordinary) differential equation. They involve

- casting of the original wave equation to its *weak* form to obtain a single functional incorporating the conditions imposed by the wave equation and the boundary conditions.

- tessellation of the computational domain allowing for a discretization of the *weak* form to a linear system of equations element by element.
- assembly of the element equation and imposition of the boundary conditions to obtain the final linear system of equations.

In this section we carry out the first two steps, and in the subsequent section we consider the assembly of elemental equations to solve for the fields and eigenvalues associated with a metallic waveguide (closed domain problem).

4.3.1 Weak Form of the Wave Equation

The pertinent residual of (4.29) is

$$R(x, y) = \nabla \cdot p(\mathbf{r})\nabla U(\mathbf{r}) + k_0^2\, q(\mathbf{r})U(\mathbf{r}) - f(\mathbf{r}) \tag{4.30}$$

where as usual $\mathbf{r} = x\hat{x} + y\hat{y}$ denotes the position vector. To derive the weak form, we multiply (4.30) by a weighting or testing function $W(x, y)$ and enforce $R(\mathbf{r}) = 0$ over the domain of each element rather than point by point. This gives

$$\iint_{\substack{\text{Domain of}\\ W(\mathbf{r})}} W(\mathbf{r})\, R(\mathbf{r})\, dx\, dy = 0 \tag{4.31}$$

which is the generalization of (3.9) for the one-dimensional case. As discussed in Section 3.4, the weighting function must again be compatible with the boundary conditions and be square integrable over the domain. Its derivative must also be square integrable.

To reduce the order of the derivatives in the residual and introduce the boundary terms, we must make use of the identities

$$W\nabla \cdot p\nabla U = \nabla \cdot (pW\nabla U) - p\nabla W \cdot \nabla U \tag{4.32}$$

$$\iint_\Omega \nabla \cdot (pW\nabla U)\, ds = \oint_C pW(\nabla U \cdot \hat{n})\, dl \tag{4.33}$$

in which Ω denotes the pertinent computational domain, C is the contour enclosing Ω as illustrated in Fig. 4.4, and $\hat{n}$ refers to the outward directed unit normal vector to the contour C. We note that the latter is the divergence theorem, which can be considered as the generalization of integration by parts to two dimensions.

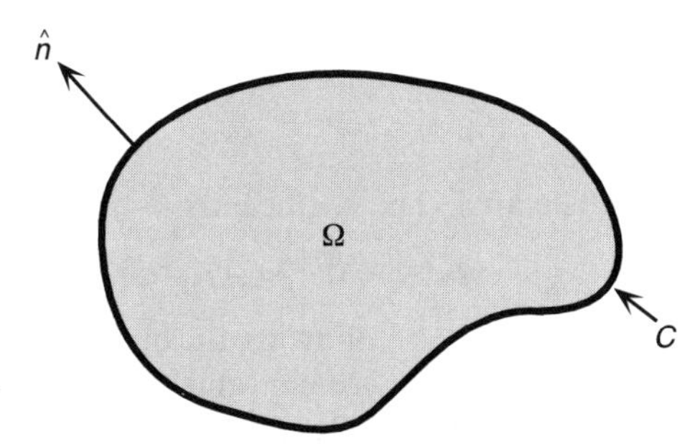

Figure 4.4 Parameters for the application of the divergence theorem.

Making use of (4.32) and (4.33) into the weighted residual equation (4.31) yields

$$\iint_{\Omega} [-p(\mathbf{r})\nabla W(\mathbf{r}) \cdot \nabla U(\mathbf{r}) + k_0^2 q(\mathbf{r}) W(\mathbf{r}) U(\mathbf{r}) - W(\mathbf{r}) f(\mathbf{r})]\, ds + \oint_C p(\mathbf{r}) W(\mathbf{r}) [\hat{n} \cdot \nabla U(\mathbf{r})]\, dl = 0 \quad (4.34)$$

This is the *weak form* of the two-dimensional scalar wave equation and should be compared to the one-dimensional weak form in (3.11). Again, we note the presence of the boundary integral over C which allows for the imposition of the boundary conditions. Thus, (4.34) provides a single statement incorporating the conditions implied by the wave equation and the pertinent boundary conditions. As noted for the one-dimensional case, this is at the heart of the finite element method.

4.3.2 Discretization of the Weak Wave Equation

To discretize (4.34) we proceed by introducing a discrete representation of the field $U(x, y)$ and making appropriate choices for the weighting function $W(x, y)$. As a first step, we consider a tessellation of the computational domain Ω in small triangular elements (see Fig. 4.1 to Fig. 4.3). We next choose to approximate the fields in each triangle as a linear function and in so doing, we can choose the expansion or basis functions $N_i^e(x, y)$ given in Section 2.3.2 to represent $U(x, y)$ over Ω. Specifically, we expand $U(x, y)$ as

$$U(x, y) = \sum_{e=1}^{N_e} \sum_{i=1}^{3} U_i^e N_i^e(x, y) \quad (4.35)$$

where U_i^e are the unknown coefficients of the expansion and represent the field or potential values at the nodes of each triangle. This representation is therefore referred to as a node-based expansion. As usual, N_e denotes the number of elements used for tessellating the domain. The procedure of tessellation is referred to as meshing and typically each side of the triangle is chosen to be less than 1/10 of a wavelength.

From Chapter 2, the explicit form of the shape functions $N_i^e(x, y)$ is

$$N_i^e(x, y) = \begin{cases} (1/2\Delta^e)(a_i^e + b_i^e x + c_i^e y) & \mathbf{r} \in \Omega^e \\ 0 & \text{otherwise} \end{cases} \quad (4.36)$$

where

$$\Delta^e = \frac{1}{2} \det \begin{bmatrix} 1 & x_1^e & y_1^e \\ 1 & x_2^e & y_2^e \\ 1 & x_3^e & y_3^e \end{bmatrix} = \frac{1}{2}[(x_2^e - x_1^e)(y_3^e - y_1^e) - (x_3^e - x_1^e)(y_2^e - y_1^e)] \quad (4.37)$$

is the triangle area. The coefficients in (4.36) are given by

$$a_i^e = x_j^e y_k^e - x_k^e y_j^e, \quad b_i^e = y_j^e - y_k^e, \quad c_i^e = x_k^e - x_j^e \quad (4.38)$$

with the indices (i, j, k) following the cyclical rule. That is, $(i, j, k) = (1, 2, 3), (2, 3, 1)$, or $(3, 1, 2)$ for the first, second, and third local nodes, respectively. The shape functions $N_i^e(x, y)$ are pictorially illustrated in Fig. 4.5. They are equal to unity at the ith

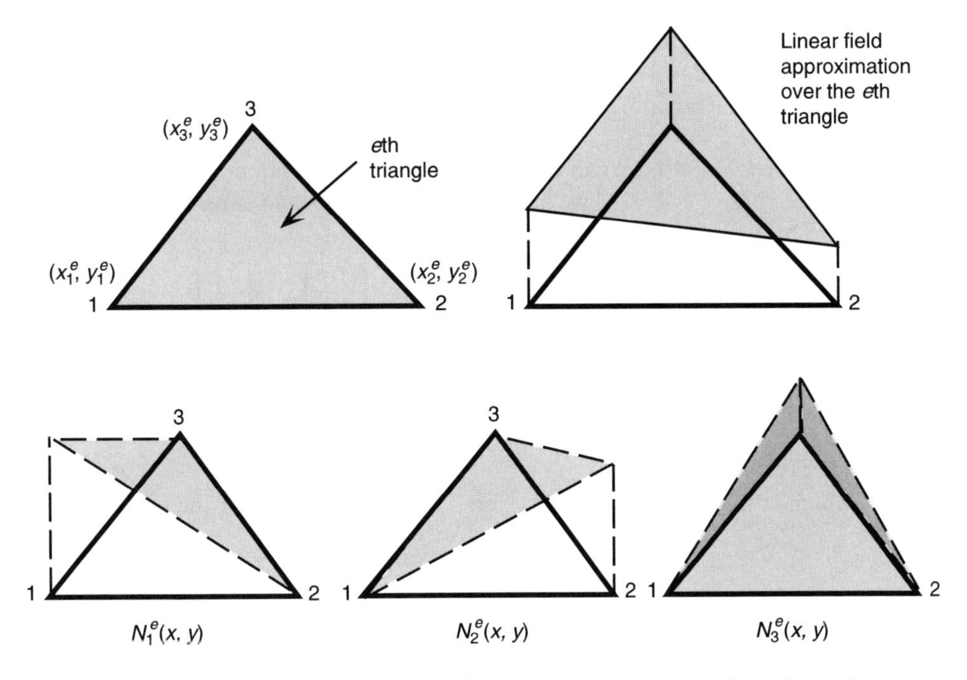

Figure 4.5 Node coordinates for the eth triangle and illustration of the node-based expansion functions $N_i^e(x, y)$.

node of the eth element and taper linearly to zero at the other two nodes. Consequently, the eth element expansion $\sum_{i=1}^{3} U_i^e N_i^e(x, y)$ scales the shape functions depending on the value of the field or potential at the nodes of the eth element.

Substituting the expansion (4.35) for $U(x, y)$ into the weak wave equation (4.34) yields

$$\sum_{e=1}^{N_e} \sum_{i=1}^{3} U_i^e \iint_{\Omega^e} [-p(\mathbf{r})\nabla W(\mathbf{r}) \cdot \nabla N_i^e(\mathbf{r}) + k_0^2 q(\mathbf{r}) W(\mathbf{r}) N_i^e(\mathbf{r})]\, dx\, dy$$
$$+ \int_C p(\mathbf{r}) W(\mathbf{r}) \hat{n} \cdot \nabla U(\mathbf{r})\, dl = \iint_{\Omega^e} W(\mathbf{r}) f(\mathbf{r})\, dy \qquad (4.39)$$

Notice that we have not used the expansion (4.35) to approximate $\hat{n} \cdot \nabla U = \frac{\partial U}{\partial n}$ on the boundary C since the behavior of $\frac{\partial U}{\partial n}$ on C must be provided through the boundary conditions.

A linear set of equations can now be obtained by employing Galerkin's method where we choose the weighting function $W(x, y)$ equal to the expansion basis $N_i^e(x, y)$, $i = 1, 2, 3$. Doing so yields

$$\sum_{i=1}^{3} U_i^e \iint_{\Omega^e} [-p(\mathbf{r})\nabla N_j^e(\mathbf{r}) \cdot \nabla N_i^e(\mathbf{r}) + k_0^2 q(\mathbf{r}) N_j^e(\mathbf{r}) N_i^e(\mathbf{r})]\, dx\, dy$$
$$+ \int_{C_s} p(\mathbf{r}) N_j^e(\mathbf{r}) \hat{n} \cdot \nabla U(\mathbf{r})\, dl = \iint_{\Omega^e} N_j^e(\mathbf{r}) f(\mathbf{r})\, dx\, dy, \qquad j = 1, 2, 3 \qquad (4.40)$$

where we have temporarily dropped the sum over the elements since $N_j^e(x, y)$ is nonzero only over the eth element. We will later perform the summation over all elements during the assembly process of the matrix system. Also, the presence of the boundary integral is required only if the eth element has an edge bordering the contour C. The contour segment C_s refers to the edge of the eth triangle which is part of C.

From (4.40), we can obtain a 3×3 system of equations by running through all choices of $j = 1, 2, 3$. If we assume that on C the field satisfies the Neumann boundary condition $\hat{n} \cdot \nabla U = \frac{\partial U}{\partial n} = 0$, we have

$$\begin{bmatrix} A_{11}^e & A_{12}^e & A_{13}^e \\ A_{21}^e & A_{22}^e & A_{23}^e \\ A_{31}^e & A_{32}^e & A_{33}^e \end{bmatrix} \begin{Bmatrix} U_1^e \\ U_2^e \\ U_3^e \end{Bmatrix} = \begin{Bmatrix} b_1^e \\ b_2^e \\ b_3^e \end{Bmatrix} \tag{4.41}$$

or

$$[A^e]\{U^e\} = \{b^e\}$$

which is the element matrix system. The explicit form of the matrix entries is

$$A_{ij}^e = -p^e \iint_{\Omega^e} \nabla N_i^e(\mathbf{r}) \cdot \nabla N_j^e(\mathbf{r})\, dx\, dy$$

$$+ k_0^2 q^e \iint_{\Omega^e} N_i^e(\mathbf{r})\, N_j^e(\mathbf{r})\, dx\, dy, \quad i = 1, 2, 3, \quad j = 1, 2, 3 \tag{4.42}$$

$$b_j^e = \iint_{\Omega^e} N_j^e(\mathbf{r}) f(\mathbf{r})\, dx\, dy, \quad j = 1, 2, 3 \tag{4.43}$$

where we assumed $p(\mathbf{r}) \approx p^e$ and $q(\mathbf{r}) \approx q^e$ are constant over each element to permit a closed form evaluation of the integrals. Since $N_i^e(\mathbf{r})$ are linear functions, the evaluation of the $[A^e]$ matrix entries can be done in closed form. We have

$$[A^e] = [K_\nabla^e] + k_0^2 [K^e]$$

where ($\{u\}$ imply column vectors)

$$[K_\nabla^e] = \left[-p^e \iint_{\Omega^e} \nabla N_i^e \cdot \nabla N_j^e\, dx\, dy \right]$$

$$- p^e \left[\iint_{\Omega^e} \left\{ \frac{\partial N_i^e}{\partial x} \right\} \left\{ \frac{\partial N_j^e}{\partial x} \right\}^T dx\, dy + \iint_{\Omega^e} \left\{ \frac{\partial N_i^e}{\partial y} \right\} \left\{ \frac{\partial N_j^e}{\partial y} \right\}^T \right] dx\, dy$$

$$[K^e] = \left[q^e \iint_{\Omega^e} N_i^e N_j^e\, dx\, dy \right] = q^e \iint_{\Omega^e} \{N_i^e\}\{N_j^e\}^T\, dx\, dy$$

Evaluating the entries of the submatrices $[K_\nabla^e]$ and $[K^e]$ yields [note that the constants b_i^e below are those in (4.38) and are not related to the excitation column in (4.41)]

$$\left[\iint_{\Omega^e} \nabla N_i^e \cdot \nabla N_j^e \right] = \frac{1}{4\Delta^e} \begin{bmatrix} (b_1^e)^2 + (c_1^e)^2 & b_1^e b_2^e + c_1^e c_2^e & b_1^e b_3^e + c_1^e c_3^e \\ b_2^e b_1^e + c_2^e c_1^e & (b_2^e)^2 + (c_2^e)^2 & b_2^e b_3^e + c_2^e c_3^e \\ b_3^e b_1^e + c_3^e c_1^e & b_3^e b_2^e + c_3^e c_2^e & (b_3^e)^2 + (c_3^e)^2 \end{bmatrix} \tag{4.44}$$

$$\left[\iint_{\Omega^e} N_i^e N_j^e \right] = \frac{\Delta^e}{12} \begin{bmatrix} 2 & 1 & 1 \\ 1 & 2 & 1 \\ 1 & 1 & 2 \end{bmatrix} \tag{4.45}$$

The latter is independent of the triangle coordinates (except for the area multiplier) and is referred to as a universal element matrix [7]. By making use of (4.44) and (4.45) in (4.42), the $[A^e]$ matrix entries can be more compactly written as

$$A_{ij}^e = K_{\nabla_{ij}}^e + k_0^2 K_{ij}^e = \frac{-p^e}{4\Delta^e}(b_i^e b_j^e + c_i^e c_j^e) + k_0^2 q^e \frac{\Delta^e}{12}(1 + \delta_{ij}) \tag{4.46}$$

where

$$\delta_{ij} = \begin{cases} 1 & i = j \\ 0 & \text{otherwise} \end{cases}$$

Thus, for the chosen linear triangular tessellation, the matrix entries are given in closed form and can be easily evaluated. Depending on the form of the function $f(x, y)$, the excitation column entries b_j^e may need to be computed numerically.

4.3.3 Assembly of Element Equations

As was the case for the one-dimensional solutions, the next step in the finite element procedure is the assembly of the element equations (4.41). This refers to the procedure of carrying out the sum

$$\sum_{e=1}^{N_e} [A^e]\{U^e\} = \sum_{e=1}^{N^e} \{b^e\} \tag{4.47}$$

as implied by the original discrete form of the "weak" wave equation (4.39). Since several of the elements may share the same node, the sum or assembly (4.47) consolidates the surrounding elements to yield a single equation. This is simply the sum of the element equations from all of the surrounding elements. Specifically, let us consider node 1 which is shared by five elements, as shown in Fig. 4.6.

For this case, the sum (4.47) produces the equation

$$\left(\sum_{i=0}^{4} A_{11}^{e'+i}\right) U_1^{e'} + \sum_{i=0}^{4} (A_{12}^{e'+i} U_2^{e'+i} + A_{13}^{e'+i} U_3^{e'+i}) = \sum_{i=0}^{4} b_1^{e'+i} \tag{4.48}$$

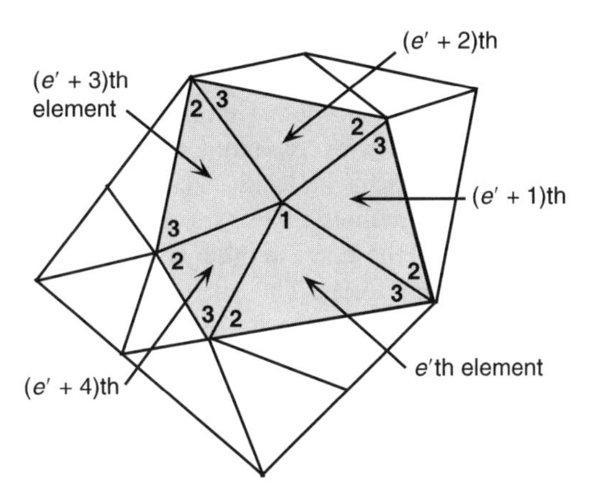

Figure 4.6 Illustration of a node shared by five elements (triangles).

for node 1 with the matrix entries as given in (4.46). The corresponding equations for the other nodes are very similar, except for differences in the superscripts/subscripts and the order of the sum. We should remark that the assembled equation (4.48) for node 1 is the same regardless of the number of elements/nodes contained in the computational domain. That is, even if the entire domain contains thousands of nodes, (4.48) will still involve only six nodal fields, implying that the corresponding row of the assembled matrix system $[A]\{U\} = \{b\}$ will contain only six nonzero entries even though the rank of $[A]$ is in the thousands. Thus, the assembled finite element matrix is always very sparse and this is a major advantage and characteristic of the FEM. The bandwidth and structure of $[A]$ is determined by the connectivity of the nodes, a result of the tessellation scheme. As can be understood, the bandwidth of the matrix $[A]$ is strongly dependent on the node numbering scheme. We can reduce the bandwidth by numbering the adjacent nodes using consecutive numbers. This is difficult to achieve, but sparse matrix storage schemes as those presented in Chapter 8 can be used to maintain their efficiency for matrices of the same sparsity but different bandwidths. Parallel computing architectures take advantage of sparsity but in the case of vector processors, narrow matrix bandwidths must also be maintained for substantial efficiency improvements [8].

A key issue in performing the assembly as dictated by (4.47) is the transformation from local to global nodes. This was discussed for the one-dimensional analysis. However, because of the easily predictable connectivity of the elements (i.e., each element was sequentially numbered and each node was shared by the two adjacent segments) the local to global transformation was not an issue for the one-dimensional case. The issue of node numbering becomes apparent when we look at the assembled equation (4.48).

Since the unknowns U_i^e must be eventually put into a single column (with one subscript), it is necessary to have a readily available mapping between the local and global nodes which are associated with the eth element. Thus, in addition to the node geometry data provided to the finite element program, we must also provide information about the local and global node numbering schemes. Four tables may be required before carrying out the matrix assembly routine:

- Node Location Table

  A listing of all mesh nodes (interior and boundary nodes) using global numbers and their corresponding (x, y) coordinates. This table specifies the geometry of the input configuration.
- Triangle Connectivity Table

  The global nodes comprising each triangle are given by this table. For example, by referring to Fig. 4.7 we observe that element #3 ($e = 3$) is formed by nodes 3, 5, and 2, as given in line 3 of the table. Basically, the table defines three arrays: $n(1, e)$, $n(2, e)$, and $n(3, e)$. The first of these provides the correspondence between local node 1 of the eth element and the global nodes. The other two arrays provide the same correspondence information for the other two nodes of the triangle. Let's say for example that we are working with the local nodes of element $e = 4$ and want to get the corresponding global nodes of that element. Then the value of $n(1, 4)$ will give the global number of local node 1, $n(2, 4)$ will provide the global number of local node 2, and so on.

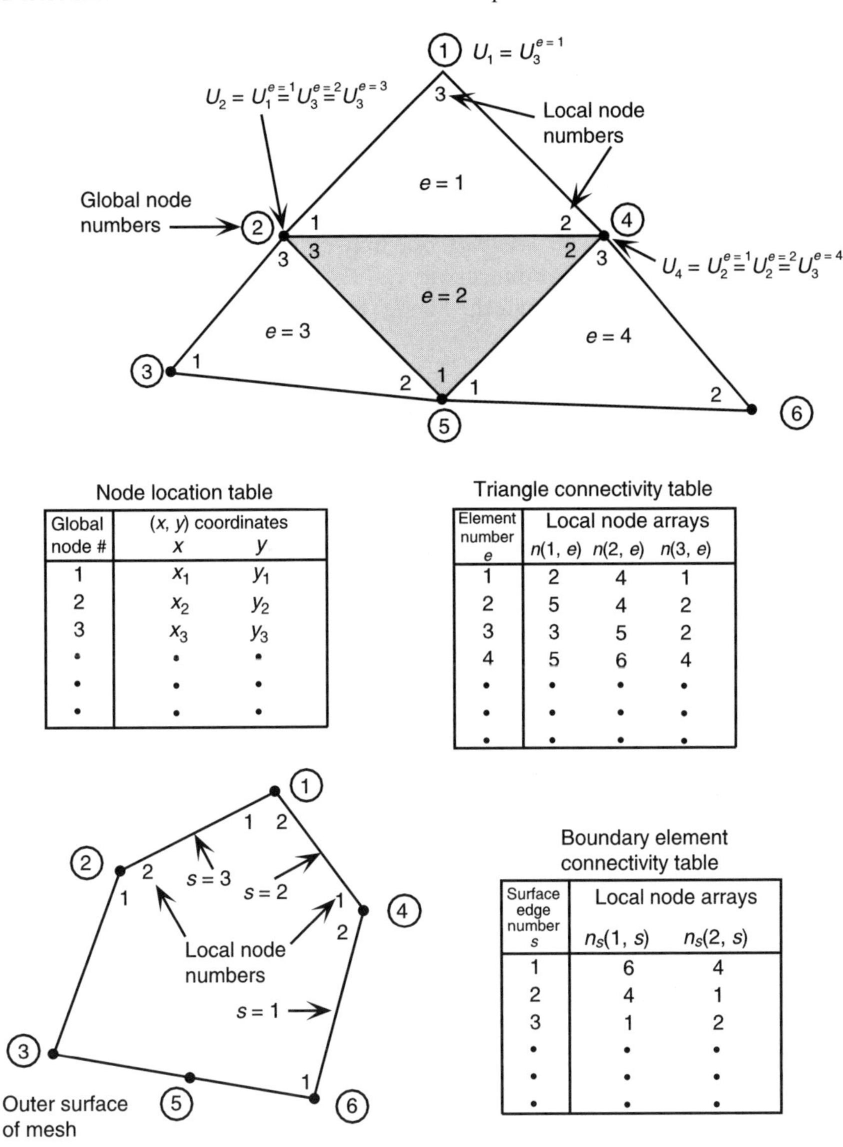

Node location table

Global node #	(x, y) coordinates x	y
1	x_1	y_1
2	x_2	y_2
3	x_3	y_3
•	•	•
•	•	•
•	•	•

Triangle connectivity table

Element number e	Local node arrays n(1, e)	n(2, e)	n(3, e)
1	2	4	1
2	5	4	2
3	3	5	2
4	5	6	4
•	•	•	•
•	•	•	•
•	•	•	•

Boundary element connectivity table

Surface edge number s	Local node arrays $n_s(1, s)$	$n_s(2, s)$
1	6	4
2	4	1
3	1	2
•	•	•
•	•	•
•	•	•

Figure 4.7 Geometry and connectivity data tables required for matrix assembly.

- Boundary Element Table

  To impose the boundary conditions it is necessary to identify the surface edges (line elements) on the outer boundary of the mesh and their associated nodes. This table can be generated using the data in the previous two tables. The data manipulations required to generate the surface node and element information is typically part of the data preprocessor and is an important step before assembling the final system. An example of such a boundary element table is given in Fig. 4.7. For the two-dimensional case, it suffices to identify all segments which bound the mesh and the nodes which form those elements. Again, the listed arrays provide the correspondence between the local surface element nodes and the global nodes.

- Material Group Table

  This is a look-up table for the material specification of each element. In practice, the same material covers blocks or sections of the domain and it is therefore not necessary to specify the material parameters for each individual element. Instead, one may choose to attach a material code column to the element connectivity table. That is, the material of each element is specified through the "code" which is in turn associated with specific values of ε_r and μ_r.

The first two of the above tables are always required but the latter two may or may not be needed depending on the application at hand. Also, it may be convenient to introduce other tables when different elements are used or more complex geometries are modeled.

4.3.4 Assembly Example: Waveguide Eigenvalues

To show how the assembly is performed, let us do a specific example. For illustrative purposes, consider the 18-element rectangular waveguide cross section shown in Fig. 4.8. The node location and element connectivities are given in the accompanying tables. If we assume that $U = H_z$, then $\frac{\partial U}{\partial n} = \hat{n} \cdot \nabla U = 0$ at the boundary nodes of the rectangular metallic waveguide. Thus, the boundary integral in (4.40) vanishes and the FEM system for the node fields is obtained by performing the assembly

$$\sum_{e=1}^{18} [A^e]\{U^e\} = [A]\{U\} = 0 \tag{4.49}$$

where the dimension of $\{U\}$ is 16 and $\{b^e\}$ was set to zero since no excitation is assumed. A procedure for carrying out the assembly is as follows:

Note the correspondence between the local and global nodes. For example, $U_1^{e=10} = U_6$, where, as usual, the single subscript refers to the global node 6. Thus, the element matrix for $e = 10$ is

$$\begin{bmatrix} A_{11}^{10} & A_{12}^{10} & A_{13}^{10} \\ A_{21}^{10} & A_{22}^{10} & A_{23}^{10} \\ A_{31}^{10} & A_{32}^{10} & A_{33}^{10} \end{bmatrix} \begin{Bmatrix} U_6 \\ U_{11} \\ U_{10} \end{Bmatrix} = 0 \tag{4.50}$$

If we are interested in obtaining the assembled equation for global node 6, we must also consider all of the elements sharing this node. In addition to element 10, node 6 is shared by elements 2, 4, 9, and 7. The corresponding element equations for each of these are

$$\begin{bmatrix} A_{11}^{2} & A_{12}^{2} & A_{13}^{2} \\ A_{21}^{2} & A_{22}^{2} & A_{23}^{2} \\ A_{31}^{2} & A_{32}^{2} & A_{33}^{2} \end{bmatrix} \begin{Bmatrix} U_2 \\ U_6 \\ U_5 \end{Bmatrix} = 0 \tag{4.51}$$

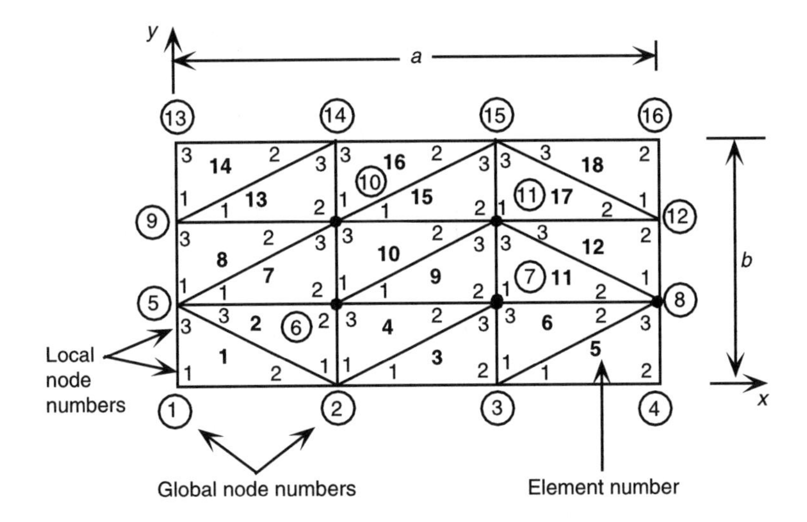

Triangle connectivity table

Element number e	Local node arrays $n(1, e)$	$n(2, e)$	$n(3, e)$
1	1	2	5
2	2	6	5
3	2	3	7
4	2	7	6
5	3	4	8
6	3	8	7
7	5	6	10
8	5	10	9
9	6	7	11
10	6	11	10
11	7	8	11
12	8	12	11
13	9	10	14
14	9	14	13
15	10	11	15
16	10	15	14
17	11	12	15
18	12	16	15

Node location table

Global node number	x	y
1	0	0
2	0.5	0
3	1.0	0
4	1.5	0
5	0	0.25
6	0.5	0.25
7	1.0	0.25
8	1.5	0.25
9	0	0.5
10	0.5	0.5
11	1.0	0.5
12	1.5	0.5
13	0	0.75
14	0.5	0.75
15	1.0	0.75
16	1.5	0.75

Figure 4.8 Geometry, node location, connectivity, and boundary node data tables required for matrix assembly.

$$\begin{bmatrix} A_{11}^4 & A_{12}^4 & A_{13}^4 \\ A_{21}^4 & A_{22}^4 & A_{23}^4 \\ A_{31}^4 & A_{32}^4 & A_{33}^4 \end{bmatrix} \begin{Bmatrix} U_2 \\ U_7 \\ U_6 \end{Bmatrix} = 0 \tag{4.52}$$

$$\begin{bmatrix} A_{11}^9 & A_{12}^9 & A_{13}^9 \\ A_{21}^9 & A_{22}^9 & A_{23}^9 \\ A_{31}^9 & A_{32}^9 & A_{33}^9 \end{bmatrix} \begin{Bmatrix} U_6 \\ U_7 \\ U_{11} \end{Bmatrix} = 0 \tag{4.53}$$

$$\begin{bmatrix} A_{11}^7 & A_{12}^7 & A_{13}^7 \\ A_{21}^7 & A_{22}^7 & A_{23}^7 \\ A_{31}^7 & A_{32}^7 & A_{33}^7 \end{bmatrix} \begin{Bmatrix} U_5 \\ U_6 \\ U_{10} \end{Bmatrix} = 0 \tag{4.54}$$

Adding the element equations in (4.50) to (4.54) which refer to testing at a common node, yields the system

$$\begin{bmatrix} A_{11}^2 + A_{11}^4 & A_{13}^2 & A_{12}^2 + A_{13}^4 & A_{12}^4 & 0 & 0 \\ A_{31}^2 & A_{11}^7 + A_{33}^2 & A_{12}^7 + A_{32}^2 & 0 & A_{13}^7 & 0 \\ A_{21}^2 + A_{31}^4 & A_{23}^2 + A_{21}^7 & \begin{matrix} A_{11}^{10} + A_{22}^2 + \\ A_{33}^4 + A_{11}^9 + A_{22}^7 \end{matrix} & A_{32}^4 + A_{12}^9 & A_{13}^{10} + A_{23}^7 & A_{12}^{10} + A_{13}^9 \\ A_{21}^4 & 0 & A_{23}^4 + A_{21}^9 & A_{22}^4 + A_{22}^9 & 0 & A_{23}^9 \\ 0 & A_{31}^7 & A_{31}^{10} + A_{32}^7 & 0 & A_{33}^{10} + A_{33}^7 & A_{32}^{10} \\ 0 & 0 & A_{21}^{10} + A_{31}^9 & A_{32}^9 & A_{23}^{10} & A_{33}^9 + A_{22}^{10} \end{bmatrix}$$

$$\times \begin{Bmatrix} U_2 \\ U_5 \\ U_6 \\ U_7 \\ U_{10} \\ U_{11} \end{Bmatrix} = 0 \tag{4.55}$$

We can continue assembling element equations onto this system until all elements have been accounted for. However, the third equation in (4.55), referring to testing at node 6, will not change through the remaining assembly process. Clearly, this equation is no different than the generic equation (4.48). Thus, we have established a pattern for assembling the final system of equations. Specifically, note that in the final assembled system, the entries will be given by the sum

$$A_{n',m'} = \sum A_{ij}^{e_1}, \quad \text{for all} \quad n(i, e_1) = n', \quad n(j, e_1) = m' \tag{4.56}$$

The index (see Fig. 4.8) e_1 of the sum must be kept identical in $n(i, e_1)$ and $n(j, e_1)$ when carrying out the sum over e_1, i and j. For example, $A_{66} = A_{11}^{10} + A_{22}^2 + A_{33}^4 + A_{11}^9 + A_{22}^7$, $A_{62} = A_{21}^2 + A_{31}^4$, $A_{44} = A_{22}^5$, $A_{55} = A_{11}^7 + A_{11}^8 + A_{33}^2 + A_{33}^1$, $A_{52} = A_{32}^1 + A_{31}^2$, and so on. A computer (MATLAB) routine for carrying out the assembly is given in the appendix where the coefficients are adjusted for the chosen polarization. For a wavenumber of $k_0 = 2\pi$ (i.e., $\lambda = 1$), the numerical values of the assembled $[A]$ matrix are given on the next page and can be used in conjunction with a given excitation vector $\{b\}$ to obtain the waveguide fields across its cross section.

$$
[A] = \begin{bmatrix}
-0.8388 & 0.4556 & 0 & 0 & 1.2056 & 0 & 0 & 0 & 0 & 0 & 0 & 0 & 0 & 0 & 0 & \\
0.4556 & -0.8551 & 0.4556 & 0 & 0.4112 & 2.4112 & 0.4112 & 0 & 0 & 0 & 0 & 0 & 0 & 0 & 0 & 0 \\
0 & 0.4556 & -1.2663 & 0.4556 & 0 & 0 & 2.4112 & 0.4112 & 0 & 0 & 0 & 0 & 0 & 0 & 0 & 0 \\
0 & 0 & 0.4556 & -0.8388 & 0 & 0 & 0 & 1.2056 & 0 & 0 & 0 & 0 & 0 & 0 & 0 & 0 \\
1.2056 & 0.4112 & 0 & 0 & -0.8551 & 0.9112 & 0 & 0 & 1.2056 & 0 & 0 & 0 & 0 & 0 & 0 & 0 \\
0 & 2.4112 & 0 & 0 & 0.9112 & -2.9438 & 0.9112 & 0 & 0 & 2.4112 & 0.4112 & 0 & 0 & 0 & 0 & 0 \\
0 & 0.4112 & 2.4112 & 0 & 0 & 0.9112 & -2.9438 & 0.9112 & 0 & 0 & 2.4112 & 0 & 0 & 0 & 0 & 0 \\
0 & 0 & 0.4112 & 1.2056 & 0 & 0 & 0.9112 & -0.8551 & 0 & 0 & 0.4112 & 1.2056 & 0 & 0 & 0 & 0 \\
0 & 0 & 0 & 0 & 1.2056 & 0 & 0 & 0 & -1.2663 & 0.9112 & 0 & 0 & 1.2056 & 0.4112 & 0 & 0 \\
0 & 0 & 0 & 0 & 0.4112 & 2.4112 & 0 & 0 & 0.9112 & -2.5326 & 0.9112 & 0 & 0 & 2.4112 & 0.4112 & 0 \\
0 & 0 & 0 & 0 & 0 & 0.4112 & 2.4112 & 0.4112 & 0 & 0.9112 & -2.5326 & 0.9112 & 0 & 0 & 2.4112 & 0 \\
0 & 0 & 0 & 0 & 0 & 0 & 0 & 1.2056 & 0 & 0 & 0.9112 & -1.2663 & 0 & 0 & 0.4112 & 1.2056 \\
0 & 0 & 0 & 0 & 0 & 0 & 0 & 0 & 1.2056 & 0 & 0 & 0 & -0.8388 & 0.4556 & 0 & 0 \\
0 & 0 & 0 & 0 & 0 & 0 & 0 & 0 & 0.4112 & 2.4112 & 0 & 0 & 0.4556 & -1.2663 & 0.4556 & 0 \\
0 & 0 & 0 & 0 & 0 & 0 & 0 & 0 & 0 & 0.4112 & 2.4112 & 0.4112 & 0 & 0.4556 & -0.8551 & 0.4556 \\
0 & 0 & 0 & 0 & 0 & 0 & 0 & 0 & 0 & 0 & 0 & 1.2056 & 0 & 0 & 0.4556 & -0.8388
\end{bmatrix}
$$

4.3.4.1 TE Modes (Neumann Boundary Conditions). One way to verify the correct evaluation and implementation of the FEM matrix is to use it in computing the known capacitance of a given transmission line such as the shielded microstrip line in Fig. 4.1. In this case, the wavenumber k_0 is set to zero and thus only the $[K_\nabla^e]$ matrix is used to evaluate the capacitance C_0, as in Fig. 4.1. A situation where both submatrices $[K_\nabla]$ and $[K]$ [obtained from the assembly of $[K_\nabla^e]$ and $[K^e]$ given in (4.46)] are used is that of computing the waveguide cutoff wavenumbers. Upon comparing the wave equations (4.17) and (4.29), we observe that the cutoff wavenumbers are obtained by solving the generalized eigenvalue problem

$$-[K_\nabla]\{H_z\} = \gamma^2[K]\{H_z\} \tag{4.57}$$

where γ^2 are the eigenvalues and γ refer to the cutoff wavenumbers. Considering a rectangular waveguide (see Fig. 4.8) with dimensions $a/b = 2$, the computed eigenvalues are given in Table 4.1 and the corresponding mode field distributions (eigenvectors) are shown in Fig. 4.9 for TE_{10} and TE_{11}. These calculations were carried out by Reddy et al. [9] using 400 triangular elements over the waveguide cross section. As seen, they are within 0.4% of the exact [10] eigenvalues given by

$$\gamma = \sqrt{\left(\frac{m\pi}{a}\right)^2 + \left(\frac{n\pi}{a/2}\right)^2}$$

for the TE_{mn} ($m \neq 0$ or $n \neq 0$) and the TM_{mn} ($m \neq 0$ and $n \neq 0$) modes. We remark that the accuracy of the calculated eigenvalues deteriorates for the higher order modes since the latter require a finer tessellation due to their more complex mode structure.

We can solve the eigenvalue problem (4.57) with equal ease for a coaxial cable of inner radius r_1 and outer radius r_2, as shown in Fig. 4.10. The typical triangular

TABLE 4.1 Cutoff Wavenumbers for a Rectangular Waveguide

		γa	$(a/b = 2)$
TE	TM	Analytical [10]	FEM Calculation
10		3.142	3.144
20		6.285	6.308
01		6.285	6.308
11	11	7.027	7.027
12	12	12.958	13.201
21	21	8.889	8.993

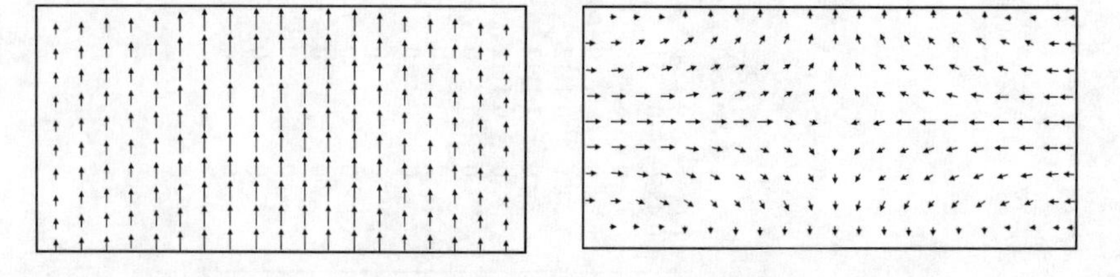

Figure 4.9 Calculated TE_{10} and TE_{11} mode electric fields in a rectangular waveguide with $a/b = 2$. [*Courtesy of Reddy et al. [9].*]

mesh for this type of cross section is illustrated in Fig. 4.3. Calculations were carried out by Reddy et al. [9] using 340 elements to model the cross section between the inner and outer conductor for $r_2/r_1 = 4$. The first few eigenvalues (cutoff wave numbers) and eigenvectors are given in Table 4.2 and Fig. 4.11, respectively. For the shown TE mode values, the agreement between the FEM calculated and analytical [11] wavenumbers is within 0.8%.

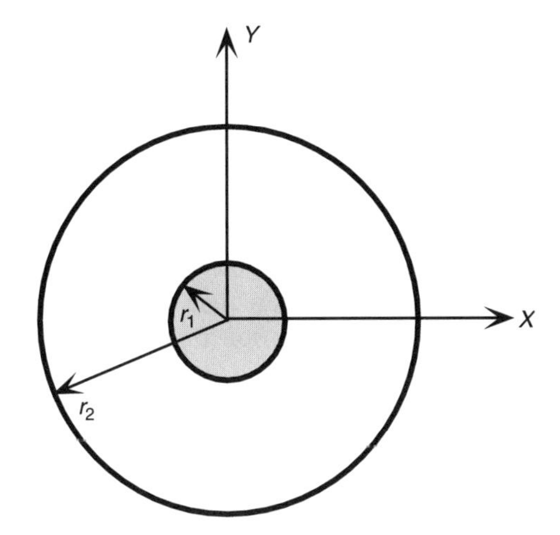

Figure 4.10 Coaxial waveguide geometry.

TABLE 4.2 Cutoff Wavenumbers for the Three Lowest Order TE Modes and Two Lowest Order TM Modes of the Coaxial Cable (the TEM mode is excluded)

	γr_1 $(r_2/r_1 = 4)$	
Mode	Analytical [11]	FEM Calculation
TE_{11}	0.411	0.412
TE_{21}	0.752	0.754
TE_{31}	1.048	1.055
TM_{01}	1.024	1.030
TM_{11}	1.112	1.120

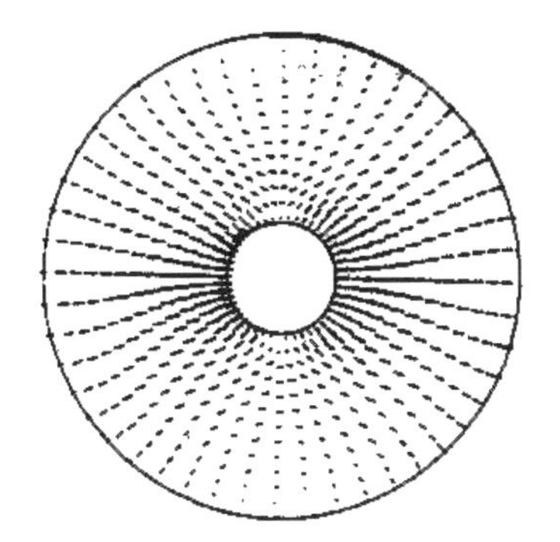

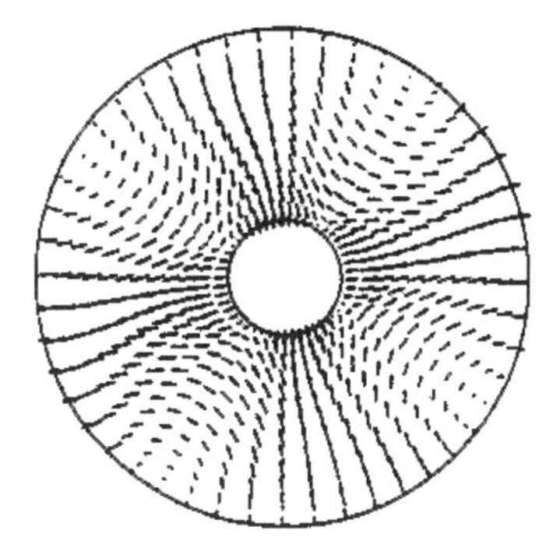

Figure 4.11 Calculated TE_{11} and TE_{21} mode electric fields in a coaxial line with $r_2/r_1 = 4$. [*Courtesy of Reddy et al. [9].*]

4.3.4.2 TM Modes (Dirichlet Boundary Conditions). For the TM modes, $U = E_z$ and $E_z = 0$ on the boundary. However, $\partial U/\partial n = \hat{n} \cdot \nabla E_z = \partial E_z/\partial n$ is non-zero on the boundary and it is therefore necessary to treat $\Psi = \partial E_z/\partial n$ as an additional unknown on the boundary. The contribution of the boundary integral then plays the same role as the [endpoints] for one-dimensional problems (see Chapter 3). As discussed in Section 3.7, since the field values at the boundary nodes are zero, it is not necessary to test at these nodes. Instead, we set the boundary node fields to zero whenever they appear in the system. By avoiding testing (or weighting) at the boundary nodes (or elements), the integrals over C_s do not enter in the construction of the final system and can thus be neglected altogether. We also remark that the choice of omitting testing at the boundary nodes is equivalent to setting the weighting functions to zero when testing at these nodes.

Although the above arguments are sufficient to proceed with the FEM matrix assembly while neglecting the presence of the boundary integral, they are nevertheless difficult to visualize without going through some of the details. Therefore, below we will (for a moment) proceed with the assumption that the boundary integral contribution is needed. We begin with the discretization of the boundary integral by introducing the expansion

$$\Psi(\mathbf{r}) = \frac{\partial E_z(\mathbf{r})}{\partial n} = \sum_{s=1}^{N_s} \sum_{i=1}^{2} \Psi_i^s L_i^s(\mathbf{r})|_{\mathbf{r} \in C_s} \tag{4.58}$$

where N_s denotes the number of boundary edges (12, for example, in Fig. 4.8) and Ψ_i^s carries the usual notation. That is, Ψ_i^s refers to the value of $\partial E_z/\partial n$ at the ith local node of the sth segment of the boundary (see Fig. 4.7). The expansion bases are linear interpolation functions between the node values of $\Psi(\mathbf{r})$. They are of the same form as those discussed in Chapter 2 (see Section 2.3.1) and Chapter 3. For a constant value of x or y, the two-dimensional shape functions reduce to the same linear one-dimensional expansion functions given in (2.3). Thus, we can write

$$L_1^s(\mathbf{r}) = \begin{cases} \dfrac{x_2^s - x}{x_2^s - x_1^s} & \text{on } y = \text{constant boundaries} \\ \dfrac{y_2^s - y}{y_2^s - y_1^s} & \text{on } x = \text{constant boundaries} \\ 0 & \text{outside the } s\text{th segment} \end{cases}$$
$$L_2^s(\mathbf{r}) = 1 - L_1^s(\mathbf{r}), \quad \mathbf{r} \in C_s \tag{4.59}$$

These expansion bases are applicable to piecewise rectangular boundaries, as in Fig. 4.8. In the case of circular boundaries, as in Fig. 4.10, an appropriate set of linear basis functions is

$$L_1^s(\phi) = \begin{cases} \dfrac{\phi_2^s - \phi}{\phi_2^s - \phi_1^s} & \phi_1^s < \phi < \phi_2^s \\ 0 & \text{otherwise} \end{cases}$$
$$L_2^s = 1 - L_1^s(\phi) \tag{4.60}$$

where ϕ is the angular variable ranging from $\phi = 0$ to $\phi = 2\pi$.

To obtain the element equations for the TM excitation, we substitute (4.58) along with (4.35) into (4.39) and proceed as done for the TE case. The resulting element system is

$$[A^e]\{U^e\} + [B^s]\{\Psi^s\} = \{b^e\} \tag{4.61}$$

where the entries of $[A^e]$ and $\{b^e\}$ are given by (4.42) and (4.43). Those of $[B^s]$ are computed from

$$B_{ij}^s = p^e \int_{C_s} N_i^e(\mathbf{r}) L_j^s(\mathbf{r})\, dl \tag{4.62}$$

or

$$B_{ij}^s = p^s \int_{C_s} L_i^s(\mathbf{r}) L_j^s(\mathbf{r})\, dl \tag{4.63}$$

($dl = dx$ or dy for rectangular boundaries) and are associated with the last left hand side of (4.39). The latter expression (4.63) for B_{ij}^s results from the identification that the sth surface or boundary segment must be an edge of the eth element for a nonzero value of B_{ij}^s. It is also understood that the boundary matrix $[B^s]$ will be nonzero only when the eth element is associated with a pair of nodes on the outer surface/boundary of the computational domain.

The assembly of the elemental equation (4.61) again amounts to summing the equations from each element weighting at the same global node. For the specific rectangular waveguide example shown in Fig. 4.8, the resulting assembled system will be of the form

$$[A]\{U\} + [B]\{\Psi\} = \{b\} \tag{4.64}$$

The expanded version of this matrix system is

$$\begin{bmatrix} A_{11} & A_{12} & A_{13} & \cdots & A_{1,16} \\ A_{21} & A_{22} & A_{23} & \cdots & A_{2,16} \\ \vdots & & & \vdots & \\ A_{n1} & A_{n2} & A_{n3} & \cdots & A_{n,16} \\ \vdots & & & \vdots & \\ A_{16,1} & A_{16,2} & A_{16,3} & \cdots & A_{16,16} \end{bmatrix} \begin{Bmatrix} U_1 \\ U_2 \\ \vdots \\ U_n \\ \vdots \\ U_{16} \end{Bmatrix} +$$

$$
\begin{bmatrix}
B_{11} & B_{12} & 0 & 0 & B_{15} & 0 & 0 & \cdots & & & & & & & \cdots & 0 \\
B_{21} & B_{22} & B_{23} & 0 & 0 & \cdots & & & & & & & & & \cdots & 0 \\
0 & B_{32} & B_{33} & B_{34} & 0 & 0 & \cdots & & & & & & & & \cdots & 0 \\
0 & 0 & B_{43} & B_{44} & 0 & 0 & 0 & B_{48} & 0 & 0 & \cdots & & & & \cdots & 0 \\
B_{51} & 0 & 0 & 0 & B_{55} & 0 & 0 & 0 & B_{59} & 0 & \cdots & & & & \cdots & 0 \\
0 & 0 & 0 & 0 & \cdots & & & & & & & & & & \cdots & 0 \\
0 & 0 & 0 & 0 & \cdots & & & & & & & & & & \cdots & 0 \\
0 & 0 & 0 & B_{84} & 0 & 0 & 0 & B_{88} & 0 & 0 & 0 & B_{8,12} & 0 & 0 & 0 & 0 \\
0 & 0 & 0 & 0 & B_{95} & 0 & 0 & 0 & B_{99} & 0 & 0 & 0 & B_{9,13} & 0 & 0 & 0 \\
0 & 0 & 0 & \cdots & & & & & & & & & & & \cdots & 0 \\
0 & 0 & 0 & \cdots & & & & & & & & & & & \cdots & 0 \\
0 & 0 & 0 & 0 & 0 & 0 & 0 & B_{12,8} & 0 & 0 & 0 & B_{12,12} & 0 & 0 & 0 & B_{12,16} \\
0 & 0 & 0 & 0 & 0 & 0 & 0 & 0 & B_{13,9} & 0 & 0 & 0 & B_{13,13} & B_{13,14} & 0 & 0 \\
0 & 0 & 0 & 0 & 0 & 0 & 0 & 0 & 0 & 0 & 0 & 0 & B_{14,13} & B_{14,14} & B_{14,15} & 0 \\
0 & 0 & 0 & 0 & 0 & 0 & 0 & 0 & 0 & 0 & 0 & 0 & 0 & B_{15,14} & B_{15,15} & B_{15,16} \\
0 & 0 & 0 & 0 & 0 & 0 & 0 & 0 & 0 & 0 & 0 & B_{16,12} & 0 & 0 & B_{16,15} & B_{16,16}
\end{bmatrix}
$$

$$
\begin{Bmatrix}
\Psi_1 \\ \Psi_2 \\ \Psi_3 \\ \Psi_4 \\ \Psi_5 \\ \Psi_6 \\ \Psi_7 \\ \Psi_8 \\ \Psi_9 \\ \Psi_{10} \\ \Psi_{11} \\ \Psi_{12} \\ \Psi_{13} \\ \Psi_{14} \\ \Psi_{15} \\ \Psi_{16}
\end{Bmatrix}
=
\begin{Bmatrix}
b_1 \\ b_2 \\ b_3 \\ b_4 \\ b_5 \\ b_6 \\ b_7 \\ b_8 \\ b_9 \\ b_{10} \\ b_{11} \\ b_{12} \\ b_{13} \\ b_{14} \\ b_{15} \\ b_{16}
\end{Bmatrix}
$$

in which Ψ_n denotes the outward directed normal derivative of E_z at the nth node. The values of Ψ at the interior nodes are irrelevant because they are associated with all zero rows, included for the proper addition of the $[A]$ and $[B]$ matrices. It is understood that $[A]$ is a very sparse matrix, as discussed earlier in the chapter.

We observe that the above system involves 12 nontrivial unknowns from the $\{\Psi\}$ column plus 16 unknowns from the $\{U\}$ column for a total of 28 unknowns. Clearly, this number of unknowns is much greater than the available 16 equations. For a solution for $\{U\}$ and $\{\Psi\}$ we must add 12 more equations or conditions on the values of $\{U\}$ and $\{\Psi\}$. This is done through the introduction of the Dirichlet boundary conditions satisfied by $\{U\} = \{E_z\}$ on the boundary nodes. The procedure is referred to as *condensation of boundary conditions* and was discussed in Chapter 3. To see how it can be done in a consistent manner, let us divide the $\{U\}$ column into two parts, one for the interior nodes and another for the boundary or surface nodes. That is, we write $\{U\}$ as

$$\{U\} = \begin{Bmatrix} \{U^I\} \\ \{U^S\} \end{Bmatrix}$$

where in the case of Fig. 4.8, $\{U^I\} = \{U_6, U_7, U_{10}, U_{11}\}^T$ are the interior node fields and $\{U^S\}$ contains the boundary node variables. Also, we formally define the column $\{\Psi\}$ as $\{\Psi\} = \{\Psi_1, \Psi_2, \Psi_3, \Psi_4, \Psi_5, \Psi_8, \Psi_9, \Psi_{12}, \Psi_{13}, \Psi_{14}, \Psi_{15}, \Psi_{16}\}^T$ which excludes all interior node values of Ψ. Using this notation, we can rewrite the system (4.64) as

$$\begin{bmatrix} [A^{II}] & [A^{IS}] \\ [A^{SI}] & [A^{SS}] \end{bmatrix} \begin{Bmatrix} \{U^I\} \\ \{U^S\} \end{Bmatrix} + \begin{bmatrix} 0 & 0 \\ 0 & [B] \end{bmatrix} \begin{Bmatrix} 0 \\ \{\Psi\} \end{Bmatrix} = \begin{Bmatrix} \{b^I\} \\ \{b^S\} \end{Bmatrix} \tag{4.65}$$

in which $[A^{II}]$ refers to the submatrix of $[A]$ containing the interactions among the interior nodes, $[A^{IS}]$ and $[A^{SI}]$ are associated with interactions among exterior and interior nodes, and $[A^{SS}]$ refers to the interactions among the boundary or outer surface nodes. Similarly to $\{U^I\}$ and $\{U^S\}$, the excitation subvectors $\{b^I\}$ and $\{b^S\}$ are associated with the interior and boundary nodes, respectively.

When we invoke the Dirichlet boundary condition

$$\{U^S\} = \{0\}$$

the system (4.65) can be decomposed into two independent subsystems

$$[A^{II}]\{U^I\} = \{b^I\} \tag{4.66}$$

and

$$[A^{SI}]\{U^I\} + [B]\{\Psi\} = \{b^S\} \tag{4.67}$$

The interior node fields are now decoupled from $\{\Psi\}$ and we can therefore proceed to solve (4.66) without a need to consider the solution of $\{\Psi\}$. Thus, the boundary integral can be neglected when assembling the FEM system provided U or its normal derivative are zero on the boundary C. After $\{U^I\}$ is found, we can return to (4.67) and determine $\{\Psi\}$, if necessary, by solving the system

$$[B]\{\Psi\} = \{\tilde{b}\} \tag{4.68}$$

where

$$\{\tilde{b}\} = \{b^S\} - [A^{SI}]\{U^I\}$$

In the case of an eigenvalue problem, the excitation column $\{b\}$ is set to zero and (4.66) is written as

$$-[K_\nabla^{II}]\{E_z\} = \gamma^2[K^{II}]\{E_z\} \tag{4.69}$$

where $[K_\nabla^{II}]$ and $[K^{II}]$ are identical to those in (4.57) except that only the entries associated with the interior nodes are kept and that p^e/q^e are defined for the TM polarization case. Some values of γ for the TM modes, obtained from (4.69), are given in Tables 4.1 and 4.2 for the rectangular and coaxial waveguides. Also, Fig. 4.12 displays the fields of the lowest order TM mode for each of these empty waveguides. The analysis can be carried out using the MATLAB program in the appendix. For the mesh in Fig. 4.8, the corresponding $[A^{II}]$, $[K_\nabla^{II}]$ and $[K^{II}]$ are

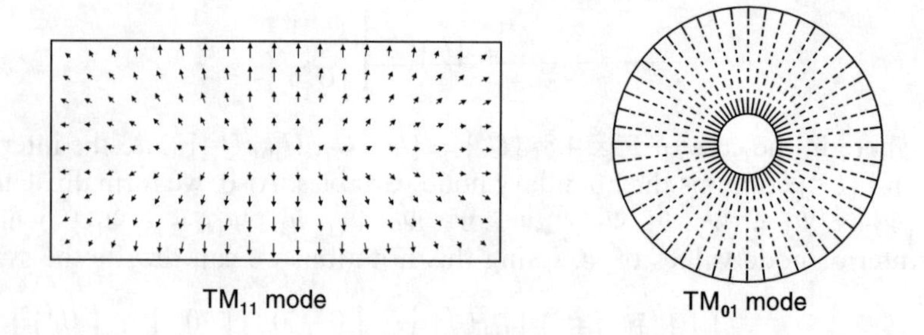

Figure 4.12 Calculated fields for the lowest TM modes of the rectangular ($a/b = 2$) and coaxial ($r_2/r_1 = 4$) waveguides. [*Courtesy of Reddy et al. [9].*]

$$[A^{II}] = \begin{bmatrix} -2.9438 & 0.9112 & 2.4112 & 0.4112 \\ 0.9112 & -2.9438 & 0 & 2.4112 \\ 2.4112 & 0 & -2.5326 & 0.9112 \\ 0.4112 & 2.4112 & 0.9112 & -2.5326 \end{bmatrix}$$

$$[K^{II}] = \begin{bmatrix} 0.521 & 0.0104 & 0.0104 & 0.0104 \\ 0.0104 & 0.0521 & 0 & 0.0104 \\ 0.0104 & 0 & 0.0625 & 0.0104 \\ 0.0104 & 0.0104 & 0.0104 & 0.0625 \end{bmatrix}$$

$$[K^{II}_{\nabla}] = \begin{bmatrix} 5.0000 & -0.5000 & -2.0000 & 0 \\ -0.5000 & 5.0000 & 0 & -2.0000 \\ -2.0000 & 0 & 5.0000 & -0.5000 \\ 0 & -2.0000 & -0.5000 & 5.0000 \end{bmatrix}$$

The above TM mode analysis confirms that the boundary integral in (4.39) could be neglected from the start when dealing with metallic domain enclosures. The final system for the TE and TM analysis is then obtained by enforcing the boundary conditions on $U(\mathbf{r})$ only. For the TE case, testing is imposed on the boundary and interior nodes without any specification for the boundary values of $U(\mathbf{r})$. However, for TM analysis, the boundary values of $U(\mathbf{r})$ are set to zero a priori. Thus testing at the boundary nodes is avoided and the final TM system is smaller than that corresponding to the TE case.

4.4 TWO-DIMENSIONAL SCATTERING

As another example application of solving the two-dimensional wave equation, we consider the scattering by a cylinder shown in Fig. 4.2. We shall assume H_z incidence as given by (4.6) and as based on the discussion in Section 4.2.2, it is best to work with the scattered rather than the total field. From (4.11), (4.29), and (4.34), the pertinent weak form of the wave equation is

$$\iint_{\Omega}\left[-\frac{1}{\varepsilon_r}\nabla W(\mathbf{r})\cdot\nabla H_z^{\text{scat}}(\mathbf{r})+k_0^2\mu_r W(\mathbf{r})H_z^{\text{scat}}(\mathbf{r})\right]ds$$
$$+\frac{1}{\varepsilon_r}\int_{C^{\text{outer}}+C^{\text{inner}}}W(\mathbf{r})[\hat{n}\cdot\nabla H_z^{\text{scat}}(\mathbf{r})]\,dl=\iint_{\Omega}W(\mathbf{r})f(\mathbf{r})\,ds \tag{4.70}$$

where

$$f(\mathbf{r})=-\nabla\cdot\left(\frac{1}{\varepsilon_r}\nabla H_z^i\right)-k_0^2\mu_r H_z^i+j\omega\varepsilon_0 M_{iz}$$

is the excitation function and $C^{\text{inner, outer}}$ represent the closure boundaries of the computation domain, as depicted in Fig. 4.13. For $H_z^i=e^{jk_0(x\cos\phi_0+y\sin\phi_0)}$ and $M_{iz}=0$, it follows that

$$f^e=k_0^2\left(\frac{1}{\varepsilon_r^e}-\mu_r^e\right)e^{jk_0(x\cos\phi_0+y\sin\phi_0)}=k_0^2\left(\frac{1}{\varepsilon_r^e}-\mu_r^e\right)H_z^i \tag{4.71}$$

in the eth element, where we approximated ε_r and μ_r by their average values ε_r^e and μ_r^e, respectively, over the eth element. Also, we note the presence of the boundary integral over $C=C^{\text{outer}}+C^{\text{inner}}$ and from (4.33) it is necessary that C encloses the entire computational domain (see Fig. 4.13 for the directions of the normals on each contour).

To discretize (4.70), we expand the scattered field as

$$H_z^{\text{scat}}(x,y)=\sum_{e=1}^{N_e}H_z^{se}=\sum_{e=1}^{N_e}\sum_{i=1}^{3}H_{zi}^{se}N_i^e(x,y) \tag{4.72}$$

where H_{zi}^{se} denotes the unknown scattered field values at the nodes and $N_i^e(x,y)$ is given by (4.36). Subsequently, on choosing Galerkin's testing (i.e., $W=N_j^e$), we obtain the element equations

$$\sum_{e=1}^{N^e}\sum_{i=1}^{3}H_{zi}^{se}\iint_{\Omega_e}\left[-\frac{1}{\varepsilon_r}\nabla N_j^e\cdot\nabla N_i^e+k_0^2\mu_r N_j^e N_i^e\right]dx\,dy$$
$$+\sum_{s_1=1}^{N_{s_1}}\int_{C_{s_1}^{\text{outer}}}\frac{1}{\varepsilon_r}N_j^e[\hat{n}\cdot\nabla H_z^{\text{scat}}]\,dl+\sum_{s_2=1}^{N_{s_2}}\int_{C_{s_2}^{\text{inner}}}\frac{1}{\varepsilon_r}N_j^e[\hat{n}\cdot\nabla H_z^{\text{scat}}]\,dl$$
$$=\sum_{e=1}^{N_e}\iint_{\Omega_e}N_j^e f^e\,dx\,dy \tag{4.73}$$

It has been assumed that the interior domain (dielectric and free space) bounded by C^{inner} and C^{outer} has been subdivided into triangles, whereas the contours C^{outer} and C^{inner} have been subdivided into N_{s_1} and N_{s_2} line segments, respectively. Note also that the excitation function f^e will be nonzero only when ε_r and μ_r are not equal to unity, i.e., only if the element is within the material region Ω_d, illustrated in Fig. 4.13.

4.4.1 Treatment of Metallic Boundaries

In the previous section, we eliminated the boundary integral over C^{inner} since $\hat{n}\cdot\nabla H_z^{\text{scat}}$ was zero on the metal boundary. However, this is not the case here because

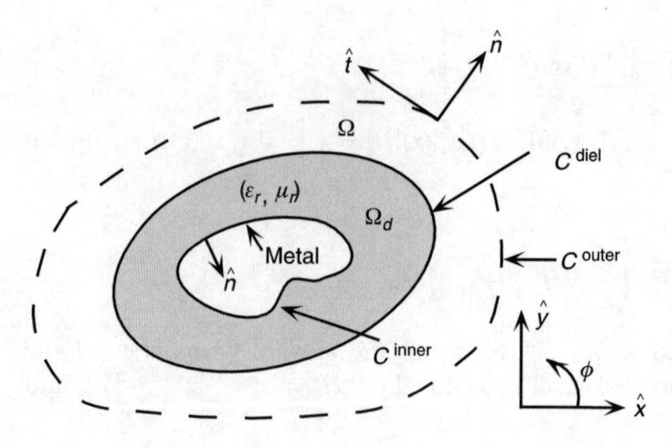

Figure 4.13 Illustration of the contours C^{outer} and C^{inner} as well as boundary vectors for the scattering geometry.

H_z^{scat} represents the scattered and not the total magnetic field. Since $H_z = H_z^i + H_z^{\text{scat}}$, it follows that

$$\hat{n} \cdot (\nabla H_z^i + \nabla H_z^{\text{scat}}) = \hat{n} \cdot \nabla H_z = 0, \qquad \mathbf{r} \in C^{\text{inner}} \tag{4.74}$$

and thus

$$\hat{n} \cdot \nabla H_z^{\text{scat}} = -\hat{n} \cdot \nabla H_z^i, \qquad \mathbf{r} \in C^{\text{inner}} \tag{4.75}$$

which is an inhomogeneous Neumann boundary condition, where $\hat{n} \cdot \nabla H_z^i$ is a known quantity everywhere on C^{inner}. An alternative form of (4.75) can be obtained by noting that

$$\hat{t} \cdot \mathbf{E} = \frac{jZ_0}{k_0} \frac{1}{\varepsilon_r} \hat{t} \cdot (\hat{z} \times \nabla H_z) = \frac{jZ_0}{k_0} \frac{1}{\varepsilon_r} \nabla H_z \cdot (\hat{t} \times \hat{z})$$

(see also (1.76)) or

$$\mathbf{E}_{\tan} = \frac{jZ_0}{k_0} \frac{1}{\varepsilon_r} \nabla H_z \cdot \hat{n} \tag{4.76}$$

where $\mathbf{E}_{\tan}$ refers to the total tangential electric field on C^{inner}. Similarly, for the scattered field

$$\frac{jk_0}{Z_0} \mathbf{E}_{\tan}^{\text{scat}} = -\frac{1}{\varepsilon_r} \nabla H_z^{\text{scat}} \cdot \hat{n} \tag{4.77}$$

and since $\mathbf{E}_{\tan} = \hat{t} \cdot (\mathbf{E}^i + \mathbf{E}^{\text{scat}}) = 0$ on conducting surfaces, it follows that

$$\frac{1}{\varepsilon_r} \nabla H_z^{\text{scat}} \cdot \hat{n} = +\frac{jk_0}{Z_0} \mathbf{E}_{\tan}^i \qquad \mathbf{r} \in C^{\text{inner}} \tag{4.78}$$

with $\mathbf{E}_{\tan}^i = Z_0 \hat{t} \cdot (\hat{x} \sin \phi_0 - \hat{y} \cos \phi_0) e^{jk_0(x \cos \phi_0 + y \sin \phi_0)}$. Thus, the boundary integral over C^{inner} can be moved to the right hand side of (4.73) to be included as part of the excitation column. This type of detail in treating a boundary condition (or constraint) demonstrates how a knowledge of the field over a boundary becomes an equivalent source excitation. It was discussed in Section 3.7 for one-dimensional applications and is referred to as condensation of boundary conditions.

4.4.2 Absorbing Boundary Conditions

Before casting (4.73) onto a matrix system, we must also consider the boundary conditions on C^{outer}. So far, no information has been given with regard to the boundary condition that must be satisfied by H_z^{scat} on C^{outer}. For scattering problems, the field continues to propagate to infinity and at large distances from the two-dimensional scatterer it has the form $Ae^{-jkr}/\sqrt{r}$, where r denotes the radial distance from the origin and A is a constant. Thus, since

$$\frac{\partial}{\partial r}\left[\frac{e^{-jk_0 r}}{\sqrt{r}}\right] = -jk_0\frac{e^{-jk_0 r}}{\sqrt{r}} - \frac{1}{2}\frac{e^{-jk_0 r}}{r^{3/2}}$$

H_z^{scat} satisfies the condition

$$\frac{\partial H_z^{\text{scat}}}{\partial r} + \left(jk_0 + \frac{1}{2r}\right)H_z^{\text{scat}} = 0, \qquad r \to \infty \tag{4.79}$$

This is the well-known first-order Bayliss-Turkel [12] absorbing boundary condition (ABC) and, by its nature, it can only be enforced on circular boundaries such as that shown in Fig. 4.14. In this case, $\hat{r}$ coincides with the normal $\hat{n}$, and we then rewrite (4.79) as

$$\frac{\partial H_z^{\text{scat}}}{\partial n} + \left(jk_0 + \frac{1}{2r_0}\right)H_z^{\text{scat}} = 0 \tag{4.80}$$

where r_0 is the radius of the circular boundary C^{outer}. We observe that (4.80) is identical in form to the impedance boundary condition given by (3.42) and its treatment in the FEM solution is therefore straightforward.

The purpose of the ABC is to create a boundary which does not perturb the field that is incident upon it—in effect, to simulate a surface which is actually not there. A measure on how well an ABC simulates a nonreflecting boundary can be obtained by examining the reflection field due to a plane wave impinging upon the ABC surface (see Fig. 4.14). For the Bayliss-Turkel ABC (4.80), it will be seen that its approximation of a nonreflecting surface may be poor unless it is placed far (1 to 2 wavelengths) from the scatterer. The latter is highly undesirable because the

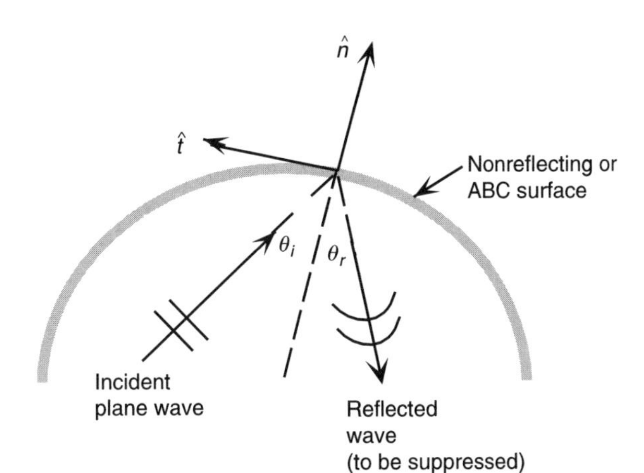

Figure 4.14. Illustration of a circular ABC surface and how to measure its effectiveness.

computational domain is enlarged substantially leading to unnecessarily large matrix systems. In the 1980s (see Senior and Volakis [13] for a review of ABCs) much work was carried out, aimed at deriving ABCs which provide a better simulation of non-reflecting surfaces even when placed at a fraction of a wavelength from the scatterer [12], [14], [15], [16]. These improved ABCs are associated with higher order tangential derivatives. For example, (4.79) is referred to as a first-order ABC because it involves a single derivative (with respect to the tangent) of the field. The general form of the second-order ABC is

$$\frac{\partial U}{\partial n} = \tilde{\alpha} U + \frac{\tilde{\beta}}{r_0^2} \frac{\partial^2 U}{\partial \phi^2} = \tilde{\alpha} U + \tilde{\beta} \frac{\partial^2 U}{\partial t^2} \tag{4.81}$$

where t and n are shown in Fig. 4.14. For the second-order Bayliss-Turkel ABC [12], the coefficients $\tilde{\alpha}$ and $\tilde{\beta}$ are given by

$$\begin{aligned} \tilde{\alpha} &= -jk_0 \left(1 + \frac{3}{2jk_0 r_0} - \frac{3}{8(k_0 r_0)^2}\right)\left(1 + \frac{1}{jk_0 r_0}\right)^{-1} \\ \tilde{\beta} &= -\frac{j}{2k_0}\left(1 + \frac{1}{jk_0 r_0}\right)^{-1} \end{aligned} \tag{4.82}$$

where again r_0 is interpreted as the radius of the boundary contour C^{outer}. Based on their derivation, C^{outer} must actually be circular for the application of the second-order Bayliss-Turkel ABC. However, it can be specialized to the case of planar boundaries by letting $r_0 \to \infty$. On constant x boundaries, as in Fig. 4.15, (4.81) becomes

$$\frac{\partial U}{\partial x} = \tilde{\alpha} U + \tilde{\beta} \frac{\partial^2 U}{\partial y^2} \tag{4.83}$$

with

$$\tilde{\alpha} = -jk_0, \qquad \tilde{\beta} = \frac{-j}{2k_0}$$

and a similar expression is obtained on constant y boundaries. This simplified second-order ABC (4.83) is the Engquist and Majda [14] ABC and is extensively used in connection with Finite Difference–Time Domain solutions.

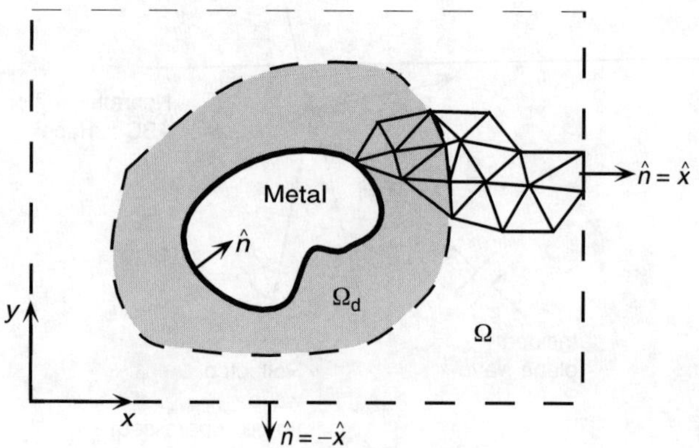

Figure 4.15 Piecewise planar boundary for mesh truncation.

Regardless of the type of ABC being used to simulate a nonreflecting surface at C^{outer}, mathematically the ABC replaces the normal field derivatives with tangential ones. In contrast to normal derivatives, tangential derivatives can be carried out using the approximate field expansion since there is no requirement for information outside the computational domain. Specifically, we have

$$\int_{C^{\text{outer}}} \frac{1}{\varepsilon_r} N_j^e [\hat{n} \cdot \nabla H_z^{\text{scat}}]\, dt = \int_{C^{\text{outer}}} \frac{1}{\varepsilon_r} N_j^e \left(\tilde{\alpha} H_z^{\text{scat}} + \tilde{\beta} \frac{\partial^2 H_z^{\text{scat}}}{\partial t^2} \right) dt$$

$$= \int_{C^{\text{outer}}} \frac{\tilde{\alpha}}{\varepsilon_r} N_j^e H_z^{\text{scat}}\, dt - \int_{C^{\text{outer}}} \frac{\tilde{\beta}}{\varepsilon_r} \frac{\partial N_j^e}{\partial t} \frac{\partial H_z^{\text{scat}}}{\partial t}\, dt \tag{4.84}$$

where we used integration by parts to transfer one of the derivatives from the field to the testing function as was done by obtaining the weak form of the wave equation. To proceed with the discretization of (4.84) we must introduce an expansion for the field on the boundary C^{outer}. Choosing the linear expansion basis (4.58) or (4.59), we have

$$H_z^{\text{scat}} = \sum_{s_1=1}^{N_{s_1}} \sum_{i=1}^{2} H_{zi}^{\text{scat}(s_1)} L_i^{s_1}(\mathbf{r})$$

and upon substitution into (4.84) we obtain

$$\int_{C^{\text{outer}}} \frac{1}{\varepsilon_r} N_j^e [\hat{n} \cdot \nabla H_z^{\text{scat}}]\, dl$$

$$= \sum_{s_1=1}^{N_{s_1}} \sum_{i=1}^{2} H_{zi}^{\text{scat}(s_1)} \frac{1}{\varepsilon_r^{s_1}} \left[\tilde{\alpha} \int_{C_{s_1}^{\text{outer}}} L_j^{s_1} L_i^{s_1}\, dt - \tilde{\beta} \int_{C_{s_1}^{\text{outer}}} \frac{\partial L_j^{s_1}}{\partial t} \frac{\partial L_i^{s_1}}{\partial t}\, dt \right] \tag{4.85}$$

As noted before, $L_i^{s_1}(\mathbf{r}) = N_j^e(\mathbf{r})$ when $\mathbf{r} \in C^{\text{outer}}$ and s_1 is an edge on C^{outer} belonging to the eth element, as depicted in Fig. 4.7.

With the evaluation of the boundary integral over C^{inner} as given by (4.78) and the discretization of the other over C^{outer} as given by (4.85), we are ready to cast (4.73) into a linear system of equations. The element equations are

$$\sum_{e=1}^{N_e} [A^e]\{H_z^{\text{scat}}\}^e + \sum_{s_1=1}^{N_{s_1}} [B^{s_1}]\{H_z^{\text{scat}}\}^{s_1} = \sum_{e=1}^{N_e} \{b^e\} \tag{4.86}$$

The entries of the $[A^e]$ matrix are again given by [see (4.46)]

$$A_{ij}^e = -\frac{1}{4\varepsilon_r \Delta^e} (b_i^e b_j^e + c_i^e c_j^e) + k_0^2 \mu_r \frac{\Delta^e}{12} (1 + \delta_{ij}) \tag{4.87}$$

and

$$B_{ij}^{s_1} = \frac{\tilde{\alpha}}{\varepsilon_r^{s_1}} \int_{C_{s_1}^{\text{outer}}} L_i^{s_1} L_j^{s_1}\, dt - \frac{\tilde{\beta}}{r_0 \varepsilon_r^{s_1}} \int_{C_{s_1}^{\text{outer}}} \frac{\partial L_i^{s_1}}{\partial \phi} \frac{\partial L_j^{s_1}}{\partial \phi}\, d\phi \tag{4.88}$$

in which the angular cylindrical variable ϕ has been used in the integral to emphasize that the employed ABC is only applicable to circular enclosures C^{outer} (note: $dt = r_0\, d\phi$).

The excitation column entries consist of two components—one from the excitation function $f(\mathbf{r})$ and another from the boundary condition on C^{inner}. Specifically [again, these entries are not related to the b_i^e in (4.87)],

$$b_i^e = \iint_{\Omega^e} N_i^e(\mathbf{r}) f^e \, dx\, dy - \frac{jk_0}{Z_0} \int_{C_{s_2}^{\text{inner}}} N_i^e(\mathbf{r}) \, (\mathbf{E}_{\tan}^i \cdot \hat{t}) \, dt \tag{4.89}$$

where s_2 is a segment on C^{inner} belonging to the eth element. Substituting for f^e and $\mathbf{E}_{\tan}^i$, a more explicit expression for b_i^e is

$$\begin{aligned} b_i^e = k_0^2 \left(\frac{1}{\varepsilon_r^e} - \mu_r^e \right) \iint_{\Omega^e} N_i^e(\mathbf{r}) \, e^{jk_0(x\cos\phi_0 + y\sin\phi_0)} \, dx\, dy \\ - jk_0 \int_{C_{s_2}^{\text{outer}}} N_i^e(\mathbf{r}) (\hat{x}\sin\phi_0 - \hat{y}\cos\phi_0) \cdot \hat{t} e^{jk_0(x\cos\phi_0 + y\sin\phi_0)} \, dt \end{aligned} \tag{4.90}$$

which can be evaluated in closed form for each of the line segments and triangles.

The assembly of the elemental equations (4.86) is carried out in the same manner as done for the TM waveguide mode analysis. The resulting global system will be of the form

$$\begin{bmatrix} [A^{II}] & [A^{IS}] \\ [A^{SI}] & [A^{SS}] \end{bmatrix} \begin{Bmatrix} H_z^{\text{scat}^{\text{interior}}} \\ H_z^{\text{scat}^{\text{boundary}}} \end{Bmatrix} + \begin{bmatrix} 0 & 0 \\ 0 & [B] \end{bmatrix} \begin{Bmatrix} H_z^{\text{scat}^{\text{interior}}} \\ H_z^{\text{scat}^{\text{boundary}}} \end{Bmatrix} = \{b\} \tag{4.91}$$

where we used the superscripts "interior" and "boundary" to indicate the separation of the node field column as done in (4.65).

This system is similar in all respects to (4.65) for the TM mode analysis. However, there is a major difference in that $\{\Psi\}$ has now been replaced with the field itself on the boundary. Thus, the convenient decomposition to a pair of smaller systems is no longer possible, nor is it needed. Since the ABC permitted the elimination of $\{\Psi\}$, no additional equations are required for the solution of $\{H_z^{\text{scat}}\}$. Addition of the two left hand matrices gives

$$\begin{bmatrix} [A^{II}] & [A^{IS}] \\ [A^{SI}] & [A^{SS}] + [B] \end{bmatrix} \{H_z^{\text{scat}}\} = \{b\} \tag{4.92}$$

where the entire matrix is sparse and the system can be solved using an iterative solver (see Chapter 9).

4.4.3 Scattered Field Computation

Once H_z^{scat} or E_z^{scat} is computed from a solution of (4.92), the scattered field outside the computational domain and echowidth of the scatterer are two observables of interest. The echowidth is obtained from the far zone scattered field using the relation

$$\sigma_{2D} = \lim_{r\to\infty} 2\pi r \frac{|H_z^{\text{scat}}|^2}{|H_z^i|^2} \tag{4.93}$$

The field outside the computational domain is obtained by application of the surface equivalence principle or Kirchhoff's integral equation. In either case, the pertinent expression is (see Chapter 1, Section 1.4)

$$U^{\text{scat}}(\mathbf{r}) = -\oint_{C_K} \{\hat{n}' \cdot \nabla' U(\mathbf{r}')\, G^{2D}(\mathbf{r}, \mathbf{r}') - [\hat{n}' \cdot \nabla' G^{2D}(\mathbf{r}, \mathbf{r}')]\, U(\mathbf{r}')\}\, dl' \tag{4.94}$$

where U is equal to E_z or H_z, depending on the excitation. The contour C_K can be of any shape or form and can be located at any distance from the scatterer. Also, $G^{2D}(\mathbf{r}, \mathbf{r}')$ is the two-dimensional Green's function

$$G^{2D}(\mathbf{r}, \mathbf{r}') = -\frac{j}{4} H_0^{(2)}(k_0|\mathbf{r} - \mathbf{r}'|) \tag{4.95}$$

where $H_0^{(2)}$ denotes the zeroth-order Hankel function of the second kind. This Green's function can be interpreted as the field generated by a line source at $\mathbf{r}'$ since it satisfies the differential equation

$$\nabla^2 G^{2D} + k_0^2 G^{2D} = -\delta(\mathbf{r} - \mathbf{r}') \tag{4.96}$$

The scattered field representation (4.94) is identical to that given in Chapter 1. Also, (4.94) can be related to the radiated fields due to equivalent current sources. For example, assuming $U = E_z$, from Maxwell's equations [see (1.76)]

$$\mathbf{J} = \hat{n} \times \mathbf{H} = -\frac{j\omega\varepsilon_0}{k_0^2}\, \hat{n} \times [\hat{z} \times \nabla E_z] \tag{4.97}$$

and since $\nabla E_z = \hat{n}(\partial E_z/\partial n) + \hat{t}(\partial E_z/\partial t)$, it follows that $(\hat{n} \times \hat{t} = \hat{z})$

$$\mathbf{J} = \hat{n} \times \mathbf{H} = -\hat{z}\, \frac{j}{k_0 Z_0} \frac{\partial E_z}{\partial n} = -\hat{z}\, \frac{j}{k_0 Z_0}\, \hat{n} \cdot \nabla U \tag{4.98}$$

Also, the equivalent magnetic current is given by

$$\mathbf{M} = \mathbf{E} \times \hat{n} = \hat{t} E_z = \hat{t} M_t \tag{4.99}$$

and therefore (4.94) can be rewritten as

$$E_z^{\text{scat}}(\mathbf{r}) = \oint_{C_K} \{-jk_0 Z_0 J_z(\mathbf{r}) G^{2D}(\mathbf{r}, \mathbf{r}') + M_t[\hat{n}' \cdot \nabla' G^{2D}(\mathbf{r}, \mathbf{r}')]\}\, dl' \tag{4.100}$$

Although C_K can be arbitrary, it is best to choose it so that the integrated fields are most accurate. For purely metallic scatterers, the computed field and its derivative is most accurate near the metallic surface. Thus, it is appropriate to choose C_K to be near to or coincide with the metallic surface of the scatterer. For TM incidence, when C_K is coincident with the metallic surface, $U(\mathbf{r}) = 0$ for $\mathbf{r} \in C_K$, and thus

$$E_z^{\text{scat}}(\mathbf{r}) = \frac{j}{4} \oint_{C_K} [\hat{n}' \cdot \nabla' E_z(\mathbf{r}')] H_0^{(2)}(k_0|\mathbf{r} - \mathbf{r}'|)\, dl' \tag{4.101}$$

Likewise, for TE incidence, $\hat{n} \cdot \nabla U(\mathbf{r}) = 0$ on C_K and (4.94) reduces to

$$H_z^{\text{scat}}(\mathbf{r}) = -\frac{j}{4} \oint_{C_K} H_z(\mathbf{r}')[\hat{n}' \cdot \nabla' H_0^{(2)}(k_0|\mathbf{r} - \mathbf{r}'|)]\, dl' \tag{4.102}$$

However, when dealing with coated metallic or purely dielectric scatterers, (4.101) and (4.102) cannot be used. In this case, the contour C_K is placed above the outer surface of the dielectric (see Fig. 4.16) and the scattered must be computed from (4.100) or its dual.

To evaluate the scattered field using (4.94) or (4.100), we proceed by first discretizing C_K into N_s segments each less than 1/10 of wavelength or as dictated by the original FEM tessellation of the domain. For simplicity, let us choose C_K to be a circle of radius $|\mathbf{r}'| = r_K$. Then

$$U^{\text{scat}}(\mathbf{r}) = I_1(\mathbf{r}) + I_2(\mathbf{r}) \tag{4.103}$$

with

$$\begin{aligned} I_1(\mathbf{r}) &= +\frac{j}{4} r_K \int_0^{2\pi} \Psi(\mathbf{r}') H_0^{(2)}\left(k_0\sqrt{r^2 + r_K^2 - 2rr_K\cos(\phi-\phi')}\right) d\phi' \\ \Psi(\mathbf{r}) &= \hat{n}\cdot\nabla U(\mathbf{r}) \end{aligned} \tag{4.104}$$

$$\begin{aligned} I_2(\mathbf{r}) = \frac{jk_0}{4} r_K \int_0^{2\pi} U(\mathbf{r}') \frac{H_1^{(2)}\left(k_0\sqrt{r^2 + r_K^2 - 2rr_K\cos(\phi-\phi')}\right)}{\sqrt{r^2 + r_K^2 - 2rr_K\cos(\phi-\phi')}} \\ \cdot [r_K - r\cos(\phi-\phi')]\, d\phi' \end{aligned} \tag{4.105}$$

and in (4.105) we used the simplifications[2]

$$\begin{aligned} \hat{r}'\cdot\nabla' H_0^{(2)}(k_0|\mathbf{r}-\mathbf{r}'|) &= \frac{\hat{r}'\cdot(\mathbf{r}'-\mathbf{r})}{|\mathbf{r}-\mathbf{r}'|} k_0 H_0^{(2)'}(k_0|\mathbf{r}-\mathbf{r}'|) \\ &= -k_0 \frac{H_1^{(2)}(k_0|\mathbf{r}-\mathbf{r}'|)}{|\mathbf{r}-\mathbf{r}'|}[r_K - r\cos(\phi-\phi')] \end{aligned} \tag{4.106}$$

in which $H_1^{(2)}$ denotes the first-order Hankel function of the second kind. Also, ϕ' refers to the angle between $\mathbf{r}'$ and the x-axis, as shown in Fig. 4.16.

For far zone computations (i.e., $r \to \infty$), we can simplify the above integrands by introducing the large argument approximations of the Hankel functions. Specifically,

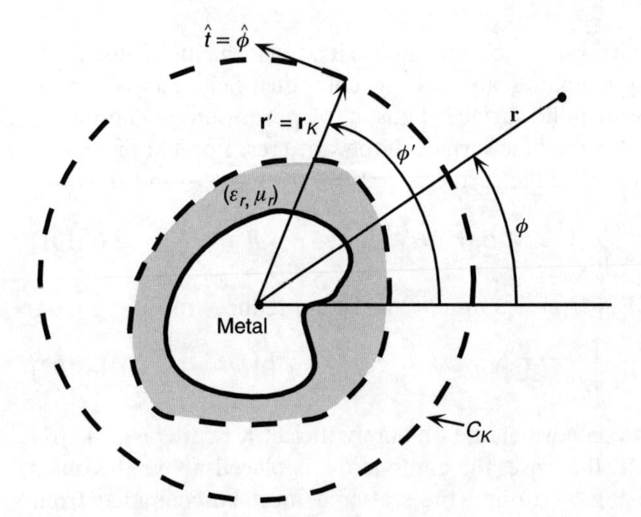

Figure 4.16 Illustration of the integration contour C_K for computing the scattered field.

[2]Note: $\partial G^{2D}/\partial n = \hat{n}\cdot\nabla G^{2D} = \frac{-jk_0}{4}(\hat{n}\cdot\nabla R)\, H_0^{(2)'}(k_0 R) = \frac{jk_0}{4}(\hat{n}\cdot\hat{R})\, H_1^{(2)}(k_0 R)$, where $\nabla R = \hat{R} = (\mathbf{r}-\mathbf{r}')/|\mathbf{r}-\mathbf{r}'|$.

$$H_n^{(2)}(k_0 r)|_{r\to\infty} \sim \sqrt{\frac{2j}{\pi k_0 r}}\, j^n e^{-jk_0 r} \tag{4.107}$$

and we approximate the radical as

$$|\mathbf{r}-\mathbf{r}'| = \sqrt{r^2 + r_K^2 - 2 r r_K \cos(\phi-\phi')} \approx \begin{cases} r & \text{for amplitude terms} \\ r - r_K\cos(\phi-\phi') & \text{for phase terms} \end{cases} \tag{4.108}$$

Substituting (4.107) and (4.108) into (4.104) and (4.105) and discretizing the integral, yields the far zone approximations

$$I_1(\mathbf{r}) \approx j\frac{r_K\Delta\phi}{4}\sqrt{\frac{2j}{\pi k_0}}\frac{e^{-jk_0 r}}{\sqrt{r}}\sum_{n=1}^{N_s}\Psi_n e^{jk_0 r_K\cos(\phi-\phi_n)} \tag{4.109}$$

$$I_2(\mathbf{r}) \approx \frac{r_K k_0\Delta\phi}{4}\sqrt{\frac{2j}{\pi k_0}}\frac{e^{-jk_0 r}}{\sqrt{r}}\sum_{n=1}^{N_s}U_n\cos(\phi-\phi_n)e^{jk_0 r_K\cos(\phi-\phi_n)} \tag{4.110}$$

In these, $\Delta\phi = 2\pi/N_s$ denotes the angular extent of each discrete arc segment and ϕ_n is the value of ϕ at the midpoint of the nth element. For this case, U_n and Ψ_n refer to the average value of the field and its normal derivative at the midpoint of the nth segment.

From (4.93), the echowidth can be computed from

$$\sigma_{2D} = 2\pi\left(\frac{r_K\Delta\phi}{4}\right)^2\left(\frac{2}{\pi k_0}\right) \cdot\left[\left|j\sum_{n=1}^{N_s}\Psi_n e^{jk_0 r_K\cos(\phi-\phi_n)} + k_0\sum_{n=1}^{N_s}U_n\cos(\phi-\phi_n)e^{jk_0 r_K\cos(\phi-\phi_n)}\right|^2\right] \tag{4.111}$$

Another expression for the far zone field using equivalent surface electric and magnetic currents is

$$E_z^{\text{scat}}(\mathbf{r}) = \sqrt{\frac{k_0}{8\pi}}\frac{e^{-j(k_0 r-\pi/4)}}{\sqrt{r}}\oint_{C_K}\hat{z}\cdot[\hat{r}\times\mathbf{M}(\mathbf{r}') + Z_0\hat{r}\times\hat{r}\times\mathbf{J}(\mathbf{r}')]\,e^{jk_0\mathbf{r}'\cdot\hat{r}}\,dl' \tag{4.112}$$

for TM incidence and

$$H_z^{\text{scat}}(\mathbf{r}) = \sqrt{\frac{k_0}{8\pi}}\frac{e^{-j(k_0 r-\pi/4)}}{\sqrt{r}}\oint_{C_K}\hat{z}\cdot\left[-\hat{r}\times\mathbf{J}(\mathbf{r}') + \frac{1}{Z_0}\hat{r}\times\hat{r}\times\mathbf{M}(\mathbf{r}')\right]e^{jk_0\mathbf{r}'\cdot\hat{r}}\,dl' \tag{4.113}$$

for TE incidence. These are obtained from (4.100) once (4.107) and (4.108) are used. For this case $\mathbf{M} = \hat{t}E_z$ and $\mathbf{J} = -\hat{z}j\Psi/(k_0 Z_0)$.

4.4.4 Scattering Example Using ABCs

A classic problem is that of scattering from a circular cylinder. As an example application of the finite element method let us consider the scattering by the dielectrically coated circular cylinder shown in Fig. 4.17. The metallic cylinder is coated with a 0.05λ dielectric layer having ($\varepsilon_r = 4$, $\mu_r = 1$). As a first step, our goal is to determine a suitable ABC radius r_0 leading to a sufficiently accurate solution. This

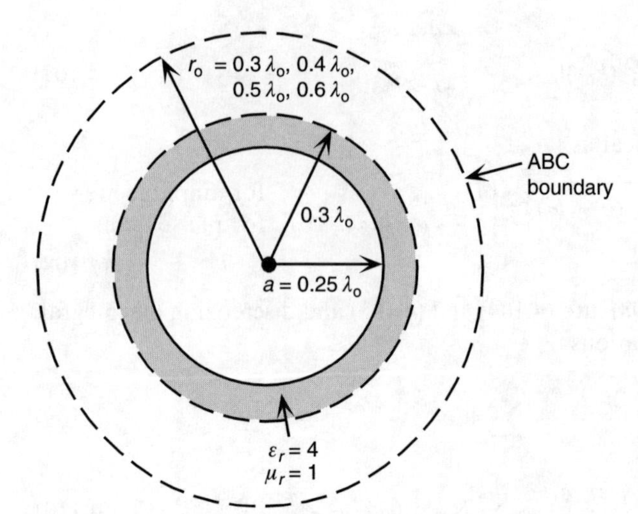

Figure 4.17 Geometry of the coated circular cylinder and illustration of the ABC boundary.

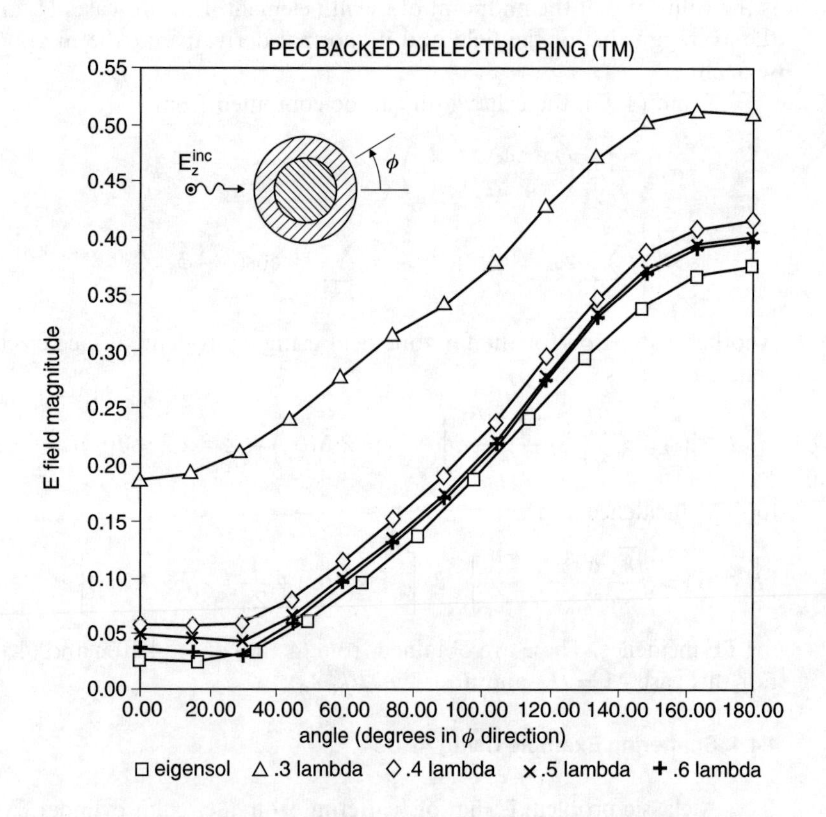

Figure 4.18 Finite element solution of the near-zone TM incidence scattered field measured at $r = 0.275\lambda_0$. The geometry is shown in Fig. 4.17 and the four curves refer to the radii at which the ABC was placed ($r_0 = 0.3\lambda_0$, $0.4\lambda_0$, $0.5\lambda_0$, and $0.6\lambda_0$). [*After Peterson and Castillo [17],* © *IEEE, 1989.*]

specific example was considered by Peterson and Castillo [17] and was used to assess the accuracy of the second order ABC (4.81) by comparison with the exact eigenfunction solution [18]. In Fig. 4.18 we show the near zone field E_z^{scat} due to a plane TM incidence for different values of the ABC radius r_0. As expected, the computed E_z^{scat} field is quite inaccurate when $r_0 = 0.3\lambda_0$ since the ABC boundary is then coincident with the outer boundary of the scattering geometry. By its derivation, the second-order ABC is valid for large r_0 since it neglects [13] terms beyond $O(r^{-9/2})$ as well as nonradial waves. Clearly, the choice of $r_0 = 0.3\lambda_0$ violates this assumption. However, the accuracy of the solution improves substantially when r_0 is increased to $0.4\lambda_0$, and continues to improve as r_0 is increased. Typically, it may be necessary to increase r_0 as much as $2\lambda_0$ to obtain very accurate results. This is especially true for the TE incidence where nonspecular fields caused by traveling and surface waves are of importance.

Another example application of the finite element method with ABCs for a noncircular cylinder is shown in Fig. 4.19. The geometry is a metallic triangular

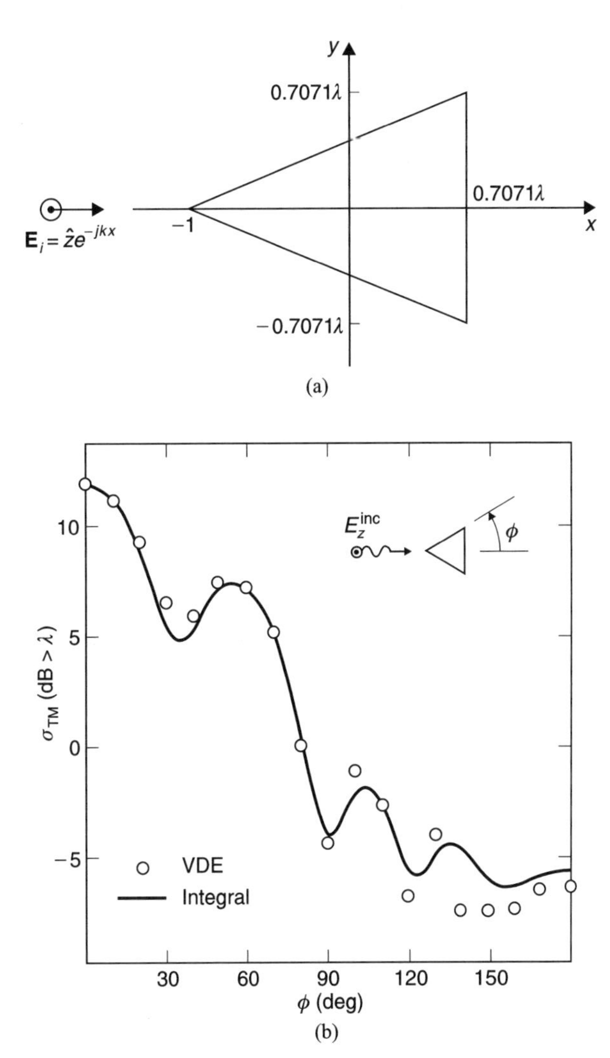

Figure 4.19 Geometry and TM incidence bistatic echowidth for a triangular metallic cylinder. The ABC was placed on a circle of radius $2\lambda_0$ and centered at the origin. The solid line gives the reference data based on the rigorous integral equation solution and the circles correspond to data from the finite element–ABC method (labeled as VDE). [*After Peterson and Castillo [17], © IEEE, 1989.*]

cylinder whose maximum length is $1.7071\lambda_0$. The second-order Bayliss-Turkel ABC was used again for mesh truncation with the ABC boundary being a circle of radius $r_0 = 2\lambda_0$ centered at the origin. The resulting mesh consisted of 1599 nodes and 2952 triangular elements [17]. For a TM plane wave impinging from the negative x-axis the echowidth is shown in Fig. 4.19 as a function of the angle ϕ measured from the x-axis. For this bistatic pattern, $\phi = 0^\circ$ corresponds to the direction of forward scattering and $\phi = 180^\circ$ refers to the backscatter direction (because of symmetry, only the pattern from $\phi = 0^\circ$ to $\phi = 180^\circ$ is shown). Two peaks (lobes) are characteristic to this pattern, one occurring at $\phi = 0^\circ$ and the other at $\phi = 60^\circ$. The latter is due to the specular scattering from the upper (nonvertical) side of the triangular cylinder and the $\phi = 0^\circ$ peak is due to the coherent addition of the diffracted fields from the two back edges of the triangle. The echowidth is the lowest (20 dB below the peak value) along the backscatter direction where the contributions are due to diffraction from the three visible edges of the triangle. Because of the lower level of these diffraction contributions and their nonspherical wave character, the accuracy of the simulation is least accurate there. Higher accuracy may be achieved by placing the ABC surface further from the scatterer. This would, of course, result in higher computational demands. Eventually, numerical errors caused by the larger mesh would limit the accuracy of the solution unless the machine precision is increased.

4.4.5 Artificial Absorbers for Mesh Truncation

In the above section we introduced ABCs for finite element mesh truncation (this approach is referred to as the finite element–ABC method). As discussed, the purpose of the ABC is to eliminate wave reflections from entering back into the computational domain. That is, the ABC serves as an absorber of the scattered waves (see Fig. 4.14) and therefore a true absorber can instead be used to absorb these waves. Indeed, material absorbers are used in anechoic chambers to eliminate wall reflections and simulate an otherwise open space, free of nearby obstacles which can interfere with antenna or scattering measurements. In practice, material absorbers are shaped as cones and are loaded with carbon or other particles to simulate a lossy environment (see, for example, Fig. 4.20). The tapered shape provides a better impedance matching, whereas the loss in the material causes absorption of the entering waves.

In a finite element analysis, we can also use material absorbers for mesh truncation and this approach will be referred to as the finite element–artificial absorber (FE–AA) method. For numerical simulations, it is not necessary to make use of material parameters or profiles which are physical. Instead, we can use any fictitious (i.e., artificial) material profile and employ it for mesh truncation, as illustrated in Fig. 4.21. The shown absorber can be curved, if necessary, to minimize the computational volume, depending on the scatterer's or radiator's shape.

Not being restricted by the material choices, an optimizer such as the simplex method can be used to determine the material parameters μ_r, ε_r, and thickness t to minimize reflections over all visible incidence angles. This approach was used by Özdemir and Volakis [19] to obtain the parameters given in Fig. 4.21. A homogeneous absorber cross section was assumed in this design for simplicity. However, improved artificial absorbing layers have been recently developed on the assumption of inhomogeneity and even anisotropy in the material. In the latter case,

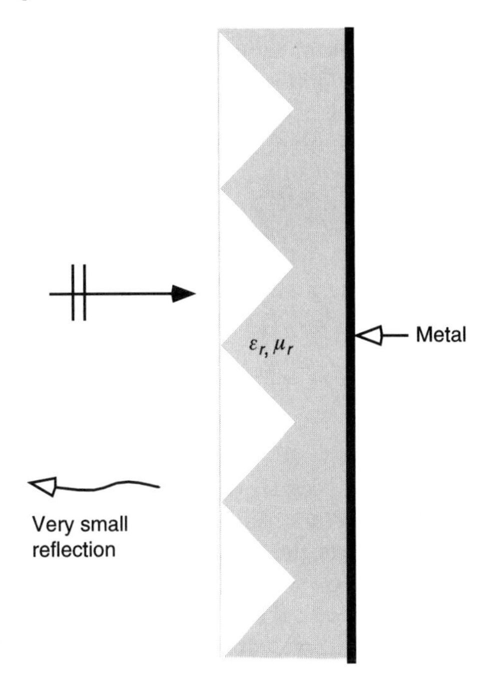

Figure 4.20 Practical configuration of a material absorber.

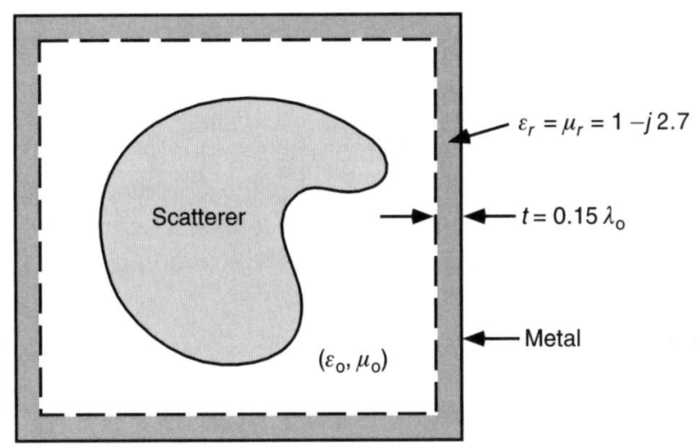

Figure 4.21 Use of a homogeneous absorber for mesh truncation. [*The shown parameters are based on a design recommended by Özdemir and Volakis* [*19*].]

it is possible to theoretically design layers which exhibit no reflection at the dielectric interface over all incidence angles. These are referred to as perfectly matched layers (PMLs) and their material parameters are given by Sacks et al. [20]

$$\bar{\bar{\varepsilon}} = \bar{\bar{\mu}} = \begin{bmatrix} c & 0 & 0 \\ 0 & c & 0 \\ 0 & 0 & 1/c \end{bmatrix}$$

in which $c = \alpha - j\beta$ is a constant. A reasonable choice for c is $1 - j1$ but other choices can be employed depending on the thickness of the layer and the discretization in the numerical implementation. A study of the PML parameters for optimal absorption is given by Legault et al. [21].

An example of using the artificial absorber for mesh truncation is shown in Fig. 4.22. This configuration is a rectangular groove situated in an otherwise flat metallic plane (ground plane). We are interested in computing the scattered field due to a TE plane wave excitation. For our formulation, the excitation is simulated by setting

$$\mathbf{E}_{\tan}^{\text{scat}} = 0 \qquad \text{on the absorber metal backing}$$

$$\mathbf{E}_{\tan}^{\text{scat}} = -\mathbf{E}_{\tan}^{i} \qquad \text{on cavity and ground plane metallic surfaces}$$

and the resulting matrix system is identical to that in (4.91) except that [B] is set to zero and

$$\{b_i^e\} = -\frac{jk_0}{Z_0}\int_{\substack{\text{cavity,}\\ \text{ground plane}}} N_i^e(\mathbf{r})\,(\mathbf{E}_{\tan}^i \cdot \hat{t})\,dt + \iint_{\Omega_e} N_i^e(\mathbf{r}) f(\mathbf{r})\,dx\,dy$$

in which $f(\mathbf{r})$ is the source function given in (4.11) with $M_{iz} = 0$. Upon completion of the solution, the scattered field is obtained by integrating equivalent currents over the cavity's aperture using (4.100) or (4.112).

To compute the scattered field we invoke surfacc equivalence theory. A convenient surface on which to place the equivalent (electric and magnetic) currents is that coincident with the ground plane ($y = 0$). Since we are interested in the fields in the $y > 0$ region, we can arbitrarily set those below the aperture to zero for the application of the surface equivalence principle. The surface magnetic currents are then given by

$$\mathbf{M} = \begin{cases} \mathbf{E} \times \hat{n} = \mathbf{E} \times \hat{y} & 0 < x < w, y = 0 \\ 0 & \text{otherwise} \end{cases} \tag{4.114}$$

(see Fig. 4.23) and $\mathbf{J} = \hat{n} \times \mathbf{H} = \hat{y} \times \mathbf{H}$, everywhere on the $y = 0$ plane. However, because of the ground plane's presence, image theory can be invoked. This leads to the introduction of new equivalent currents which provide us with the same radiated field as the original ones which radiated in the presence of the ground plane. The new equivalent currents are

$$\mathbf{M}_{\text{equ}} = \begin{cases} 2\mathbf{E} \times \hat{n} = 2\mathbf{E} \times \hat{y} & 0 < x < w, y = 0 \\ 0 & \text{otherwise} \end{cases} \tag{4.115}$$

and $\mathbf{J} = 0$, and these radiate in free space. The fact that the electric current vanishes is a substantial simplification since the integration of the radiation integral is limited over the aperture of the groove. Specifically, from (4.113)

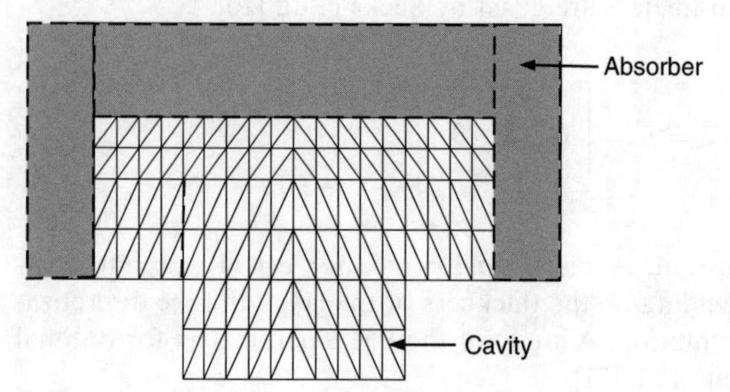

Figure 4.22 Illustration of a groove recessed in a perfectly conducting ground plane. The artificial absorber in Fig. 4.21 is used for mesh truncations.

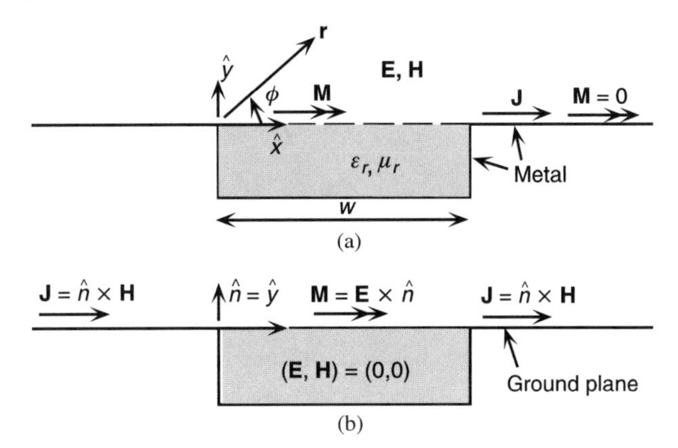

Figure 4.23 Illustration of the surface equivalence principle for computing the scattered fields from a groove: (a) original geometry; (b) setup for surface equivalence; (c) equivalent currents after application of image theory.

$$H_z^{\text{scat}}(\mathbf{r}) = \sqrt{\frac{k_0}{8\pi}} \frac{e^{-j(k_0 r - \pi/4)}}{\sqrt{r}} \int_0^w \hat{z} \cdot \left[\frac{1}{Z_0} \hat{r} \times \hat{r} \times \mathbf{M}_{\text{equ}}(\mathbf{r}') \right] e^{jk_0 x' \hat{x} \cdot \hat{r}} dx' \tag{4.116}$$

and from the finite element solution, the magnetic current is given by

$$\begin{aligned} \mathbf{M}_{\text{equ}} &= 2(\mathbf{E} \times \hat{y}) = 2\hat{z}(E_x^i + E_x^{\text{scat}}) \\ &= \hat{z}2\left[Z_0 \sin\phi_0 e^{jk_0(x\cos\phi_0 + \sin\phi_0)} + \frac{Z_0}{2jk_0\Delta^e} \sum_{i=1}^{3} c_i^e H_{z_i}^e \right] \end{aligned} \tag{4.117}$$

and we should emphasize that H_{zi}^e refers to the scattered field at the ith node of the eth element. The latter sum is over the three nodes of the element bordering the aperture at the computation point for $\mathbf{M}_{\text{equ}}$. From (4.93), the corresponding echowidth is

$$\sigma_{2D} = \frac{k_0}{4(Z_0)^2} \left| \int_0^w (\mathbf{M}_{\text{equ}} \cdot \hat{z})\, e^{jk_0 x' \cos\phi}\, dx' \right|^2 \tag{4.118}$$

Bistatic echowidth calculations for the rectangular groove depicted in Fig. 4.22 are given in Fig. 4.24. The curve corresponds to a groove width of $W = 2.5\lambda_0$ and a depth of $d = 0.2\lambda_0$. The incident plane wave was incoming at an angle of 70° from the face of the ground plane and the absorber was placed $0.15\lambda_0$ from the top of the groove. As seen, the echowidth computations using the FE–AA method is in good agreement with those based on the rigorous FE–BI method discussed next (see [22], [23]). However, care must be given when using artificial absorbers for mesh truncation. The accuracy of the results is not assured, and this is more so for the near zone fields. Also, the convergence rate of the iterative solver may deteriorate for certain absorber parameters.

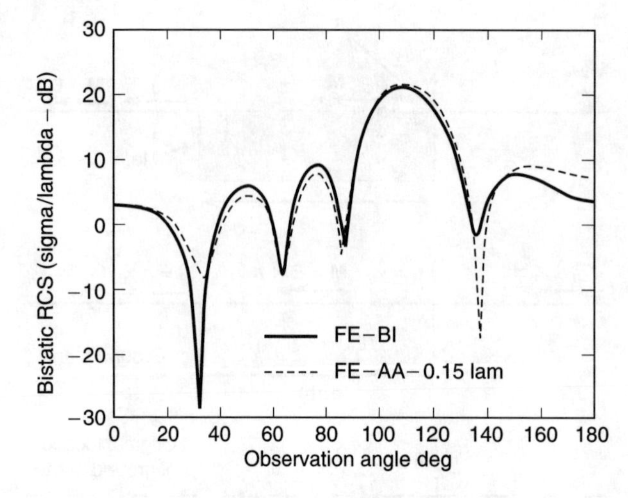

Figure 4.24 Bistatic echowidth (TE plane wave incidence at an angle of $\phi_0 = 70°$ from the face of the ground plane) for a $2.5\lambda_0$ wide and $0.2\lambda_0$ deep groove recessed in a ground plane, as illustrated in Fig. 4.22. Comparison of the finite element–artificial absorber and finite element–boundary integral [22] methods. The absorber was placed $0.15\lambda_0$ from the groove's aperture. [*Courtesy of S. Bindiganavale.*]

4.4.6 Boundary Integral Mesh Truncation

Absorbing boundary conditions provide an approximation of the relationship between the normal and tangential field derivatives on the mesh truncation boundary C^{outer} (see Fig. 4.13). Instead, we could use the exact integral equation (4.94)

$$H_z(\mathbf{r})|_{\mathbf{r}\in C^{\text{outer}}} = H_z^i - \oint_{C^{\text{outer}}} \left\{ \frac{\partial H_z(\mathbf{r}')}{\partial n'} G^{2D}(\mathbf{r},\mathbf{r}') - [\hat{n}' \cdot \nabla' G^{2D}(\mathbf{r},\mathbf{r}')] H_z(\mathbf{r}') \right\} dl' \quad (4.119)$$

where $H_z = H_z^i + H_z^{\text{scat}}$. Here, the quantities $\partial H_z/\partial n$ and H_z are not related through an explicit expression as is the case with the ABC (4.80) or (4.81). Because of the integral and the nonvanishing Green's function (4.95), all values of H_z and $\partial H_z/\partial n$ on the boundary C^{outer} are now interrelated. The relationship between H_z and $\partial H_z/\partial n$ becomes clearer when we carry out the discretization of (4.119). To do so, we may employ the linear expansions (4.58)–(4.60) for $\Psi = \partial H_z/\partial n$ and a similar one for the total field H_z. Galerkin's method, may then be used to generate a linear system which relates to the nodal values of Ψ and H_z on C^{outer}.

Because of the nonpolynomial form of $G^{2D}(\mathbf{r},\mathbf{r}')$ in (4.119), application of Galerkin's method with linear basis to discretize (4.119) leads to rather complex integrals for the matrix entries, albeit more rigorous [24]. To illustrate the procedure, we will instead employ the piecewise constant expansion

$$\Psi = \sum_{s_1=1}^{N_{s_1}} \left(\frac{\Psi_1^{s_1} + \Psi_2^{s_1}}{2} \right) P_{\Delta\phi}(\phi - \phi_{s_1}) \quad (4.120)$$

$$H_z = \sum_{s_1=1}^{N_{s_1}} \left(\frac{H_{z_1}^{s_1} + H_{z_2}^{s_1}}{2} \right) P_{\Delta\phi}(\phi - \phi_{s_1}) \quad (4.121)$$

in which

$$P_{\Delta\phi}(\phi - \phi_{s_1}) = \begin{cases} 1 & |\phi - \phi_{s_1}| < \dfrac{\Delta\phi}{2} \\ 0 & \text{otherwise} \end{cases} \quad (4.122)$$

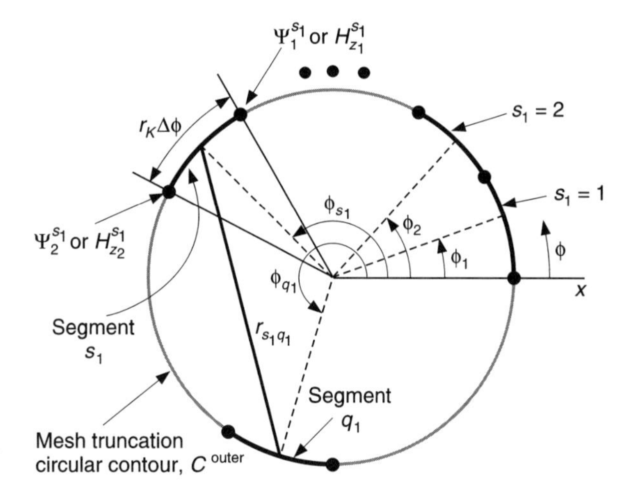

Figure 4.25 Discretization of the mesh truncation circular contour.

is the pulse function and we have assumed a circular contour C^{outer}, as shown in Fig. 4.25.

Basically, the above expansions (4.120) and (4.121) approximate the values of H_z and $\partial H_z/\partial n$ over each boundary segment by a constant value equal to the average of their values at the bordering nodes. They are less accurate than the linear expansions but provide substantial simplification in discretizing Kirchhoff's boundary integral (4.119). Substituting (4.120) and (4.121) into (4.119) yields

$$H_z(\mathbf{r}) + \sum_{s_1=1}^{N_{s_1}} \left\{ \left(\frac{\Psi_1^{s_1} + \Psi_2^{s_1}}{2} \right) \int_{C_{s_1}^{\text{outer}}} G^{2D}(\mathbf{r}, \mathbf{r})\, dl' \right.$$
$$\left. - \left(\frac{H_{z_1}^{s_1} + H_{z_2}^{s_1}}{2} \right) \int_{C_{s_1}^{\text{outer}}} \hat{n}' \cdot \nabla' G^{2D}(\mathbf{r}, \mathbf{r})\, dl' \right\} - H_z^i(\mathbf{r}) = R(\mathbf{r}), \quad \mathbf{r} \in C^{\text{outer}} \tag{4.123}$$

in which $R(\mathbf{r})$ is again the residual. To generate a linear system of equations, we proceed by applying the weighted residual method, viz.,

$$\int_{C^{\text{outer}}} R(\mathbf{r}) W(\mathbf{r})\, dl = 0$$

Choosing $W(\mathbf{r}) = \delta(\phi - \phi_{q_1})$, $q_1 = 1, 2, \ldots, N_{s_1}$, implying point matching, yields the system

$$\frac{H_{z_1}^{q_1} + H_{z_2}^{q_1}}{2} + \sum_{s_1=1}^{N_{s_1}} \left[\left(\frac{\Psi_1^{s_1} + \Psi_2^{s_1}}{2} \right) G_{s_1 q_1} - \left(\frac{H_{z_1}^{s_1} + H_{z_2}^{s_1}}{2} \right) G_{\nabla_{s_1 q_1}} \right]$$
$$= H_z^i(\mathbf{r}_{q_1}), \quad q_1 = 1, 2, \ldots, N_{s_1} \tag{4.124}$$

in which $\mathbf{r}_{q_1}$ denotes the location of the midpoint in the q_1th testing segment. The entries $G_{s_1 q_1}$ and $G_{\nabla_{s_1 q_1}}$ are the integrals

$$G_{s_1q_1} = \int_{\substack{s_1\text{th}\\\text{segment}}} G^{2D}(\mathbf{r}_{q_1}, \mathbf{r}')\, dl'$$

$$= \frac{-jr_K}{4}\int_{\phi_{s_1}-\Delta\phi/2}^{\phi_{s_1}+\Delta\phi/2} H_0^{(2)}\left(\left|2k_0 r_K \sin\left(\frac{\phi_{q_1}-\phi'}{2}\right)\right|\right) d\phi' \tag{4.125}$$

$$G_{\nabla_{s_1q_1}} = \int_{\substack{s_1\text{th}\\\text{segment}}} \hat{n}'\cdot\nabla' G^{2D}(\mathbf{r}_{q_1}, \mathbf{r}')\, dl'$$

$$= +\int_{\substack{s_1\text{th}\\\text{segment}}} \frac{dG^{2D}}{dn'}\, dl' = -\frac{d}{dn}\int_{\substack{s_1\text{th}\\\text{segment}}} G^{2D}(\mathbf{r}, \mathbf{r}')|_{\mathbf{r}=\mathbf{r}_{q_1}}\, dl' \tag{4.126}$$

in which r_K denotes the radius of C^{outer}. For $q_1 \neq s_1$, the evaluation of these integrals can be carried out via the simple midpoint method since $r_K \Delta\phi$ is typically very small. However, when $q_1 = s_1$, the argument of the Hankel function vanishes. Noting that

$$H_0^{(2)}(z) \approx A_1 - j\frac{2}{\pi}\ln z + A_2 z^2 + j\frac{z^2}{2\pi}\ln z + O(z^4, z^4 \ln z) \tag{4.127}$$

with $A_1 = 1 - j\frac{2}{\pi}(\ln\gamma - \ln 2)$, $A_2 = -\frac{1}{2} + j\frac{1}{2\pi}(\ln\gamma - 1 - \ln 2)$ and $\gamma = 1.781072418$, it is clear that the integrands in (4.125) and (4.126) become singular when $q_1 = s_1$. However, the evaluation of $G_{s_1s_1}$ can be readily carried out by substituting (4.127) into (4.125) to get

$$G_{s_1s_1} = -\frac{jr_K}{4}\int_{-\Delta\phi/2}^{\Delta\phi/2} H_0^{(2)}\left(\left|2k_0 r_K \sin\frac{u}{2}\right|\right) du$$

$$\approx -\frac{jr_K}{4}\,\Delta\phi\left\{1 - j\frac{2}{\pi}\left[\ln\left(\frac{\gamma k_0 r_K \Delta\phi}{4}\right) - 1\right]\right\} \tag{4.128}$$

In the case of $G_{\nabla_{s_1s_1}}$, the integral is improper and, to evaluate it, it is necessary to work with the rightmost expression in (4.126). That is, we first integrate the small argument expansion (4.127) and then perform the differentiation before setting $r = r_K$. Doing so yields

$$G_{\nabla_{s_1s_1}} = -\frac{d}{dr}\int_{\phi_{s_1}-\Delta\phi/2}^{\phi_{s_1}+\Delta\phi/2}\left[-\frac{j}{4}H_0^{(2)}\left(\left|2k_0 r \sin\left(\frac{\phi_{s_1}-\phi'}{2}\right)\right|\right) r_K\, d\phi'\right]_{r=r_K}$$

$$= +\tfrac{1}{2} + \frac{jk_0 r_K}{4}\left[k_0 r_K\left(\frac{\Delta\phi}{2} - \sin\frac{\Delta\phi}{2}\right) + j\frac{\Delta\phi}{\pi k_0 r_K}\right] \tag{4.129}$$

In the above, the 1/2 term can be also viewed as a result of the identity

$$\oint_C \Psi(\mathbf{r}')\frac{\partial G^{2D}(\mathbf{r}^{\pm}, \mathbf{r}')}{\partial n'}\, dl' = \pm\tfrac{1}{2}\Psi(\mathbf{r}) + ⨏_C \Psi(\mathbf{r}')\frac{\partial G^{2D}(\mathbf{r}^{\pm}, \mathbf{r}')}{\partial n'}\, dl'$$

where the bar through the integral implies principal value and $r^{\pm}$ denotes that the evaluation point may be just inside (+) or just outside (−) the contour C. However, as pointed out by M. Sancer (see Chapter 7), it is not appropriate to introduce the principal value concept for evaluating these integrals since the 1/2 term can be extracted in the limit as the testing point approaches the integration surface or as done above where the differentiation is carried out after the integration.

We can now approximate $G_{s_1q_1}$ and $G_{\nabla_{s_1q_1}}$ as

$$G_{s_1q_1} = \begin{cases} -\frac{jr_K}{4}\,\Delta\phi\left\{1 - j\,\frac{2}{\pi}\left[\ln\left(\frac{\gamma k_0 r_K \Delta\phi}{4}\right) - 1\right]\right\}, & s_1 = q_1 \\ -\frac{jr_K}{4}\,\Delta\phi H_0^{(2)}\left(2k_0 r_K\left|\sin\left(\frac{\phi_{s_1} - \phi_{q_1}}{2}\right)\right|\right), & s_1 \neq q_1 \end{cases} \tag{4.130}$$

$$G_{\nabla_{s_1q_1}} = \begin{cases} +\frac{1}{2} + \frac{jk_0 r_K}{4}\left[k_0 r_K\left(\frac{\Delta\phi}{2} - \sin\frac{\Delta\phi}{2}\right) + j\,\frac{\Delta\phi}{\pi k_0 r_K}\right], & s_1 = q_1 \\ \frac{jk_0 r_K \Delta\phi}{4}\,H_1^{(2)}\left(2k_0 r_K\left|\sin\left(\frac{\phi_{s_1} - \phi_{q_1}}{2}\right)\right|\right)\left|\sin\left(\frac{\phi_{s_1} - \phi_{q_1}}{2}\right)\right|, & s_1 \neq q_1 \end{cases} \tag{4.131}$$

To solve for the nodal values of H_z and Ψ, it is necessary to cast (4.124) into a matrix system of the form

$$[\tilde{G}_\nabla]\{H_z^{\text{boundary}}\} + [\tilde{G}]\{\Psi\} = \{V\} \tag{4.132}$$

Unlike the earlier systems, here the matrices $[\tilde{G}_\nabla]$ and $[\tilde{G}]$ are fully populated. Their entries $\tilde{G}_{\nabla_{s_1q_1}}$ and $\tilde{G}_{s_1q_1}$ are easily extracted from (4.124) in terms of $G_{\nabla_{s_1q_1}}$ and $G_{s_1q_1}$, respectively, and $V_{q_1} = H_z^i(r_{q_1})$, $q_1 = 1, 2, \ldots, N_{s_1}$. When this boundary element system (4.132) is combined with finite element equations

$$\begin{bmatrix} [A^{II}] & [A^{IS}] \\ [A^{SI}] & [A^{SS}] \end{bmatrix}\begin{Bmatrix} H_z^{\text{interior}} \\ H_z^{\text{boundary}} \end{Bmatrix} + \begin{bmatrix} 0 & 0 \\ 0 & [B] \end{bmatrix}\begin{Bmatrix} 0 \\ \{\Psi\} \end{Bmatrix} = \{0\} \tag{4.133}$$

we obtain a total of $N + N_s$ equations for the N node field values and N_s values of Ψ on C^{outer}. The entries of $[A]$ and $[B]$ are identical to those given in reference to (4.65). Although more cumbersome, this approach is rigorous and would be exact apart from the numerical approximation required in obtaining the linear systems (4.132) and (4.133). It is commonly referred to as the finite element–boundary integral (FE–BI) method, and scattering results based on it were used for reference in Fig. 4.24.

It should be noted that the boundary integral subsystem (4.132) is associated with possible fictitious resonances [25, 26] and the solution fails when the resonances are excited. To suppress them, one can simply introduce a small imaginary part in k_0 [27] or employ the combined field formulation [28].

4.5 EDGE ELEMENTS

In previous sections, FEM solutions were carried out using *node-based* scalar basis functions based on expanding the unknown quantity in terms of its nodal values, i.e., its values at element nodes. Such an expansion is suitable for modeling a scalar quantity. This is indeed the case for static potentials or a single field component as is the case for homogeneous waveguides and 2D scattering. For inhomogeneous waveguides, it is necessary to work with the vector wave equations (4.23)–(4.26) requiring an expansion of the transverse vector field component $\mathbf{E}_t$ or $\mathbf{H}_t$. However, it has been found that node-based expansions are not ideal for representing the vector nature of an electromagnetic field. Node-based expansions require specification of field values at element nodes where the field may not be defined (corners). Also, the implementation of boundary conditions occurring in electromagnetics (tangential field continuity) is a challenging task. Further, as noted in Chapter 2, false solutions (also referred to as 'spurious modes') to eigenvalue problems can be generated. An example of an erroneous mode field solution is shown in Fig. 4.26 [29] for a metallic cylinder enclosing two dissimilar dielectric media. The cylinder has a radius of $R = 25\,\text{cm}$ and one of the dielectric regions is an offset cylinder of radius $a = 10\,\text{cm}$. Mode field solutions are given when the wavenumber of the inner offset cylinder is $k^2_{\text{inner}} = 2000$ and that of the remaining region is $k^2_{\text{outer}} = (196.1, 39.22)$. As seen, the mode field solution using the node-based elements is substantially in error and corrective approaches have been extensively studied [29], [30], [31]. Initially, an

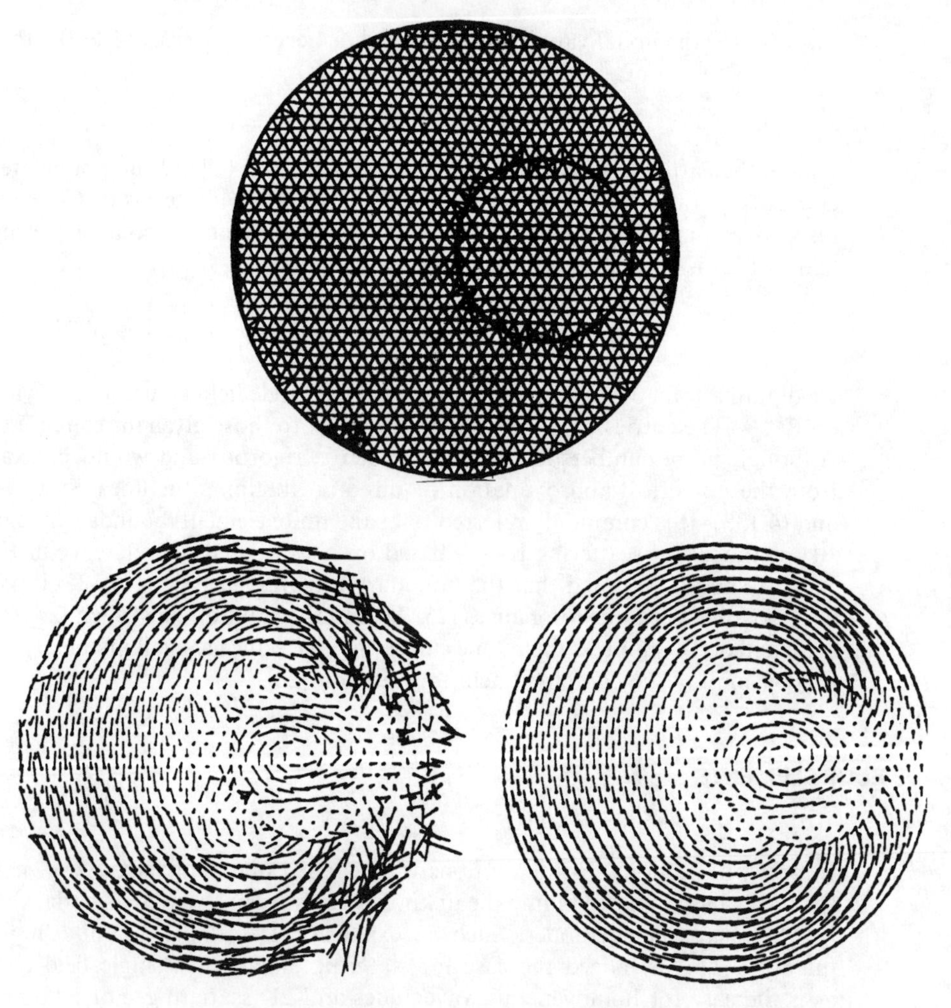

Figure 4.26 Failure of the node-based FEM implementation to predict the correct mode fields inside a coaxial guide with an offset center conductor. Top: mesh interior to the conducting cylinder—offset inner cylinder material wavenumber is $k^2_{\text{inner}} = 2000$, and that of the remaining region is $k^2_{\text{outer}} = (196.1,\ 39.22)$; bottom left: mode solution using node-based elements; bottom right: reference mode fields. [*After Paulsen and Lynch, © IEEE, 1991.*]

approach referred to as the penalty method [32] was employed to reformulate the weak wave equation (or variational functional) in conjunction with the node-based elements. However, in recent years it has been recognized that edge-based elements or Tangential Vector Finite Elements (TVFEs) remove the shortcomings of node-based elements.

As discussed in Chapter 2, TVFEs are based on expanding the unknown quantity in terms of its average values along element edges. The corresponding basis functions are vector basis functions as opposed to scalar basis functions (separate expansion for each component) when node-based finite elements are applied. TVFEs enforce tangential field continuity along element boundaries but allow for normal field discontinuities and have been shown to be free of the shortcomings of node-based elements [2, 33, 34, 35]. By using TVFEs, field values are not specified where the field is not defined, spurious modes can automatically be eliminated, and Dirichlet boundary conditions are easily imposed.

To describe the Whitney element at some greater detail (see also Chapter 2), let us consider the rectangular (x, y) coordinate system shown in Fig. 4.27. As usual, we denote the coordinates of the first, second, and third node of a triangular element by (x_1, y_1), (x_2, y_2), and (x_3, y_3), respectively. Also, we denote the edge from node 1 to node 2 as edge #1, the edge from node 2 to node 3 as edge #2, and the edge from node 3 to node 1 as edge #3. The length of the kth edge will be denoted ℓ_k, and the unit vector directed from node i toward node j will be referred to as $\hat{e}_{ij}$ or $\hat{e}_k$. We will assume that a point P internal to the triangle has the coordinates (x, y). The geometry is illustrated in Fig. 4.27.

Next, we define A as the area of $\triangle 123$, A_1 as the area of $\triangle P23$, A_2 as the area of $\triangle P31$, and A_3 as the area of $\triangle P12$. Using the simplex or area coordinates L_1, L_2, and

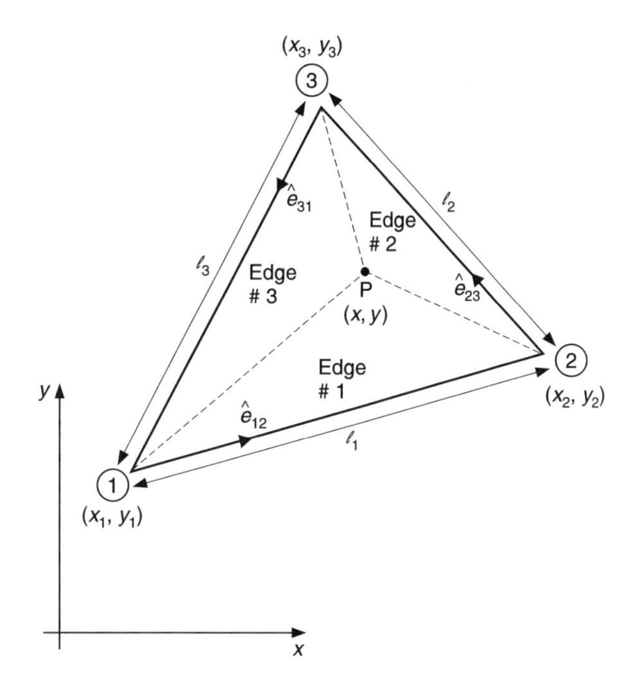

Figure 4.27 Geometry of triangular element and definition of parameters for TVFEs.

L_3 (given by $L_1 = N_1^e = A_1/A$, $L_2 = N_2^e = A_2/A$, and $L_3 = N_3^e = A_3/A$) defined in Chapter 2, we introduce the vector basis functions $\mathbf{W}_1^e$, $\mathbf{W}_2^e$, and $\mathbf{W}_3^e$ by

$$\mathbf{W}_1^e = \mathbf{N}_{12}^e = \ell_1(L_1\nabla L_2 - L_2\nabla L_1) \tag{4.134}$$

$$\mathbf{W}_2^e = \mathbf{N}_{23}^e = \ell_2(L_2\nabla L_3 - L_3\nabla L_2) \tag{4.135}$$

$$\mathbf{W}_3^e = \mathbf{N}_{31}^e = \ell_3(L_3\nabla L_1 - L_1\nabla L_3) \tag{4.136}$$

Following the usual notation, the basis function $\mathbf{W}_k^e$ is associated with the kth edge of the eth element. As noted in Chapter 2, the ℓ_k factor serves as a normalization parameter. Each basis function can be shown to be divergence free (i.e., $\nabla \cdot \mathbf{W}_k^e = 0$), to ensure tangential field continuity across element boundaries and to allow normal field variation across element boundaries. The field $\mathbf{F}^e$ (either an electric field $\mathbf{E}^e$ or a magnetic field $\mathbf{H}^e$) in the eth element is expanded as

$$\mathbf{F}^e = \sum_{k=1}^{e} F_k^e \mathbf{W}_k^e \tag{4.137}$$

where the unknown coefficient F_k^e represents the average field value along edge #k of the eth element. That is, the field along the kth edge is expanded in the direction of the unit vector introduced for the kth edge. In the case of an electromagnetic scattering problem, the scattered field can be expanded via (4.137) and the known incident field can then be added to form the total field.

For the triangle in Fig. 4.28 for which $(x_1, y_1) = (0, 0)$, $(x_2, y_2) = (1, 0)$ and $(x_3, y_3) = (0, 1)$, the three vector basis functions $\mathbf{W}_1^e$, $\mathbf{W}_2^e$, and $\mathbf{W}_3^e$ associated with edge #1, #2, and #3 are plotted in Fig. 4.29 to Fig. 4.31.

The vector basis function $\mathbf{W}_k^e$ provides a constant tangential component (with unit magnitude) along edge #k and zero tangential component along the two other edges. The normal component of $\mathbf{W}_k^e$, however, is varying linearly along all three edges.

Previously, a FEM analysis using node-based triangular finite elements was carried out and we ended up with element matrix entries involving the integrals

$$\iint_{\Omega^e} \nabla N_i^e \cdot \nabla_j^e \, dy, \qquad \iint_{\Omega^e} N_i^e N_j^e \, dy$$

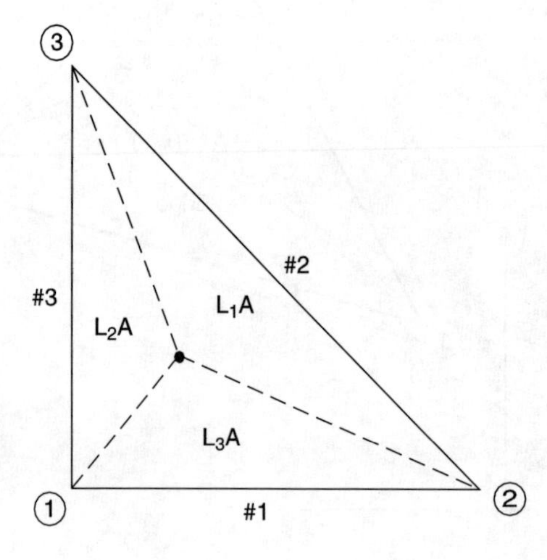

Figure 4.28 Triangle for which vector basis functions are plotted.

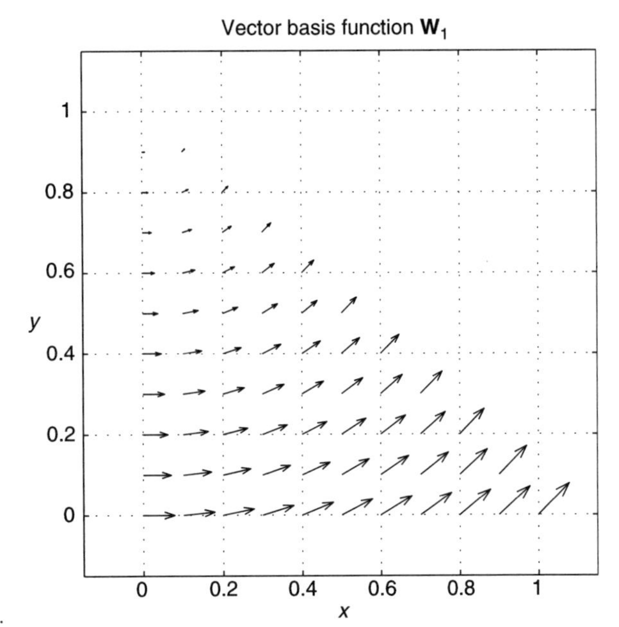

Figure 4.29 Vector basis function $\mathbf{W}_1$.

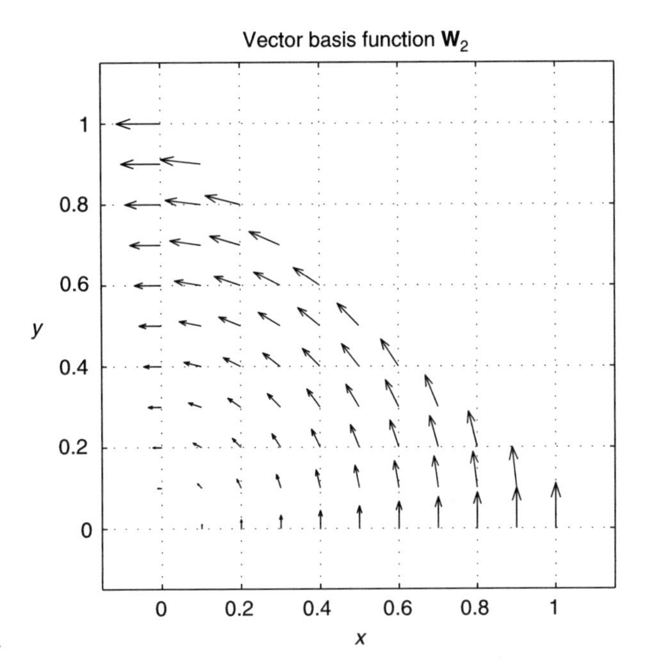

Figure 4.30 Vector-basis function $\mathbf{W}_2$.

with N_i^e $(i = 1, 2, 3)$ and N_j^e $(j = 1, 2, 3)$ being shape functions for the triangular element and Ω^e denoting the surface of eth element. For the standard linear shape functions $N_i^e = L_i$, closed-form expressions for these integrals were given in (4.44) and (4.45). Alternatively, they could be evaluated numerically.

For TVFEs, a similar analysis can be carried out. When using TVFEs, we can proceed with the construction of the weak vector wave equation by working directly with the vector wave equation (4.19) for homogeneous waveguides or 2D scattering

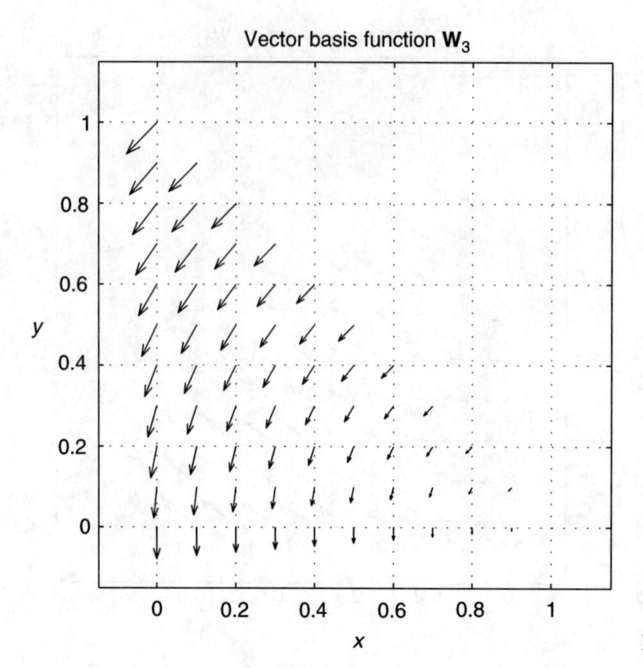

Figure 4.31 Vector basis function $\mathbf{W}_3$.

problems (since $E_z = H_z = 0$). For inhomogeneous waveguides, it is necessary to work with variations of the PDEs (4.23), (4.24), and (4.26) or their dual. Since our goal here is to introduce TVFEs we will work with the simpler vector wave equation (4.19). To construct the weak form of

$$\nabla_t \times \left(\frac{1}{\mu_r} \nabla_t \times \mathbf{E}_t\right) - \gamma^2 \mathbf{E}_t = 0$$

we must employ the identities

$$\mathbf{A} \cdot \nabla_t \times (\nabla_t \times \mathbf{B}) = (\nabla_t \times \mathbf{A}) \cdot (\nabla_t \times \mathbf{B}) - \nabla_t \cdot [\mathbf{A} \times (\nabla_t \times \mathbf{B})] \tag{4.138}$$

$$\iint_\Omega \nabla_t \cdot (\mathbf{A} \times \mathbf{B})\, ds = \oint_C (\mathbf{A} \times \mathbf{B}) \cdot \hat{n}\, dl \tag{4.139}$$

$$-\iint_\Omega \mathbf{A} \cdot \nabla_t(\nabla_t \cdot \mathbf{B})\, ds = \iint_\Omega (\nabla_t \cdot \mathbf{A})(\nabla_t \cdot \mathbf{B})\, ds - \oint_C (\nabla_t \cdot \mathbf{B})(\mathbf{A} \cdot \hat{n})\, dl \tag{4.140}$$

(see Fig. 4.4 for a definition of $\hat{n}$, Ω, and C) in conjunction with the weighted residual equation

$$\iint_\Omega \mathbf{T}_t \cdot \left[\nabla_t \times \left(\frac{1}{\mu_r} \nabla_t \times \mathbf{E}_t\right) - \gamma^2 \mathbf{E}_t\right] dx\, dy = 0, \qquad n = 1, 2, 3 \tag{4.141}$$

where $\mathbf{T}_t$ is the weighting function (we did not use the usual notation to avoid confusion with the TVFE basis notation). The resulting weak equation becomes

$$\iint_\Omega \left[(\nabla_t \times \mathbf{T}_t) \cdot \left(\frac{1}{\mu_r} \nabla_t \times \mathbf{E}_t\right) - \gamma^2 \mathbf{T}_t \cdot \mathbf{E}_t\right] dx\, dy - \oint_C \mathbf{T}_t \cdot \left[\frac{1}{\mu_r} (\nabla_t \times \mathbf{E}_t) \times \hat{n}\right] dl = 0 \tag{4.142}$$

As before, the boundary conditions must be introduced to evaluate the line integral over C. In case of metallic enclosures, the boundary integral is eliminated since $\mathbf{E}_t \times \hat{n}$ vanishes on the enclosures and thus these edges/unknowns need not be considered.

We can proceed to discretize (4.142) by introducing the expansion (4.137) and choosing $\mathbf{T}_t = \mathbf{W}_n^e$, i.e., Galerkin testing. This gives the element equations

$$[K_\nabla^e]\{E_n^e\} = \gamma^2[K^e]\{E_n^e\} \tag{4.143}$$

for the eigenvalue problem. The column $\{E_n^e\}$ now refers to the edge field values of the eth element, and from (4.142)

$$K_{\nabla,mn}^e = \frac{1}{\mu_r^e}\iint_{\Omega^e} (\nabla_t \times \mathbf{W}_m^e)\cdot(\nabla_t \times \mathbf{W}_n^e)\, dx\, dy \tag{4.144}$$

$$K_{mn}^e = \iint_{\Omega^e} \mathbf{W}_m^e \cdot \mathbf{W}_n^e\, dx\, dy \tag{4.145}$$

These can be evaluated in closed form. Specifically [9],

$$K_{\nabla,mn}^e = \frac{1}{\mu_r^e}\frac{\ell_m \ell_n}{4(\Delta^e)^3} D_m D_n \tag{4.146}$$

$$K_{mn}^e = \frac{\ell_m \ell_n}{16(\Delta^e)^3}(I_1 + I_2 + I_3 + I_4 + I_5) \tag{4.147}$$

$$I_1 = A_m A_n + C_m C_n \tag{4.148}$$

$$I_2 = \frac{1}{\Delta^e}(C_m D_n + C_n D_m)\iint_{\Omega^e} x\, dx\, dy \tag{4.149}$$

$$I_3 = \frac{1}{\Delta^e}(A_m B_n + A_n B_m)\iint_{\Omega^e} y\, dx\, dy \tag{4.150}$$

$$I_4 = \frac{1}{\Delta^e} B_m B_n \iint_{\Omega^e} y^2\, dx\, dy \tag{4.151}$$

$$I_5 = \frac{1}{\Delta^e} D_m D_n \iint_{\Omega^e} x^2\, dx\, dy \tag{4.152}$$

in which

$$A_m = a_i^e b_j^e - a_j^e b_i^e \tag{4.153}$$

$$B_m = c_i^e b_j^e - c_j^e b_i^e \tag{4.154}$$

$$C_m = a_i^e c_j^e - a_j^e c_i^e \tag{4.155}$$

$$D_m = -B_m \tag{4.156}$$

and the remaining parameters are given by (4.38). We note that the subscripts m and n refer to edge numbers, whereas the subscripts i and j are associated with the node numbers as specified in Fig. 4.27. The above closed form expressions are not necessarily less expensive to evaluate than a direct numerical evaluation of the integrals (4.144) and (4.145). Thus, in practice, one may simply opt to use Gaussian integration for matrix element evaluation.

The assembly of the element equations (4.143) is carried out in the usual manner. For TVFEs, each edge (or unknown) is shared by two elements only (unless the edge is on C) and thus the resulting assembled global matrix has greater sparsity than the corresponding global matrix for node-based elements. This makes up for the larger number of unknowns associated with TVFEs and typically the storage requirements are about the same for the two types of elements. However, the pre-processing stage of writing a FEM code based on TVFEs is more involved. In addition to the geometry and connectivity tables needed for node-based elements (see Fig. 4.7), tables describing edge to node correspondence, groups of edges on conducting and dielectric boundaries, and unique directions for all edges (unknowns) must be generated since these are used during the element matrix construction and assembly process.

4.5.1 Example 1: Propagation Constants of a Homogeneously Filled Waveguide

As an example application of TVFEs, we again consider a rectangular waveguide with PEC walls. The waveguide is assumed to be homogeneously filled with a material of permittivity $\epsilon = \epsilon_r \epsilon_0$ and permeability $\mu = \mu_r \mu_0$. For TE polarization, $\mathbf{F}_t$ in (4.137) is an electric field $\mathbf{E}_t$, whereas for TM polarization $\mathbf{F}_t$ is the magnetic field $\mathbf{H}_t$. In either case, we are interested in determining the eigenvalues γ^2 from which we can then obtain the transverse propagation constants $\beta = \sqrt{\gamma^2 - \epsilon_r \mu_r k_0^2}$ of the guide.

Assembly of the element equations yields the global matrix equation system

$$[K_\Delta]\{F\} = \gamma^2 [K]\{F\} \tag{4.157}$$

which is identical (in form) to (4.57) but where $\{F\}$ now represents the field values along element edges rather than at element nodes. For TM polarization the number of unknowns equals the number of global edges, whereas for TE polarization the number of unknowns equals the number of global edges minus the number of boundary edges (since the field here is known a priori to be zero).

A FEM computer code was written to again model the rectangular waveguide shown in Fig. 4.8. The geometry was discretized using 261 nodes forming 720 edges and 440 elements. Of these, 60 nodes and 60 edges were on the PEC boundary. Analytical and numerical values for γa are compared in Tables 4.3 and 4.4 for TE and TM polarization, and we observe that the agreement between analytical and numerical values is good. In fact, the TVFE simulation yields slightly more accurate results for comparable discretization than the simulation for node-based elements.

TABLE 4.3 Analytical and Numerical γa Values for TE Modes Using TVFEs

Mode	Analytical Result	FEM Result
TE_{10}	3.141593	3.141520
TE_{20}	6.283185	6.282485
TE_{01}	6.283185	6.283132
TE_{11}	7.024818	7.024096
TE_{21}	8.885766	8.884424
TE_{30}	9.424778	9.426045

TABLE 4.4 Analytical and Numerical γa Values for TM Modes Using TVFEs

Mode	Analytical Result	FEM Result
TM_{11}	7.024818	7.023634
TM_{21}	8.885766	8.884080
TM_{31}	11.327173	11.326886
TM_{12}	12.953118	12.947988
TM_{22}	14.049629	14.041209
TM_{32}	15.707963	15.698376

4.5.2 Example 2: Scattering by a Square-Shaped Material Coated Cylinder

As another application, we consider a square metallic cylinder situated in free space and coated with a dielectric shell of thickness d, as shown in Fig. 4.32. The cylinder is assumed to be coated by a material layer of relative permittivity ϵ_r and relative permeability μ_r. It is illuminated by an incident TE or TM polarized plane wave as defined by (4.6) and (4.7).

As in the case of node-based elements, we will work with the scattered field as the unknown quantity for this application. For TM polarization, the resulting element equations will be of the form

$$[A^e]\{H_e^{\text{scat}}\} = \{b^e\} \tag{4.158}$$

with the column vector $\{H_e^{\text{scat}}\}$ representing the unknowns scattered magnetic field components along the edges of the triangular mesh. Following the analysis in Section 4.4, we readily find

$$A_{mn}^e = \iint_{\Omega^e} \left[(\nabla_t \times \mathbf{W}_m^e) \cdot \left(\frac{1}{\epsilon_r} \nabla_t \times \mathbf{W}_n^e \right) - k_0^2 \mu_r \mathbf{W}_m^e \cdot \mathbf{W}_n^e \right] d\Omega \tag{4.159}$$

$$\begin{aligned} b_m^e = & \int_{C^{\text{inner}}, C^{\text{diel}}} \mathbf{W}_m^e \cdot \left[\hat{n} \times \left(\frac{1}{\epsilon_r} \nabla_t \times \mathbf{H}^i \right) \right] dl \\ & - \iint_{\Omega^e \in \Omega^d} \mathbf{W}_m^e \cdot \left[\nabla_t \times \frac{1}{\epsilon_r} \nabla_t \times \mathbf{H}^i - k_0^2 \mu_r \mathbf{H}^i \right] dx\, dy \end{aligned} \tag{4.160}$$

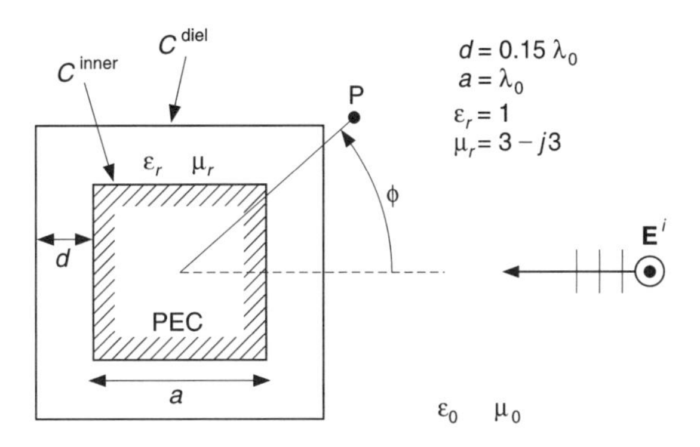

Figure 4.32 Coated square cylinder illuminated by TM polarized plane wave.

in which Ω^d refers to the region occupied by the dielectric coating and C^{diel} is the contour on the outer boundary of the dielectric as depicted in Fig. 4.32 (see also Fig. 4.16). The computational domain will be terminated by a metal-backed artificial absorber and therefore we can neglect the presence of any boundary integrals. The absorber was $0.5\lambda_0$ thick and placed at a distance $0.5\lambda_0$ from the boundary between the coating and free space. The absorber's relative permeability and permittivity was set to $\epsilon_a/\epsilon_0 = \mu_a/\mu_0 = 2 - j7$.

For our scattering calculations, we choose a PEC cylinder of side length $a = \lambda_0$. The coating has the uniform thickness $d = 0.15\lambda_0$, a relative permittivity $\epsilon_r = 1$ and a relative permeability $\mu_r = 3 - j3$. The FEM simulation of this domain used 1264 nodes forming 3620 edges and 2356 elements. Of these, 40 nodes and edges were on the PEC boundary C^{inner}, 52 nodes and edges were on the material/free space boundary C^{diel} and 132 nodes and edges were on the C^{outer} which coincides with the metal boundary backing the absorber. The final number of unknowns was thus 3620 for TM polarization.

Once the FEM solution is carried out and the fields are found everywhere in the domain, the next step is to compute the far zone scattered field. An intermediate part of this process involves the application of the surface equivalence principle (see Chapter 1) to determine equivalent surface currents on a surface enclosing the scatterer (such as C^{diel} or any other contour of choice). For TM polarization, the equivalent electric and magnetic surface currents are given by

$$\begin{aligned}\mathbf{J}^B &= \hat{n} \times \mathbf{H} = (n_x\hat{x} + n_y\hat{y}) \times (H_x\hat{x} + h_y\hat{y}) \\ &= (n_x H_y - n_y H_x)\hat{z} = J_z\hat{z} \end{aligned} \tag{4.161}$$

$$\begin{aligned}\mathbf{M}^B &= \mathbf{E} \times \hat{n} = (E_z\hat{z}) \times (n_x\hat{x} + n_y\hat{y}) \\ &= -n_y E_z\hat{x} + n_x E_z\hat{y} = M_x\hat{x} + M_y\hat{y} \end{aligned} \tag{4.162}$$

in which $\mathbf{E}$ is found from $\mathbf{E} = \frac{\nabla\times\mathbf{H}}{j\omega\epsilon}$. These currents can be integrated using the radiation integrals to give the far zone scattered field. For TM incidence, we find

$$E_z^{\text{scat}} = \sqrt{\frac{k_0}{8\pi}} \frac{e^{-j(k_0 r - \pi/4)}}{\sqrt{r}} \int_{C^{\text{diel}}} e^{jk_0\hat{r}\cdot\mathbf{r}'} [-Z_0 J_z(\mathbf{r}') - M_x(\mathbf{r}')\sin\phi' + M_y(\mathbf{r}')\cos\phi']\, dl' \tag{4.163}$$

where the reader is referred to Fig. 4.16 for a definition of the primed parameters.

In Figs. 4.33 to 4.36, the computed equivalent currents $\mathbf{J}^B$ and $\mathbf{M}^B$ on C^{diel} are compared with those obtained using a moment method (denoted as MoM on the figures) analysis [36] with the incident wave impinging at $\phi_0 = 0°$. As seen, the two analyses give nearly identical results (for magnitude as well as phase). The corresponding bistatic echo width as computed from (4.163) and the dual of (4.93) is given in Fig. 4.37. Again, the FEM results are in agreement with the moment method data. The scattering patterns show distinctively that the largest scattering occurs in the forward ($\phi = 180°$) and backscatter ($\phi = 0°$) directions. The beamwidth of these forward and backscatter lobes decreases inversely with the size of the cylinder, whereas the number of side lobes increase with increasing cylinder size.

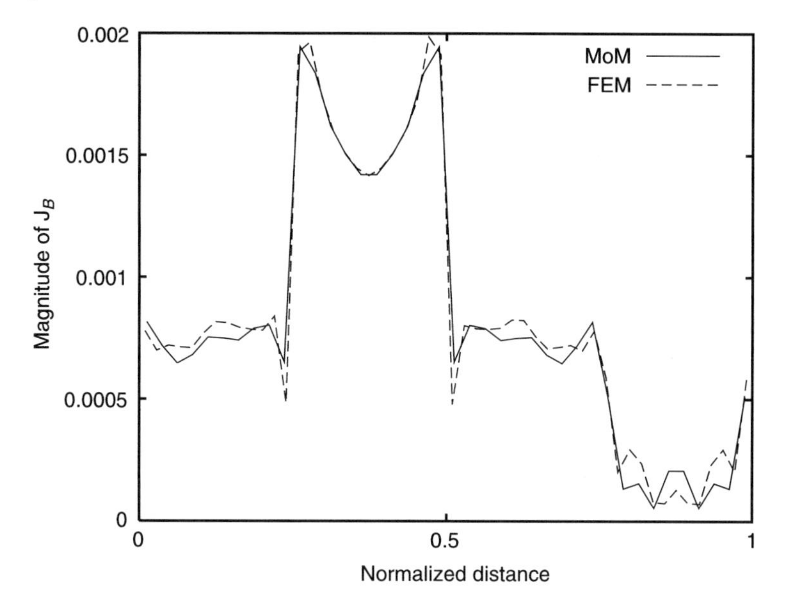

Figure 4.33 $|\mathbf{J}^B|$ on C^{diel} in Fig. 4.32.

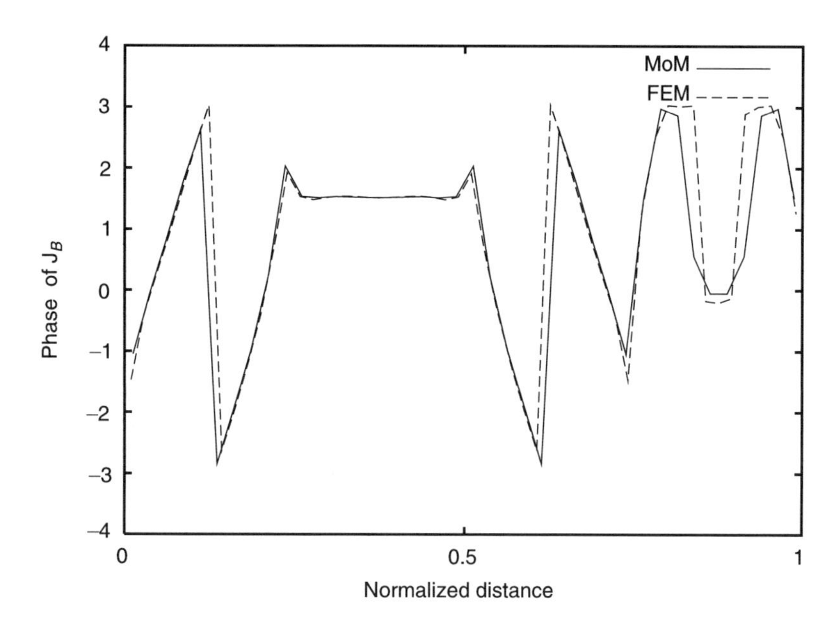

Figure 4.34 $\angle\mathbf{J}^B$ on C^{diel} in Fig. 4.32.

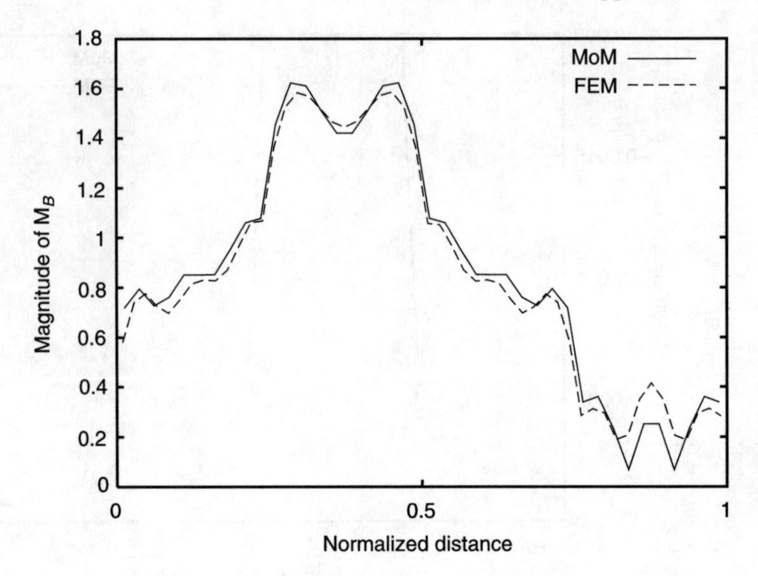

Figure 4.35 $|\mathbf{M}^B|$ on C^{diel} in Fig. 4.32.

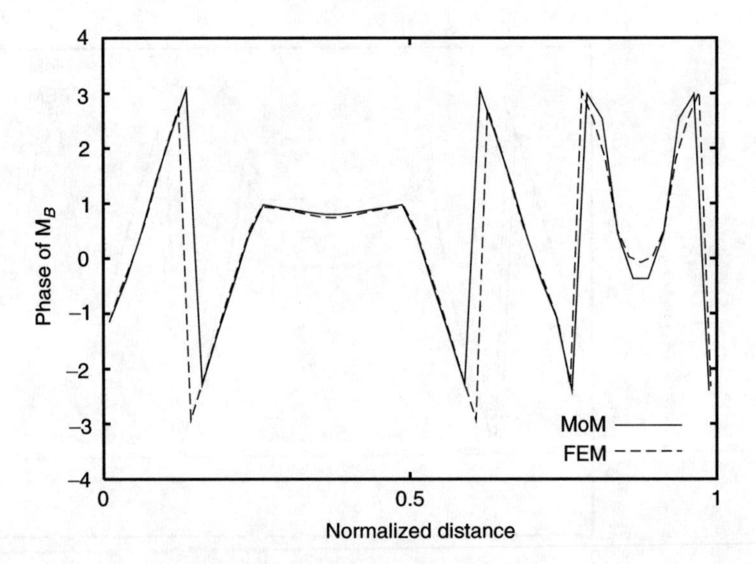

Figure 4.36 $\angle\mathbf{M}^B$ on C^{diel} in Fig. 4.32.

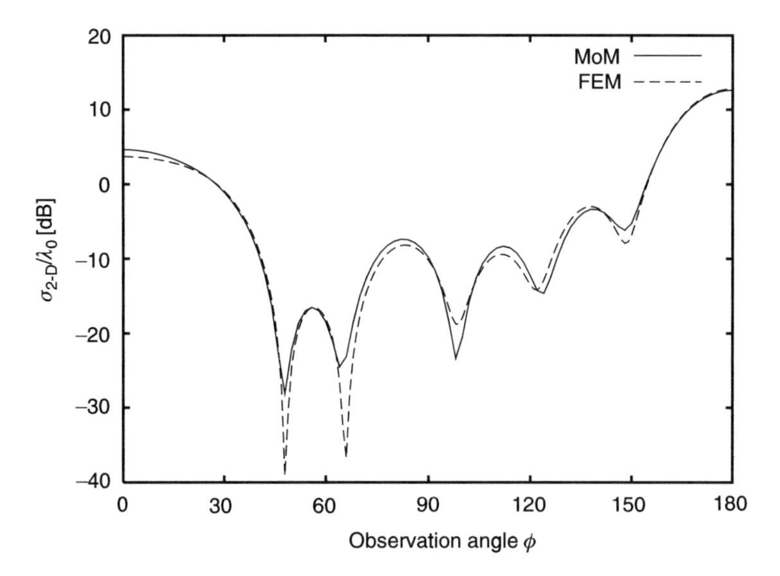

Figure 4.37 Bistatic RCS of cylinder in Fig. 4.32. (*Results in Figs. 4.33 to 4.37 are courtesy of L. S. Andersen.*) The RCS was computed by integrating equivalent electric and magnetic currents $(\mathbf{J}^B, \mathbf{M}^B)$ on a contour a small distance from C^{diel}.

APPENDIX 1: ELEMENT MATRIX FOR NODE-BASED BILINEAR RECTANGLES

The bilinear node-based expansion for the rectangle (right-angled quadrilateral) is given by (2.7) in Chapter 2, and has four degrees of freedom. Thus, the element matrix will have 4 × 4 entries, viz.

$$[A^e] = \begin{bmatrix} A_{11}^e & A_{12}^e & A_{13}^e & A_{14}^e \\ A_{21}^e & A_{22}^e & A_{23}^e & A_{24}^e \\ A_{31}^e & A_{32}^e & A_{33}^e & A_{34}^e \\ A_{41}^e & A_{42}^e & A_{43}^e & A_{44}^e \end{bmatrix}$$

where the expression for A_{ij}^e is given in (4.159). For the rectangle in Fig. 4.38 (see also Fig. 2.1), the resulting entries are

$$A_{ii}^e = \frac{1}{3\epsilon_r^e}\left(\frac{h_y^e}{h_x^e} + \frac{h_x^e}{h_y^e}\right) - k_0^2\mu_r^e\,\frac{h_x^e h_y^e}{9}, \qquad i = 1, 2, 3, 4$$

$$A_{12}^e = A_{21}^e = A_{34}^e = A_{43}^e = \frac{1}{3\epsilon_r^e}\left(-\frac{h_y^e}{h_x^e} + \frac{h_x^e}{2h_y^e}\right) - k_0^2\mu_r^e\,\frac{h_x^e h_y^e}{18}$$

$$A_{23}^e = A_{32}^e = A_{41}^e = A_{14}^e = \frac{1}{3\epsilon_r^e}\left(\frac{h_y^e}{2h_x^e} - \frac{h_x^e}{h_y^e}\right) - k_0^2\mu_r^e\,\frac{h_x^e h_y^e}{18}$$

$$A_{24}^e = A_{42}^e = A_{31}^e = A_{13}^e = -\frac{1}{6\epsilon_r^e}\left(\frac{h_y^e}{h_x^e} + \frac{h_x^e}{h_y^e}\right) - k_0^2\mu_r^e\,\frac{h_x^e h_y^e}{36}$$

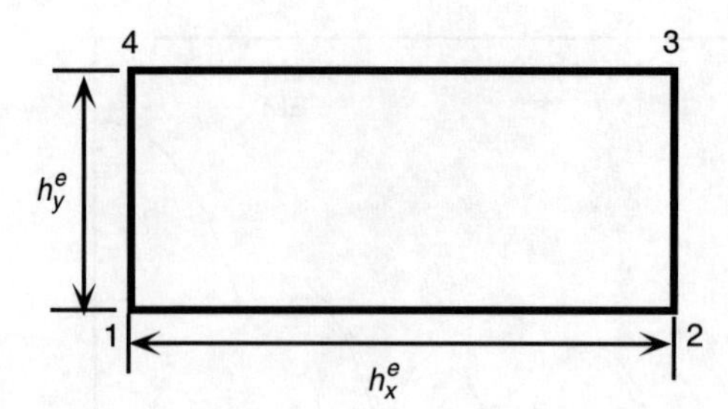

Figure 4.38 Node numbering and geometry of the rectangle.

APPENDIX 2: SAMPLE MATLAB CODE FOR IMPLEMENTING THE MATRIX ASSEMBLY

```
function[K_TE,Kdel_TE, A_TE,A,K,Kdel,K_TM,Kdel_TM, A_TM,x,y,xnodes,ynodes]=
fem_2d(lambda);

% Generation of the Node Connectivity Table:
% ==========================================
%This MATLAB code is courtesy of Y. Botros

N_elements=18;
n_1st_local_node=[1 2 2 2  3  3 5  5 6  6 7  8  9  9 10 10 11  12];
n_2nd_local_node=[2 6  3 7 4 8 6 10 7 11 8 12 10 14 11 15 12 16];
n_3rd_local_node=[ 5 5 7 6 8 7 10 9 11 10 11 11 14 13 15 14 15 15];
mu=ones(1:N_elements);
eps=ones(1:N_elements);
ko=2*pi/lambda;

for e=1:N_elements;

n(1,e)=n_1st_local_node(e);
n(2,e)=n_2nd_local_node(e);
n(3,e)=n_3rd_local_node(e);
end;

non_cond=[6 7 10 11];
% Nodes locations table:
%=======================
xnodes=[0 .5 1 1.5 0 .5 1 1.5 0 .5 1 1.5 0 .5 1 1.5 ];
ynodes=[0 0  0  0 .25 .25 .25 .25  .5 .5 .5 .5 .75 .75 .75 .75 ];

%Initialization Process:
%=======================

A=zeros(N_elements);
Kdel=zeros(N_elements);
K=zeros(N_elements);
Kedel=zeros(3);
```

```
%Loop through all elements:
%===========================

for e=1:N_elements;

%E_z/TM polarization :
%====================

pe=1/mu((e));
qe=eps((e));

% H_z/TE polarization :
% =========================

pe=1/eps((e));
qe=mu((e));

% Coordinates of the Element nodes:
%==================================

for i=1:3;
x(i)=xnodes(n(i,e));
y(i)=ynodes(n(i,e));
end;

% Compute the Element matrix entries:
%================================

Area=.5*abs((x(2)-x(1))*(y(3)-y(1))-x(3)-x(1))*y(2)-y(1)));

% A(n(i,e),n(i,j))=0;
% Kdel(n(i,e),n(i,j))=0;
% K(n(i,e),n(i,j))=0;

for i=1:3;
i1=0;
if i==3;
i1=3;
end;
ip1=(i+1)=i1;
i2=0;
if ip1==3;
i2=3;
end;
ip2=ip1+1-i2;
bi=y(ip1)-y(ip2);
ci=x(ip2)-x(ip1);
```

```
for j=1:3;
j1=0;
if j==3;
j1=3;
end;
jp1=(j+1)-j1;
j2=0;
if jp1==3;
j2=3;
end;
jp2=(jp1+1)-j2;
bj=y(jp1)-y(jp2);
cj=x(jp2)-x(jp1);

Ae(i,j)=-(pe*(bi*bj+ci*cj))/(4*Area);

Kedel(i,j)=-Ae(i,j);


if i==j;
Ke(i,j)=qe*(Area/6);
Ae(i,j)=Ae(i,j)+(ko^2)*qe*(Area/6);
else;
Ke(i,j)=qe*(Area/12);
Ae(i,j)=Ae(i,j)+(ko^2)*qe*(Area/12);
end;

% Assemble the Element matrices into the Global FEM System:
%=========================================================

A(n(i,e),n(j,e))=A(n(i,e),n(j,e))+Ae(i,j);
Kdel(n(i,e),n(j,e))=Kdel(n(i,e),n(j,e))+Kedel(i,j);
K(n(i,e),n(j,e))=K(n(i,e),n(j,e))+Ke(i,j);
end;
end;
end;

K_TE=K;
Kdel_TE=Kdel;
A_TE=A;

K_TM=K(non_cond,non_cond);
Kdel_TM=Kdel(non_cond,non_cond);
A_TM=A(non_cond,non_cond);

eig_squares_TE=eig(Kdel_TE,K_TE);
eig_values_indices_TE=find(eig_squares_TE >= 0);
eig_values_TE=sqrt(eig_squares_TE(eig_values_indices_TE));
eig_values_TE=sort(eig_values_TE);
```

```
figure(1)
plot (eig_values_TE,'*');
xlabel('The Modes (TE case)')
ylabel('The eigen_values');
title('Eigen values for TE modes in a rectangular waveguide');
grid;
legend('a=1.5 cm and b=.75 cm');

figure(2)
eig_squares_TM=eig(Kdel_TM,K_TM);
eig_values_indices_TM=find(eig_squares_TM >= 0);
eig_values_TM=sqrt(eig_squares_TE(eig_values_indices_TM));
eig_values_TM=sort(eig_values_TM);
plot( eig_values_TM,'o');
xlabel('The Modes (TM case)')
ylabel('The eigen_values');
title('Eigen values for TM modes in a rectangular waveguide');
grid;
legend('a=1.5 cm and b=.75 cm');

e_no=1:16;
table=[e_no' xnodes' ynodes'];

diary data.data

% The General A matrix
A
% The General K matrix
K
% The General del matrix
Kdel
%*****************
% The TE A matrix
A_TE
% The TE K matrix
K_TE
% The TE Kdel matrix
Kdel_TE
%*****************
% The TM A matrix
A_TM
% The TM K matrix
K_TM
% The TM Kdel matrix
Kdel_TM
%*****************
% The next table indicates the node coordinates.
% First column ====> Node Numbers.
```

```
% Second column ===> x coord.
% Third column  ===> y coord.
table
diary
```

REFERENCES

[1] B. M. A. Rahman and J. B. Davies. Finite element analysis of optical and microwave waveguide problems. *IEEE Trans. Microwave Theory Tech.*, MTT-32:20–28, January 1984.

[2] J. F. Lee, D. K. Sun, and Z. J. Cendes. Full-wave analysis of dielectric waveguides using tangential vector finite elements. *IEEE Trans. Microwave Theory Tech.*, MTT-39(8):1262–1271, August 1991.

[3] J. van Bladel. *Electromagnetic Fields*. Hemisphere Publishing Corp., New York, 1985.

[4] Y. Lu and F. A. Fernandez. An efficient finite element solution of inhomogeneous anisotropic and lossy dielectric waveguides. *IEEE Trans. Microwave Theory Tech.*, 41(6/7):1215–1223, June/July 1993.

[5] J.-F Lee. Finite element analysis for lossy dielectric waveguides. *IEEE Trans. Microwave Theory Tech.*, 42(6):1025–1031, June 1994.

[6] J. B. Davies. Complete modes in uniform waveguides. In T. Itoh, G. Pelosi, and P. P. Silvester, editors, *Finite Element Software for Microwave Engineering*, Chapter 1. Wiley, New York, 1996.

[7] P. Silvester. Construction of triangular finite element universal matrices. *Int. J. Numer. Methods Eng.*, 12(2):237–244, 1978.

[8] A Chatterjee, J. L. Volakis, and L. C. Kempel. Optimization issues in finite element codes for solving open 3D electromagnetic scattering and conformal antenna problems. *Int. J. Num. Modeling: Electr. Net. Dev. and Fields*, 9:335–344, 1996.

[9] C. J. Reddy, M. D. Deshpande, C. R. Cockrell, and F. B. Beck. Finite element method for eigenvalue problems. Technical report, NASA Technical Paper 3485, NASA Langley Research Center, December 1994.

[10] R. F. Harrington. *Time-Harmonic Electromagnetic Fields*. McGraw-Hill, New York, 1961.

[11] N. Marcuvitz. *Waveguide Handbook*. Peter Peregrinus, London, 1986. Originally published by McGraw-Hill Co., 1951.

[12] A. Bayliss and E. Turkel. Radiation boundary conditions for wave-like equations. *Comm. Pure Appl. Math.*, 33:707–725, 1980.

[13] T. B. A. Senior and J. L. Volakis. *Approximate Boundary Conditions in Electromagnetics*. IEE Press, London, 1995.

[14] B. Engquist and A. Majda. Absorbing boundary conditions for the numerical simulation of waves. *Math. Comp.*, 31:629-651, 1977.

[15] L. N. Trefethen and L. Halpern. Wide-angle one-way wave equations. *J. Acoust. Soc. Amer.*, 84:1397–1404, October 1988.

[16] R. L. Higdon. Absorbing boundary conditions for acoustic and elastic waves in stratified media. *J. Comp. Physics*, 101:386–418, 1992.

[17] A. F. Peterson and S. P. Castillo. A frequency-domain differential equation formulation for electromagnetic scattering from inhomogeneous cylinders. *IEEE Trans. Antennas Propagat.*, 37(5):601–607, May 1989.

[18] G. T. Ruck, D. E. Barrick, W. D. Stuart, and C. K. Krichbaum. *Radar Cross Section Handbook*. Plenum Press, New York, 1970, 1970.

[19] T. Özdemir and J. L. Volakis. A comparative study of an absorber boundary condition and an artificial absorber for truncating finite element meshes. *Radio Science*, 29:1255–1263, September–October, 1994.

[20] Z. J. Sacks, D. M. Kingsland, R. Lee, and J.-F. Lee. A perfectly matched anisotropic-absorber for use as an absorbing boundary condition. *IEEE Trans. Antennas Propagat.*, 43:1460–1463, 1995.

[21] S. R. Legault, T. B. A. Senior, and J. L. Volakis. Design of planar absorbing layers for domain truncation in FEM applications. *Electromagnetics*, 16(4):451–464, July–August 1996.

[22] J. M. Jin and J. L. Volakis. TE scattering by an inhomogeneously filled aperture in a thick conducting plane. *IEEE Trans. Antennas Propagat.*, 38:1280–1286, August 1990.

[23] J. M. Jin and J. L. Volakis. TM scattering by an inhomogeneously filled aperture in a thick ground plane. *IEE Proc. H.*, 137(3):153–159, June 1990.

[24] J. M. Jin, J. L. Volakis, and J. D. Collins. A finite element-boundary integral method for scattering and radiation by two- and three-dimensional structures. *IEEE Antennas Propagat. Soc. Mag.*, 33(3):22–32, June 1991.

[25] J. R. Mautz and R. F. Harrington. H-field, E-field and combined-field solutions for conducting bodies of revolution. *Arch. Elek. Übertragung*, 32:157–163, 1978.

[26] D. R. Wilton and J. E. Wheeler III. Comparison of convergence rates of the conjugate gradient method applied to various integral equation formulations. In T. K. Sarkar, editor, *Application of Conjugate Gradient Method to Electromagnetics and Signal Analysis*, Chapter 5. Elsevier, New York, 1990.

[27] J. D. Collins, J. M. Jin, and J. L. Volakis. Eliminating interior cavity resonances in FE–BI methods for scattering. *IEEE Trans. Antennas Propagat.*, 40:1583–1585, December 1992.

[28] P. L. Huddleston, L. N. Medgyesi-Mitschang, and J. M. Putnam. Combined field integral formulation for scattering by dielectrically coated bodies. *IEEE Trans. Antennas Propagat.*, AP-34:510–520, 1986.

[29] K. D. Paulsen and D. R. Lynch. Elimination of vector parasites in finite element Maxwell solutions. *IEEE Trans. Microwave Theory Tech.*, 39(3):395–404, March 1991.

[30] D. Sun, J. Manges, X. Yuan, and Z. Cendes. Spurious modes in finite-element methods. *IEEE Trans. Antennas Propagat.*, 37(5):12–24, October 1995.

[31] J. R. Winkler and J. B. Davies. Elimination of spurious modes in finite element analysis. *J. Comput. Physics*, 56:1–14, 1984.

[32] B. M. A. Rahman and J. B. Davies. Penalty function improvement of waveguide solution by finite elements. *IEEE Trans. Microwave Theory Tech.*, 32:922–928, August 1984.

[33] J. M. Jin and J. L. Volakis. Electromagnetic scattering by and transmission through a three-dimensional slot in a thick conducting plane. *IEEE Trans. Antennas Propagat.*, 39(4):543–550, April 1991.

[34] J. P. Webb. Edge elements and what they can do for you. *IEEE Trans. Magnetics*, 29:1460–1465, 1993.

[35] I. V. Yioultsis and T. D. Tsiboukis. Vector finite element analysis of waveguide discontinuities involving anisotropic media. *IEEE Trans. Magnetics*, MAG-31, pp. 1550–1553, May 1995.

[36] L. S. Andersen. Scattering by non-perfectly conducting structures. Master's thesis, Electromagnetics Institute, Technical University of Denmark, Lyngby, No. E544, August 1995.

5

Three-Dimensional Problems: Closed Domain

5.1 INTRODUCTION

Finite elements have been used extensively to model open- and closed-domain electromagnetic problems in scalar form in two and three dimensions [1], [2], [3]. However, the true power of the finite element method is revealed in three-dimensional volume formulations since surface-based integral equation methods have great difficulty in dealing with material and structural inhomogeneities. As explained in earlier chapters, finite elements do not suffer from these shortcomings. But for a long time [4], reliable full vector formulation proved to be extremely difficult to implement. Discretization of the curl-curl version of the wave equations (1.30) usually resulted in the appearance of nonphysical or *spurious* modes. The cause of the problem is the traditional nodal basis functions that are used to discretize the unknown field variable.[1] The reasons for the failure of node-based elements in modeling the vector wave equation will be discussed in a later section. Fortunately, a novel remedy was found by assigning the degrees of freedom to the edges rather than to the nodes of elements. These types of elements had been described by Whitney [5] in terms of geometrical forms about 35 years back and were revived by Nedelec [6] in 1980. In recent years, Bossavit [7] and others [8], [9], [10], [11] applied these edge-based finite elements successfully to model full three-dimensional vector problems. In all these works, edge elements were seen to be devoid of the shortcomings commonly experienced with node-based elements. In this chapter, we examine the application of edge basis functions to closed domain problems as found in packaged microwave circuits and cavity resonators.

[1]See Chapter 2 for a presentation of the node-based and edge-based elements. Also, see the last section of Chapter 4 for a two-dimensional formulation using edge-based elements along with a discussion on the shortcomings of node-based elements.

The first part of this chapter describes the variational formulation for the closed domain problem in terms of field intensity. As noted in Chapter 1 (Section 1.11.1), the variational formulation leads to the same system of equations as the weighted residual method employed in Chapters 3 and 4. We also formulate the problem in terms of vector potentials. The field formulation and the potential formulation are equivalent; however, each has its pros and cons. After the formulation to obtain the linear system of equations, we briefly describe the problem of *spurious* solutions encountered with node-based elements. Generation and assembly of the finite element coefficient matrix using tetrahedrals and bricks are given and issues related to modeling sources for circuit problems are discussed. This is very critical for the computation of the scattering parameters (S parameters) in a circuit. We end the chapter by presenting a few applications of the finite element method pertaining to cavity resonators and packaged circuit configurations.

5.2 FORMULATION

The geometry of the problem is illustrated in Fig. 5.1. The structure of interest is a three-dimensional inhomogeneous body occupying the volume V that may include embedded perfect magnetic and resistive sheet surfaces as well as metallic sheets. We shall assume that V is bounded by perfect electric walls.

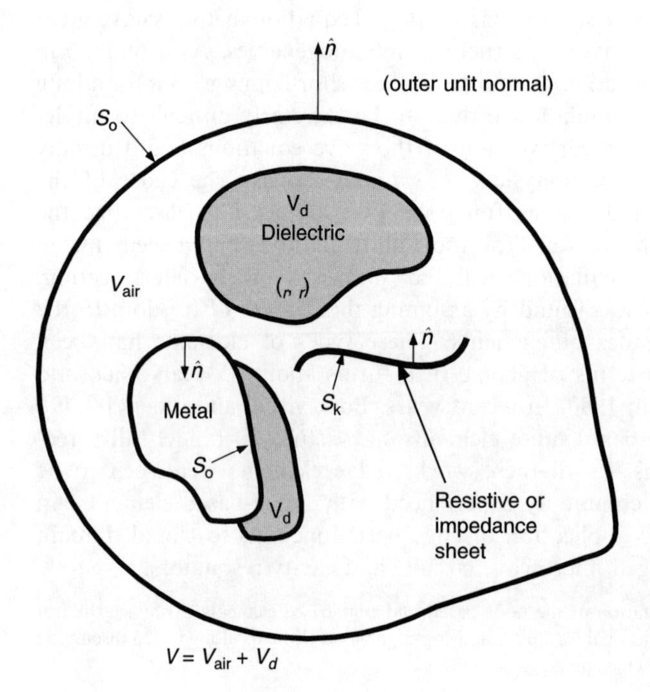

Figure 5.1 Inhomogeneous structure enclosed by a mesh termination surface S_0 assumed to be a perfect electric conductor (PEC).

5.2.1 Field Formulation

As done in the previous chapters, the problem statement is to satisfy the vector wave equation

$$\nabla \times \frac{1}{\mu_r} \nabla \times \mathbf{E} - k_0^2 \epsilon_r \mathbf{E} = -jk_0 Z_0 \mathbf{J}^i - \nabla \times \left(\frac{1}{\mu_r} \mathbf{M}^i \right) \tag{5.1}$$

throughout the volume V subject to a given set of boundary conditions on the surface S_0 enclosing the volume V. Here, $\mathbf{J}^i$ and $\mathbf{M}^i$ are the electric and magnetic current sources, respectively, contained within the volume. These current sources are usually known a priori and form the excitation for the problem. In this chapter, we take the variational approach to formulating the finite element solution. This approach is often employed in the literature to construct the linear system, but as discussed in Chapter 1, it typically leads to a system that is identical to that obtained from the weighted residual method. The variational formulation is therefore another method for finding the solution of a given boundary value problem.

The calculus of variations originates from a generalization of the elementary theory of maxima and minima of functions. In the variational technique, we strive to find the extrema of functionals (see Chapter 1). The functional can loosely be taken to mean a function which depends on the entire course of one or more functions within the domain of interest. For the wave equation, we can express the functional for the total electric field as

$$\begin{aligned} F(\mathbf{E}) = & \frac{1}{2} \int_V \mathbf{E} \cdot \left[\nabla \times \frac{1}{\mu_r} (\nabla \times \mathbf{E}) - k_0^2 \epsilon_r \mathbf{E} \right] dV \\ & + \int_V \mathbf{E} \cdot \left[jk_0 Z_0 \boldsymbol{J}^i + \nabla \times \left(\frac{1}{\mu_r} \boldsymbol{M}^i \right) \right] dV \end{aligned} \tag{5.2}$$

where V is the entire computational domain. For the sake of simplicity, we will assume a source-free volume domain and consider that the only sources used to excite the circuit are coupled through the ports of the geometry. Using the vector identity

$$\mathbf{A} \cdot \nabla \times \mathbf{B} = \mathbf{B} \cdot \nabla \times \mathbf{A} - \nabla \cdot (\mathbf{A} \times \mathbf{B})$$

and the divergence theorem

$$\int_V \nabla \cdot (\mathbf{A} \times \mathbf{B}) \, dV = \oint_{S_0} (\mathbf{A} \times \mathbf{B}) \cdot \hat{n} \, dS \tag{5.3}$$

on the double curl in the right hand side of (5.2), we get

$$\begin{aligned} F(\mathbf{E}) = & \frac{1}{2} \int_V \left[\frac{1}{\mu_r} (\nabla \times \mathbf{E}) \cdot (\nabla \times \mathbf{E}) - k_0^2 \epsilon_r \mathbf{E} \cdot \mathbf{E} \right] dv \\ & + \frac{1}{2} \int_{S_0} [\mathbf{E} \cdot (\hat{n} \times \nabla \times \mathbf{E})] \, dS + \int_V \mathbf{E} \cdot \mathbf{f}^i \, dV \end{aligned} \tag{5.4}$$

In (5.4), $\mathbf{f}^i$ is the source function given by

$$\mathbf{f}^i = jk_0 Z_0 \mathbf{J}^i + \nabla \times \left(\frac{1}{\mu_r} \right) \mathbf{M}^i \tag{5.5}$$

and S_0 denotes the surfaces within V for which the tangential component of $\mathbf{E}$ and/or $\mathbf{H}$ is discontinuous. We remark that the corresponding formulation based on the weighted residual method would lead to the weak form of the vector wave equation

$$\langle \mathbf{R}, \mathbf{W} \rangle = \int_V \left[\frac{1}{\mu_r} (\nabla \times \mathbf{E}) \cdot (\nabla \times \mathbf{W}) - k_0^2 \epsilon_r \mathbf{E} \cdot \mathbf{W} \right] dV + \int_{S_0} [\mathbf{W} \cdot (\hat{n} \times \nabla \times \mathbf{E})] \, dS + \int_V \mathbf{W} \cdot \mathbf{f}^{\mathrm{i}} \, dV \tag{5.6}$$

where $\mathbf{R}$ is the residual vector resulting from the finite-dimensional approximation to the exact solution and $\mathbf{W}$ is the weighting function.

The above functional can be generalized to account for the presence of impedance and resistive sheets or other discontinuous boundaries. In the case of a resistive card, the transition condition [12] (see Section 1.6)

$$\hat{n}_K \times (\hat{n}_K \times \mathbf{E}) = -R \hat{n}_K \times (\mathbf{H}^+ - \mathbf{H}^-) \tag{5.7}$$

must be enforced, where $\mathbf{H}^{\pm}$ denotes the total magnetic field above and below the sheet, R is the resistivity in Ohms per square meter and $\hat{n}_K$ is the unit normal vector to the sheet pointing in the upward direction (+ side). For an impenetrable impedance surface, the appropriate boundary condition is (see Section 1.6)

$$\hat{n}_K \times (\hat{n}_K \times \mathbf{E}) = -\eta \hat{n}_K \times \mathbf{H} \tag{5.8}$$

where $\hat{n}_K = -\hat{n}$ is the unit normal vector to the surface and η is the surface impedance ($\hat{n}_K$ points into the computational domain and as usual $\hat{n}$ points away from the computational domain). Taking into consideration these boundary/transition conditions, the functional for the total electric field can be more explicitly written as

$$F(\mathbf{E}) = \int_V \left[\frac{1}{\mu_r} (\nabla \times \mathbf{E}) \cdot (\nabla \times \mathbf{E}) - k_0^2 \epsilon_r \mathbf{E} \cdot \mathbf{E} \right] dV + jk_0 Z_0 \int_{S_k} \frac{1}{K} (\hat{n} \times \mathbf{E}) \cdot (\hat{n} \times \mathbf{E}) \, dS + \int_{S_0} [\mathbf{E} \cdot P(\mathbf{E})] \, dS \tag{5.9}$$

where K is the surface resistivity of a resistive card and equals the surface impedance for an impedance sheet. It has also been assumed that the relation between tangential $\mathbf{H}$ and tangential $\mathbf{E}$ on the surface S_0 can be expressed in terms of the differential equation

$$\hat{n} \times \nabla \times \mathbf{E} = P(\mathbf{E}) \tag{5.10}$$

Note that the factor of $\frac{1}{2}$ in (5.9) was dropped on the assumption that in carrying out the extremization of $F(\mathbf{E})$, the differentiation will be applied to $P(\mathbf{E})$ as well. Again, we note that the corresponding weighted residual equation which would lead to the same linear system is [13]

$$\int_V \left[\frac{1}{\mu_r} (\nabla \times \mathbf{E}) \cdot (\nabla \times \mathbf{W}) - k_0^2 \epsilon_r \mathbf{E} \cdot \mathbf{W} \right] dV + jk_0 Z_0 \int_{S_K} \frac{(\hat{n} \times \mathbf{E}) \cdot (\hat{n} \times \mathbf{W})}{K} \, dS + \int_{S_0} \mathbf{W} \cdot P(\mathbf{E}) \, dS = 0 \tag{5.11}$$

because of the following identity:

$$\mathbf{E} \cdot (\hat{n} \times \nabla \times \mathbf{E}) = -jk_0 Z_0 \mathbf{E} \cdot (\hat{n} \times \mathbf{H}) = -jk_0 Z_0 \mathbf{H} \cdot (\mathbf{E} \times \hat{n})$$

Therefore, the integral over S_0 in (5.9) vanishes for PEC and PMC surfaces comprising the inner boundaries of the volume V. Thus, the surface integral over S_0 reduces to the integral over the outer boundary or the mesh termination surface. However, for packaged structures which are bounded with electric walls (PEC surfaces), the integral expression over the outer boundary vanishes too. For open structures, this is not the case and leads to the use of absorbing boundary conditions (ABCs) which are explained in detail in Chapters 4 and 6.

To deal with anisotropic structures, the functional (5.9) undergoes a slight modification since the material properties of the object (permeability and permittivity) are now second-rank tensors rather than scalars. Equation (5.9) can therefore be written as

$$F(\mathbf{E}) = \int_V \left[(\nabla \times \mathbf{E}) \cdot \left\{ [\overline{\overline{\mu}}]^{-1} \cdot (\nabla \times \mathbf{E}) \right\} - k_0^2 \mathbf{E} \cdot \{ \overline{\overline{\epsilon}} \cdot \mathbf{E} \} \right] dV + jk_0 Z_0 \int_{S_k} \frac{1}{K} (\hat{n} \times \mathbf{E}) \cdot (\hat{n} \times \mathbf{E}) \, dS + \int_{S_0} \mathbf{E} \cdot P(\mathbf{E}) \, dS \tag{5.12}$$

where

$$\overline{\overline{\mu}} = \begin{bmatrix} \mu_{xx} & \mu_{xy} & \mu_{xz} \\ \mu_{yx} & \mu_{yy} & \mu_{yz} \\ \mu_{zx} & \mu_{zy} & \mu_{zz} \end{bmatrix} \tag{5.13}$$

and

$$\overline{\overline{\epsilon}} = \begin{bmatrix} \epsilon_{xx} & \epsilon_{xy} & \epsilon_{xz} \\ \epsilon_{yx} & \epsilon_{yy} & \epsilon_{yz} \\ \epsilon_{zx} & \epsilon_{zy} & \epsilon_{zz} \end{bmatrix} \tag{5.14}$$

The symmetry of the final system of equations now depends on the symmetry of $\overline{\overline{\mu}}$ and $\overline{\overline{\epsilon}}$. For packaged structures, when the surface integral over the outer mesh boundary S_0 vanishes, we are left with the functional

$$F(\mathbf{E}) = \int_V \left[(\nabla \times \mathbf{E}) \cdot \left\{ [\overline{\overline{\mu}}]^{-1} \cdot (\nabla \times \mathbf{E}) \right\} - k_0^2 \mathbf{E} \cdot \{ \overline{\overline{\epsilon}} \cdot \mathbf{E} \} \right] dV + jk_0 Z_0 \int_{S_k} \frac{1}{K} (\hat{n} \times \mathbf{E}) \cdot (\hat{n} \times \mathbf{E}) \, dS \tag{5.15}$$

The functional representation for open boundaries is the subject of the next chapter. The extremum of the functional can be found by using the Rayleigh-Ritz minimization procedure which amounts to differentiating F with respect to $\mathbf{E}$ and then setting it to zero, as explained in Chapter 1. In practice, the differentiation is done after introducing the expansion for $\mathbf{E}$. By setting the derivative with respect to each coefficient of the expansion to zero, a set of equations is obtained which is said to be stationary with respect to the first variation in F.

5.2.2 Potential Formulation

The finite element problem can also be formulated in terms of vector and scalar potentials. The potential formulation is attractive for obtaining reliable solutions throughout the frequency spectrum. It is well known that the field formulation is unstable for very low frequencies where the system of equations becomes nearly singular. In circuit and waveguide applications, characterization of low frequency design behavior could have significant importance. The potentials used as the solution variables are the magnetic vector potential **A** and the electric scalar potential Φ. The potentials are defined in terms of electric and magnetic fields as [14]

$$\nabla \times \mathbf{A} = \mathbf{B} \tag{5.16}$$

$$-j\omega(\nabla\Phi + \mathbf{A}) = \mathbf{E} \tag{5.17}$$

where **B** is the magnetic field density and ω is the angular frequency. Again, assuming a source-free domain and using Maxwell's equations, the curl-curl wave equation (5.1) can be rewritten as

$$\nabla \times \left(\frac{1}{\mu}\nabla \times \mathbf{A}\right) - \omega^2\epsilon(\mathbf{A} + \nabla\Phi) = 0 \tag{5.18}$$

The boundary conditions for perfect electric and perfect magnetic materials are a little more complicated than the field formulation. The perfect electric boundary conditions in terms of potentials are

$$\begin{aligned} \mathbf{A} \times \hat{n} &= \frac{j}{\omega}(\mathbf{E} \times \hat{n}) = 0 \\ \Phi &= 0 \end{aligned} \tag{5.19}$$

while the perfect magnetic condition reduces to

$$\left(\frac{1}{\mu}\nabla \times \mathbf{A}\right) \times \hat{n} = \mathbf{H} \times \hat{n} = 0 \tag{5.20}$$

The corresponding functional can then be formulated as before to yield

$$\begin{aligned} F(\mathbf{A}, \Phi) = {} & \frac{1}{2}\int_V \left[\frac{1}{\mu}\nabla \times \mathbf{A} \cdot \nabla \times \mathbf{A} - \omega^2\epsilon(\mathbf{A} + \nabla\Phi) \cdot \mathbf{A}\right] dV \\ & - \frac{1}{2}\int_V [\omega^2\epsilon(\mathbf{A} + \alpha^2\nabla\Phi) \cdot \nabla\Phi]\, dV \\ & - \frac{1}{2}\int_{S_0} (\mathbf{H} \times \hat{n}) \cdot \mathbf{A} + (\mathbf{H} \times \hat{n}) \cdot \nabla\Phi\, dS \end{aligned} \tag{5.21}$$

where α is a scalar parameter used to provide a gauge. The formulation is ungauged for $\alpha^2 = 1$. The gauge varies between 0 and 1 and is guaranteed to provide the same field values as the field formulation. However, it may be used to improve the numerical stability of the algorithm.

As in the field formulation, the surface integral over S_0 vanishes for PEC and PMC boundaries as well as for interior surfaces without transition conditions. The sheet boundary condition can also be applied rather easily as in the field formulation by setting

$$\hat{n} \times \frac{1}{K} \hat{n} \times \mathbf{A} = \frac{1}{jk_0 Z_0 \mu_r} \hat{n} \times \nabla \times \mathbf{A}$$
$$\Phi = 0 \tag{5.22}$$

where K is defined as the surface impedance. The functional given by (5.21) is discretized by using edge basis functions for representing the vector potential and nodal basis functions for the scalar potential. Use of the edge-based vector potential eliminates spurious modes (to be discussed in the next section) in the solution spectrum and also helps in the enforcement of boundary conditions on edges and corners of the desired structure. It should be noted that the potential formulation results in a larger solution space for an identical mesh topology than the field formulation. This is due to the extra nodal unknowns required to discretize the scalar potential Φ. However, as noted earlier in this section, the formulation allows robust analysis even for the lower frequencies in the analysis spectrum [15]. The solution for (5.21) is obtained by extremizing $F(\mathbf{A}, \Phi)$ with respect to $\mathbf{A}$ and Φ.

5.3 ORIGIN OF SPURIOUS SOLUTIONS

Conventional finite element basis functions give rise to spurious solutions when (5.31) is solved. As Wong and Cendes point out in [16], the origin of these spurious solutions lies in the infinitely degenerate eigenvalue $k = 0$ in the eigenvalue spectrum. Given the eigenvalue matrix system along with the PEC condition $\hat{n} \times \mathbf{E} = 0$ on the boundary, there exists an infinite number of scalar functions Φ such that $\Phi = 0$ on the boundary. Then $\mathbf{E} = -\nabla\Phi$ is a permitted eigenfunction corresponding to the eigenvalue $k = 0$. If the discretization scheme fails to model this infinite dimensional nullspace of the curl operator properly, spurious solutions to the eigenvalue problem will appear. This is also the reason why spurious solutions do not occur when the scalar wave equation is discretized since the nullspace of the Helmholtz operator is the trivial eigensolution $\mathbf{E} = 0$.

One way to get rid of spurious modes is to formulate the eigenvalue problem such that $k = 0$ is no longer a permissible eigenvalue. This is achieved by enforcing

$$\nabla \cdot \mathbf{E} = 0 \tag{5.23}$$

exactly everywhere in the source-free solution region. Then the only solution corresponding to the $k = 0$ eigenvalue is the trivial one $\mathbf{E} = 0$. In finite elements, solving an eigenproblem along with a constraint (5.23) is well known [17]. Researchers have mostly tried the penalty function approach of constrained minimization [18], [19] since it is simple to implement. However, the penalty approach is a mere *fix* and not a *cure* for the problem. Since the spurious eigenmodes are now shifted far into the visible spectrum, they are not completely eliminated and are dependent on a user-defined parameter which specifies how strongly the divergenceless condition (5.23) is imposed. The spurious modes do not really go away since the nullspace is still being modeled with very few variables when nodal elements are used.

Other than the penalty method, derivative continuous finite elements (C^1 elements) have also been proposed [16] to alleviate this problem. In this method, an auxiliary vector field ζ is introduced such that

$$\mathbf{E} = \nabla \times \zeta \tag{5.24}$$

Since substitution of (5.24) for **E** into (5.15) results in second derivatives, we need to construct first derivative continuous elements or C^1 elements. As shown in [16], discretization of **E** using node-based C^1 elements eliminates the problem of spurious solutions since the nullspace of the curl operator is modeled exactly. The added constraint of normal derivative continuity across inter-element boundaries provides the excess degrees of freedom to increase the size of the $\nabla\Phi$ subspace. However, C^1 elements are not commonly found in finite elements and may need to be explicitly derived for the problem at hand.

Another method of eliminating spurious modes, without getting rid of the eigenvalue $k = 0$, is by using edge elements [20]. Webb in [21] provides an elegant rationale as to why spurious modes do not appear with edge-based elements and why they are likely to be present with node-based vectorial elements.

Let us consider a single tetrahedral element. With a vector first-order node-based formulation, the element has 12 degrees of freedom (three each for the four vertices), i.e., the bases span a 12-dimensional space V. Since we are solving the curl-curl equation, the space of $\nabla\Phi$ must contain all vectors of polynomial order less than or equal to one. Therefore, Φ must be a second-order polynomial which needs ten degrees of freedom to be complete. However, discounting the constant term which has zero gradient, the dimension of the $\nabla\Phi$ subspace is nine. In a general mesh, the degrees of freedom belonging to the $\nabla\Phi$ subspace must satisfy both tangential continuity in Φ and normal continuity in $\nabla\Phi$. This requirement is usually not met in general unless the mesh is a specially constructed C^1 mesh. Thus, the elements belonging to the $\nabla\Phi$ subspace are relatively few.

In edge elements, however, the picture is different. The space V has dimension 6. More importantly, the dimension of the $\nabla\Phi$ subspace is only 3 since Φ needs to be only a first-order polynomial. Hence, the elements of the subspace of $\nabla\Phi$ must now satisfy *only* tangential continuity which is achieved by making Φ continuous across element boundaries. This is automatically satisfied by edge elements and, therefore, the $\nabla\Phi$ subspace is fully represented by the mesh entities. In fact, the dimension of the $\nabla\Phi$ subspace is one less than the total number of internal nodes in the mesh. Thus, the null space $\nabla\Phi$ of the curl operator is modeled more completely using edge elements rather than using node-based elements.

The above discussion conclusively proves that the root cause of spurious modes is the improper modeling of the nullspace of the curl operator. Any basis function which approximates it fully will be stable and free of spurious modes. As it turns out, conventional Lagrangian finite elements are unsuitable; either C^1 node-based elements or edge-based elements of any order can be used to obtain the true solutions.

5.4 MATRIX GENERATION AND ASSEMBLY

To discretize the electric field **E** within this volume, we subdivide the volume into small tetrahedra, rectangular bricks or any suitable three-dimensional element (see Fig. 3.3) each occupying the volume $V^e (e = 1, 2, \ldots, M)$, where M is the total number of elements. As in the previous chapters, we expand the discretization variable within each finite element in terms of polynomial basis functions. In this section, we will use the electric field as the unknown variable. Other electromagnetic quantities like currents and potentials can be expanded similarly.

Let us introduce the expansion (valid within the eth volume element)

$$\mathbf{E} = \sum_{j=1}^{m} E_j^e \mathbf{W}_j^e \tag{5.25}$$

where $\mathbf{W}_j^e$ are the edge-based vector basis functions and E_j^e denotes the expansion coefficients of the basis $\mathbf{W}_j^e$. The upper summation index m represents the number of edges comprising the element, and the superscript e stands for the element number. On substituting the expansion (5.25) into (5.15) and setting $\partial F(\mathbf{E})/\partial E_j = 0$, we obtain the system of equations

$$\sum_{e=1}^{M}[A^e]\{E^e\} - k_0^2 \sum_{e=1}^{M}[B^e]\{E^e\} - \sum_{e=1}^{M}[C^e]\{E^e\} = \{0\} \tag{5.26}$$

where

$$A_{ij}^e = \int_{V^e} \frac{1}{\mu_r} (\nabla \times \mathbf{W}_i^e) \cdot (\nabla \times \mathbf{W}_j^e)\, dv \tag{5.27}$$

$$B_{ij}^e = \int_{V^e} \epsilon_r \mathbf{W}_i^e \cdot \mathbf{W}_j^e\, dv \tag{5.28}$$

$$C_i^e = jk_0 Z_0 \left[\oint_{S^e} \mathbf{W}_i^e \cdot (\hat{\mathbf{n}} \times \mathbf{H})\, ds - \int_{S_k^e} \frac{1}{K} (\hat{n} \times \mathbf{W}_i^e) \cdot (\hat{n} \times \mathbf{W}_j^e)\, dS \right] \tag{5.29}$$

In the preceding equations, all matrices and vectors following the summation sign have been augmented using global numbers.

The surface area S^e indicates the boundary of element e. As mentioned earlier, all free faces, i.e., faces on the boundary, have zero contribution to the surface integral since we consider only packaged structures in this chapter. This reduces the original unknown count and eliminates the need to generate equations for those edges/unknowns which would otherwise have to be included in the solution. We can further simplify the surface integral contribution to $[C^e]$ by taking advantage of the inter-element continuity afforded by finite element basis functions. Due to the continuity of tangential $\mathbf{H}$ at the interface between two elements, an element face lying inside the body does not contribute to the integral over S^e in the final assembly of the element equations. As a result, the last term of (5.26) merely reduces to the integral contribution over imperfectly conducting or impedance surfaces (S_k).

For simplicity, let us assume that no impedance surfaces exist in the structure. As will become clear in the derivation below, it is difficult to solve the problem in the presence of impedance structures. With all surface integrals reduced to zero, we are left with the equation

$$[A]\{E\} - k_0^2[B]\{E\} = 0 \tag{5.30}$$

which corresponds to the generalized eigenvalue problem for finding cavity resonances. The matrices $[A]$ and $[B]$ are $N \times N$ symmetric, sparse matrices with N being the total number of edges resulting from the subdivision of the structure excluding the edges on the boundary. Their entries are given by (5.27) and (5.28). As usual, $\{E\}$ is a column vector of order N denoting the coefficients of edge bases and $\lambda = k_0^2$ is the eigenvalues of the system.

For mathematical treatment, we rewrite (5.30) as

$$[A]\{E\} = \lambda[B]\{E\} \tag{5.31}$$

whose solution will yield the resonant field distribution $\{E\}$ and the corresponding wavenumber k_0. The inclusion of impedance structures in (5.31) would require us to solve a nonlinear eigenproblem given by

$$[A]\{E\} = \{k_0[R] + k_0^2[B]\}\{E\} \tag{5.32}$$

where k_0 is the desired eigenvalue and $[R]$ is the contribution from the surface impedance terms. This problem is usually solved by obtaining an initial guess from the solution of the linear eigenproblem and subsequently using this solution to find the true solution iteratively. For more discussion on such a solution process, the interested reader is referred to [22].

As mentioned in Chapters 3 and 4, the finite element matrices are sparse and can be filled very quickly. If the element shapes used for discretization are simple, then the local element matrices A^e and B^e can be derived analytically and used to fill the global system after taking the proper numbering into account. For tetrahedra using $H_0(curl)$ elements (see Chapter 2), the element matrices are given by

$$[A^e] = \frac{1}{\mu_r}\frac{1}{9V^e}\begin{bmatrix} (\mathbf{b}_6, \mathbf{b}_6) & -(\mathbf{b}_6, \mathbf{b}_5) & (\mathbf{b}_6, \mathbf{b}_4) & (\mathbf{b}_6, \mathbf{b}_3) & -(\mathbf{b}_6, \mathbf{b}_2) & (\mathbf{b}_6, \mathbf{b}_1) \\ -(\mathbf{b}_5, \mathbf{b}_6) & (\mathbf{b}_5, \mathbf{b}_5) & -(\mathbf{b}_5, \mathbf{b}_4) & -(\mathbf{b}_5, \mathbf{b}_3) & (\mathbf{b}_5, \mathbf{b}_2) & -(\mathbf{b}_5, \mathbf{b}_1) \\ (\mathbf{b}_4, \mathbf{b}_6) & -(\mathbf{b}_4, \mathbf{b}_5) & (\mathbf{b}_4, \mathbf{b}_4) & (\mathbf{b}_4, \mathbf{b}_3) & -(\mathbf{b}_4, \mathbf{b}_2) & (\mathbf{b}_4, \mathbf{b}_1) \\ (\mathbf{b}_3, \mathbf{b}_6) & -(\mathbf{b}_3, \mathbf{b}_5) & (\mathbf{b}_3, \mathbf{b}_4) & (\mathbf{b}_3, \mathbf{b}_3) & -(\mathbf{b}_3, \mathbf{b}_2) & (\mathbf{b}_3, \mathbf{b}_1) \\ -(\mathbf{b}_2, \mathbf{b}_6) & (\mathbf{b}_2, \mathbf{b}_5) & -(\mathbf{b}_2, \mathbf{b}_4) & -(\mathbf{b}_2, \mathbf{b}_3) & (\mathbf{b}_2, \mathbf{b}_2) & -(\mathbf{b}_2, \mathbf{b}_1) \\ (\mathbf{b}_1, \mathbf{b}_6) & -(\mathbf{b}_1, \mathbf{b}_5) & (\mathbf{b}_1, \mathbf{b}_4) & (\mathbf{b}_1, \mathbf{b}_3) & -(\mathbf{b}_1, \mathbf{b}_2) & (\mathbf{b}_1, \mathbf{b}_1) \end{bmatrix} \tag{5.33}$$

where $(\mathbf{b}_i, \mathbf{b}_j)$ denotes the dot product of the edge vectors $\mathbf{b}_i$ and $\mathbf{b}_j$ of the tetrahedral element. Referring to Fig. 2.6 and Table 2.4 as an example, we note that

$$\mathbf{b}_1 = (x_2 - x_1)\hat{x} + (y_2 - y_1)\hat{y} + (z_2 - z_1)\hat{z}$$
$$\mathbf{b}_2 = (x_3 - x_1)\hat{x} + (y_3 - y_1)\hat{y} + (z_3 - z_1)\hat{z}$$

The element matrix $[B^e]$ for a linear edge basis expansion on a tetrahedron is given by [23]

$$[B^e] = \frac{\epsilon_r}{180 V^e}$$

$$\begin{bmatrix} 2(I_{11} - I_{12} + I_{22}) & I_{11} - I_{12} - I_{13} + 2I_{23} & I_{11} - I_{12} - I_{14} + 2I_{24} & I_{12} - I_{22} - 2I_{13} + I_{23} & I_{12} - I_{22} - 2I_{14} + I_{24} & I_{13} - I_{23} - I_{14} + I_{24} \\ I_{11} - I_{12} - I_{13} + 2I_{23} & 2(I_{11} - I_{13} + I_{33}) & I_{11} - I_{13} - I_{14} + 2I_{34} & 2I_{12} - I_{23} - I_{13} + I_{33} & I_{12} - I_{23} - I_{14} + I_{34} & I_{13} - I_{33} - 2I_{14} + I_{34} \\ I_{11} - I_{12} - I_{14} + 2I_{24} & I_{11} - I_{13} - I_{14} + 2I_{34} & 2(I_{11} - I_{14} + I_{44}) & I_{12} - I_{24} - I_{13} + I_{34} & 2I_{12} - I_{24} - I_{14} + I_{44} & 2I_{13} - I_{34} - I_{14} + I_{44} \\ I_{12} - I_{22} - 2I_{13} + I_{23} & 2I_{12} - I_{23} - I_{13} + I_{33} & I_{12} - I_{24} - I_{13} + I_{34} & 2(I_{22} - I_{23} + I_{33}) & I_{22} - I_{23} - I_{24} + 2I_{34} & I_{23} - I_{33} - 2I_{24} + I_{34} \\ I_{12} - I_{22} - 2I_{14} + I_{24} & I_{12} - I_{23} - I_{14} + I_{34} & 2I_{12} - I_{24} - I_{14} + I_{44} & I_{22} - I_{23} - I_{24} + 2I_{34} & 2(I_{22} - I_{24} + I_{44}) & 2I_{23} - I_{34} - I_{24} + I_{44} \\ I_{13} - I_{23} - I_{14} + I_{24} & I_{13} - I_{33} - 2I_{14} + I_{34} & 2I_{13} - I_{34} - I_{14} + I_{44} & I_{23} - I_{33} - 2I_{24} + I_{34} & 2I_{23} - I_{34} - I_{24} + I_{44} & 2(I_{33} - I_{34} + I_{44}) \end{bmatrix} \tag{5.34}$$

with $I_{ij} = \mathbf{F}_i \cdot \mathbf{F}_j$. Here, $\mathbf{F}_i$ is the inward oriented normal vector to the tetrahedron's face opposite to node i and has an amplitude equal to the area of the triangular face.

The vector edge-based expansion functions for rectangular bricks were presented in [24], [25]. These basis functions were reviewed in Chapter 2 and consist of 12 unknowns for each of the 12 edges in the rectangular brick element. They are rather simple to derive analytically and are presented here for the sake of completeness. The $[A^e]$ elemental matrix is given by

$$[A^e] = \frac{V^e}{6\mu_r} \begin{bmatrix} \frac{[K_1]}{h_y^2} + \frac{[K_2]}{h_z^2} & -\frac{h_z}{V^e}[K_3]^T & -\frac{h_y}{V^e}[K_3]^T \\ -\frac{h_z}{V^e}[K_3]^T & \frac{[K_1]}{h_z^2} + \frac{[K_2]}{h_x^2} & -\frac{h_x}{V^e}[K_3] \\ -\frac{h_y}{V^e}[K_3] & -\frac{h_x}{V^e}[K_3]^T & \frac{[K_1]}{h_x^2} + \frac{[K_2]}{h_y^2} \end{bmatrix} \tag{5.35}$$

where the square matrices $[K_1]$, $[K_2]$, and $[K_3]$ are

$$[K_1] = \begin{bmatrix} 2 & -2 & 1 & -1 \\ -2 & 2 & -1 & 1 \\ 1 & -1 & 2 & -2 \\ -1 & 1 & -2 & 2 \end{bmatrix} \tag{5.36}$$

$$[K_2] = \begin{bmatrix} 2 & 1 & -2 & -1 \\ 1 & 2 & -1 & -2 \\ -2 & -1 & 2 & 1 \\ -1 & -2 & 1 & 2 \end{bmatrix} \tag{5.37}$$

$$[K_3] = \begin{bmatrix} 2 & 1 & -2 & -1 \\ -2 & -1 & 2 & 1 \\ 1 & 2 & -1 & -2 \\ -1 & -2 & 1 & 2 \end{bmatrix} \tag{5.38}$$

and the notation $[\,]^T$ denotes the matrix transpose (see Fig. 2.11 for the definition of parameters and node specification).[2] Also, $h_{x,y,z}$ denotes the length of the edges as shown in Fig. 2.11. Substituting the values of $[K_1]$, $[K_2]$, and $[K_3]$ in (5.35) yields a (12×12) system for the $[A^e]$ element matrix. The values for the matrix $[B^e]$ are given by

$$[B^e] = \frac{\epsilon_r V^e}{36} \begin{bmatrix} [L_1] & 0 & 0 \\ 0 & [L_1] & 0 \\ 0 & 0 & [L_1] \end{bmatrix} \tag{5.39}$$

where

$$[L_1] = \begin{bmatrix} 4 & 2 & 2 & 1 \\ 2 & 4 & 1 & 2 \\ 2 & 1 & 4 & 2 \\ 1 & 2 & 2 & 4 \end{bmatrix} \tag{5.40}$$

In general, the implementation of the finite element discretization involves two numbering systems, and thus some unique global edge direction must be defined to ensure the continuity of $\hat{\mathbf{n}} \times \mathbf{E}$ across all edges [26]. Here, we choose this direction to be coincident with the edge vector pointing from the smaller to the larger global node number. This convention ensures that the edge directions are consistent with the local and global numbering of the unknowns.

5.5 SOURCE MODELING

To model microwave circuits, we usually need sources at the input ports to excite the circuit. The source modeling issue is critical since it provides the input boundary condition for the problem. A small error in the source model could lead to large errors in the 3D simulation of circuit parameters. In providing the source model for a microwave circuit, it is assumed that the field is already accurately solved (stabilized) throughout the volume of the configuration. A rather simplistic way of establishing the stabilized mode pattern would be to use a voltage source excitation at the input

[2]The reader is cautioned that the matrix given in [25] contains misprints which have been corrected in (5.35).

port and use a feeder transmission line of suitable length to excite the actual circuit. The disadvantage of this method is that the feeder structure increases problem size and wastes valuable computer resources. Moreover, complicated transmission line configurations cannot be adequately modeled and evanescent modes cannot be included. In this section, we will be talking about the transfinite element method in modeling the source excitation for circuit applications [27], [28].

The transfinite element method matches the projection of the desired modal fields at the port with the fields obtained from the three-dimensional solution inside the structure. Thus, it allows the specification of the desired modal excitations—propagating as well as evanescent—for obtaining the circuit parameters in a systematic manner. The modal patterns are obtained from an eigensolution of the 2.5D problem, as discussed in Chapter 4. Since the dominant eigenvalues for the ports can be obtained relatively cheaply and the feed line of the source model mentioned earlier can be eliminated completely, the transfinite element method is very efficient numerically. The following derivation for modeling the source excitation borrows heavily from [28].

The problem at hand is to characterize the scattering parameters of an N-port circuit. At first, we assume for simplicity that port 1 is excited with the dominant mode. We also assume that the normal direction of propagation is the z-axis. The electric field at each port surface can then be described as

$$\mathbf{E}^{(1)} = \mathbf{E}^{\text{inc}} + \sum_{j=1}^{\infty} a_{1j} \boldsymbol{M}_{1j} e^{-\gamma_{1j} z} \tag{5.41}$$

$$\mathbf{E}^{(i)} = \sum_{j=1}^{\infty} a_{ij} \boldsymbol{M}_{ij} e^{-\gamma_{ij} z}, \qquad \text{for } i = 2, \ldots, N \tag{5.42}$$

where

$$\mathbf{E}^{\text{inc}} = \boldsymbol{M}_{11} e^{\gamma_{11} z} \tag{5.43}$$

since only the dominant mode excitation is being considered for now. In (5.42), $\boldsymbol{M}_{ij}$ is the modal pattern (eigenvector) for mode j at port i, γ_{ij} is the propagation constant of the jth mode in the ith port and a_{ij} are the scattering coefficients to be determined in the analysis. As noted earlier, the modal information is obtained from a 2.5D analysis of the port eigenvalue problem.

It remains to match the fields at the ports to the edge-based volume finite elements in the interior of the domain. Using the continuity condition for tangential electric fields at material interfaces, we set the projection of the modal approximation equal to projections on the port plane from the finite element basis functions inside the computational volume. Tangential continuity of edge-based finite elements across inter-element boundaries simplifies the field matching condition. Moreover, continuity of the tangential magnetic field at the port surface is automatically satisfied since it is a natural boundary condition of the electric field formulation. The field matching condition is imposed through the surface integrals on the port planes. Assuming unit power at each port, we can normalize the surface integrals for each mode on port surfaces such that

$$-\int_{S_i} \frac{1}{j\omega\mu} \mathbf{E}_{ij} \times \nabla \times \mathbf{E}_{ij} \bigg|_{z=0} \cdot \hat{z}\, dS = s_{ij} \tag{5.44}$$

where S_i is the aperture surface at the ith port and

$$s_{ij} = \begin{cases} 1 & \text{for propagating modes} \\ j = \sqrt{-1} & \text{for evanescent modes} \end{cases} \tag{5.45}$$

On substituting (5.44) into the surface integral of the electric field functional, we obtain

$$\int_{S_i} \frac{1}{\mu_r} (\mathbf{E} \times \nabla \times \mathbf{E}) \cdot \hat{z}\, dS = -j\omega\mu_0 \sum_{j=1}^{\text{No. of Modes}} s_{ij}(a_{ij})^2 \tag{5.46}$$

$$\int_{S_1} \frac{1}{\mu_r} (\mathbf{E} \times \nabla \times \mathbf{E}) \cdot \hat{z}\, dS = j\omega\mu_0 \left(1 - \sum_{j=1}^{\text{No. of Modes}} s_{ij}(a_{ij})^2\right) \tag{5.47}$$

Thus the boundary integrals over the port surfaces are obtained analytically. The obvious advantage of this technique is that the scattering parameters—a_{ij}—are obtained as a result of the solution process and are, therefore, variational. Moreover, all the unknowns on the port collapse to the desired scattering parameters for the 3D solution. Of course, the modal solution needs to be computed initially from a 2.5D analysis of the port eigenproblem.

The functional obtained by enforcing the mode matching condition at the ports is outlined in [28]. The resulting system of equations is derived by separating the interior domain from the exterior and coupling the known modal solutions at the ports through the surface integrals shown in (5.47). The final system of equations is given by extremizing the functional by the Rayleigh-Ritz technique and is expressed in matrix form as

$$\begin{bmatrix} [A^{\text{int}}] & [A_{iS_1}]\{M_1\} & [A_{iS_2}]\{M_2\} \\ \{M_1\}[A_{iS_1}] & \{M_1\}[A_{S_1S_1}]\{M_1\} + j\omega\mu[s]_1 & \{M_1\}[A_{S_1S_2}]\{M_2\} \\ \{M_2\}[A_{iS_2}] & \{M_2\}[A_{S_2S_1}]\{M_1\} & \{M_2\}[A_{S_2S_2}]\{M_2\} + j\omega\mu[s]_2 \end{bmatrix} \cdot \begin{Bmatrix} \{E\} \\ \{a_1\} \\ \{a_2\} \end{Bmatrix} = \begin{Bmatrix} [A_{S_1}]\{M_{11}\} \\ \{M_{11}\}[A_{S_1S_1}]\{M_{11}\} + j\omega\mu_0 \\ \{M_{21}\}[A_{S_2S_1}]\{M_{11}\} \end{Bmatrix} \tag{5.48}$$

In defining the matrix entries, the subscript i denotes an interior edge and the subscripts S_1 and S_2 denote edges on ports 1 and 2, respectively. The matrix entries of $[A^{\text{int}}]$ are the usual volume integral representation given by

$$A_{jk}^{\text{int}} = \int_V \left(\frac{1}{\mu_r} \nabla \times \mathbf{W}_j \cdot \nabla \times \mathbf{W}_k - k_0^2 \epsilon_r \mathbf{W}_j \cdot \mathbf{W}_k\right) dV \tag{5.49}$$

The column $\{E\}$ represents the interior volume field coefficients as in (5.31), whereas $\{a_1\}$ and $\{a_2\}$ stand for the unknown modal coefficients of the expansions (5.41) and (5.42). Thus, the transfinite method gives rise to a partly sparse and a partly dense matrix. However, the dimension of the dense part is limited to the sum of the number of modes per port and the number of ports in the structure. It is, therefore, always much smaller than the sparse portion of the matrix which represents the volume unknowns in the geometry. Other circuit excitations, like a voltage or a current source, can be modeled in the usual way by placing a voltage or current element and integrating over the volume of the element or elements occupied by the source.

More details on source modeling are given in Chapter 7 for antenna and scattering applications. The integrals over the sources are given in (5.4) and (5.5).

5.6 APPLICATIONS

Finite elements in closed domains have a variety of applications, especially for analyzing cavity resonances, modeling packaging discontinuities and transitions, characterizing shielded circuit configurations, and in analyzing inhomogeneous waveguides with arbitrary material filling. In the subsequent sections, a few examples are mentioned to illustrate the highlights of the technique.

5.6.1 Cavity Resonators

Solving Maxwell's equations for the resonances of a closed cavity is important in understanding and controlling the operation of many devices, including particle accelerators, microwave filters, and microwave ovens.

The rectangular cavity is a simple structure, but is widely used for feeding complex microwave devices. In Table 5.1, we present a comparison of the percentage error in the computation of eigenvalues for a 1 cm × 0.5 cm × 0.75 cm rectangular cavity using edge-based rectangular bricks and tetrahedra. The edge-based formulation using tetrahedral elements predicts the first six distinct non-trivial eigenvalues with less than four percent error and is seen to provide better accuracy than rectangular brick elements. Both the tetrahedral and the brick elements used in the computation are H^0 (*curl*) elements. The maximum edge length for the rectangular brick elements is 0.15 cm whereas that for the tetrahedral elements is 0.2 cm. Figure 5.2 shows that the tetrahedral elements have slightly less error when the same number of unknowns with bricks are used for modeling the rectangular cavity. However, it cannot be categorically stated that the brick elements are better than tetrahedrals for modeling rectangular structures. The primary advantage of tetrahedra lies in their generality in being able to automatically mesh arbitrary structures. This is

TABLE 5.1 Eigenvalues (k_0, cm^{-1}) for an Empty 1 cm × 0.5 cm × 0.75 cm Rectangular Cavity

Mode	Analytical	Computed (bricks) 270 Unknowns	Computed (tetra.) 260 Unknowns	Error (%) (bricks)	Error (%) (tetra.)
TE_{101}	5.236	5.307	5.213	−1.36	.44
TM_{110}	7.025	7.182	6.977	−2.23	.70
TE_{011}	7.531	7.725	7.474	−2.58	1.00
TE_{201}		7.767	7.573	−3.13	−.56
TM_{111}	8.179	8.350	7.991	−2.09	2.29
TE_{111}		8.350	8.122	−2.09	.70
TM_{210}	8.886	9.151	8.572	−2.98	3.53
TE_{102}	8.947	9.428	8.795	−5.38	1.70

Source: After Chatterjee et al. [29], © IEEE, 1992.

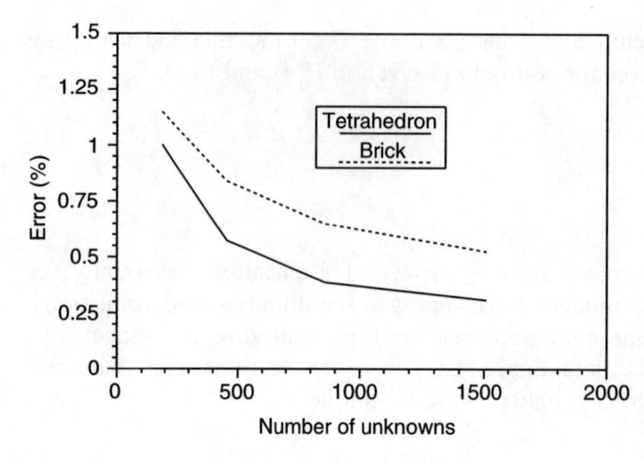

Figure 5.2 Performance comparison of rectangular bricks and tetrahedrals. [*After Chatterjee et al.* [*29*], © *IEEE, 1992.*]

not true for rectangular bricks unless staircasing is permitted to model curved surfaces as is done in the finite difference method. Bricks are used primarily for convenience in meshing and can lead to a reduction in unknowns in solids with rectangular cross sections.

For our next example, we compute the eigenvalues of a rectangular cavity, half of which is filled with a dielectric with permittivity $\epsilon_r = 2$. In Table 5.2 we compare the exact eigenvalues with those computed using edge-based tetrahedral finite elements. The exact eigenvalues of the half-filled cavity as described in Table 5.2 are computed by solving the transcendental equation obtained upon matching the tangential electric and magnetic fields at the air-dielectric interface. As seen, these results agree with those predicted by the finite element solution to within one percent.

TABLE 5.2 Eigenvalues (k_0, cm^{-1}) for a Half-Filled 1 cm × 0.1 cm × 1 cm Rectangular Cavity Having a Dielectric Filling of $\epsilon_r = 2$ Extending from $z = 0.5$ cm to $z = 1.0$ cm

Mode	Analytical	Computed 192 Unknowns	Error (%)
TEz_{101}	3.538	3.534	.11
TEz_{201}	5.445	5.440	.10
TEz_{102}	5.935	5.916	.32
TEz_{301}	7.503	7.501	.04
TEz_{202}	7.633	7.560	.97
TEz_{103}	8.096	8.056	.50

Finally, Table 5.3 presents the eigenvalues of the geometry illustrated in Fig. 5.3. This is a closed metallic cavity with a ridge along one of its faces. Note that even with a relatively coarse initial mesh (267 unknowns), the dominant eigenvalues are recovered with less than two percent error. However, a much finer mesh is needed to obtain a reasonable approximation to the modal field pattern or the eigenvector of the geometry.

As the degeneracy of the eigenvalues increases, the eigenvalue problem becomes increasingly ill-conditioned and the numerical solution is correspondingly

TABLE 5.3 Ten Lowest Nontrivial Eigenvalues (k_0, cm^{-1}) for the Ridged Waveguide Geometry: (a) 267 Unknowns; (b) 671 Unknowns

No.	(a)	(b)
1	4.941	4.999
2	7.284	7.354
3	7.691	7.832
4	7.855	7.942
5	8.016	7.959
6	8.593	8.650
7	8.906	8.916
8	9.163	9.103
9	9.679	9.757
10	9.837	9.927

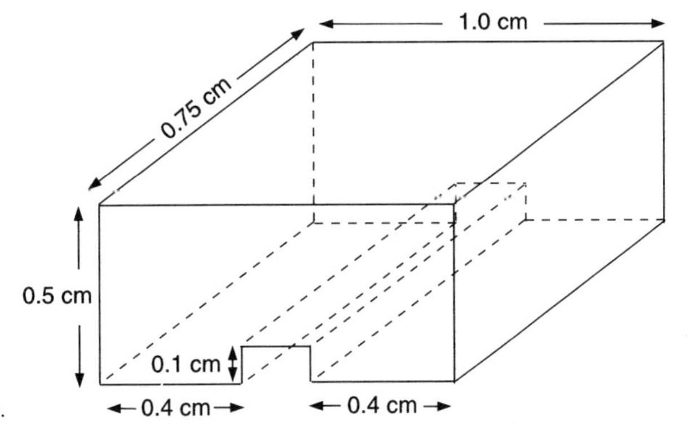

Figure 5.3 Geometry of ridged cavity.

less accurate [30]. Therefore, for the partially filled rectangular cavity, the absence of degenerate modes gives results which are accurate to within one percent of the exact eigensolutions. As expected, the solution yields a set of eigenvalues equal to the degrees of freedom (unknowns). Of these, there is an inherent presence of zero eigenvalues the number of which equals the number of internal nodes. These eigenvalues correspond to the dimension of the nullspace of the curl–curl operator and were explained in an earlier section. The zero eigenvalues are easily identifiable, and because they do not correspond to physical modes, they are always discarded.

5.6.2 Circuit Applications

In this section, we present two examples of using the closed domain formulation for modeling circuit transitions and packaging in microwave circuits. In Fig. 5.4, a 50 Ω coplanar waveguide (CPW) to a 50 Ω microstrip line transition is modeled in a packaged environment. The transition uses both surfaces of the GaAs substrate to print the microstrip and CPW metallizations. This is done for achieving better packing density in high frequency design. To reduce cross-talk, one of the substrate surfaces hosts CPW components and the other surface contains microstrip elements.

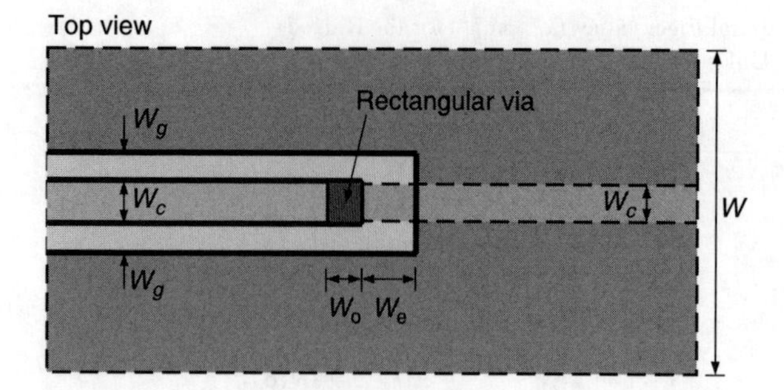

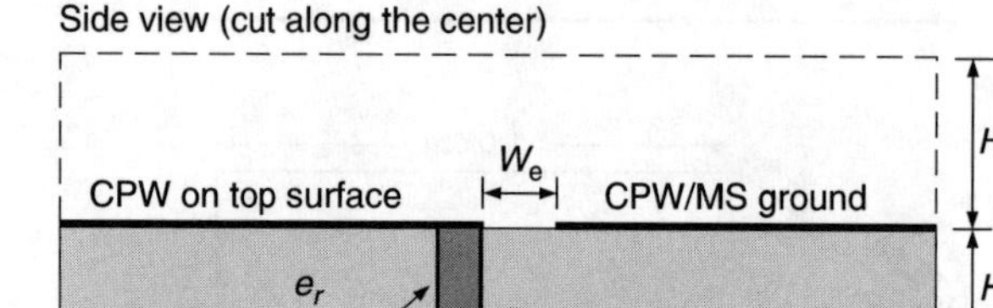

Figure 5.4 Geometry of two-layer microstrip-CPW transition on GaAs substrate ($\epsilon_r = 12.9$) with a rectangular via hole. Dimensions are as follows: $W_g = 50\,\mu$m; $W_c = W_0 = 75\,\mu$m; $W_e = 200\,\mu$m; $W = 1000\,\mu$m; $H_1 = 100\,\mu$m; $H_2 = H_3 = 400\,\mu$m. [*After Yook et al.* © *IEEE* [*31*].]

This configuration serves two purposes: first, the CPW ground serves as the microstrip ground and second, the CPW aperture and the top conductor in the microstrip run in parallel with the separation to reduce cross-talk. Without the rectangular via hole, the transition geometry has a significant cross-talk of about −15 dB at 20 GHz. However, as the frequency increases, the transmission coefficient (S_{21}) increases to about −4 dB due to radiation effects from the open ends of the microstrip and the CPW. On drilling a via hole connecting the CPW and the microstrip, the crosstalk levels stay below 20 dB for a wider frequency range (from 20 GHz to 50 GHz), as shown in Fig. 5.5. This is just one of a wide range of applications where a full-wave three-dimensional solution can be used to meet design specifications for critical parts of a complicated circuit.

Another type of common inter-chip feed-through is the hermetic bead transition. In Fig. 5.6, a coax-to-microstrip transition is modeled by approximating the circular coax cross section with a rectangular stripline configuration. Dimensions are chosen such that the low-frequency stripline impedance is approximately 50 Ω. The dielectric filling of the bead has the same permittivity ($\epsilon_r = 10.8$) as the substrate. The dielectric bead is hermetically enclosed within a PEC wall of thickness 1.5 mm and the air gap spacing is chosen to be 0.4 mm. Predictably, the air gap thickness governs the insertion loss at lower frequencies with the loss increasing as the gap spacing is enlarged. At higher frequencies, impedance mismatch will further degrade interconnect performance.

The reflection and insertion losses for the hermetic transition were computed by Yook et al. [32] using the finite element method with first order edge-based

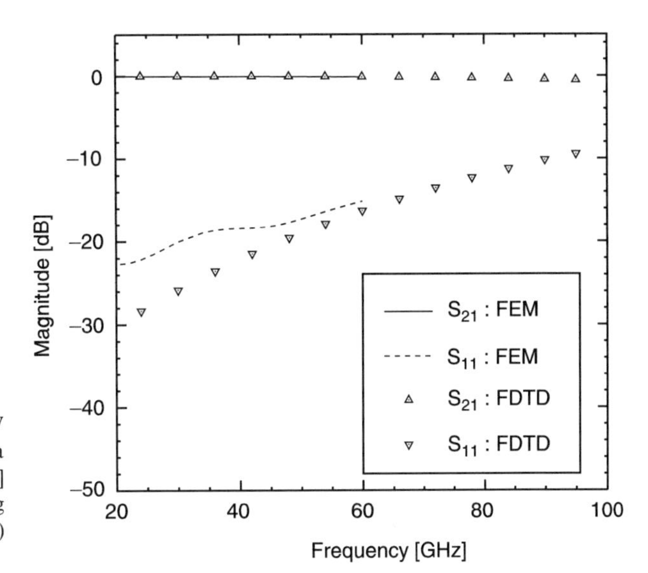

Figure 5.5 Scattering parameters of the CPW to microstrip transition with rectangular via connection. [*After Yook et al.* © *IEEE* [*31*].] Symbols are reference data computed using the Finite Difference–Time Domain (FDTD) Method.

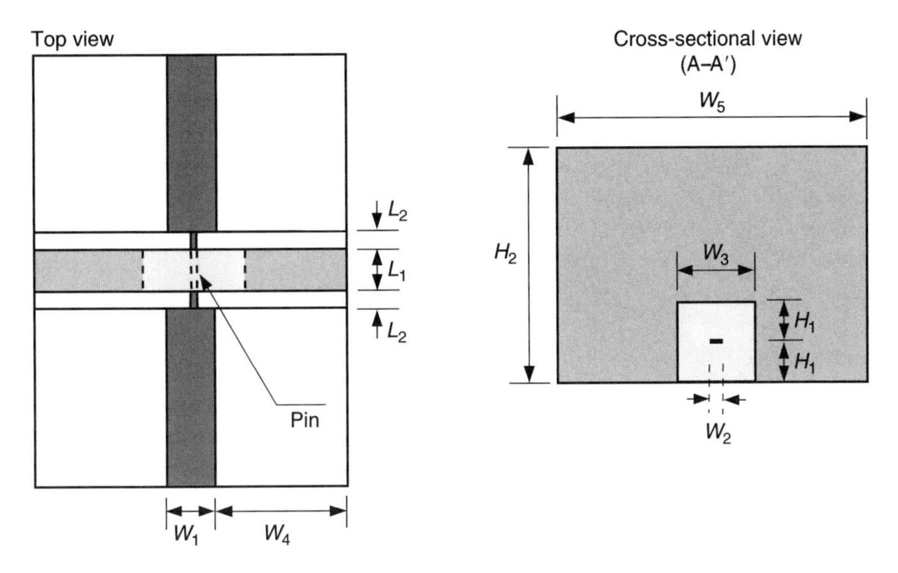

Figure 5.6 Hermetic bead transition in electronic packaging. Dimensions are as follows: $W_1 = 0.55\,\text{mm}$; $W_2 = 0.21\,\text{mm}$; $W_3 = 1.27\,\text{mm}$; $W_4 = 2.225\,\text{mm}$; $W_5 = 5\,\text{mm}$; $H_1 = 0.635\,\text{mm}$; $H_2 = 4\,\text{mm}$; $L_1 = 1.5\,\text{mm}$; $L_2 = 0.4\,\text{mm}$; $\epsilon_r = 10.8$. [*After Yook et al.* © *IEEE* [*32*].]

tetrahedral elements. The results are given in Fig. 5.7 and as expected the insertion loss increases from close to 0 dB at 10 GHz to nearly −3 dB at 25 GHz. To lower the insertion loss, the air gap spacing can be decreased and the geometrical parameters can be optimized to reduce the impedance mismatch at higher frequencies.

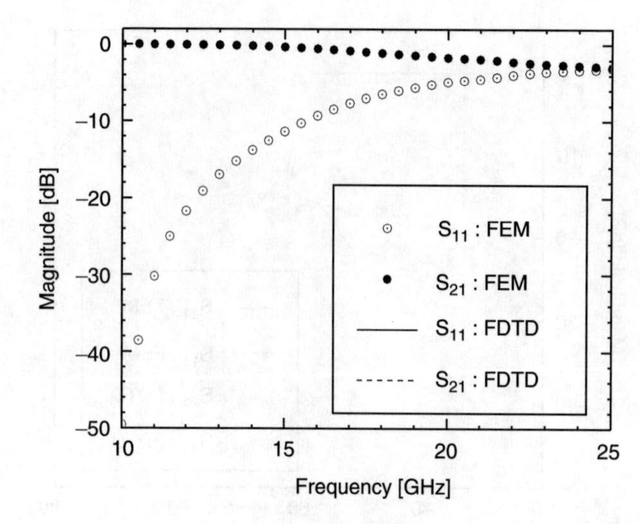

Figure 5.7 Scattering parameters of the hermetic bead transition shown in Fig. 5.6. [*After Yook et al.* © *IEEE* [32].]

APPENDIX: EDGE-BASED RIGHT TRIANGULAR PRISMS

In this appendix we present the matrix entries needed for finite element analysis using right triangular prisms (see Chapter 2 for their basis functions). Figure 5.8 shows a right triangular prism as an edge-based vector finite element [33], [34]. The top and bottom surfaces are identical and parallel to each other, while the vertical arms are perpendicular to the base of the prism. The vector electric field inside the element is

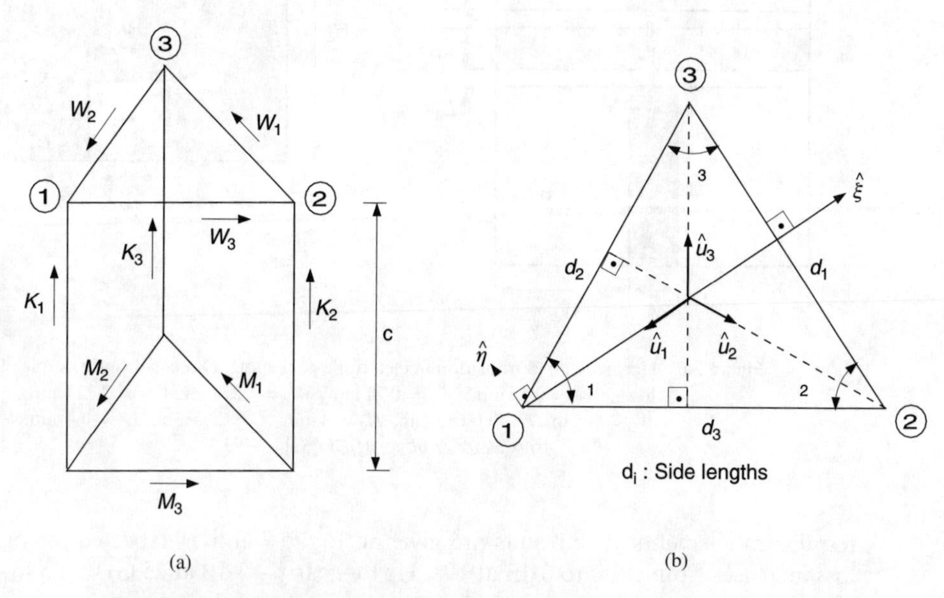

Figure 5.8 Right-prism with edge-based unknowns: (a) perspective view; (b) top view. [*Courtesy of T. Özdemir.*]

an interpolation among the nine vector unknowns each parallel to and constant along a particular edge of the prism.

The prism is specified by its height, the lengths of, and the angle between two sides of its triangular surface, namely, c, d_2, d_3, and α_1, respectively. First, we need to compute some scalar and vector quantities that will be used later in computing the matrix elements. These are illustrated in Fig. 5.8 and given by

$$d_1 = \sqrt{d_2^2 + d_3^2 - 2d_2 d_3 \cos\alpha_1} \tag{5.50}$$

$$\alpha_2 = \cos^{-1}\left(\frac{d_1^2 + d_3^2 - d_2^2}{2d_1 d_3}\right) \tag{5.51}$$

$$\alpha_3 = \cos^{-1}\left(\frac{d_1^2 + d_2^2 - d_3^2}{2d_1 d_2}\right) \tag{5.52}$$

$$h_1 = d_2 \sin\alpha_3 \tag{5.53}$$

$$h_2 = d_3 \sin\alpha_1 \tag{5.54}$$

$$h_3 = d_1 \sin\alpha_2 \tag{5.55}$$

$$\hat{u}_1 = -\hat{\xi} \tag{5.56}$$

$$\hat{u}_2 = \cos\alpha_3 \hat{\xi} - \sin\alpha_3 \hat{\eta} \tag{5.57}$$

$$\hat{u}_3 = \cos\alpha_2 \hat{\xi} + \sin\alpha_2 \hat{\eta} \tag{5.58}$$

The edge-based basis functions in their final forms are given by (see also Chapter 2)

$$\mathbf{W}_1^e = d_1\left(L_2^e \frac{\hat{u}_3}{h_3} - L_3^e \frac{\hat{u}_2}{h_2}\right)(z/c) \tag{5.59}$$

$$\mathbf{W}_2^e = d_2\left(L_3^e \frac{\hat{u}_1}{h_1} - L_1^e \frac{\hat{u}_3}{h_3}\right)(z/c) \tag{5.60}$$

$$\mathbf{W}_3^e = d_3\left(L_1^e \frac{\hat{u}_2}{h_2} - L_2^e \frac{\hat{u}_1}{h_1}\right)(z/c) \tag{5.61}$$

$$\mathbf{M}_1^e = d_1\left(L_2^e \frac{\hat{u}_3}{h_3} - L_3^e \frac{\hat{u}_2}{h_2}\right)(1 - z/c) \tag{5.62}$$

$$\mathbf{M}_2^e = d_2\left(L_3^e \frac{\hat{u}_1}{h_1} - L_1^e \frac{\hat{u}_3}{h_3}\right)(1 - z/c) \tag{5.63}$$

$$\mathbf{M}_3^e = d_3\left(L_1^e \frac{\hat{u}_2}{h_2} - L_2^e \frac{\hat{u}_1}{h_1}\right)(1 - z/c) \tag{5.64}$$

$$\mathbf{K}_1^e = \hat{z}\, L_1^e \tag{5.65}$$

$$\mathbf{K}_2^e = \hat{z}\, L_2^e \tag{5.66}$$

$$\mathbf{K}_3^e = \hat{z}\, L_3^e \tag{5.67}$$

where

$$L_1^e(\xi, \eta) = 1 - \frac{1}{h_1}\,\xi$$
$$L_2^e(\xi, \eta) = \frac{\cos\alpha_3}{h_2}\,\xi - \frac{\sin\alpha_3}{h_2}\,\eta$$
$$L_3^e(\xi, \eta) = \frac{\cos\alpha_2}{h_3}\,\xi + \frac{\sin\alpha_2}{h_3}\,\eta$$

are the usual nodal shape functions for a triangle in the (ζ, η) coordinate frame as defined in Fig. 5.8(b).

The relevant quantities for constructing the element matrix for the prism are

$$\begin{aligned} EWWC_{i\ell} &= \iiint_{V^e} (\nabla \times \mathbf{W}_i^e) \cdot (\nabla \times \mathbf{W}_\ell^e)\, dV \\ &= \frac{d_i d_\ell}{c}\left(\frac{\cos\beta_{kn}}{h_k h_n}\,\chi_{jm} + \frac{\cos\beta_{jm}}{h_j h_m}\,\chi_{kn} - \frac{\cos\beta_{km}}{h_k h_m}\,\chi_{jn} - \frac{\cos\beta_{jn}}{h_j h_n}\,\chi_{km}\right. \\ &\quad \left. + \frac{2}{3}\,\frac{c^2 h_1 d_1}{h_j h_k h_m h_n}\sin\beta_{jk}\sin\beta_{mn}\right) \end{aligned} \tag{5.68}$$

$$\begin{aligned} EWMC_{i\ell} &= \iiint_{V^e} (\nabla \times \mathbf{W}_i^e) \cdot (\nabla \times \mathbf{M}_\ell^e)\, dV \\ &= \frac{d_i d_\ell}{c}\left(-\frac{\cos\beta_{kn}}{h_k h_n}\,\chi_{jm} - \frac{\cos\beta_{jm}}{h_j h_m}\,\chi_{kn} + \frac{\cos\beta_{km}}{h_k h_m}\,\chi_{jn} + \frac{\cos\beta_{jn}}{h_j h_n}\,\chi_{km}\right. \\ &\quad \left. + \frac{1}{3}\,\frac{c^2 h_1 d_1}{h_j h_k h_m h_n}\sin\beta_{jk}\sin\beta_{mn}\right) \end{aligned} \tag{5.69}$$

$$\begin{aligned} EWKC_{i\ell} &= \iiint_{V^e} (\nabla \times \mathbf{W}_i^e) \cdot (\nabla \times \mathbf{K}_\ell^e)\, dV \\ &= \frac{h_1 d_1}{6}\, d_i\left(\frac{\cos\beta_{j\ell}}{h_j h_\ell} - \frac{\cos\beta_{k\ell}}{h_k h_\ell}\right) \end{aligned} \tag{5.70}$$

$$\begin{aligned} EMMC_{i\ell} &= \iiint_{V^e} (\nabla \times \mathbf{M}_i^e) \cdot (\nabla \times \mathbf{M}_\ell^e)\, dV \\ &= EWWC_{i\ell} \end{aligned} \tag{5.71}$$

$$\begin{aligned} EMKC_{i\ell} &= \iiint_{V^e} (\nabla \times \mathbf{M}_i^e) \cdot (\nabla \times \mathbf{K}_\ell^e)\, dV \\ &= -EWKC_{i\ell} \end{aligned} \tag{5.72}$$

$$\begin{aligned} EKKC_{i\ell} &= \iiint_{V^e} (\nabla \times \mathbf{K}_i^e) \cdot (\nabla \times \mathbf{K}_\ell^e)\, dV \\ &= c\,\frac{h_1 d_1}{2}\,\frac{\cos\beta_{i\ell}}{h_i h_l} \end{aligned} \tag{5.73}$$

$$\begin{aligned} EWWD_{i\ell} &= \iiint_{V^e} \mathbf{W}_i^e \cdot \mathbf{W}_\ell^e\, dV \\ &= c\,\frac{d_i d_\ell}{3}\left(\frac{\cos\beta_{kn}}{h_k h_n}\,\chi_{jm} + \frac{\cos\beta_{jm}}{h_j h_m}\,\chi_{kn} - \frac{\cos\beta_{km}}{h_k h_m}\,\chi_{jn} - \frac{\cos\beta_{jn}}{h_j h_n}\,\chi_{km}\right) \end{aligned} \tag{5.74}$$

$$\begin{aligned} EWMD_{i\ell} &= \iiint_{V^e} \mathbf{W}_i^e \cdot \mathbf{M}_\ell^e\, dV \\ &= \tfrac{1}{2}\, EWWD_{i\ell} \end{aligned} \tag{5.75}$$

$$EWKD_{i\ell} = \iiint_{V^e} \mathbf{W}_i^e \cdot \mathbf{K}_\ell^e \, dV$$
$$= 0 \tag{5.76}$$

$$EMMD_{i\ell} = \iiint_{V^e} \mathbf{M}_i^e \cdot \mathbf{M}_\ell^e \, dV$$
$$= EWWD_{i\ell} \tag{5.77}$$

$$EMKD_{i\ell} = \iiint_{V^e} \mathbf{M}_i^e \cdot \mathbf{K}_\ell^e \, dV$$
$$= 0 \tag{5.78}$$

$$EKKD_{i\ell} = \iiint_{V^e} \mathbf{K}_i^e \cdot \mathbf{K}_\ell^e \, dV$$
$$= c\chi_{i\ell} \tag{5.79}$$

where V^e is the volume of the prism. Also

$$\beta_{rs} = \begin{cases} 0 & \text{if } r = s \\ \alpha_r + \alpha_s & \text{otherwise} \end{cases}$$

$$\chi_{rs} = \frac{h_1 d_1}{2} \left\{ w_r w_s + \frac{h_1}{3} [(\cot\alpha_3 - \cot\alpha_2)(\eta_s w_r + \eta_r w_s) + 2(\xi_s w_r + \xi_r w_s)] \right.$$
$$+ \frac{h_1^2}{12} [3(\cot\alpha_3 - \cot\alpha_2)(\eta_s \xi_r + \eta_r \xi_s) + 2\eta_r \eta_s (\cot^2\alpha_2 - \cot\alpha_2 \cot\alpha_3 + \cot^2\alpha_3)$$
$$\left. + 6\xi_r \xi_s] \right\}$$

in which $r, s = 1, 2, 3$ and w_r, ξ_r, η_r are given by

$$w_1 = 1, \quad \xi_1 = -\frac{1}{h_1}, \quad \eta_1 = 0$$
$$w_2 = 0, \quad \xi_2 = \frac{\cos\alpha_3}{h_2}, \quad \eta_2 = -\frac{\sin\alpha_3}{h_2}$$
$$w_3 = 0, \quad \xi_3 = \frac{\cos\alpha_2}{h_3}, \quad \eta_3 = \frac{\sin\alpha_2}{h_3}$$

In all the above formulae the indices i, j, k and ℓ, m, n follow the cyclic rule given by the following table:

i	j	k	ℓ	m	n
1	2	3	1	2	3
2	3	1	2	3	1
3	1	2	3	1	2

As a test, we compute the eigenvalues of the rectangular cavity shown in Fig. 5.9. The cavity was first discretized using bricks, and these prisms were formed by slicing each brick diagonally. Note that the bricks at two corners were sliced along different diagonals from the bricks at the other two corners.

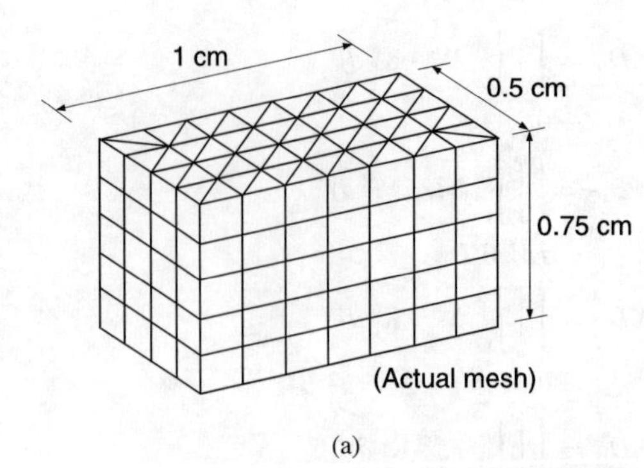

(a)

Mode	k, cm^{-1} (Exact)	% Error Prism	% Error Brick	% Error Tetra.
TE_{101}	5.236	0.73	-1.36	0.44
TM_{110}	7.025	2.32	-2.23	0.70
TE_{011}	7.531	0.53	-2.58	1.00
TE_{201}	7.531	0.64	-3.13	-0.56
TM_{111}	8.179	0.22	-2.09	2.29

(b)

Figure 5.9 Eigenvalues of a rectangular cavity: (a) discretization of the rectangular cavity; (b) comparison of eigenvalues using bricks and tetrahedrals. [*After Özdemir and Volakis, © IEEE, 1997.*]

The percent error for triangular prisms along with the results for bricks and tetrahedrals [29] is given in the table next to the cavity mesh in Fig. 5.9 and should be compared with Table 5.1. The number of segments along the x-, y-, and z-directed edges were seven, four, and five, respectively, for both the triangular prism and the brick discretizations resulting in 382 edge unknowns in the triangular prism case and 270 edge unknowns in the brick case. The tetrahedral discretization, on the other hand, resulted in 260 unknowns. As seen, the performance of the triangular prisms is comparable to that of bricks and tetrahedrals.

As a second test, we consider the eigenvalues of a cylindrical-circular cavity with metal walls, as shown in Fig. 5.10. The table accompanying the figure shows the percentage error in calculating the first five eigenvalues. Note that the prism modeling is quite good given that the discretization results in only four edges along the radius as well as the axis of the cylinder. This example shows the advantage of the triangular elements over rectangular ones in being able to model cavities with arbitrary cross section.

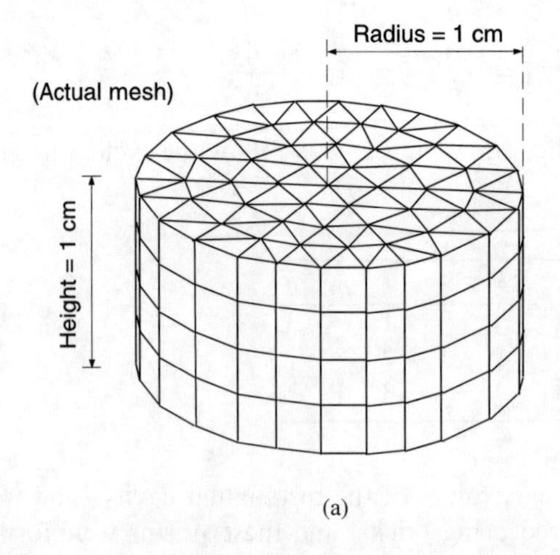

(a)

Mode	k, cm^{-1} (Exact)	% Error (Computed)
TM_{010}	2.405	1.29
TE_{111}	3.640	2.17
TM_{110}	3.830	-2.90
TM_{011}	3.955	0.81
TE_{211}	4.380	-8.97

(b)

Figure 5.10 Accuracy of triangular prisms in calculating the eigenvalues of a cylindrical-circular cavity of 1 cm radius and height. [*Courtesy of T. Özdemir.*]

REFERENCES

[1] P. P. Silvester and R. L. Ferrari. *Finite Elements for Electrical Engineers*. Cambridge Univ. Press, second edition, 1990.

[2] A. F. Peterson and S. P. Castillo. A frequency-domain differential equation formulation for electromagnetic scattering from inhomogeneous cylinders. *IEEE Trans. Antennas Propagat.*, 37(5):601–607, May 1989.

[3] R. Mittra and O. Ramahi. Absorbing boundary conditions for the direct solution of partial differential equations arising in electromagnetic scattering problems. In M. A. Morgan, editor, *Finite Element and Finite Difference Method in Electromagnetic Scattering*, Chapter 4. Elsevier, New York, 1990.

[4] Z. J. Cendes and P. Silvester. Numerical solution of dielectric loaded waveguides: I—Finite element analysis. *IEEE Trans. Microwave Theory Tech.*, 118:1124–1131, 1970.

[5] H. Whitney. *Geometric Integration Theory*. Princeton Univ. Press, NJ, 1957.

[6] J. C. Nedelec. Mixed finite elements in R^3. *Numer. Math.*, 35:315–341, 1980.

[7] A. Bossavit and J. C. Verite. A mixed FEM-BIEM method to solve 3D eddy current problems. *IEEE Trans. Magnetics*, 18:431–435, March 1982.

[8] M. Hano. Finite element analysis of dielectric-loaded waveguides. *IEEE Trans. Microwave Theory Tech.*, 32:1275–1279, October 1984.

[9] G. Mur and A. T. de Hoop. A finite element method for computing three-dimensional electromagnetic fields in inhomogeneous media. *IEEE Trans. Magnetics*, 21:2188–2191, November 1985.

[10] J. S. van Welij. Calculation of eddy currents in terms of H on hexahedra. *IEEE Trans. Magnetics*, 21:2239–2241, November 1985.

[11] M. L. Barton and Z. J. Cendes. New vector finite elements for three-dimensional magnetic field computation. *J. Appl. Phys.*, 61(8):3919–3921, April 1987.

[12] T. B. A. Senior. Combined resistive and conductive sheets. *IEEE Trans. Antennas Propagat.*, 33:577–579, 1985.

[13] J. L. Volakis, A. Chatterjee, and L. Kempel. Review of the finite element method for three-dimensional electromagnetic scattering. *J. Opt. Soc. Am. A*, 11(4):1422–1432, April 1994.

[14] R. F. Harrington. *Time Harmonic Electromagnetic Fields*. McGraw-Hill, New York, 1987.

[15] R. Dyczij-Edlinger and O. Biro. A joint vector and scalar potential formulation for driven high frequency problems using hybrid edge and nodal finite elements. *IEEE Trans. Microwave Theory Tech.*, 44(1):15–23, January 1996.

[16] S. H. Wong and Z. J. Cendes. Combined finite element-modal solution of three-dimensional eddy current problems. *IEEE Trans. Magnetics*, 24(6), November 1988.

[17] O. C. Zienkiewicz. *The Finite Element Method*. McGraw-Hill, New York, third edition, 1979.

[18] B. M. A. Rahman and J. B. Davies. Penalty function improvement of waveguide solution by finite elements. *IEEE Trans. Microwave Theory Tech.*, 32:922–928, August 1984.

[19] J. P. Webb. Finite element analysis of dispersion in waveguides with sharp metal edges. *IEEE Trans. Microwave Theory Tech.*, 36(12):1819–1824, December 1988.

[20] A Bossavit. Solving Maxwell's equations in a closed cavity, and the question of spurious modes. *IEEE Trans. Magnetics*, 26(2):702–705, March 1990.

[21] J. P. Webb. Edge elements and what they can do for you. *IEEE Trans. Magnetics*, 29:1460–1465, 1993.

[22] B. R. Crain and A. F. Peterson. Analysis of propagation on open microstrip lines using mixed-order covariant projection vector finite elements. *Int. J. Microwave and Millimeter Wave CAE*, 5(2):59–67, March 1995.

[23] Jin-Fa Lee and R. Mittra. A note on the application of edge elements for modeling three-dimensional inhomogeneously filled cavities. *IEEE Trans. Microwave Theory Tech.*, 40:1767–1773, 1992.

[24] J. M. Jin and J. L. Volakis. Electromagnetic scattering by and transmission through a three-dimensional slot in a thick conducting plane. *IEEE Trans. Antennas Propagat.*, 39(4):543–550, April 1991.

[25] J. L. Volakis, J. Gong, and A. Alexanian. A finite element boundary integral method for antenna RCS analysis. *Electromagnetics*, 14(1):63–85, 1994.

[26] X. Yuan. Three-dimensional electromagnetic scattering from inhomogeneous objects by the hybrid moment and finite element method. *IEEE Trans. Antennas Propagat.*, 38:1053–1058, 1990.

[27] Z. J. Cendes and J. F. Lee. The transfinite method for modeling MMIC devices. *IEEE Trans. Microwave Theory Tech.*, 36:1639–1649, 1988.

[28] J. F. Lee. Analysis of passive microwave devices by using three-dimensional tangential vector finite elements. *Int. J. Num. Modeling*, 3:235–246, 1990.

[29] A Chatterjee, J. M. Jin, and J. L. Volakis. Computation of cavity resonances using edge-based finite elements. *IEEE Trans. Microwave Theory Tech.*, 40(11):2106–2108, November 1992.

[30] G. H. Golub and C. F. Van Loan. *Matrix Computations*. Johns Hopkins Univ. Press, Baltimore, MD, 1983.

[31] J. G. Yook, N. I. Dib, and L. P. B. Katehi. Characterization of high frequency interconnects using finite difference time domain and finite element methods. *IEEE Trans. Microwave Theory Tech.*, 42(9):1727–1736, September 1994.

[32] J. G. Yook, N. I. Dib, E. Yasan, and L. P. B. Katehi. A study of hermetic transitions for microwave packages. In *IEEE MTT-S Int. Microwave Symp.*, pages 1579–1582, 1995.

[33] T. Özdemir and J. L. Volakis. Triangular prisms for edge-based vector finite element antenna analysis. *IEEE Trans. Antennas Propagat.*, pages 788–797, May 1997.

[34] Z. S. Sacks and J. F. Lee. A finite element time domain method using prism elements for microwave cavities. *IEEE Trans. Electromagnetic Compatibility*, November 1995.

6

Three-Dimensional Problems: Radiation and Scattering

6.1 INTRODUCTION

In the previous chapter, we outlined methods to deal with packaged three-dimensional (3D) structures in electromagnetics. However, the solution of packaged or bounded structures is only one of the problems that finite elements can be applied to. Open problems, like radiation and scattering, present a unique challenge to finite domain methods. Since the mesh of the computational domain cannot be extended to infinity, boundary conditions must be applied to simulate the effect of the infinite domain. This was discussed in detail in Chapter 4 for two-dimensional (2D) geometries. In this chapter, we concentrate on solving the radiation and scattering problem for 3D structures using vector absorbing boundary conditions (ABCs).

As described in Chapter 4, absorbing boundary conditions were derived for finite element solutions of 2D scattering problems. However, the method's implementation and performance for scattering by three-dimensional geometries has become acceptable only recently [1]. The 3D implementations of the FEM for radiation and scattering have been limited primarily to a hybrid solution using the boundary integral (BI) technique [2], [3], [4] and those incorporating ABCs [1], [5], [6], [7]. The finite element–boundary integral (FE–BI) method will be discussed in the next chapter. This chapter focuses on ABCs and artificial absorbers to truncate the finite element mesh for applications to radiation and scattering.

The motivation for using ABCs comes from the localized nature of its effect which preserves the $O(N)$ storage advantage of the finite element method. This feature permits scalability to large 3D problems. The boundary integral is equivalent to employing a global boundary condition for terminating the mesh and consequently leads to a full submatrix, restricting the method's utility to small or regular geometries. Recently, artificial absorbers [8] and coordinate stretching methods [9] have been used to absorb the outgoing waves and minimize non-physical reflections back into the computational domain. These techniques maintain sparsity but may worsen convergence properties for iterative equation solvers. Even for direct solvers, the ill conditioning of the matrix could lead to unstable or incorrect solutions.

In the first part of this chapter, we present a survey of the more popular vector ABCs and artificial absorbers. Detailed derivations for the ABCs along with recent advances made in understanding their behavior are included. In the following sections, we formulate the open domain problem in terms of the linear functional and incorporate the ABC into the finite element system. The scattered and total field functionals are then presented. In the last section of the chapter, we include examples of various applications solved using finite elements and absorbing boundary conditions or artificial absorbers.

6.2 SURVEY OF VECTOR ABCs

The motivation for applying ABCs to simulate open domain problems was discussed in detail in Chapter 4. In three dimensions, the advantages of locality and subsequent scalability are even clearer. All 3D finite element formulations rely on a vector representation of the underlying variable to maintain generality over a wide class of problems. ABCs for 3D problems must, therefore, be expressed in terms of vectors. Three-dimensional ABCs usually start with the Wilcox representation [10] of the radiating field variable, which is the electric or the magnetic field. The tangential component of the curl of the radiating function is then expressed in terms of the tangential components of the function and its higher-order derivatives on the ABC surface. The order of the derivative usually determines the order of the ABC and the degree of its accuracy. The higher the order, the better the accuracy of the boundary condition. Of course, this accuracy is achieved at the expense of increased complexity in modeling the boundary condition. In the next section, we outline the derivation of the ABCs for vector formulations.

6.2.1 Three-Dimensional Vector ABCs

Consider the scattering volume V_d enclosed by a fictitious surface S_0, as shown in Fig. 6.1. We shall assume that the immediate surrounding of the surface S_0 is free space and the fields near S_0 are thus governed by the vector wave equation

$$\nabla \times \nabla \times \mathbf{E} - k_0^2 \mathbf{E} = 0 \tag{6.1}$$

where k_0 is the free-space wave number. We also assume that the field has a well-defined phase front in the region under consideration. This is a valid assumption when the ABC boundary is placed far enough from the radiator or the scatterer. Since we are concerned only with local behavior, we can assume that the phase fronts can be treated as parallel regions. Consequently, the surface describing the phase fronts can be specified by a net of coordinate curves orthogonal to each other denoted by t_1 and t_2 and a third variable n denotes the coordinate along the normal to the phase front (see Fig. 6.2).

The point of observation in the Dupin coordinate system [11] can now be defined as

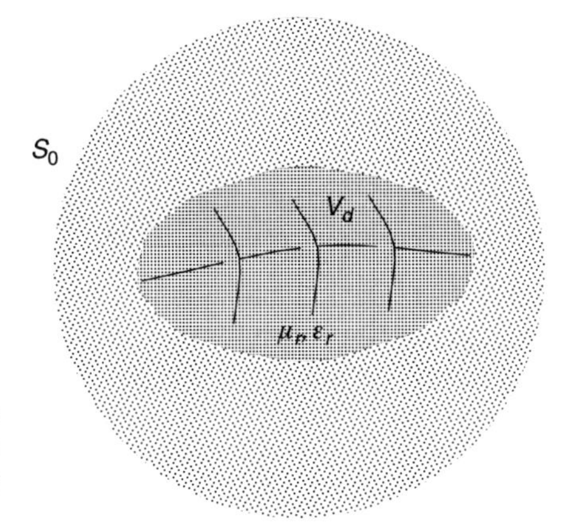

Figure 6.1 Illustration of scattering structure V_d enclosed by an artificial mesh termination surface, S_0, on which the absorbing boundary condition is imposed.

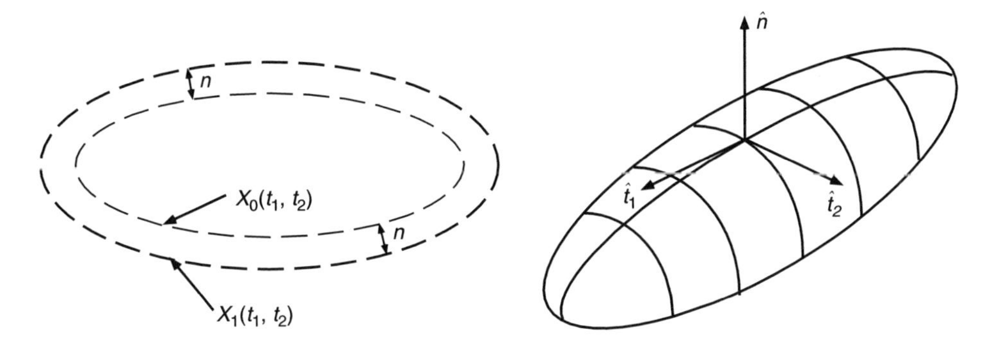

Figure 6.2 Dupin coordinate system and related parameters.

$$\mathbf{x} = n\hat{n} + \mathbf{x}_0 (t_1, t_2) \tag{6.2}$$

where $\hat{n}$ is the unit normal and $\mathbf{x}_0 (t_1, t_2)$ denotes the surface of the reference phase front. The curl of a vector in the above coordinate system is given by

$$\nabla \times \mathbf{E} = \nabla_T \times \mathbf{E} + \hat{n} \times \frac{\partial \mathbf{E}}{\partial n} \tag{6.3}$$

where $\nabla_T \times \mathbf{E}$ is called the surface curl involving only the tangential derivatives and is defined in [12], [13] as

$$\nabla_T \times \mathbf{E} = -\hat{n} \times \nabla E_n + \hat{t}_2 \kappa_1 E_{t_1} - \hat{t}_1 \kappa_2 E_{t_2} + \hat{n} \nabla \cdot (\mathbf{E} \times \hat{n}) \tag{6.4}$$

In (6.4), κ_1 and κ_2 denote the principal curvatures of the surface under consideration, E_{t_1}, E_{t_2} are the tangential components, and E_n is the normal component of the electric field on the surface. The principal curvatures are associated with the principal directions $\hat{t}_{1,2}$ of a surface and are given by [11]

$$\kappa_1 = \frac{1}{R_1} = -\frac{1}{h_1} \frac{\partial h_1}{\partial n} \tag{6.5}$$

$$\kappa_2 = \frac{1}{R_2} = -\frac{1}{h_2} \frac{\partial h_2}{\partial n} \tag{6.6}$$

where h_1, h_2 are the metric coefficients and R_1, R_2 are the principal radii of curvature.

Using the aforementioned coordinates, the Wilcox expansion for a vector radiating function can now be generalized to read

$$\mathbf{E}(n, t_1, t_2) = \frac{e^{-jk_0 n}}{4\pi\sqrt{R_1 R_2}} \sum_{p=0}^{\infty} \frac{\mathbf{E}_p(t_1, t_2)}{(\sqrt{R_1 R_2})^p} \tag{6.7}$$

where $R_i = \rho_i + n$, $i = 1, 2$ and ρ_i is the principal radius of curvature associated with the outgoing wavefront at the target. The lowest-order term in (6.7) represents the geometrical optics spread factor for a doubly curved wavefront and reduces to the standard Wilcox expansion [10] for a spherical wave. Moreover, (6.7) can be differentiated term by term any number of times and the resulting series converges absolutely and uniformly [10].

6.2.1.1 Unsymmetric ABCs. In the 3D finite element implementation using vector basis functions and the electric field as the working variable, we need to relate the tangential component of the magnetic field in terms of the electric field at any surface discontinuity. Therefore, our next task is to derive a relation between $\hat{n} \times \nabla \times \mathbf{E}$ (i.e., $\hat{n} \times \mathbf{H}$ where $\mathbf{H}$ is the magnetic field) and the tangential components of the electric field on the surface. Taking the curl of the electric field expansion given by (6.7) and crossing it with the normal vector, we have

$$\hat{n} \times \nabla \times \mathbf{E} = \frac{e^{-jk_0 n}}{4\pi} \sum_{p=0}^{\infty} \left[\frac{(jk_0 + \kappa_m - \overline{\overline{K}}\cdot)\mathbf{E}_{pt}}{u^{p+1}} + \frac{\nabla_t E_{pn} + p\kappa_m \mathbf{E}_{pt}}{u^{p+1}} \right] \tag{6.8}$$

where $u = \sqrt{R_1 R_2}$ and

$$\nabla_t E_n = -(\hat{n} \times \hat{n} \times \nabla)\, E_n$$

$$\kappa_m = \frac{\kappa_1 + \kappa_2}{2}$$

$$\overline{\overline{K}} = \kappa_1 \hat{t}_1 \hat{t}_2 + \kappa_2 \hat{t}_2 \hat{t}_2$$

Considering that E_{0n} is zero due to the divergenceless condition [10] and simplifying, we obtain the first-order absorbing boundary condition

$$\hat{n} \times \nabla \times \mathbf{E} - (jk_0 + \kappa_m - \overline{\overline{K}}\cdot)\, \mathbf{E}_t = \frac{e^{-jk_0 n}}{4\pi} \sum_{p=1}^{\infty} \frac{\nabla_t E_{pn} + p\kappa_m \mathbf{E}_{pt}}{u^{p+1}} \tag{6.9}$$

$$\text{or, } \hat{n} \times \nabla \times \mathbf{E} - (jk_o + \kappa_m - \overline{\overline{K}}\cdot)\, \mathbf{E}_t = 0 + O(n^{-3}) \tag{6.10}$$

for a conformal outer boundary. An example of a geometry enclosed in a conformal ABC boundary is shown in Fig. 6.3.

The first-order conformal ABC derived in (6.10) is identical to the impedance boundary condition for curved surfaces as derived by Rytov [14]. It should be noted that in the above equation, $\nabla_t E_n$ and κ_m are each proportional to n^{-1}. Therefore, the leading order behavior of (6.10) is $O(n^{-3})$, i.e., only the first two terms of (6.7) are exactly satisfied by (6.10). If the scattered field contains higher order terms, application of (6.10) will give rise to nonphysical reflections back into the computational domain. To reduce these spurious reflections, we need to either shift the mesh trun-

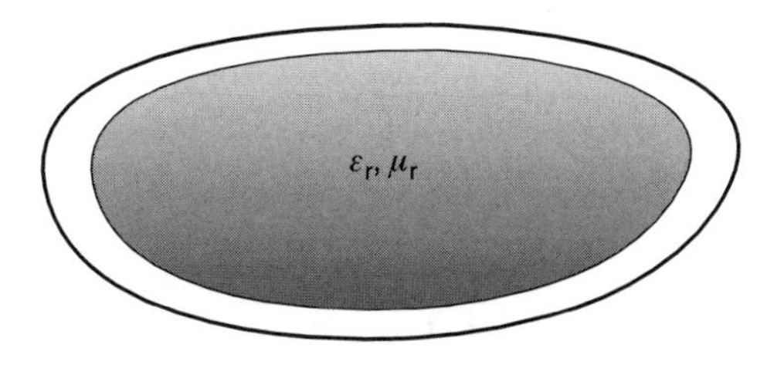

Figure 6.3 Composite structure enclosed in a conformal mesh termination boundary. The geometry of the mesh termination allows smaller problem sizes than with spherical terminations.

cation boundary farther away from the scatterer or employ higher order boundary conditions which satisfy higher order terms of (6.7).

To reduce further the order of the residual error, we include the tangential components of the curl of (6.9). This yields

$$\hat{n}\times\nabla \times [\hat{n} \times \nabla \times \mathbf{E} - (jk_0 + \kappa_m - \overline{\overline{K}}\cdot)\,\mathbf{E}_t]$$
$$= \frac{e^{-jk_0 n}}{4\pi} \sum_{p=1}^{\infty} \Bigg[\left\{ jk_0 + (p+3)\kappa_m - \frac{\kappa_g}{\kappa_m} \right\} \frac{\nabla_t E_{pn} + p\kappa_m \mathbf{E}_{pt}}{u^{p+1}}$$
$$- \left(2\kappa_m \quad \frac{\kappa_g}{\kappa_m}\right) \frac{\nabla_t E_{pn}}{u^{p+1}} - \frac{p\kappa_m \overline{\overline{K}} \cdot \mathbf{E}_{pt}}{u^{p+1}} \Bigg] \tag{6.11}$$

where $\kappa_g = \kappa_1\kappa_2$ is the Gaussian curvature. Using the result derived in (6.9) and simplifying (6.11) reduces to

$$\hat{n}\times\nabla \times [\hat{n} \times \nabla \times \mathbf{E} - (jk_0 + \kappa_m - \overline{\overline{K}}\cdot)\,\mathbf{E}_t]$$
$$= \frac{e^{-jk_0 n}}{4\pi} \sum_{p=1}^{\infty} \left[\left\{ jk_0 + (p+3)\kappa_m - \frac{\kappa_g}{\kappa_m} - \overline{\overline{K}}\cdot \right\} \frac{\nabla_t E_{pn} + p\kappa_m \mathbf{E}_{pt}}{u^{p+1}} \right]$$
$$- \left(2\kappa_m - \frac{\kappa_g}{\kappa_m} - \overline{\overline{K}}\cdot\right) \nabla_t E_n \tag{6.12}$$

If we take a closer look at the term in the square brackets on the RHS of (6.12), we find that it can be written as

$$\frac{e^{-jk_0 n}}{4\pi} \sum_{p=1}^{\infty} p\kappa_m \frac{\nabla_t E_{pn} + p\kappa_m \mathbf{E}_{pt}}{u^{p+1}}$$
$$+ \left(jk_0 + 3\kappa_m - \frac{\kappa_g}{\kappa_m} - \overline{\overline{K}}\cdot\right) \left\{ \hat{n} \times \nabla \times \mathbf{E} - \left(jk_0 + \kappa_m - \overline{\overline{K}}\cdot\right) \mathbf{E}_t \right\}$$

where we have substituted

$$\frac{e^{-jk_0 n}}{4\pi} \sum_{p=1}^{\infty} \left[\frac{\nabla_t E_{pn} + p\kappa_m \mathbf{E}_{pt}}{u^{p+1}} \right] = \hat{n} \times \nabla \times \mathbf{E} - \left(jk_0 + \kappa_m - \overline{\overline{K}}\cdot\right) \mathbf{E}_t \tag{6.13}$$

using the relation derived in (6.9).

Now the dominant terms on the RHS of (6.12) can be eliminated by considering the higher order operator

$$\left[\hat{n} \times \nabla \times -\left(jk_0 + 3\kappa_m - \frac{\kappa_g}{\kappa_m} - \overline{\overline{K}}\cdot\right)\right]\{\hat{n} \times \nabla \times \mathbf{E} - (jk_0 + \kappa_m - \overline{\overline{K}}\cdot)\,\mathbf{E}_t\}$$
$$+\left(2\kappa_m - \frac{\kappa_g}{\kappa_m} - \overline{\overline{K}}\cdot\right)\nabla_t E_n = \frac{e^{-jk_0 n}}{4\pi}\sum_{p=1}^{\infty} p\kappa_m \frac{\nabla_t E_{pn} + p\kappa_m \mathbf{E}_{pt}}{u^{p+1}} \tag{6.14}$$

The residual of (6.14) can be reduced further to yield the absorbing boundary condition of second order which satisfies (6.7) to $O(n^{-5})$, where n is the normal distance from the object surface to the phase front. This second-order ABC is found to be

$$\left[\hat{n} \times \nabla \times -\left(jk_0 + 4\kappa_m - \frac{\kappa_g}{\kappa_m} - \overline{\overline{K}}\cdot\right)\right]\{\hat{n} \times \nabla \times \mathbf{E} - (jk_0 + \kappa_m - \overline{\overline{K}}\cdot)\,\mathbf{E}_t\}$$
$$+\left(2\kappa_m - \frac{\kappa_g}{\kappa_m} - \overline{\overline{K}}\cdot\right)\nabla_t E_n = 0 \tag{6.15}$$

and the residual is equal to

$$\frac{e^{-jk_0 n}}{4\pi}\sum_{p=2}^{\infty}(p-1)\kappa_m \frac{\nabla_t E_{pn} + p\kappa_m \mathbf{E}_{pt}}{u^{p+1}} \tag{6.16}$$

The operator on the LHS of (6.15) can be applied repeatedly to obtain ABCs of increasing order; however, higher order basis functions are needed for their implementation.

After some algebraic manipulation, the terms on the LHS of (6.15) reduce to simpler ones. In addition to the wave equation, the following vector identities were utilized to carry out the simplifications and are provided below for the reader's convenience:

$$\hat{n} \times \nabla \times \mathbf{E}_t = \hat{n} \times \nabla \times \mathbf{E} - \nabla_t E_n$$
$$\hat{n} \times \nabla \times \nabla_t E_n = \nabla_t(\nabla \cdot \mathbf{E}_t) + 2\kappa_m \nabla_t E_n$$
$$\hat{n} \times \nabla \times (\hat{n} \times \nabla \times \mathbf{E}) = \nabla \times \{\hat{n}(\nabla \times \mathbf{E})_n\} - k_0^2 \mathbf{E}_t - \Delta\kappa\{(\nabla \times \mathbf{E})_{t_2}\hat{t}_1 + (\nabla \times \mathbf{E})_{t_1}\hat{t}_2\}$$

where $\Delta\kappa = \kappa_1 - \kappa_2$. The derivation of these identities is given in an appendix to this chapter. Upon simplification, the second-order ABC can be compactly written as

$$-(D - 2\kappa_m)\hat{n} \times \nabla \times \mathbf{E} + \{4\kappa_m^2 - \kappa_g + D(jk_0 - \overline{\overline{K}}\cdot) + \overline{\overline{K}} \cdot \overline{\overline{K}} \cdot + \kappa_m \Delta\kappa \overline{\overline{\xi}}\cdot\}\,\mathbf{E}_t$$
$$+\nabla \times \{\hat{n}(\nabla \times \mathbf{E})_n\} + \left(jk_0 + 3\kappa_m - \frac{\kappa_g}{\kappa_m} - 2\overline{\overline{K}}\cdot\right)\nabla_t E_n = 0 \tag{6.17}$$

in which

$$D = 2jk_0 + 5\kappa_m - \frac{\kappa_g}{\kappa_m}$$

and

$$\overline{\overline{K}} \cdot \overline{\overline{K}} \cdot \mathbf{E}_t = \kappa_1^2 E_{t_1}\hat{t}_1 + \kappa_2^2 E_{t_2}\hat{t}_2$$
$$\overline{\overline{\xi}} = \hat{t}_1\hat{t}_1 - \hat{t}_2\hat{t}_2 \tag{6.18}$$

The second-order ABC derived in [15] is recovered by setting $\kappa_1 = \kappa_2 = 1/r$.

6.2.1.2 Symmetric Approximation. It has been shown by Peterson [15] that the LHS of (6.17) when incorporated into the finite element equations gives rise to an unsymmetric matrix system in spherical coordinates. To alleviate this problem, Kanellopoulos and Webb [16] suggested an alternative derivation involving an arbitrary parameter which would lead to a symmetric matrix while sacrificing some accuracy. Below, we discuss a different approach which leads to a symmetric ABC without the introduction of an arbitrary parameter.

On considering the series expansion of the term $\hat{n} \times \nabla \times \nabla_t E_n$, we have

$$\begin{aligned}
\hat{n} \times \nabla \times \nabla_t E_n &= \frac{e^{-jk_0 n}}{4\pi} \sum_{p=1}^{\infty} \{jk_0 + (p+1)\kappa_m\} \frac{\nabla_t E_{pn}}{u^{p+1}} \\
&= jk_0 \nabla_t E_n + 2\kappa_m \nabla_t E_n + \sum_{p=2}^{\infty} (p-1)\kappa_m \frac{\nabla_t E_{pn}}{u^{p+1}} \\
&= jk_0 \nabla_t E_n + 2\kappa_m \nabla_t E_n + O(n^{-5})
\end{aligned}$$

and on making use of the vector identity

$$\nabla_t(\nabla \cdot \mathbf{E}_t) = \hat{n} \times \nabla \times \nabla_t E_n - 2\kappa_m \nabla_t E_n$$

given earlier, we arrive at the following result

$$\nabla_t(\nabla \cdot \mathbf{E}_t) = jk_0 \nabla_t E_n + O(n^{-5}) \tag{6.19}$$

Since the ABC (6.17) was derived to have a residual error of $O(n^{-5})$, we can replace $jk_0 \nabla_t E_n$ with $\nabla_t(\nabla \cdot \mathbf{E}_t)$ without affecting the order of the approximation. Doing so, the second-order ABC with a symmetric operator can be rewritten as

$$\begin{aligned}
(D - 2\kappa_m)\hat{n} \times \nabla \times \mathbf{E} &= \{4\kappa_m^2 - \kappa_g + D(jk_0 - \overline{\overline{K}}\cdot) + \overline{\overline{K}} \cdot \overline{\overline{K}} \cdot + \kappa_m \Delta\kappa \overline{\overline{\zeta}}\cdot\} \mathbf{E}_t \\
&+ \nabla \times \{\hat{n}(\nabla \times \mathbf{E})_n\} + \frac{1}{jk_0}\left(jk_0 + 3\kappa_m - \frac{\kappa_g}{\kappa_m} - 2\overline{\overline{K}}\cdot\right) \nabla_t(\nabla \cdot \mathbf{E}_t) = 0
\end{aligned} \tag{6.20}$$

in which $\Delta\kappa = \kappa_1 - \kappa_2$. It can be easily shown that the above boundary condition leads to a symmetric system of equations when incorporated into the finite element functional for surfaces having $\kappa_1 = \kappa_2$. Equations (6.10) and (6.20) reduce to the boundary conditions derived in [16] on setting $\kappa_1 = \kappa_2 = 1/r$ which have been found to work well for spherical and flat boundaries [1]. Symmetry cannot be guaranteed when $\kappa_1 \neq \kappa_2$ as explained in the next section.

6.2.1.3 Finite Element Implementation. The boundary condition outlined in equation (6.20) cannot be incorporated into the finite element equations without modification. As explained in Chapters 3 and 4, the absorbing boundary condition is implemented in the finite element system through the surface integral over the mesh termination surface S_0.

$$\int_{S_0} \mathbf{E} \cdot \hat{n} \times \nabla \times \mathbf{E} \, dS = \int_{S_0} \mathbf{E} \cdot P(\mathbf{E}) \, dS \tag{6.21}$$

where $P(\mathbf{E})$ denotes the boundary condition relating the tangential magnetic field to the tangential electric field on the surface.

Let $P_1(\mathbf{E})$ denote the first-order absorbing boundary condition given by (6.10), where the subscript represents the order of the ABC. Therefore, the surface integral contribution for the first-order ABC reduces to

$$\int_{S_0} \mathbf{E} \cdot P_1(\mathbf{E}) = (jk_0 + \kappa_m) \int_{S_0} \mathbf{E} \cdot \mathbf{E}_t \, dS - \int_{S_0} \mathbf{E} \cdot (\overline{\overline{K}} \cdot \mathbf{E}_t) \, dS \tag{6.22}$$

Using some basic vector identities and considering that $\mathbf{E}_t = -\hat{n} \times \hat{n} \times \mathbf{E}$, we deduce that

$$\int_{S_0} \mathbf{E} \cdot P_1(\mathbf{E}) \, dS = (jk_0 + \kappa_m) \int_{S_0} (E_{t_1}^2 + E_{t_2}^2) \, dS - \int_{S_0} (\kappa_1 E_{t_1}^2 + \kappa_2 E_{t_2}^2) \, dS \tag{6.23}$$

which is a readily implementable form of the first-order ABC. However, the second-order ABC does not simplify as easily. If $P_2(\mathbf{E})$ denotes the second-order ABC given by (6.20), we can rewrite it in more compact vector notation as

$$P_2(\mathbf{E}) = \overline{\overline{\alpha}} \cdot \mathbf{E}_t + \overline{\overline{\beta}} \cdot [\nabla \times \{\hat{n}(\nabla \times \mathbf{E})_n\}] + \overline{\overline{\gamma}} \cdot \{\nabla_t(\nabla \cdot \mathbf{E}_t)\} \tag{6.24}$$

where the tensors $\overline{\overline{\alpha}}$, $\overline{\overline{\beta}}$, and $\overline{\overline{\gamma}}$ are given by

$$\begin{aligned} \overline{\overline{\alpha}} = & \frac{1}{D - 2\kappa_m} \{4\kappa_m^2 - \kappa_g + D(jk_0 - \kappa_1) + \kappa_1^2 + \kappa_m \Delta\kappa\} \hat{t}_1 \hat{t}_1 \\ & + \frac{1}{D - 2\kappa_m} \{4\kappa_m^2 - \kappa_g + D(jk_0 - \kappa_2) + \kappa_2^2 - \kappa_m \Delta\kappa\} \hat{t}_2 \hat{t}_2 \end{aligned} \tag{6.25}$$

$$\overline{\overline{\beta}} = \frac{1}{D - 2\kappa_m} \{\hat{t}_1 \hat{t}_1 + \hat{t}_2 \hat{t}_2\} = \beta\{\hat{t}_1 \hat{t}_1 + \hat{t}_2 \hat{t}_2\} \tag{6.26}$$

$$\begin{aligned} \overline{\overline{\gamma}} = & \frac{1}{jk_0(D - 2\kappa_m)} \left(jk_0 + 3\kappa_m - \frac{\kappa_g}{\kappa_m} - 2\kappa_1 \right) \hat{t}_1 \hat{t}_1 \\ & + \frac{1}{jk_0(D - 2\kappa_m)} \left(jk_0 + 3\kappa_m - \frac{\kappa_g}{\kappa_m} - 2\kappa_2 \right) \hat{t}_2 \hat{t}_2 \end{aligned} \tag{6.27}$$

Substituting the second-order absorbing boundary condition in the surface integral given in (6.24), we have

$$\begin{aligned} \int_{S_0} \mathbf{E} \cdot P_2(\mathbf{E}) \, dS = & \int_{S_0} \mathbf{E} \cdot (\overline{\overline{\alpha}} \cdot \mathbf{E}_t) \, dS + \int_{S_0} \mathbf{E} \cdot \left(\overline{\overline{\beta}} \cdot \nabla \times \{\hat{n}(\nabla \times \mathbf{E})_n\} \right) dS \\ & + \int_{S_0} \mathbf{E} \cdot \{\overline{\overline{\gamma}} \cdot \nabla_t(\nabla \cdot \mathbf{E}_t)\} \, dS \\ = & I_1 + I_2 + I_3 \end{aligned}$$

Let us examine the integral I_1. Again, since $\mathbf{E}_t = -\hat{n} \times \hat{n} \times \mathbf{E}$, we have

$$I_1 = \int_{S_0} \alpha_1 E_{t_1}^2 + \alpha_2 E_{t_2}^2 \, dS \tag{6.28}$$

after employing some simple vector identities.

The other two integrals (I_2 and I_3) do not reduce as easily to simple, implementable forms. They are first simplified using basic vector and tensor identities, and then the divergence theorem is employed to eliminate one of the terms. Considering the integrand of the second integral I_2, we note that

$$\mathbf{E} \cdot \left[\overline{\overline{\beta}} \cdot (\nabla \times \hat{n}\phi) \right] = \left(\overline{\overline{\beta}} \cdot \mathbf{E} \right) \cdot (\nabla \times \hat{n}\phi)$$

where we have set $\phi = (\nabla \times \mathbf{E}) \cdot \hat{n} = (\nabla \times \mathbf{E})_n$. Using some additional vector identities and letting $\overline{\overline{\beta}} \cdot \mathbf{E} = \boldsymbol{F}$, we get

$$\begin{aligned} \boldsymbol{F} \cdot \nabla \times \hat{n}\phi &= \nabla \cdot (\phi \hat{n} \times \boldsymbol{F}) + \phi(\hat{n} \cdot \nabla \times \boldsymbol{F}) \\ &= \nabla \cdot (\phi \hat{n} \times \boldsymbol{F}) + \phi(\nabla \times \boldsymbol{F})_n \end{aligned}$$

Using the results from [11], the first term in the above identity can be further simplified to read[1]

$$\begin{aligned} \nabla \cdot (\phi \hat{n} \times \boldsymbol{F}) &= \nabla_s \cdot (\phi \hat{n} \times \boldsymbol{F}) + \frac{\partial}{\partial n} \{\phi \hat{n} \cdot (\hat{n} \times \boldsymbol{F})\} - J \{\phi \hat{n} \cdot (\hat{n} \times \boldsymbol{F})\} \\ &= \nabla_s \cdot (\phi \hat{n} \times \boldsymbol{F}) \end{aligned} \tag{6.29}$$

where ∇_s denotes the surface gradient operator and $J = \kappa_1 + \kappa_2$. The integral I_2 can now be written as

$$I_2 = \int_{S_0} \nabla_s \cdot (\phi \hat{n} \times \boldsymbol{F})\, dS + \int_{S_0} \phi (\nabla \times \boldsymbol{F})_n \, dS$$

We next apply the surface divergence theorem to the first term on the RHS of this expression to yield

$$\int_{S_0} \nabla_s \cdot (\phi \hat{n} \times \boldsymbol{F})\, dS = \int_C \phi \hat{\mathbf{m}} \cdot (\hat{n} \times \boldsymbol{F})\, dl = 0 \tag{6.30}$$

since the surface S_0 is closed. We note that $\hat{m} = \hat{l} \times \hat{n}$ and $\hat{l}$ is the unit vector along the edge of the surface element and C denotes the contour of integration (see Fig. 1.1). On the basis of (6.30) and considering that $\overline{\overline{\beta}}$ is a simple scalar, I_2 reduces to

$$I_2 = \int_{S_0} \beta [(\nabla \times \mathbf{E})_n]^2 \, dS \tag{6.31}$$

We now turn our attention to simplifying I_3 for implementation in the finite element equations. Considering the integrand of I_3, we have

$$\begin{aligned} \mathbf{E} \cdot \left\{ \overline{\overline{\gamma}} \cdot \nabla_t (\nabla \cdot \mathbf{E}_t) \right\} &= \left(\overline{\overline{\gamma}} \cdot \mathbf{E} \right) \cdot \{\nabla_t (\nabla \cdot \mathbf{E}_t)\} \\ &= \left(\overline{\overline{\gamma}} \cdot \mathbf{E} \right) \cdot \left\{ \nabla \psi - \hat{n} \frac{\partial \psi}{\partial n} \right\} \end{aligned}$$

where $\psi = \nabla \cdot \mathbf{E}_t$. Next, setting $\boldsymbol{G} = \overline{\overline{\gamma}} \cdot \mathbf{E}$, we obtain

$$\boldsymbol{G} \cdot \left\{ \nabla \psi - \hat{n} \frac{\partial \psi}{\partial n} \right\} = \nabla \cdot (\psi \boldsymbol{G}) - \psi \nabla \cdot \boldsymbol{G} - G_n \frac{\partial \psi}{\partial n} \tag{6.32}$$

The first term in the above identity can be written as

$$\nabla \cdot (\psi \boldsymbol{G}) = \nabla_s \cdot (\psi \boldsymbol{G}) + \frac{\partial}{\partial n} (\psi G_n) - J (\psi G_n)$$

and as usual $G_n = \hat{n} \cdot \boldsymbol{G}$. Also, since $\partial G_n / \partial n = \nabla \cdot \boldsymbol{G} - \nabla \cdot \boldsymbol{G}_t + J G_n$, the LHS of (6.32) reduces to

[1] The book by van Bladel [13] also contains an extensive list of identities associated with divergence and curl operators.

$$\boldsymbol{G} \cdot \left\{ \nabla\psi - \hat{n} \frac{\partial\psi}{\partial n} \right\} = \nabla_s \cdot (\psi\boldsymbol{G}) - \psi\nabla \cdot \boldsymbol{G}_t \tag{6.33}$$

We can thus replace the integrand of I_3 with the expression in (6.33) and use the divergence theorem to eliminate the first term of (6.33). Specifically,

$$\begin{aligned} I_3 &= \int_{S_0} \nabla_s \cdot (\psi\boldsymbol{G})\, dS - \int_{S_0} \psi\nabla \cdot \boldsymbol{G}_t\, dS \\ &= \int_C \hat{m} \cdot (\psi\boldsymbol{G})\, dS - \int_{S_0} \psi\nabla \cdot \boldsymbol{G}_t\, dS = -\int_{S_0} \psi\nabla \cdot \boldsymbol{G}_t\, dS \end{aligned}$$

where $\hat{m}$ has been defined earlier and the contour integral vanishes when the principal curvatures of the outer boundary are equal to zero, i.e., for a rectangular ABC surface. The integral, however, does not vanish for spherical or cylindrical boundaries, as was pointed out in [38]. In our computations, we have ignored the contribution of the contour integral that results from the non-vanishing portion of the surface integral. For further discussion on how to include the effect of the contour integral without destroying symmetry of the finite element matrix, the interested reader is referred to [38]. The integral I_3 can finally be rewritten as

$$I_3 = -\int_{S_0} (\nabla \cdot \mathbf{E}_t)(\nabla \cdot \boldsymbol{G}_t)\, dS \tag{6.34}$$

Using (6.28), (6.31), and (6.34), the complete surface integral term incorporating the conformal second-order ABC reduces to

$$\begin{aligned} \int_{S_0} \mathbf{E} \cdot P_2(\mathbf{E})\, dS = &\int_{S_0} (\alpha_1 E_{t_1}^2 + \alpha_2 E_{t_2}^2)\, dS + \int_{S_0} \beta[(\nabla \times \mathbf{E})_n]^2\, dS \\ &- \int_{S_0} (\nabla \cdot \mathbf{E}_t)\{\nabla \cdot (\overline{\overline{\gamma}} \cdot \mathbf{E})_t\}\, dS \end{aligned} \tag{6.35}$$

It remains to be seen whether the integrals in (6.35) lead to a symmetric system when incorporated into the finite element equations. With this in mind, we will examine three simple shapes and check whether they preserve symmetry of the finite element system. It will then be possible to generalize our findings to a more general mesh truncation boundary.

Let us consider the case of a sphere of radius r. Since the two principal curvatures of the sphere are identical ($\kappa_1 = \kappa_2 = 1/r$), the first-order boundary condition reduces to the simple Sommerfeld radiation condition

$$\int_{S_0} \mathbf{E} \cdot P_1(\mathbf{E})\, dS = jk_0 \int_{S_0} (E_\theta^2 + E_\phi^2)\, dS \tag{6.36}$$

On a spherical boundary, the second-order ABC also reduces to the comparatively simple form

$$\int_{S_0} \mathbf{E} \cdot P_2(\mathbf{E})\, dS = \int_{S_0} \left[jk_0 \mathbf{E}_t^2 + \frac{1}{2jk_0 + 2/r}[(\nabla \times \mathbf{E})_n]^2 - \frac{1}{2jk_0 + 2/r}(\nabla \cdot \mathbf{E}_t)^2\right] dS \tag{6.37}$$

The ABC given in (6.37) is identical to the boundary condition derived in [16] for a spherical mesh termination surface and leads to a symmetric system of equations.

Next, we consider a piecewise planar termination boundary in which case $\kappa_1 = \kappa_2 = 0$ (see Fig. 6.4). The first-order ABC then reduces to the Sommerfeld radiation condition, and the second-order ABC for a planar boundary simplifies to

$$\int_{S_0} \mathbf{E} \cdot P_2(\mathbf{E})\, dS = \int_{S_0} \left[jk_0 \mathbf{E}_t^2 + \frac{1}{2jk_0} [(\nabla \times \mathbf{E})_n]^2 - \frac{1}{2jk_0} (\nabla \cdot \mathbf{E}_t)^2 \right] dS \tag{6.38}$$

Since the planar boundary is a special case of a spherical boundary, (6.38) again reduces to a symmetric system of equations.

Now we examine the situation when the mesh termination boundary is cylindrical in shape and of radius ρ. The principal curvatures of the cylindrical surface are then $\kappa_1 = 1/\rho$ and $\kappa_2 = 0$. Since the principal curvatures are no longer identical, the tensors $\overline{\overline{\alpha}}$ and $\overline{\overline{\gamma}}$ do not reduce to simple scalars. The first-order ABC for a cylindrical outer boundary is given by

$$\int_{S_0} \mathbf{E} \cdot P_1(\mathbf{E})\, dS = \int_{S_0} \left(jk_0 + \frac{1}{\rho} \right) \left(jk_0 + \frac{1}{\rho} \right) \mathbf{E}_t^2\, dS - \int_{S_0} \frac{1}{\rho} E_\phi^2\, dS \tag{6.39}$$

and the second-order ABC gives

$$\begin{aligned} \int_{S_0} \mathbf{E} \cdot P_2(\mathbf{E})\, dS = & \int_{S_0} (\alpha_{c1} E_\phi^2 + \alpha_{c2} E_z^2)\, dS + \int_{S_0} \beta_c [(\nabla \times \mathbf{E})_\rho]^2\, dS \\ & - \int_{S_0} (\nabla \cdot \mathbf{E}_t) \{ \nabla \cdot (\overline{\overline{\gamma}}_c \cdot \mathbf{E})_t \}\, dS \end{aligned} \tag{6.40}$$

where α_{c1}, α_{c2}, β_c, and $\overline{\overline{\gamma}}_c$ are obtained by substituting $\kappa_1 = 1/\rho$ and $\kappa_2 = 0$ in the original expressions for $\overline{\overline{\alpha}}$, $\overline{\overline{\beta}}$, and $\overline{\overline{\gamma}}$. It is seen that the first-order ABC given by (6.39) leads to a symmetric matrix for a cylindrical boundary. On the other hand, the second-order ABC does not yield a symmetric matrix for an arbitrary choice of basis functions. However, the boundary condition outlined in (6.40) preserves symmetry on using linear edge-based elements for discretization.

The above discussion enables us to conclude that the first-order boundary condition leads to a symmetric system for surfaces having arbitrary principal curvatures. However, symmetry is guaranteed for the second-order ABC only when the two principal curvatures of the mesh termination boundary are identical, i.e., only when the outer boundary is limited to a planar or a spherical surface. Thus, if we want to enclose a scatterer having arbitrary shape within a conformal outer boundary, an unsymmetric system of equations will have to be solved. It should, however,

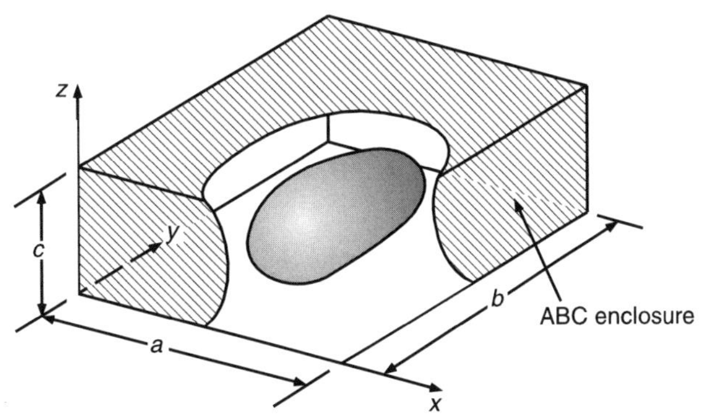

Figure 6.4 Scatterer enclosed in a piecewise planar ABC surface.

be noted that the resulting unsymmetric system will, in general, have fewer unknowns than its symmetric counterpart.

6.2.2 Artificial Absorbers

The purpose of the absorbing boundary condition is to absorb the outgoing wave so that there are no reflections back into the computational domain. An alternative to these ABCs is to use lossy layers to absorb the outgoing field [8]. The absorber proposed in [8] was a three-layer absorber with the relative permeability (μ_r) equal to the relative permittivity (ϵ_r) to preserve duality. The constitutive parameters of the medium were obtained by minimizing the reflection coefficient over a wide range of incidence angles, using a multidimensional simplex optimization algorithm. These absorbers usually consist of *nonphysical* materials to obtain the high level of absorption over the entire spectrum. The advantage of such a technique is evident: the user can specify the number of layers and their thickness and can, therefore, customize the absorber over a wide range of incidence angles and even frequencies. However, in finite element analysis, it is not yet clear whether they can be used to terminate the mesh closer than ABCs in a wide class of problems. Moreover, in three dimensions, the mesh for such structures can get increasingly complicated and reduce their viability when compared with ABCs. In the following section, we discuss a particular version of artificial absorbers called *PML* (*Perfectly Matched Layer*) that has shown considerable promise in dealing with open boundary problems.

6.2.2.1 Perfectly Matched Layer (PML). The PML concept as envisioned by Berenger [17] rests on the concept of splitting each of the electric and magnetic field components into two parts. In the case of the most general medium, Berenger defines an electric conductivity σ^e and a magnetic conductivity σ^h. However, Berenger goes one step further by introducing additional degrees of freedom in the electric and magnetic conductivity of the medium. The resulting medium, thus, becomes an artificial absorber that may no longer support Maxwellian waves. Nevertheless, the resulting solution of the modified Maxwell's equation permits the mathematical introduction of media interfaces which are completely nonreflecting for all incidence angles and material layers which are highly absorptive.

As an example, Berenger replaced the x component

$$\epsilon_0 \frac{\partial \mathcal{E}_x}{\partial t} + \sigma^e \mathcal{E}_x = \frac{\partial \mathcal{H}_z}{\partial y} - \frac{\partial \mathcal{H}_y}{\partial z} \tag{6.41}$$

of Maxwell's time domain equation by the split equations

$$\begin{aligned} \epsilon_0 \frac{\partial \mathcal{E}_{xy}}{\partial t} + \sigma_y^e \mathcal{E}_{xy} &= \frac{\partial \mathcal{H}_z}{\partial y} \\ \epsilon_0 \frac{\partial \mathcal{E}_{xz}}{\partial t} + \sigma_z^e \mathcal{E}_{xz} &= -\frac{\partial \mathcal{H}_y}{\partial z} \end{aligned} \tag{6.42}$$

With this replacement, the field component $\mathcal{E}_x$ was split into $\mathcal{E}_{xy}$ and $\mathcal{E}_{xz}$ with the corresponding introduction of the new conductivities σ_y^e and σ_z^e. Thus an additional degree of freedom was introduced in defining the conductivity of the medium leading to the appearance of the split field components. Similarly for the dual of (6.42) we have

$$
\begin{aligned}
\mu_0 \frac{\partial \mathcal{H}_{xy}}{\partial t} + \sigma_y^h \mathcal{H}_{xy} &= -\frac{\partial \mathcal{E}_z}{\partial y} \\
\mu_0 \frac{\partial \mathcal{H}_{xz}}{\partial t} + \sigma_z^h \mathcal{H}_{xz} &= \frac{\partial \mathcal{E}_y}{\partial z}
\end{aligned}
\tag{6.43}
$$

where now σ^h was generalized to σ_y^h and σ_z^h and $\mathcal{H}_x$ was split to $\mathcal{H}_{xy}$ and $\mathcal{H}_{xz}$.

In generalizing (6.42) and (6.43), Maxwell's equation $\nabla \times \mathcal{H} = \sigma\mathcal{E} + \epsilon_0 \frac{\partial \mathcal{E}}{\partial t}$ becomes [18]

$$
\begin{aligned}
\epsilon_0 \frac{\partial \mathcal{E}_{xy}}{\partial t} + \sigma_y^e \mathcal{E}_{xy} &= \frac{\partial(\mathcal{H}_{zx} + \mathcal{H}_{zy})}{\partial y} \\
\epsilon_0 \frac{\partial \mathcal{E}_{xz}}{\partial t} + \sigma_z^e \mathcal{E}_{xz} &= -\frac{\partial(\mathcal{H}_{yx} + \mathcal{H}_{yz})}{\partial z}
\end{aligned}
\tag{6.44}
$$

$$
\begin{aligned}
\epsilon_0 \frac{\partial \mathcal{E}_{yz}}{\partial t} + \sigma_z^e \mathcal{E}_{yz} &= \frac{\partial(\mathcal{H}_{xy} + \mathcal{H}_{xz})}{\partial z} \\
\epsilon_0 \frac{\partial \mathcal{E}_{yx}}{\partial t} + \sigma_x^e \mathcal{E}_{yx} &= -\frac{\partial(\mathcal{H}_{zx} + \mathcal{H}_{zy})}{\partial x}
\end{aligned}
\tag{6.45}
$$

$$
\begin{aligned}
\epsilon_0 \frac{\partial \mathcal{E}_{zx}}{\partial t} + \sigma_x^e \mathcal{E}_{zx} &= \frac{\partial(\mathcal{H}_{yx} + \mathcal{H}_{yz})}{\partial x} \\
\epsilon_0 \frac{\partial \mathcal{E}_{zy}}{\partial t} + \sigma_y^e \mathcal{E}_{zy} &= -\frac{\partial(\mathcal{H}_{xy} + \mathcal{H}_{xz})}{\partial y}
\end{aligned}
\tag{6.46}
$$

in which $\sigma_{x,y,z}^e$ denote the new electric conductivity parameters. The dual equations of (6.44)–(6.46) can be obtained by replacing $\mathcal{E}_{ab}$ with $\mathcal{H}_{ab}$, $\mathcal{H}_{ab}$ with $-\mathcal{E}_{ab}$, ϵ_0 with μ_0 and vice versa, and $\sigma_{x,y,z}^e$ with $\sigma_{x,y,z}^h$. The latter are the new magnetic conductivity components. If the condition

$$
\sigma^e/\epsilon_0 = \sigma^h/\mu_0 \tag{6.47}
$$

is satisfied, it can be shown that the medium is perfectly matched to vacuum (hence, the acronym *PML*) and plane waves propagating *normally* into a PML half-space should be completely absorbed [17]. The most powerful attribute of PMLs is that zero reflection can also be obtained irrespective of incidence angles under special conditions.

As an example application of the PML concept, let us consider the propagation of a plane wave through the PML medium (Fig. 6.5). The following derivation borrows heavily from [17]. We assume a TE plane wave of magnitude $\mathcal{E}_0$ traveling in the xy plane. The electric field and the split components of the magnetic field, H_{zx} and H_{zy}, can be expressed as

$$
\begin{aligned}
\mathcal{E}_x &= -\mathcal{E}_0 \sin\phi e^{j\omega(t-\alpha x-\beta y)} \\
\mathcal{E}_y &= \mathcal{E}_0 \cos\phi e^{j\omega(t-\alpha x-\beta y)} \\
\mathcal{H}_{zx} &= \mathcal{H}_{zx0} e^{j\omega(t-\alpha x-\beta y)} \\
\mathcal{H}_{zy} &= \mathcal{H}_{zy0} e^{j\omega(t-\alpha x-\beta y)}
\end{aligned}
\tag{6.48}
$$

In (6.48), ϕ is the angle the electric field makes with the y-axis (equal to the angle between the incident field direction and the x-axis) and α, β, $\mathcal{H}_{zx0}$, $\mathcal{H}_{zy0}$ are unknowns to be determined from the split Maxwell's equations. Substituting the

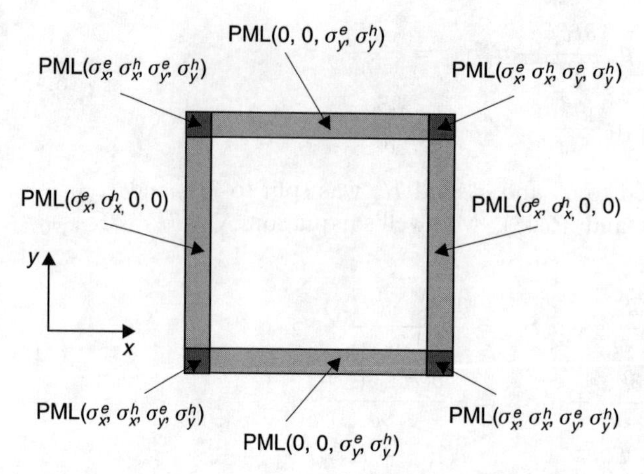

Figure 6.5 Cross section of 3D geometry with PML enclosing the outer boundary of the mesh. PML $(\sigma_x^e, \sigma_x^h, \sigma_y^e, \sigma_y^h)$ implies that the medium has a conductivity of (σ_x^e, σ_x^h) in the x direction and a conductivity of (σ_y^e, σ_y^h) along the y-axis.

values from (6.48) into (6.44) and (6.45) and the relation for H_z, after taking the necessary time derivatives, we obtain

$$
\begin{aligned}
\epsilon_0 E_0 \sin\phi - j\frac{\sigma_y^e}{\omega} E_0 \sin\phi &= \beta (H_{zx0} + H_{zy0}) \\
\epsilon_0 E_0 \cos\phi - j\frac{\sigma_x^e}{\omega} E_0 \cos\phi &= \alpha (H_{zx0} + H_{zy0}) \\
\left(\mu_0 - j\frac{\sigma_x^h}{\omega}\right) H_{zx0} &= \alpha E_0 \cos\phi \\
\left(\mu_0 - j\frac{\sigma_y^h}{\omega}\right) H_{zy0} &= \beta E_0 \sin\phi
\end{aligned}
\tag{6.49}
$$

Eliminating H_{zx0} and H_{zy0} from (6.49), we arrive at a relation between α and β.

$$
\frac{\beta}{\alpha} = \frac{\sin\phi}{\cos\phi} \frac{1 - j\sigma_y^e/(\epsilon_0\omega)}{1 - j\sigma_x^e/(\epsilon_0\omega)}
\tag{6.50}
$$

Solving for α and β and considering a positive direction of propagation, we have

$$
\begin{aligned}
\alpha &= \frac{\sqrt{\epsilon_0\mu_0}}{G}\left(1 - j\frac{\sigma_x^e}{\epsilon_0\omega}\right)\cos\phi \\
\beta &= \frac{\sqrt{\epsilon_0\mu_0}}{G}\left(1 - j\frac{\sigma_y^e}{\epsilon_0\omega}\right)\sin\phi
\end{aligned}
\tag{6.51}
$$

where $G = \sqrt{w_x \cos^2\phi + w_y \sin^2\phi}$ and

$$
\begin{aligned}
w_x &= \frac{1 - j\sigma_x^e/(\epsilon_0\omega)}{1 - j\sigma_x^h/(\mu_0\omega)} \\
w_y &= \frac{1 - j\sigma_y^e/(\epsilon_0\omega)}{1 - j\sigma_y^h/(\mu_0\omega)}
\end{aligned}
\tag{6.52}
$$

The remaining unknowns H_{zx0} and H_{zy0} can now be obtained as

$$
\begin{aligned}
H_{zx0} &= E_0 \frac{1}{Z_0 G} w_x \cos^2 \phi \\
H_{zy0} &= E_0 \frac{1}{Z_0 G} w_y \sin^2 \phi
\end{aligned} \tag{6.53}
$$

Thus, the impedance of the plane wave in the PML medium is given by

$$
Z = \frac{Z_0}{G} \tag{6.54}
$$

Assuming that both σ_x^e and σ_y^e satisfy the PML condition (6.47), the variables G, w_x, w_y become unity irrespective of frequency or angle of incidence. Consequently, the impedance of the plane wave in the PML medium reduces to the free-space impedance, and hence, the PML medium is perfectly matched to vacuum. Thus, any plane wave traveling from vacuum to a properly matched PML medium will be entirely transmitted. Another interesting thing happens to the magnitude of the propagating plane wave. The electric field of the plane wave in the PML medium can be written as $\mathbf{E} = \mathbf{E}_0 \psi$ where

$$
\psi = \psi_0 \exp\left(-j\omega \frac{x\cos\phi + y\sin\phi}{c}\right) \exp\left[-\frac{\sigma_x^e \cos\phi}{\epsilon_0 c} x\right] \exp\left[-\frac{\sigma_y^e \sin\phi}{\epsilon_0 c} y\right] \tag{6.55}
$$

where c denotes the velocity of light in free space. Therefore, the PML acts as an absorbing medium and, in the limit, will eventually absorb all propagating waves of all frequencies and incidence angles. The same results hold true for the TM case with the electric field replaced with a magnetic field. Since an arbitrary plane wave can be considered as a superposition of TE_z and TM_z modes, the above analysis holds true for all plane waves.

The story, however, is not complete. We have merely established the fact that the PML absorbs plane waves incident upon it very effectively. It is still unclear what happens when a plane wave is incident at the interface of two PML media. It has been shown in [17] that for an interface normal to the x- or y-axis and lying between two matched PML media having the same conductivity couplet (σ_y^e, σ_y^h) or (σ_x^e, σ_x^h), a plane wave is transmitted without reflection regardless of angle of incidence and frequency. This is true between vacuum and a PML medium as well because vacuum can be thought of as a zero electric and magnetic conductivity medium. This principle is illustrated in Fig. 6.5. Not considering the corners of the domain, PML media along the x-axis are given conductivity values of $(0, 0, \sigma_y^e, \sigma_y^h)$, whereas PML media parallel to the y-axis have conductivity values of $(\sigma_x^e, \sigma_x^h, 0, 0)$. The corners have conductivities which are a superposition of the intersecting layers. The corners play a very important role since they absorb the transverse components of the field entering into the PML layer [19]. The above derivation can be easily extended to three dimensions by considering a separate couplet of conductivities in the third orthogonal direction and superimposing the conductivity couplets in the two preferred directions of propagation.

In [17], Berenger terminated the PML boundary with a perfect electric conductor. Therefore, the reflection of the propagating plane wave occurs at the electric wall and is turned back into the medium. The plane wave then passes back through the PML layer, a part of it getting absorbed along the way, and then re-enters the domain. A refinement of this approach is to use a second-order ABC termination on the outer wall rather than the perfect electric condition to reduce spurious reflections.

There are some important issues raised by the PML concept.

- The PML concept essentially falls out of the split-field formulation of Maxwell's equations. Unfortunately, this also renders the PML medium non-Maxwellian with active sources embedded inside the medium [19].
- It is not very clear how to include the PML formulation for frequency domain problems since the conventional curl–curl formulation may not be valid.

We will try to answer a few of these questions in the next section. However, this topic is still an area of active research and the behavior of PML is not very well understood.

6.2.2.2 Various Interpretations of PML. In the previous section, we outlined the split-field formulation of PML. What if the fields were not split into two constituent parts? Pekel and Mittra [19] explore this concept and arrive at some interesting conclusions. They rewrite the set of PML equations (6.44)–(6.46) such that all split fields are eliminated. Assuming $U_{xy} + U_{xz} = U_x$ for all components of the electric and magnetic field, the modified equations in the frequency domain are

$$\begin{aligned}
(j\omega\epsilon_0 + \sigma_y^e)\,E_x &= -jk_yH_z + jk_zH_y - \frac{\sigma_z^e - \sigma_y^e}{j\omega\epsilon_0 + \sigma_z^e}\,jk_zH_y \\
(j\omega\epsilon_0 + \sigma_z^e)\,E_y &= -jk_zH_x + jk_xH_z - \frac{\sigma_x^e - \sigma_z^e}{j\omega\epsilon_0 + \sigma_x^e}\,jk_xH_z \\
(j\omega\epsilon_0 + \sigma_x^e)\,E_z &= -jk_xH_y + jk_yH_x - \frac{\sigma_y^e - \sigma_x^e}{j\omega\epsilon_0 + \sigma_y^e}\,jk_yH_x
\end{aligned} \tag{6.56}$$

for the electric field, where $k_{x,y,z}$ denote the (x, y, z) components of the propagation vector $\mathbf{k}$. The magnetic field equations can be obtained from (6.56) as outlined earlier. On imposing the PML condition (6.47), duality is restored between the electric and the magnetic fields. This is necessary so that electric and magnetic field formulations lead to equivalent final results. Thus, (6.56) shows that in addition to the regular terms arising from Maxwell's equations for an anisotropic medium with a uniaxial 3×3 conductivity tensor, there are excitation-dependent source terms. These source terms are proportional to the differences in the conductivities. The PML medium is thus active.

Considering z-directed propagation through the PML medium and enforcing continuity and phase matching conditions at the PML interface, it is found that the following constraints must be satisfied [19]

$$\begin{aligned}
k_{ri} &= \frac{\sigma_z^e}{\omega\epsilon_0} = \frac{\sigma_z^h}{\omega\mu_0} \\
\sigma_x^e &= \sigma_y^e = 0; \qquad \sigma_x^h = \sigma_y^h = 0
\end{aligned} \tag{6.57}$$

where $k_z = k_0(1 - jk_{ri})\cos\theta$. The σ^e values need to be permuted as the direction of propagation changes from z to x or y. This result is identical to the PML medium proposed by Berenger [17] and is illustrated in Fig. 6.5. When the unsplit PML equations (6.56) are once again rewritten with the values in (6.57), we obtain

$$\begin{aligned} j\omega\epsilon_0 E_x &= -jk_y H_z + jk_z H_y \left(\frac{1}{1 - j\sigma_z^e/\omega\epsilon_0}\right) \\ j\omega\epsilon_0 E_y &= jk_x H_z - jk_x H_z \left(\frac{1}{1 - j\sigma_z^e/\omega\epsilon_0}\right) \\ j\omega\epsilon_0 E_z &= -jk_x H_y + jk_y H_x \end{aligned} \tag{6.58}$$

for z-directed propagation. These equations are identical to the ones presented in [9], based on *coordinate stretching*. In the coordinate stretching technique, the spatial variable u is replaced by the complex spatial variable u' given by

$$u' = u\left(1 - j\frac{\sigma_u^e}{\omega\epsilon_0}\right) = u\left(1 - j\frac{\sigma_u^h}{\omega\epsilon_0}\right) \tag{6.59}$$

assuming that the wave is propagating in the u-direction. The spatial variable u can be either x, y, or z. The problem with this formulation is that the curl–curl equation must be reformulated in the PML medium to allow the fields to satisfy Maxwell's equations. However, it provides us with valuable insight into the true nature of the PML medium.

As was shown in [19], the PML medium can be thought to consist of an anisotropic material with a conductivity tensor. Along these lines, Sacks et al. [20] (see also Kingsland et al. [21]) have proposed an anisotropic absorber with perfect transmission characteristics over all incident angles and frequencies for planar surfaces. Assuming diagonal tensors, the permeability and permittivity tensors in the most general case can be written as

$$\frac{\overline{\overline{\mu}}}{\mu_0} = \frac{\overline{\overline{\epsilon}}}{\epsilon_0} = \begin{bmatrix} a & 0 & 0 \\ 0 & b & 0 \\ 0 & 0 & c \end{bmatrix} \tag{6.60}$$

The corresponding reflection coefficients for TE and TM cases are given by

$$\begin{aligned} R^{TE} &= \frac{\cos\theta_i - (\sqrt{b/a})\cos\theta_t}{\cos\theta_i + (\sqrt{b/a})\cos\theta_t} \\ R^{TM} &= \frac{(\sqrt{b/a})\cos\theta_t - \cos\theta_i}{\cos\theta_i + (\sqrt{b/a})\cos\theta_t} \end{aligned} \tag{6.61}$$

with θ_i and θ_t, as displayed in Fig. 6.6. From the phase matching condition at the interface, we also have

$$\sqrt{bc}\,\sin\theta_t = \sin\theta_i \tag{6.62}$$

In order to make the reflection coefficient independent of incidence angle, we choose $\sqrt{bc} = 1$. The zero reflection condition is achieved by setting $a = b$. Thus, we can set a to be an arbitrary complex number $\alpha - j\beta$ (for example) and rewrite the anisotropic material tensors as

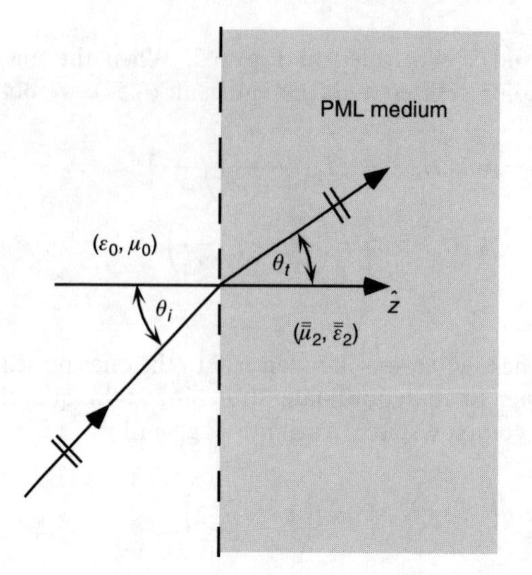

Figure 6.6 Plane wave incidence on a PML interface.

$$\frac{\bar{\bar{\mu}}}{\mu_0} = \frac{\bar{\bar{\epsilon}}}{\epsilon_0} = \begin{bmatrix} \alpha - j\beta & 0 & 0 \\ 0 & \alpha - j\beta & 0 \\ 0 & 0 & \dfrac{1}{\alpha - j\beta} \end{bmatrix} \tag{6.63}$$

The design of the anisotropic absorber, therefore, reduces to determining the values of α and β. The parameter β is more crucial since it controls the absorptivity in the PML medium. In [22], it is shown that the choice of β is critical for the performance of the absorber. If β is too small, field decay is insufficient to eliminate reflection. Too large a value of β leads to reflection since the mesh is insufficient to model the sudden jump in material property. This phenomenon is illustrated clearly in Fig. 6.7. A finer mesh will improve the situation but will perhaps never remove the problem. The value of β optimized for normal incidence is given by the relation [22]

$$\begin{aligned} \frac{\beta t}{\lambda_g} &= -0.0106|R| + 0.0433 \\ N &= 0.147 e^{[7.353\beta t/\lambda_g]} \end{aligned} \tag{6.64}$$

where $|R|$ is the desired reflection coefficient in dB, $\lambda_g = \lambda_0/\cos\theta$, t is the thickness of the PML, and N is the sampling density in the PML. Typical good choices for α and β are $\alpha \approx \beta \approx 1$.

The use of anisotropic absorbers thus simplifies the implementation of the PML to a certain degree since the curl–curl equation can now be solved within the framework of Maxwell's equations. It is without doubt that the PML is a very effective absorbing medium. In fact, it is probably the best artificial absorber known to date since it is reflectionless in the limit for all incident angles, frequencies, and polarization. But how effective is PML and how close to the target can we place it? It has been shown that the PML does not do a good job of absorbing evanescent waves [24] implying that it still needs to be placed far enough from the target for the evanescent modes to die out. This fact coupled with the convergence difficulties

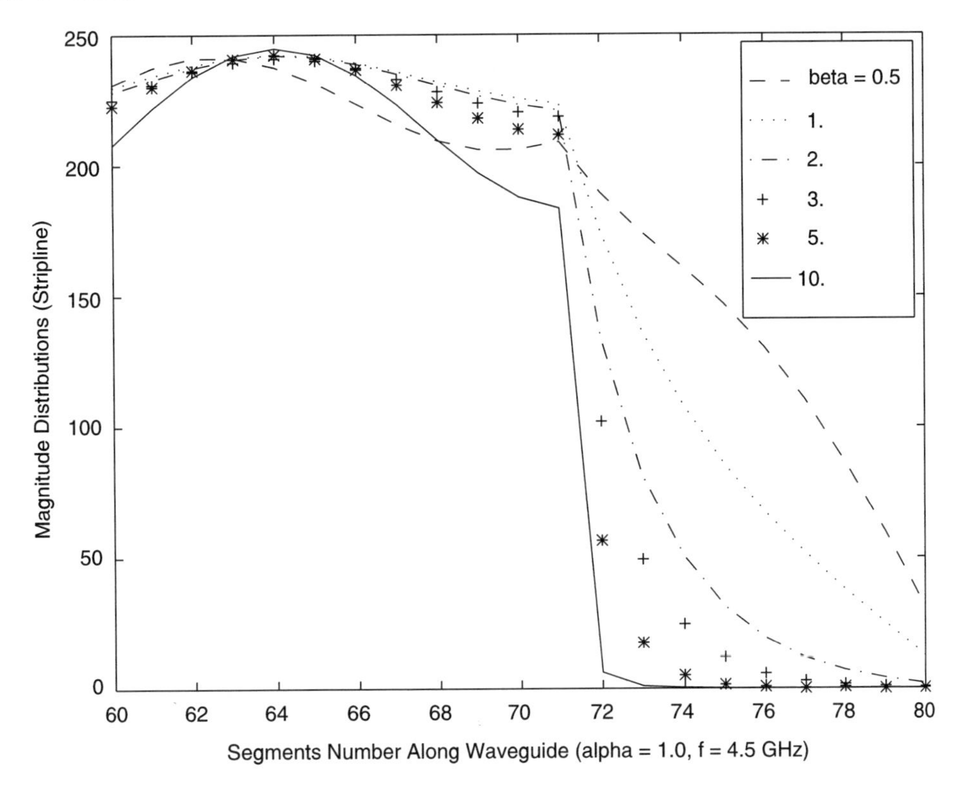

Figure 6.7 Field decay in the PML medium for different values of β and $\alpha = 1$. Curves refer to the decay of a wave in a substrate guided by a stripline. [*After Gong and Volakis, © IEE, 1995* [23].]

presented in solving for fields in an active medium indicate that further research needs to be done to determine the viability of the PML when compared to the ABCs.

6.3 FORMULATION

In the following section, the open domain problem is modeled with ABCs described earlier and is formulated in terms of the finite element functional. A Rayleigh-Ritz minimization is then carried out in the usual way to find the stationary point of the functional.

6.3.1 Scattered and Total Field Formulations

Let us consider the problem of scattering by an inhomogeneous target having possible material discontinuities. As outlined in Chapter 4 for the 2D case, it is necessary to enclose the radiator or the scatterer, embedded inside the volume V, by an artificial surface S_0 (contour in the case of 2D domains) on which the ABC is enforced (see Fig. 6.1). The general form of the ABC is as given in (6.21) and (6.37) provided $\mathbf{E}$ is interpreted as the scattered field $\mathbf{E}^{\text{scat}}$. On writing the same equation with the total field as the working variable, we get

$$\hat{n} \times \nabla \times \mathbf{E} = P(\mathbf{E}) + \mathbf{U}^{\text{inc}} \tag{6.65}$$

where

$$\mathbf{U}^{\text{inc}} = \hat{n} \times \nabla \times \mathbf{E}^{\text{inc}} - P(\mathbf{E}^{\text{inc}}) \tag{6.66}$$

and $\mathbf{E} = \mathbf{E}^{\text{scat}} + \mathbf{E}^{\text{inc}}$ is the total field with $\mathbf{E}^{\text{inc}}$ being the incident electric field. As usual, $\mathbf{E}^{\text{scat}}$ denotes the scattered field. Considering (6.65) to be the boundary condition employed at S_0, we can express the functional for the total electric field as

$$\begin{aligned} F(\mathbf{E}) = & \int_V \left[\frac{1}{\mu_r} (\nabla \times \mathbf{E}) \cdot (\nabla \times \mathbf{E}) - k_0^2 \epsilon_r \mathbf{E} \cdot \mathbf{E} \right] dV \\ & + jk_0 Z_0 \int_{S_k} \frac{1}{K} (\hat{n} \times \mathbf{E}) \cdot (\hat{n} \times \mathbf{E}) \, dS \\ & + \int_{S_0} [\mathbf{E} \cdot P(\mathbf{E}) + 2\mathbf{E} \cdot \mathbf{U}^{\text{inc}}] \, dS \end{aligned} \tag{6.67}$$

where the scalar K is the surface resistivity (R) when integrating over a resistive card and equals the surface impedance (η) for an impedance sheet (see Chapters 1 and 5).

The linear functional (6.67) is formulated in terms of the total field, but we can easily revert to a scattered field formulation by setting $\mathbf{E}^{\text{scat}} = \mathbf{E} - \mathbf{E}^{\text{inc}}$ and noting that the scattered field must satisfy the wave equation inside the domain of interest.

Let us consider the case where the computational volume V is occupied by a dielectric structure and is bounded internally by the surface of a perfect conductor and externally by the mesh termination boundary. On examining the terms inside the volume integral in (6.67), we can define

$$G(\mathbf{E}, \mathbf{E}) = \int_V \left[\frac{1}{\mu_r} (\nabla \times \mathbf{E}) \cdot (\nabla \times \mathbf{E}) - k_0^2 \epsilon_r \mathbf{E} \cdot \mathbf{E} \right] dV \tag{6.68}$$

Expressing the above relation in terms of the incident and the scattered fields, we have

$$G(\mathbf{E}, \mathbf{E}) = G(\mathbf{E}^{\text{scat}}, \mathbf{E}^{\text{scat}}) + 2G(\mathbf{E}^{\text{scat}}, \mathbf{E}^{\text{inc}}) + G(\mathbf{E}^{\text{inc}}, \mathbf{E}^{\text{inc}}) \tag{6.69}$$

The first and third terms on the RHS of (6.69) cannot be simplified any further than the form given in (6.68). The second term does, however, lend itself to more simplification. Making use of a simple vector identity and the divergence theorem, we can rewrite $G(\mathbf{E}^{\text{scat}}, \mathbf{E}^{\text{inc}})$ as

$$\begin{aligned} G(\mathbf{E}^{\text{scat}}, \mathbf{E}^{\text{inc}}) = & \int_V \mathbf{E}^{\text{scat}} \cdot \left[\nabla \times \frac{1}{\mu_r} \nabla \times \mathbf{E}^{\text{inc}} - k_0^2 \epsilon_r \mathbf{E}^{\text{inc}} \right] dV \\ & - \int_{S_0} \mathbf{E}^{\text{scat}} \cdot (\hat{n} \times \nabla \times \mathbf{E}^{\text{inc}}) \, dS \end{aligned} \tag{6.70}$$

since

$$\int_V \left[\frac{1}{\mu_r}(\nabla \times \mathbf{E}^{\text{scat}}) \cdot (\nabla \times \mathbf{E}^{\text{inc}})\right] dV$$
$$= \int_V \mathbf{E}^{\text{scat}} \cdot \left[\nabla \times \frac{1}{\mu_r} \nabla \times \mathbf{E}^{\text{inc}}\right] dV - \int_S \frac{1}{\mu_r} \mathbf{E}^{\text{scat}} \cdot (\hat{n} \times \nabla \times \mathbf{E}^{\text{inc}})\, dS \tag{6.71}$$

and the surface integral cancels out everywhere inside the computational domain except on the mesh termination boundary S_0. If we define V_d to be the volume occupied by dielectric materials, then the remaining volume ($V_0 = V - V_d$) is the volume occupied by free space. On incorporating this into (6.70), we have

$$\begin{aligned} G(\mathbf{E}^{\text{scat}}, \mathbf{E}^{\text{inc}}) = &\int_{V_0} \mathbf{E}^{\text{scat}} \cdot [\nabla \times \nabla \times \mathbf{E}^{\text{inc}} - k_0^2 \mathbf{E}^{\text{inc}}]\, dV \\ &+ \int_{V_d} \mathbf{E}^{\text{scat}} \cdot \left[\nabla \times \frac{1}{\mu_r} \nabla \times \mathbf{E}^{\text{inc}} - k_0^2 \epsilon_r \mathbf{E}^{\text{inc}}\right] dV \\ &- \int_{S_0} \mathbf{E}^{\text{scat}} \cdot (\hat{n} \times \nabla \times \mathbf{E}^{\text{inc}})\, dS \end{aligned} \tag{6.72}$$

Since the incident electric field satisfies the wave equation in free space, the first term of (6.72) is identically zero. The third term cancels exactly with the cross term $\int_{S_0} \mathbf{E} \cdot \mathbf{U}^{\text{inc}}\, dS$ in the total field functional (6.67). The second term can be simplified by employing the first vector Green's theorem to yield

$$\begin{aligned} \int_{V_d} \mathbf{E}^{\text{scat}} \cdot &\left[\nabla \times \frac{1}{\mu_r} \nabla \times \mathbf{E}^{\text{inc}} - k_0^2 \epsilon_r \mathbf{E}^{\text{inc}}\right] dV \\ &= \int_{V_d} \frac{1}{\mu_r} (\nabla \times \mathbf{E}^{\text{scat}}) \cdot (\nabla \times \mathbf{E}^{\text{inc}}) - k_0^2 \epsilon_r \mathbf{E}^{\text{scat}} \cdot \mathbf{E}^{\text{inc}}\, dV \\ &+ jk_0 Z_0 \int_{S_d} \frac{1}{\mu_r} \mathbf{E}^{\text{scat}} \cdot (\hat{n} \times \mathbf{H}^{\text{inc}})\, dS \end{aligned} \tag{6.73}$$

where the normal to S_d is directed away from V_d. The surface integral over the dielectric interface S_d occurs since the tangential component of the scattered electric field is discontinuous over the interface between two dielectrics having dissimilar permeabilities. It should be noted that (6.73) is valid even when there are multiple dielectric regions present. If the dielectric regions have the same permeability ($\mu_{r_1} = \mu_{r_2} = \cdots = \mu_{r_n} = 1$, for example) and different permittivities, the surface integral contribution over the dielectric interfaces, $S_{d_1}, \ldots, S_{d_n}$, is zero. If different permeability values are also present, then the permeability values must be substituted into the element equations and the direction of the normal for the two elements on the interface should take care of the respective signs.

Using (6.72) and (6.73), $G(\mathbf{E}^{\text{scat}}, \mathbf{E}^{\text{inc}})$ reduces to

$$
\begin{aligned}
G(\mathbf{E}^{\rm scat}, \mathbf{E}^{\rm inc}) = & \int_{V_d} \frac{1}{\mu_r} (\nabla \times \mathbf{E}^{\rm scat}) \cdot (\nabla \times \mathbf{E}^{\rm inc}) - k_0^2 \epsilon_r \mathbf{E}^{\rm scat} \cdot \mathbf{E}^{\rm inc}\, dV \\
& + jk_0 Z_0 \int_{S_d} \frac{1}{\mu_r} \mathbf{E}^{\rm scat} \cdot (\hat{n} \times \mathbf{H}^{\rm inc})\, dS \\
& - \int_{S_0} \mathbf{E}^{\rm scat} \cdot (\hat{n} \times \nabla \times \mathbf{E}^{\rm inc})\, dS
\end{aligned} \tag{6.74}
$$

The impedance and resistive sheet boundary conditions can be incorporated in a similar way into the scattered field functional. After simplification, the functional $F(\mathbf{E}^{\rm scat})$ for the scattered field is given by

$$
\begin{aligned}
F(\mathbf{E}^{\rm scat}) = & \int_V \left[\frac{1}{\mu_r} (\nabla \times \mathbf{E}^{\rm scat}) \cdot (\nabla \times \mathbf{E}^{\rm scat}) - k_0^2 \epsilon_r \mathbf{E}^{\rm scat} \cdot \mathbf{E}^{\rm scat} \right] dV \\
& + jk_0 Z_0 \int_{S_k} \frac{1}{K} (\hat{n} \times \mathbf{E}^{\rm scat}) \cdot (\hat{n} \times \mathbf{E}^{\rm scat})\, dS \\
& + \int_{S_0} \mathbf{E}^{\rm scat} \cdot P(\mathbf{E}^{\rm scat})\, dS \\
& + 2jk_0 Z_0 \int_{S_d} \frac{1}{\mu_r} \mathbf{E}^{\rm scat} \cdot (\hat{n} \times \mathbf{H}^{\rm inc})\, dS \\
& + 2 \int_{V_d} \left[\frac{1}{\mu_r} (\nabla \times \mathbf{E}^{\rm scat}) \cdot (\nabla \times \mathbf{E}^{\rm inc}) - k_0^2 \epsilon_r \mathbf{E}^{\rm scat} \cdot \mathbf{E}^{\rm inc} \right] dV \\
& + 2jk_0 Z_0 \int_{S_k} \frac{1}{K} (\hat{n} \times \mathbf{E}^{\rm scat}) \cdot (\hat{n} \times \mathbf{E}^{\rm inc})\, dS \\
& + f(\mathbf{E}^{\rm inc})
\end{aligned} \tag{6.75}
$$

where V_d is the volume occupied by the dielectric (portion of V where ϵ_r or μ_r are not unity) and S_d encompasses all dielectric interface surfaces. The function $f(\mathbf{E}^{\rm inc})$ is solely in terms of the incident electric field and vanishes when we take the first variation of $F(\mathbf{E}^{\rm scat})$.

6.4 APPLICATIONS

The open domain problem has varied applications in scattering, radiation, and microwave circuit simulations. The computation of the radar echo area from geometrically complex 3D structures is of primary interest for object detection and identification. Antenna applications are gaining widespread popularity in recent times primarily to model radiation characteristics in the presence of complicated objects. Modeling the radiation from patch antennas mounted on airborne structures or human tissue penetration of radiation from cellular phones are cases in point. In microwave circuits, calculation of power lost through radiation is often critical for meeting EMI/EMC specifications or satisfying design goals in densely-packed high frequency integrated circuits. In the following sections, we will present examples that model some of the phenomena mentioned earlier. The first few examples demonstrate the validity of the conformal ABCs; the remaining ones are progressively more complicated both in terms of modeling difficulty and structural features. They include scattering simulations as well as applications of ABCs and artificial absorbers in computing radiation from antennas and microwave circuits. It should also be mentioned that all scattering and circuit computations were done using $H_0(curl)$ elements (i.e., six unknowns per tetrahedron as mentioned in Chapter 2). The radiation problems were solved using linear basis functions on bricks and triangular prisms. An iterative solver was used for solving the final system of equations in all cases. Storage was, therefore, never a problem in any of the applications although convergence rate was geometry and excitation dependent.

6.4.1 Scattering Examples

It was discussed in Chapter 4 that of particular interest in scattering is the evaluation of radar echo width (for 2D) or echo area for three dimensions. The latter is given by

$$\sigma_{3D} = \lim_{r\to\infty} 4\pi r^2 \frac{|\mathbf{E}^{scat}|}{|\mathbf{E}^{inc}|}$$

with $\mathbf{E}^{scat}$ as described by the far zone expressions (1.63). The currents $\mathbf{J}$ and $\mathbf{M}$ now represent equivalent currents computed from (see Fig. 1.3)

$$\mathbf{J} = \hat{n} \times \mathbf{H} = \hat{n} \times \left[-\frac{\nabla \times \mathbf{E}}{jk_0 Z_0} \right]$$

$$\mathbf{M} = \mathbf{E} \times \hat{n}$$

where $\hat{n}$ denotes the outward normal of a surface S_c that encloses the scatterer. This surface can be arbitrary but for better accuracy it should be placed as close as possible to the scatterer. In the following, we will mention the $\sigma_{\theta\theta}/\sigma_{\phi\phi}$ echo area or radar cross section (RCS), as it is commonly referred to. The subscripts in these quantities simply identify the polarization of the incident and scattered fields used for the evaluation of the RCS. Specifically,

$$\sigma_{pq}^{3D} = \lim_{r\to\infty} 4\pi r^2 \frac{|\mathbf{E}^{scat} \cdot \hat{p}|^2}{|E_q^{inc}|^2}$$

implying that σ_{pq} is the measured RCS due to the pth component of the scattered field for a q-polarized incident plane wave. As usual, p and q represent either θ or ϕ spherical components with θ measured from the z-axis and ϕ is measured in the xy plane from the x-axis.

Below we present RCS calculations, using the presented finite element–ABC formulation, of cubes, cavities (inlets), plates and conesphere composite and metallic targets. These calculations were carried out using the University of Michigan code FEMATS [25].

6.4.1.1 Cube. We chose the cube as a validation example because the sharp edges and corners of this geometry make the problem very difficult to solve using node-based elements. However, by their very definition, edge elements pose no such problem in enforcing boundary conditions on the edges and corners of this structure. Figure 6.8 compares the measured [26] bistatic cross section ($\theta^{\rm inc} = 180°$, $\phi^{\rm inc} = 90°$) of a metallic cube having an edge length of 0.755λ with the corresponding pattern computed by the 3D finite element–ABC (FE–ABC) code. The results agree quite well with reference data. The second-order vector ABC derived earlier in (6.37) on a spherical mesh truncation boundary was employed and placed only 0.1λ from the edge of the cube. About 33,000 unknowns were used for the discretization of the computational domain, and the finite element matrix contained a total of 264,000 distinct nonzero entries. This compares very favorably with the approximate storage requirement of 2 million nonzero entries for a matrix obtained from the method of moments (assuming the same sampling rate as the FEM of 14 points/λ).

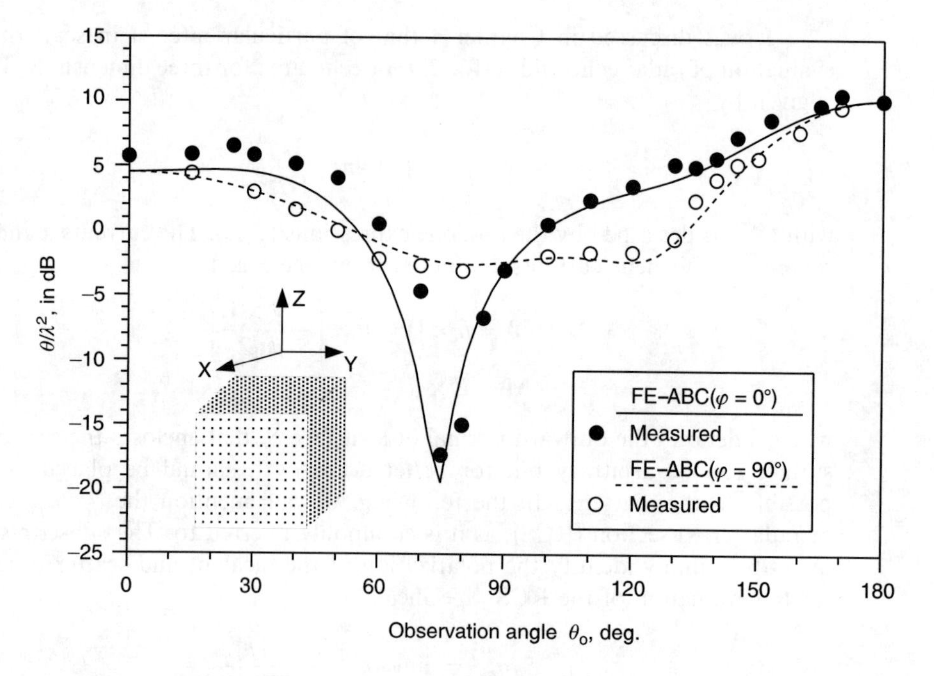

Figure 6.8 Bistatic echo-area of a perfectly conducting cube having edge length of 0.755λ. Plane wave incident from $\theta = 180°$; $\phi = 90°$.

Figure 6.9 shows the normal incidence backscatter RCS of a perfectly conducting cube as a function of its edge length. The meshes constructed for this experiment were terminated on conformal boundaries, i.e., on another cube placed a small distance (more than 0.15λ) from the scatterer. As seen, the agreement with measured data [26] is remarkably good over a 50-dB dynamic range. The interesting thing about this example is that terminating the mesh conformally did not have any effect on the accuracy of the final solution. Moreover, generalization of the Wilcox expansion to conformal structures (6.7) seems to work well, at least for planar structures.

Consider next the scattering from a variation of the cube solved earlier. The target shown in Fig. 6.10 consists of an air-filled resistive card block ($0.5\lambda \times 0.5\lambda \times 0.25\lambda$) attached to a metallic block ($0.5\lambda \times 0.5\lambda \times 0.25\lambda$). This

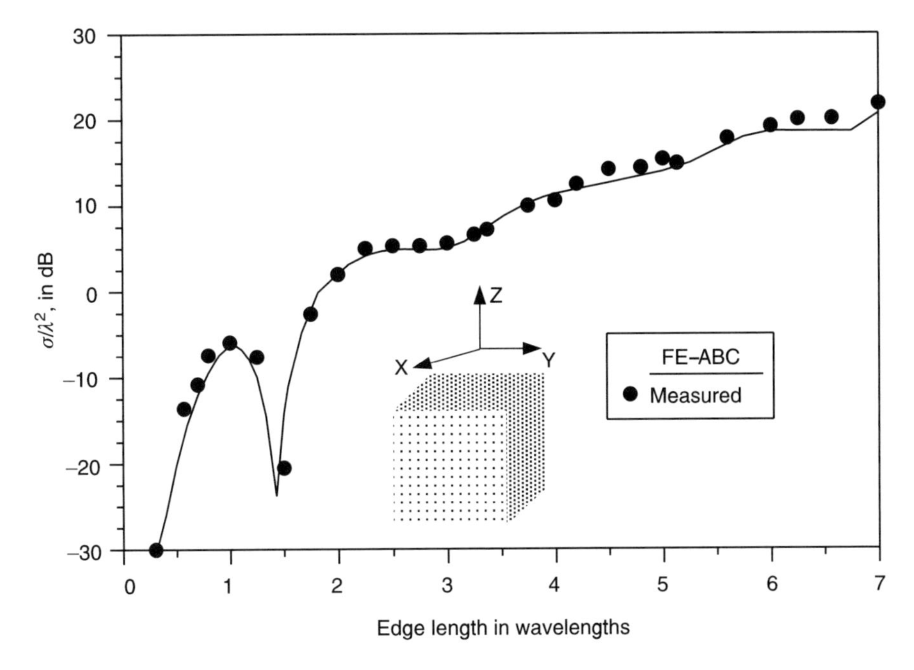

Figure 6.9 Backscatter RCS of a perfectly conducting cube at normal incidence as a function of edge length.

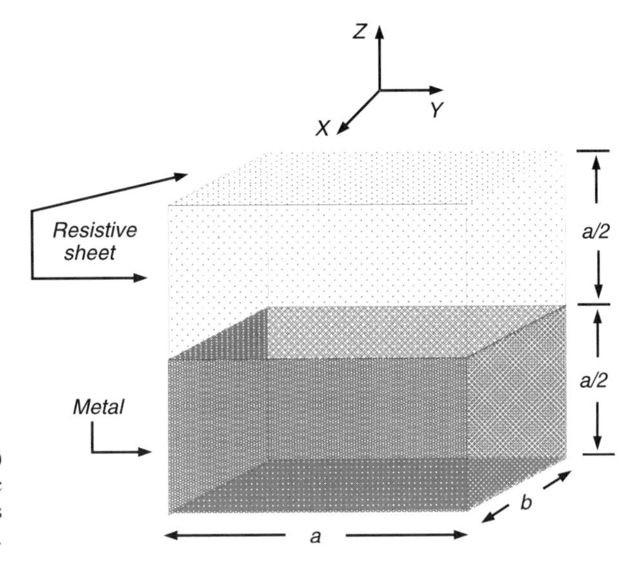

Figure 6.10 Geometry of cube ($a = b = 0.5\lambda$) consisting of a metallic section and a dielectric section ($\epsilon_r = 2 - j2$), where the latter is bounded by a resistive surface having $R = Z_0$.

example is used to validate the implementation of composite dielectric structures in the FE–ABC technique. In Fig. 6.11, we compare a principal plane backscatter pattern (both polarizations) obtained from the 3D FE–ABC implementation with data computed using a traditional moment method code.[2] The computed curve is again seen to follow the reference data closely. For the FE–ABC solution, the

[2]Courtesy of Northrop Corp., B2 Division, Pico Rivera, CA.

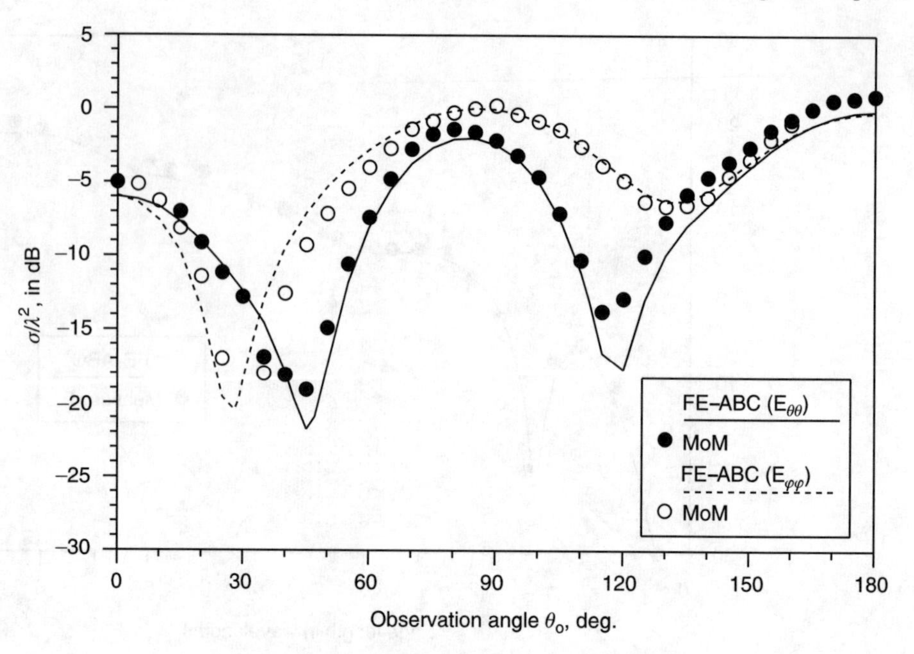

Figure 6.11 RCS pattern in the *xz* plane for the composite cube shown in Fig. 6.10. The lower half of the cube is metallic while the upper half is air-filled with a resistive card draped over it.

scatterer was enclosed within a cubical outer boundary placed only 0.3λ away from the scatterer. This resulted in a 30,000 unknown system which converged to the solution in about 400 iterations when the Sommerfeld radiation condition is employed to terminate the mesh and in 1600 iterations when the second order ABC was used. Increased iteration count for higher order ABCs is a direct result of the shift in the spectrum of the matrix. Higher order ABCs usually result in more eigenvalues of the coefficient matrix to shift toward the negative real axis. For this geometry, the second-order ABC did not provide a significant improvement in accuracy (only about 0.1 dB) over the first-order condition. However, this is not true in all cases as will be demonstrated later.

The problem size with the conformal mesh termination is much smaller than the 40,000 unknown system which results when the same target is enclosed in a spherical termination boundary. The decrease in the unknown count is even more dramatic as we go to larger scatterers. The same case was run with a higher discretization resulting in a system of 50,000 unknowns; however, there was no significant difference in the far-field values with the earlier case. The geometry for the backscatter pattern shown in Fig. 6.12 is the same as the geometry drawn in Fig. 6.10 with the air-filled section now occupied by a lossy dielectric having $\epsilon_r = 2 - j2$. The backscatter echo-area pattern for the $\phi\phi$ polarization as computed by the FE–ABC computer program is again seen to be in good agreement with corresponding moment method data.[3]

[3]Courtesy of Northrop Corp., B2 Division, Pico Rivera, CA.

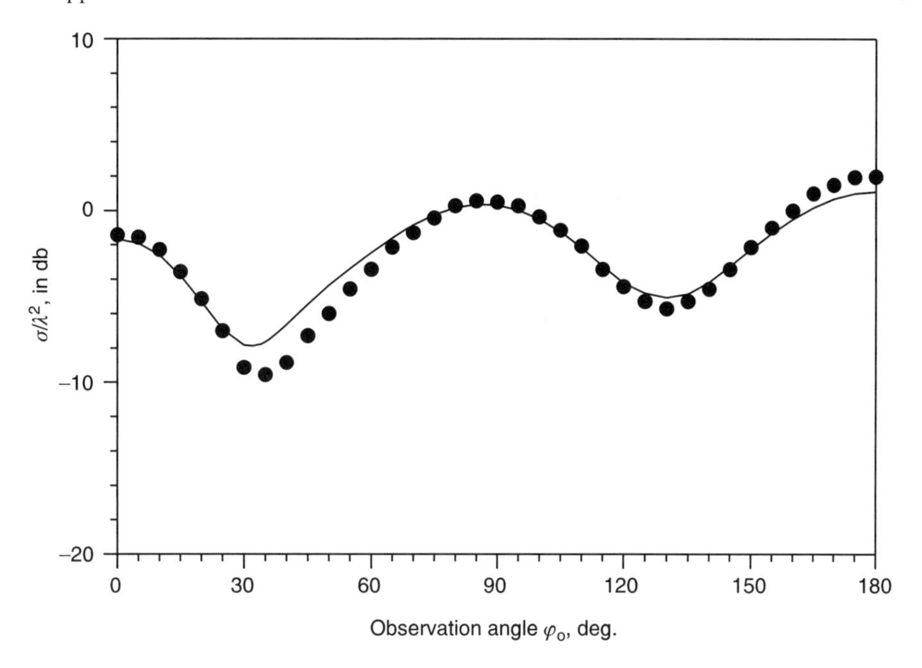

Figure 6.12 RCS pattern in the xz plane for the composite cube shown in Fig. 6.10. The composition of the cube is the same as in Fig. 6.11, except that the air-filled portion now consists of dielectric. The solid curve is the FE–ABC pattern, and the black dots are moment method (MoM) data for the $E_z^{\text{inc}} = 0$ polarization.

6.4.1.2 Inlets. As another example, we compute the scattering from perfectly conducting inlets. The aperture of an inlet usually has a large radar cross section around normal incidence. Therefore, a good understanding of its scattering characteristics is critical if measures need to be taken for reducing its echo-area. A different method for simulating electrically large jet-engine inlets can be found in [27]. An accurate computer simulation of such a geometry provides a cost-effective and ready way of allowing the designer to experiment with complex material fillings to meet design specifications. All the examples shown here are for empty inlets due to lack of reference data for more complicated structures. However, it is just as easy for finite elements to model empty inlets as arbitrarily filled ones.

Let us consider the perfectly conducting rectangular inlet with dimensions $1\lambda \times 1\lambda \times 1.5\lambda$. For the plots shown in Figs. 6.13 and 6.14, the target was enclosed within a sphere of radius 1.35λ, which is only about 0.35λ from the farthest edge of the scatterer. This resulted in a system of 224,476 unknowns and converged in an average of 785 seconds per incidence angle on a 56-processor KSR1. The computed values from the FE–ABC code agree very well with measured data for both HH and VV polarizations. We include the example of the spherical mesh boundary to illustrate how quickly a relatively small structure ($1.5\lambda^3$) can become prohibitively expensive in terms of computing resources and time.

To reduce the computational demands, the natural choice is to use the conformal mesh termination scheme formulated in the previous section and utilized in the last example. Therefore, instead of using a spherical mesh truncation surface, the

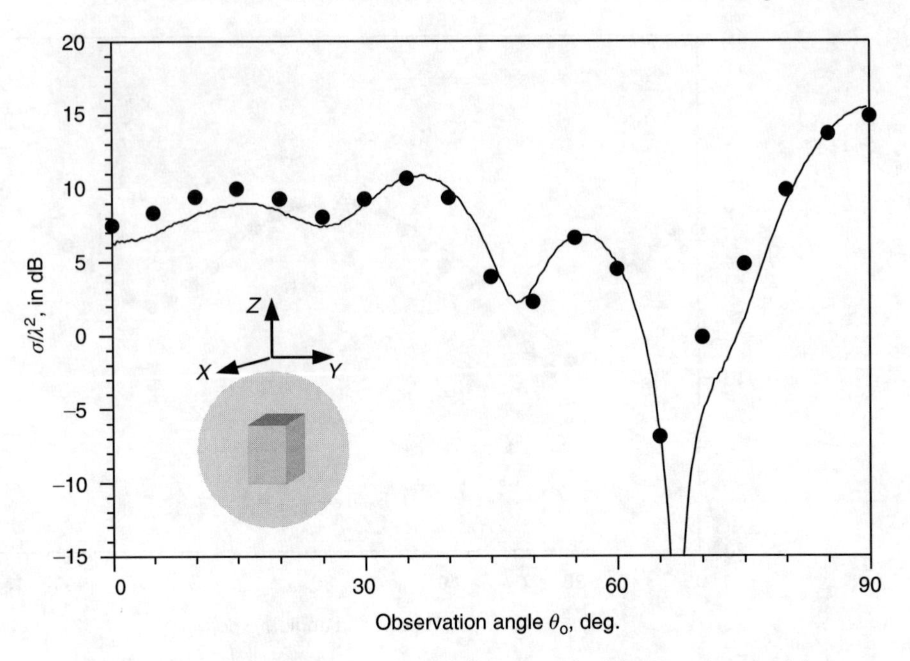

Figure 6.13 Backscatter pattern of a metallic rectangular inlet ($1\lambda \times 1\lambda \times 1.5\lambda$) for HH polarization. Black dots indicate computed values, and the solid line represents measured data [28]. Mesh termination surface is spherical.

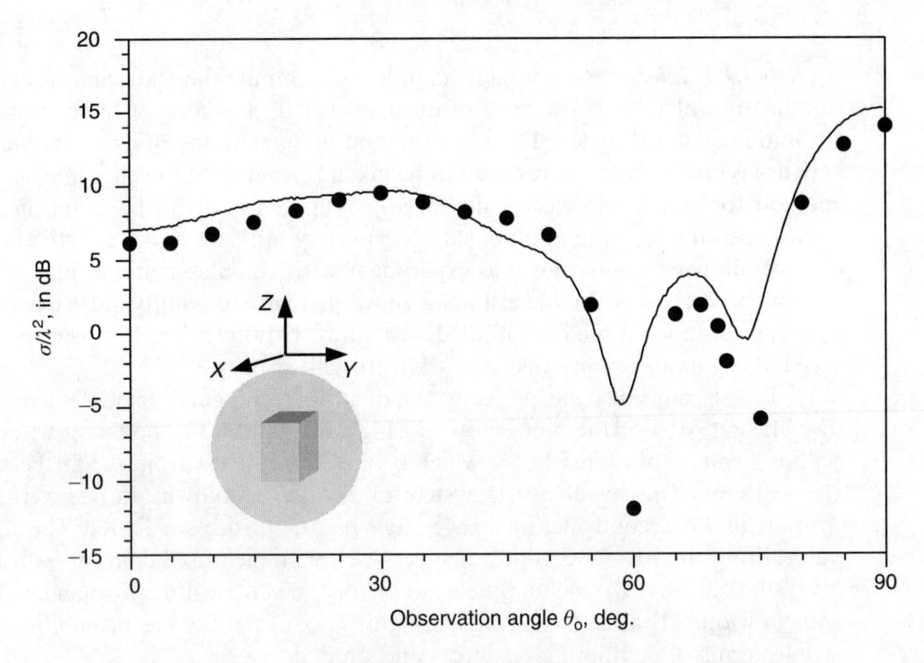

Figure 6.14 Backscatter pattern of a metallic rectangular inlet ($1\lambda \times 1\lambda \times 1.5\lambda$) for VV polarization. Black dots indicate computed values, and the solid line represents measured data [28]. Mesh termination surface is spherical.

mesh can be terminated with a rectangular box placed only 0.35λ away from the scatterer (see inset of Fig. 6.15). The problem size reduces dramatically to 145,000 unknowns, a 35% reduction over the spherical mesh termination scheme. The convergence time for each excitation vector is about 220 seconds, less than 4 minutes, when run on all 56 processors of a KSR1. The computed values are again compared with measured data for the HH polarization in Fig. 6.15; the agreement is excellent, albeit a bit worse than the spherical case. However, this observation is overshadowed by the fact that the problem size has been reduced by more than a third and computing time by about a fourth. Thus in many cases, a conformal ABC makes it possible to obtain a solution with the resident storage capacity and within a reasonable time interval. The results for the VV polarization with a rectangular mesh termination are equally accurate.

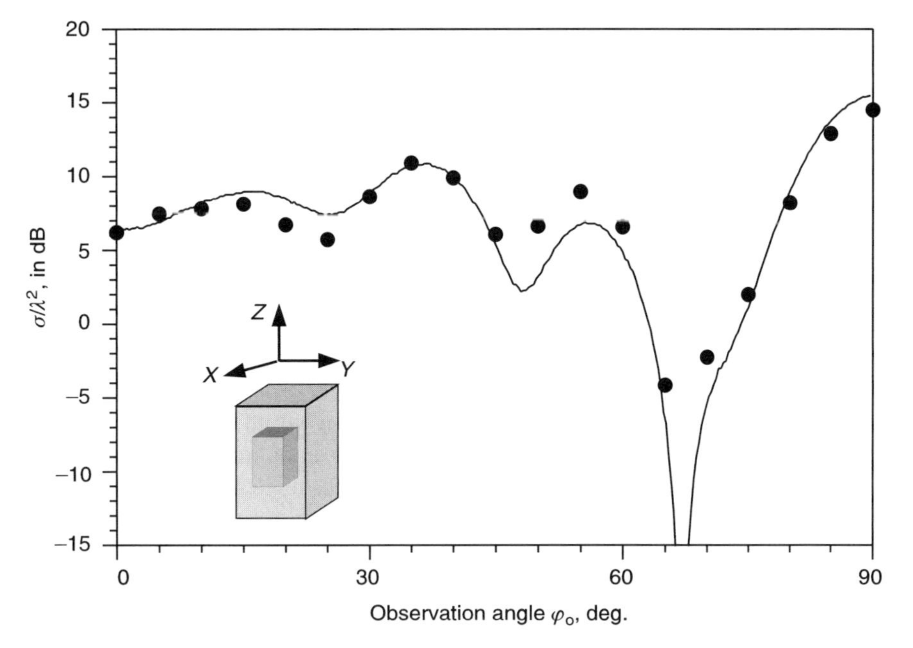

Figure 6.15 Backscatter pattern of a metallic rectangular inlet (1λ × 1λ × 1.5λ) for HH polarization. Black dots indicate computed values, and the solid line represents measured data [28]. Mesh termination surface is piecewise planar.

Next, consider the scattering from a perfectly conducting cylindrical inlet. Even though integral equation codes are more efficient for such bodies of revolution, the goal with this test is to examine the performance of the conformal absorbing boundary conditions. Moreover, the real strength of finite elements lies in its ease of handling material inhomogeneities encountered in practical structures. The target is a perfectly conducting cylindrical inlet having a diameter of 1.25λ and a height of 1.875λ. A rectangular outer boundary was first used and placed 0.45λ from the farthest edge of the target to enclose the scatterer. For this case, the radar cross section for a ϕ-polarized incident wave in the yz plane is shown in Fig. 6.16 and is compared with measured data. The agreement is quite good for all lobes except the

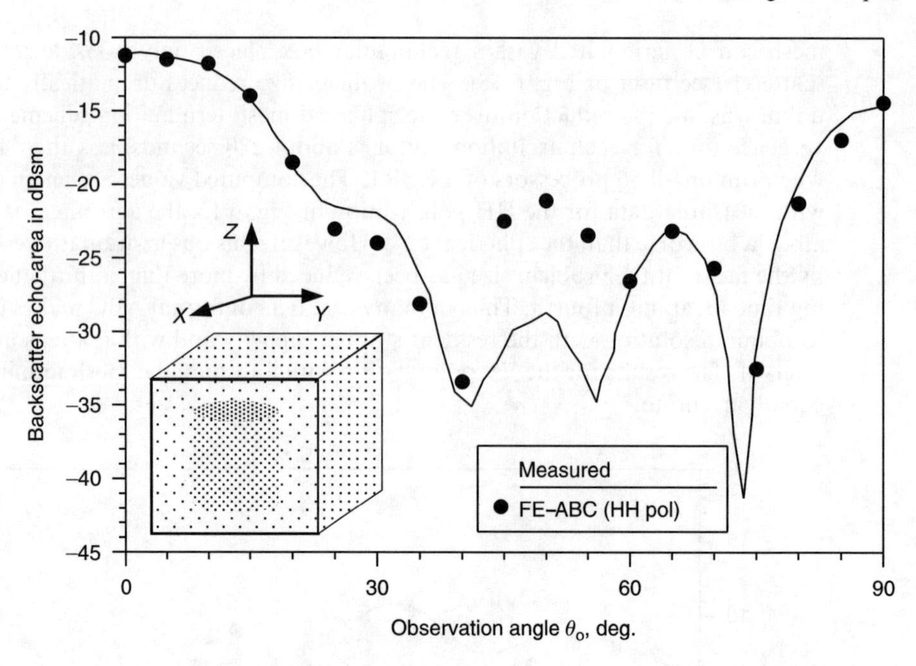

Figure 6.16 Backscatter pattern of a perfectly conducting cylindrical inlet (diameter $= 1.25\lambda$, height $= 1.875\lambda$) for HH polarization. The solid line indicates measured data [30], and the black dots indicate computed data. Mesh termination surface is a rectangular box.

third. However, the backscatter echo-area computed for the same geometry by Shankar [29] using the finite difference-time domain method agrees with the computed results via the FE–ABC for all incidence angles. In [29], the absorbing boundary was placed *a few* wavelengths from the scattering structure.

Next, we employ a conformal termination scheme with a cylindrical surface for mesh truncation. This example further demonstrates that a truly conformal mesh boundary is possible with the ABCs derived earlier in the chapter. The cylindrical outer boundary was placed about 0.45λ from the target, and the computed RCS for a ϕ-polarized incident wave is given in Fig. 6.17. For purposes of validation, the results are compared with measured data [30] and with a body of revolution code [31] (Fig. 6.17). As can be observed from Figs. 6.16 and 6.17, the far-field results for a cylindrical and a rectangular termination do not differ significantly. However, the savings in computational cost is quite impressive. The cylindrical mesh termination has only 144,392 unknowns compared to the 191,788 unknowns for a rectangular truncation scheme. A spherical mesh termination would have swelled to about 265,000 unknowns, given identical sampling density and outer boundary distance. Thus, the problem size was reduced by about 45% and the computation time by a similar, if not greater, amount. The savings in computational resources is quite significant even when we compare the rectangular and cylindrical termination schemes—a 25% reduction in problem size and a similar decrease in computation time. This phenomenon is only to be expected from the geometrical point of view and, of course, improves with the problem size.

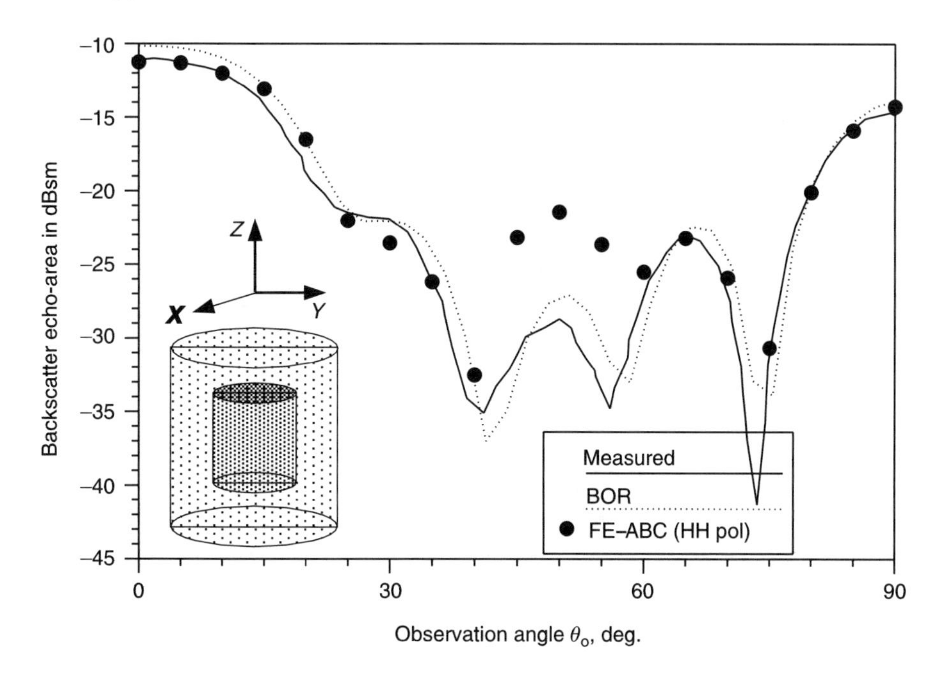

Figure 6.17 Backscatter pattern of a perfectly conducting cylindrical inlet (diameter $= 1.25\lambda$, height $= 1.875\lambda$) for HH polarization. Black dots indicate computed values, the solid line represents measured data [30], and the dotted line is the body of revolution data [31]. Mesh termination surface is a circular cylinder.

6.4.1.3 Plate. The motivation for testing the FE–ABC method for perfectly conducting plates is twofold. It is usually very difficult to model the scattering from the edges of the plate even using integral equation methods. Therefore, in this section we present examples to see how the method performs at edge-on incidence. Second, we examine the performance of termination boundaries of esoteric shapes. The first choice is to enclose the plate in a rectangular box. The second choice is to use a box with half cylinders attached to the faces normal to the plane of the plate—the reasoning being that because the edge of the plate behaves like a line source and scatters cylindrical waves, a cylindrical mesh termination is most suitable for wave absorption. Both mesh termination schemes require approximately the same number of unknowns; the superiority of one over the other is thus decided only on the basis of accuracy of the computed backscatter values.

The test case is a $3.5\lambda \times 2\lambda$ perfectly conducting rectangular plate. In Fig. 6.18, we plot the backscatter pattern for the $\phi\phi$ polarization in the xz plane, i.e., over the long side of the plate. Generally, the agreement with reference data is quite good. However, the backscatter echo-area at edge-on incidence is not calculated accurately. Thus, we need to check whether other mesh termination shapes will perform better.

In Fig. 6.19, we show the RCS of the conducting plate in the yz plane, i.e., over its short side, for the $\phi\phi$ polarization. The backscatter echo-area for edge-on incidence is picked up very well for a rectangular-cylindrical termination, whereas a rectangular truncation scheme gives completely incorrect results. These two schemes

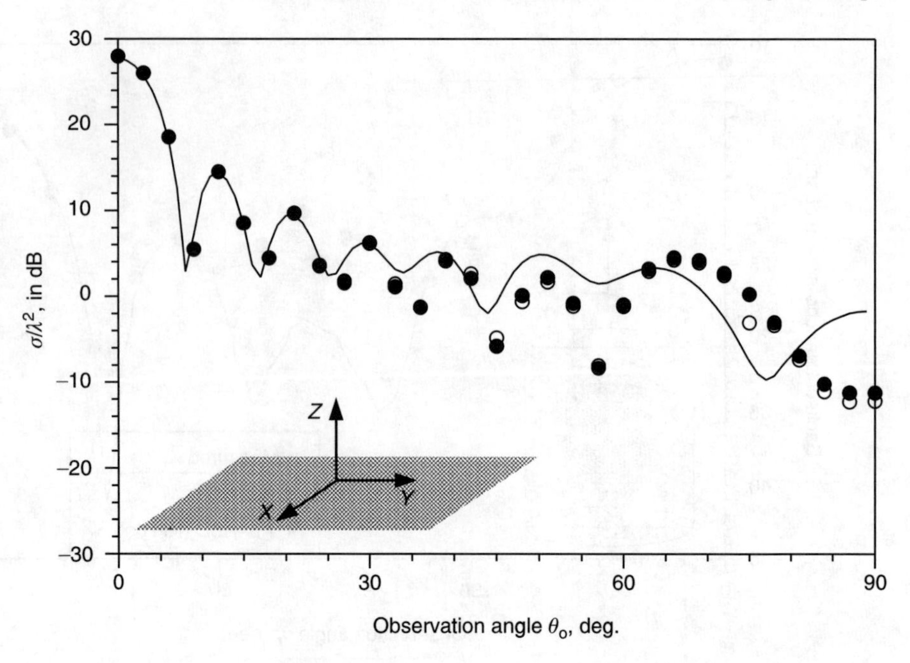

Figure 6.18 Backscatter pattern ($\sigma_{\phi\phi}$) of a $3.5\lambda \times 2\lambda$ perfectly conducting plate in the *xz* plane. The white dots indicate box termination; the black dots represent a combined box-cylinder termination.

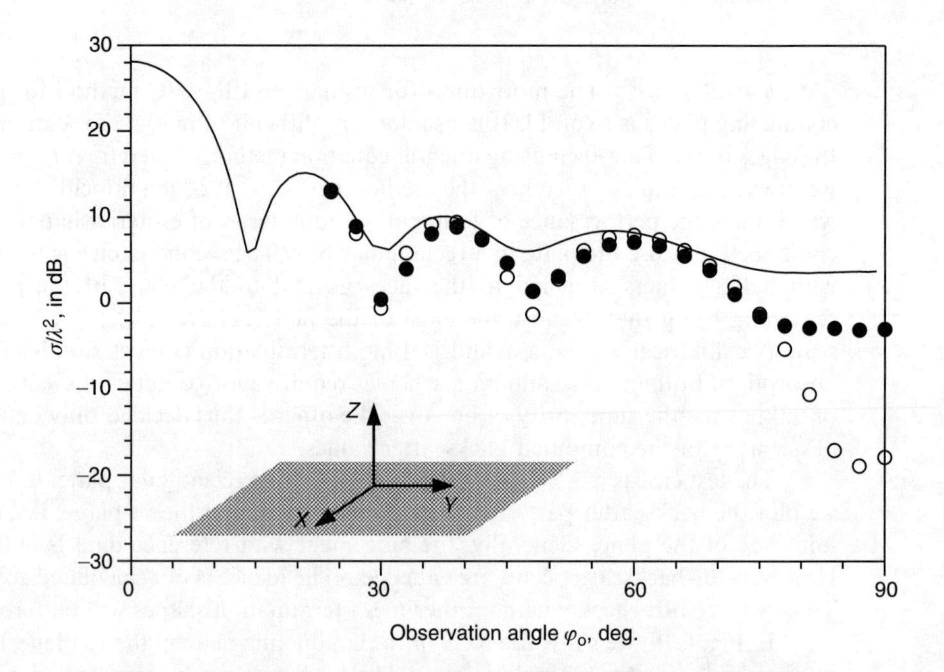

Figure 6.19 Backscatter pattern ($\sigma_{\phi\phi}$) of a $3.5\lambda \times 2\lambda$ perfectly conducting plate in the *yz* plane. The white dots indicate box termination; the black dots represent a combined box-cylinder termination.

have approximately the same storage requirement; in fact, the box-cylinder combination yields a slightly smaller system of equations. This example truly illustrates the power of a conformal truncation scheme composed of simple shapes; not only are the results far more accurate but even the storage requirement is slightly less.

In the above simulations, the boundary was terminated at 0.35λ from the flat face of the plate and 0.5λ from the edges of the plate. To test the accuracy of the ABC method as a function of mesh termination distance, we consider the backscatter patterns from the edges of the plate as the mesh termination distance is increased. Figure 6.20 shows that the backscatter values from the plate edges slowly take the shape of the reference data as the mesh truncation distance is increased. However, even though the results are seen to approach reference data as the mesh boundary is pushed farther away from the plate edge, the experiment also shows us the limitations of this technique. The FE–ABC method is a true 3D technique; therefore, although it is possible to use it in solving 2D problems, the associated computational cost makes it unjustifiably expensive. A surface formulation using integral equations or a hybrid finite element–boundary integral formulation (see Chapter 7) is more efficient for such applications.

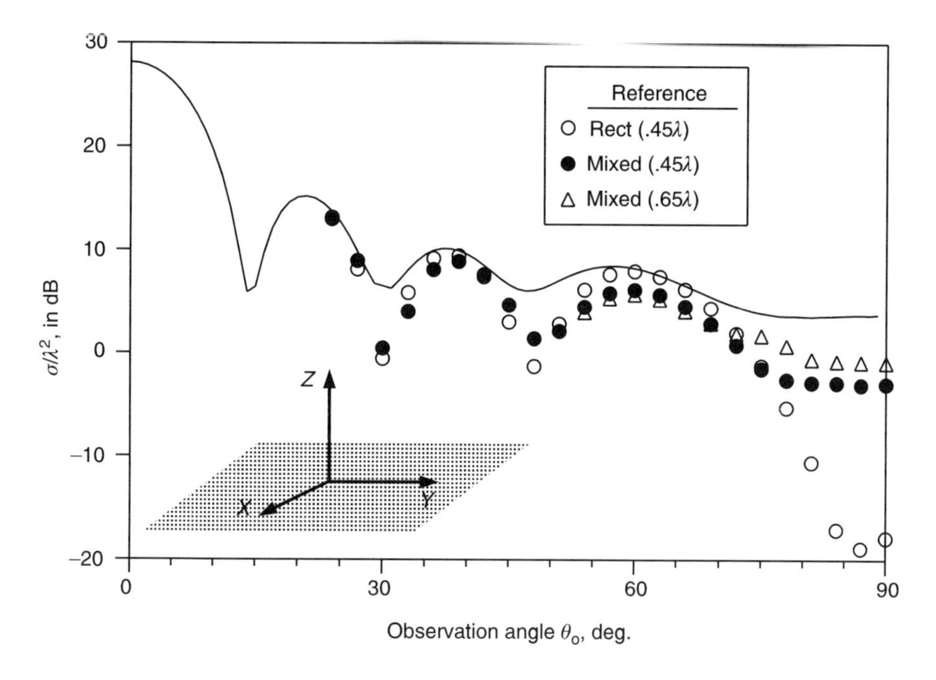

Figure 6.20 Backscatter pattern ($\sigma_{\phi\phi}$) of a 3.5λ × 2λ perfectly conducting plate in the yz plane. The numbers in the legend indicate mesh termination distance from the plate edges.

6.4.1.4 Conesphere. The last scattering example geometry presented in this chapter is unique in its own way. A conesphere is a hemisphere attached to a cone, and this shape is typical for airborne structures. It is a difficult geometry to mesh since a surface singularity exists at the tip of the cone. The singularity can be removed in two ways: (i) by creating a small region near the tip and detaching it from the surface or (ii) by chopping off a small part near the tip of the cone. The second option inevitably leads to small inaccuracies for backscatter from the conical tip; however, this option was chosen since the conical angle in our tested geometry is extremely small (around 7°) and the mesh generator fails to mesh the first case on numerous occasions. In Fig. 6.21, we plot the backscatter patterns of a 4.5λ long conesphere having a radius of 0.5λ for $\theta\theta$ and $\phi\phi$ polarizations. The mesh truncation surface is a rectangular box placed 0.4λ from the surface of the conesphere. The far-field results compare extremely well with computations from a body of revolution code [32].

In this conesphere example, the choice of a piecewise planar rectangular boundary might be questioned. A truly conformal boundary would have been a larger conesphere placed an appropriate distance from the target conesphere. However, there are two serious limitations. As mentioned in the derivation of the ABCs earlier in this chapter, ABCs need a piecewise smooth surface where the scattered or radiated field can be expressed in terms of an infinite series in $1/r$. ABCs have also been found to fail for concave and re-entrant structures. Thus, using a truly conformal ABC surface in the form of a conesphere is not a good idea. The second hurdle in using arbitrary conformal surfaces as the mesh termination boundary is the difficulty in implementation. Surfaces of arbitrary curvature will usually lead to loss of symmetry in the finite element matrix, thus resulting in a more complicated solution process for a small reduction in the size of the problem. In order to address these problems, mesh termination strategies are being investigated which use artificial absorbers instead of ABCs. Applications of artificial absorbers will be shown in the next section.

6.4.2 Antenna and Circuit Examples

This section demonstrates the application of ABCs and artificial absorbers for truncating finite element meshes in computing the radiation patterns of antennas and reflection coefficients of microwave circuits. The implementation of the ABCs is very similar to the scattering formulation except that there is the additional aspect of source modeling. This is outlined in detail in the next chapter.

6.4.2.1 Conformal Patch Antennas. Figure 6.22 illustrates the general configuration of printed antennas on conformal platforms. Various patch configurations situated on cylindrical platforms were considered for the purpose of examining the performance of ABCs for this application. Among those studied, we present the analysis of a 2 cm × 3 cm patch antenna printed atop a metallic cavity which is filled with a 5 cm × 6 cm × 0.07874 cm substrate having a dielectric constant of $\epsilon_r = 2.17$. This cavity is recessed in a metallic cylinder (whose infinite dimension is along the z-axis) with a radius of 15.28 cm, and the scattering and radiation calculations for this patch were carried out at 3 GHz. The second-order vector ABC is placed $\tau\lambda_0$ from the cavity aperture, while the lateral walls of the ABC were placed $0.5\lambda_0$ from the cavity aperture. The H-plane antenna pattern for an axially polarized patch is shown in Fig. 6.23 where the probe feed is placed at $\phi_f = 0°$ and $z_f = -0.375$ cm (with $\phi_f = 0°$ and $z_f = 0$ corresponding to the center of the patch). As in the case of

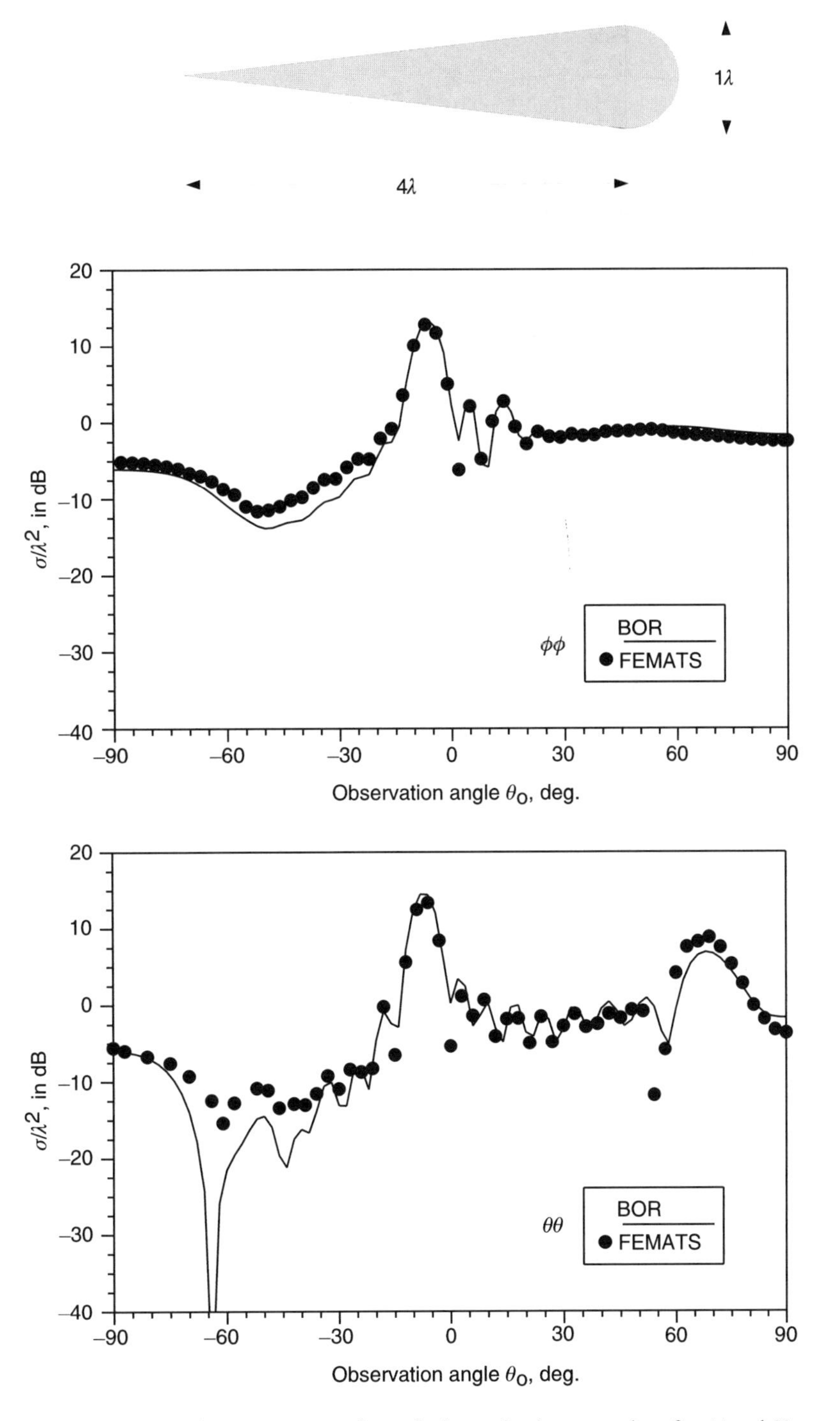

Figure 6.21 Backscatter pattern of a perfectly conducting conesphere for $\phi\phi$ and $\theta\theta$ polarizations. Black dots indicate computed values using the FE–ABC code (referred to as FEMATS), and the solid line represents data from a body of revolution code [32]. Mesh termination surface is a rectangular box.

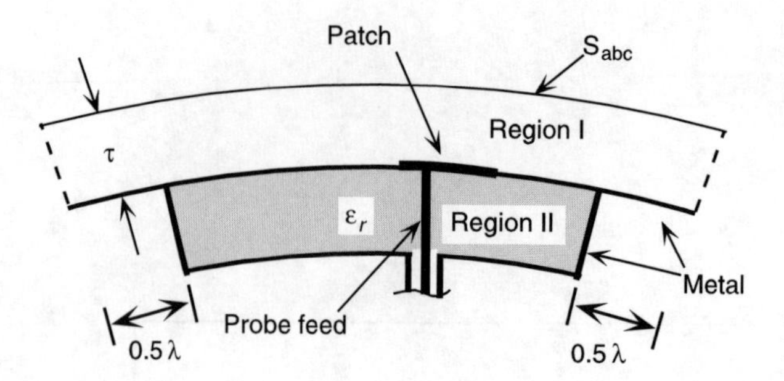

Figure 6.22 Cavity-backed patch antenna with ABC mesh termination.

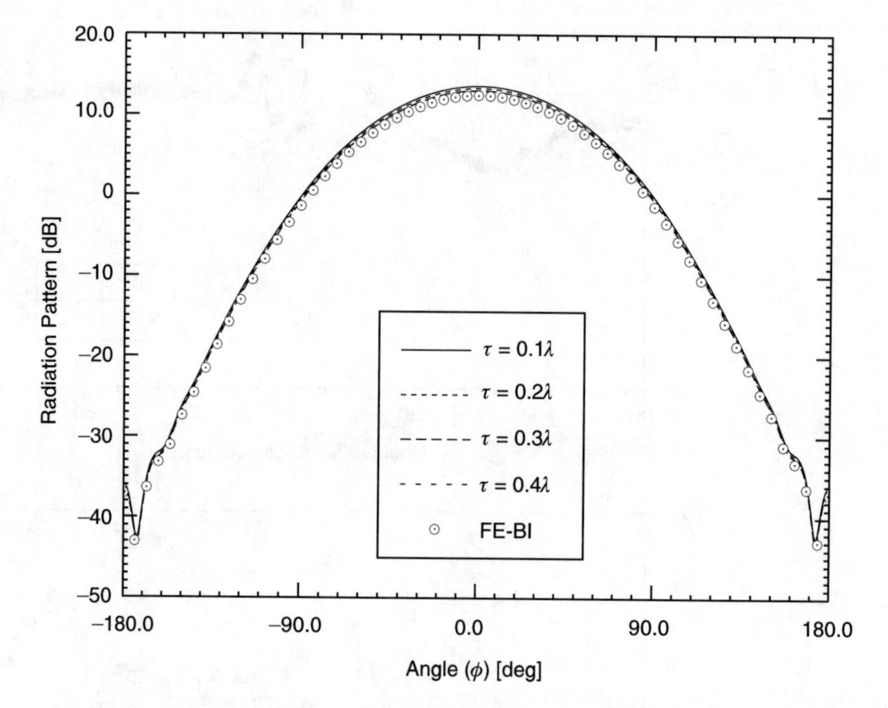

Figure 6.23 Convergence of the FE–ABC method for computing the H-plane radiation pattern of a cavity-backed axially polarized patch. The reference data is provided by a rigorous FE–BI formulation for the same cavity-backed antenna.

scattering, the radiation pattern calculated via the FE–ABC method is seen to be in excellent agreement with the pattern computed by the more rigorous finite element–boundary integral (FE–BI) [33] approach even when the mesh is terminated only 0.3λ from the aperture.

6.4.2.2 Patch Antenna on Ogive. In this example, an artificial absorber is used for terminating the radiation boundary. Artificial absorbers are especially useful for such situations when a truly conformal ABC may be much more difficult to implement. Figure 6.24(a) shows the setup, where a 2 cm × 3 cm rectangular patch

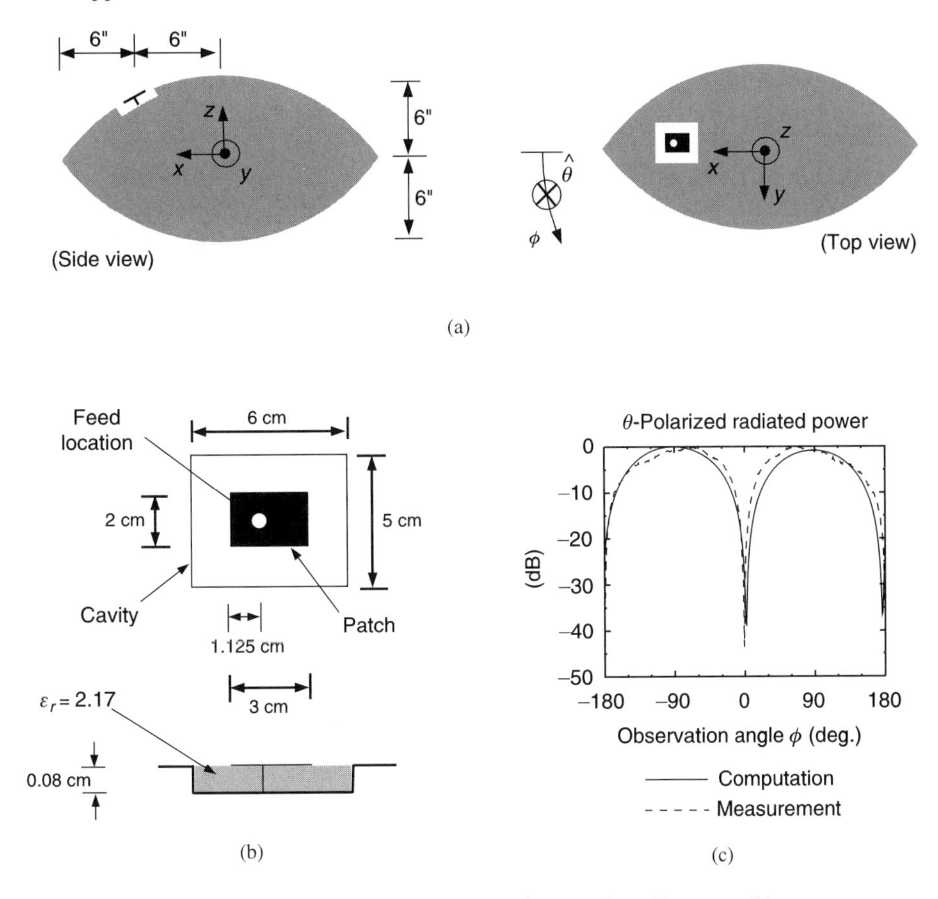

Figure 6.24 Cavity-backed rectangular patch on ogive: (a) setup, (b) antenna dimensions, (c) comparison with the measurement. [*After Özdemir and Volakis*, © *IEEE, 1997* [*34*].]

is placed on the aperture of a 5 cm × 6 cm × 0.08 cm rectangular cavity recessed in the ogive's surface. Also, Fig. 6.25 shows the artificial absorber used for terminating the mesh around the patch. The ogive is electrically much larger than the cavity (ten times the maximum cavity dimension). The θ-polarized radiation does

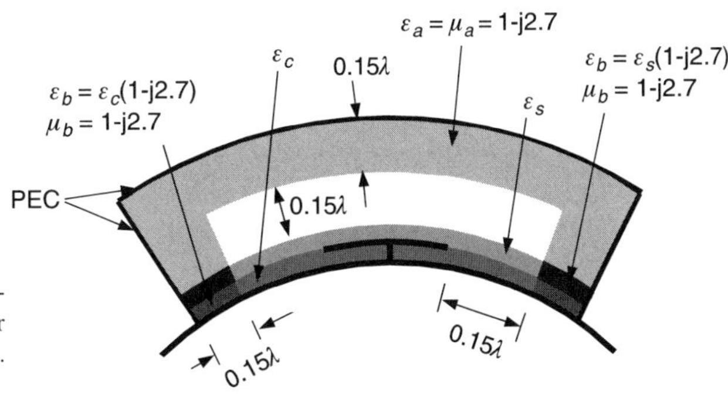

Figure 6.25 Illustration of the mesh termination approach using artificial absorbers for antenna analysis on doubly curved surfaces. [*Courtesy of T. Özdemir.*]

not interact with the edges of the ogive and can therefore be computed by localizing the mesh near the cavity on the ogive. Thus, the radiated power pattern accounts only for the antenna (and the curvature of the platform), but does not include interactions with the ogive's tips. Figure 6.24(c) shows the computed θ-polarized radiation as compared to the measurement [35]. The agreement with measured data is very good for this polarization. However, predicting the ϕ-polarized radiation (not shown) requires modeling the entire ogive as this particular polarization has a vertical surface field component which is known to cause diffraction from the ogive's tips. A way to account for such secondary diffractions is to interface the finite element–artificial absorber (FE–AA) method with a high-frequency technique, and an encouraging study in this direction has been carried out in [36].

6.4.2.3 Cylindrical Via. For all practical problems in circuit design, a full wave analysis of the circuit components can be carried out only for small parts of the circuit. Therefore, in analyzing microwave circuit problems, ABCs may be required to predict radiation loss from circuit elements, for analyzing circuit discontinuities or for modeling small critical paths in a large, complicated integrated circuit design. Figure 6.26 shows a 0.77-mm radius cylindrical via discontinuity connecting

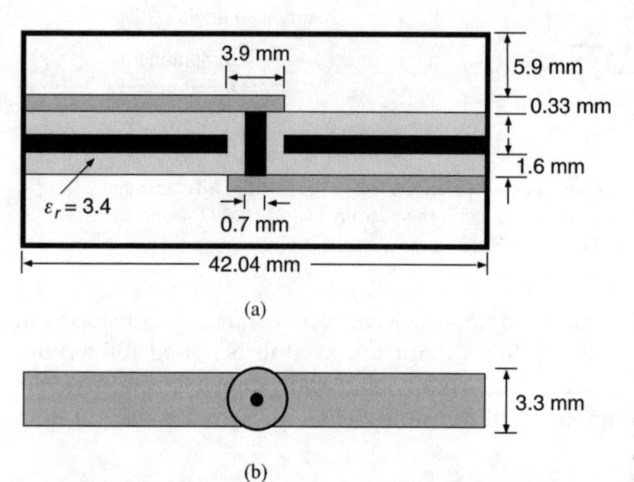

Figure 6.26 (a) Side view of cylindrical via connecting two microstrip lines. (b) Top view. [*After Wang et al.* © *IEEE, 1994* [37].]

two striplines, each 3.3 mm by 0.33 mm. The metallization in the dielectric serves as a finite thickness (0.33 mm) ground plane. In [37], the authors analyze the structure with open and closed side walls of the top microstrip. As seen in Fig. 6.27, the power loss due to the open walls at frequencies below 10 GHz is negligible. However, for frequencies above 10 GHz there is a significant amount of radiation loss from the via.

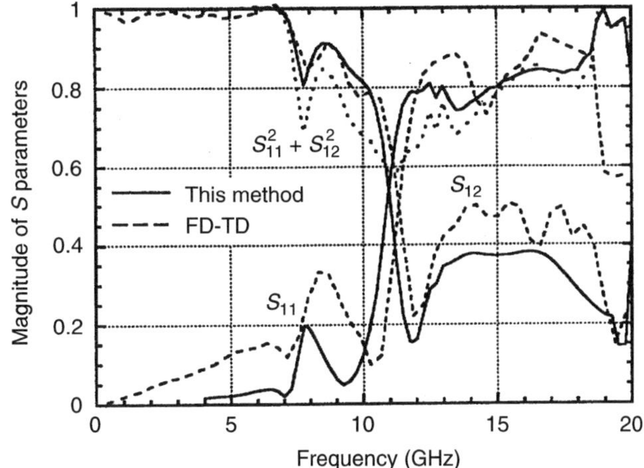

Figure 6.27 Comparison of scattering parameters of cylindrical via for open and closed walls of top microstrip. [*After Wang et al.* © *IEEE, 1994* [37].]

APPENDIX: DERIVATION OF SOME VECTOR IDENTITIES

The curl of a vector in the Dupin coordinate system is given by

$$\nabla \times \mathbf{E} = \nabla_T \times \mathbf{E} + \hat{\boldsymbol{n}} \times \frac{\partial \mathbf{E}}{\partial n} \tag{6.76}$$

where $\nabla_T \times \mathbf{E}$ is called the surface curl involving only the tangential derivatives and is defined as

$$\nabla_T \times \mathbf{E} = -\hat{\boldsymbol{n}} \times \nabla E_n + \hat{\mathbf{t}}_2 \kappa_1 E_{t_1} - \hat{\mathbf{t}}_1 \kappa_2 E_{t_2} + \hat{\boldsymbol{n}} \nabla \cdot (\mathbf{E} \times \hat{\boldsymbol{n}}) \tag{6.77}$$

As before, κ_1 and κ_2 denote the principal curvatures of the surface under consideration, E_{t_1}, E_{t_2} are the tangential components, and E_n is the normal component of the vector $\mathbf{E}$ on the surface.

We are interested in the evaluation of the three vector identities given earlier in the chapter. Let us consider simplifying the tangential components of the curl of a vector, $\mathbf{E}$ in this case. Using the definition of the curl given above, we have

$$\begin{aligned} \hat{\boldsymbol{n}} \times \nabla \times \mathbf{E} &= -(\hat{\boldsymbol{n}} \times \hat{\boldsymbol{n}} \times \nabla) E_n - \kappa_1 E_{t1} \hat{\mathbf{t}}_1 - \kappa_2 E_{t2} \hat{\mathbf{t}}_2 + \hat{\boldsymbol{n}} \times \left(\hat{\boldsymbol{n}} \times \frac{\partial \mathbf{E}}{\partial n} \right) \\ &= \nabla_t E_n - \left[\left(\frac{\partial E_{t1}}{\partial n} + \kappa_1 E_{t1} \right) \hat{\mathbf{t}}_1 + \left(\frac{\partial E_{t2}}{\partial n} + \kappa_2 E_{t2} \right) \hat{\mathbf{t}}_2 \right] \\ &= \nabla_t E_n + \hat{\boldsymbol{n}} \times \nabla \times \mathbf{E}_t \end{aligned} \tag{6.78}$$

where $-(\hat{\boldsymbol{n}} \times \hat{\boldsymbol{n}} \times \nabla) = \nabla_t$. The first vector identity is, therefore, easily proved.

Next, we will prove the second of the three identities. We start with the term $\hat{\boldsymbol{n}} \times \nabla \times \nabla_t E_n$ and simplify it using the definition of the curl of a vector given above.

$$\begin{aligned}\hat{\boldsymbol{n}} \times \nabla \times \nabla_t E_n &= \hat{\boldsymbol{n}} \times \nabla \times \left[\nabla E_n - \hat{\boldsymbol{n}}\frac{\partial E_n}{\partial n}\right] \\ &= -\hat{\boldsymbol{n}} \times \nabla \times \hat{\boldsymbol{n}}\frac{\partial E_n}{\partial n} \\ &= (\hat{\boldsymbol{n}} \times \hat{\boldsymbol{n}} \times \nabla)\frac{\partial E_n}{\partial n} \\ &= -\nabla_t\left(\frac{\partial E_n}{\partial n}\right) \end{aligned} \tag{6.79}$$

Since $\nabla \cdot \mathbf{E} = \nabla \cdot \mathbf{E}_t + (\nabla \cdot \hat{\boldsymbol{n}})E_n + \partial E_n/\partial n$, we can simplify the above relation even further by substituting the appropriate expression for the normal derivative of the normal component of the electric field and using the fact that the electric field is divergence-free in a source-free region.

$$\begin{aligned}\hat{\boldsymbol{n}} \times \nabla \times \nabla_t E_n &= \nabla_t(\nabla \cdot \mathbf{E}_t) + (\nabla \cdot \hat{\boldsymbol{n}})\nabla_t E_n \\ &= \nabla_t(\nabla \cdot \mathbf{E}_t) + 2\kappa_m \nabla_t E_n \end{aligned} \tag{6.80}$$

where $\kappa_m = (\kappa_1 + \kappa_2)/2$ is the mean curvature.

The proof of the third identity is more complicated because it involves two curl operations on the electric field. We first need to switch the positions of the outermost $\hat{\boldsymbol{n}}\times$ and the $\nabla\times$ operators to arrive at a simplified form of the rather complex expression. Therefore,

$$\begin{aligned}\hat{\boldsymbol{n}} \times \nabla \times (\hat{\boldsymbol{n}} \times \nabla \times \mathbf{E}) &= \nabla \times \{\hat{\boldsymbol{n}} \times \hat{\boldsymbol{n}} \times \nabla \times \mathbf{E}\} - \hat{\boldsymbol{n}}\nabla_t \cdot (\hat{\boldsymbol{n}} \times \nabla \times \mathbf{E}) \\ &\quad - \Delta\kappa\{(\nabla \times \mathbf{E})_{t_2}\hat{\mathbf{t}}_1 + (\nabla \times \mathbf{E})_{t_1}\hat{\mathbf{t}}_2\} \\ &= -\nabla \times \{\nabla \times \mathbf{E} - \hat{\boldsymbol{n}}(\nabla \times \mathbf{E})_n\} - \hat{\boldsymbol{n}}\nabla_t \cdot (\hat{\boldsymbol{n}} \times \nabla \times \mathbf{E}) \\ &\quad - \Delta\kappa\{(\nabla \times \mathbf{E})_{t_2}\hat{\mathbf{t}}_1 + (\nabla \times \mathbf{E})_{t_1}\hat{\mathbf{t}}_2\} \end{aligned} \tag{6.81}$$

Now we use the fact that the electric field satisfies the wave equation to reduce the expression even further.

$$\begin{aligned}\hat{\boldsymbol{n}} \times \nabla \times (\hat{\boldsymbol{n}} \times \nabla \times \mathbf{E}) &= -k_0^2\mathbf{E} + \nabla \times \{\hat{\boldsymbol{n}}(\nabla \times \mathbf{E})_n\} + k_0^2\hat{\boldsymbol{n}}E_n \\ &\quad - \Delta\kappa\{(\nabla \times \mathbf{E})_{t_2}\hat{\mathbf{t}}_1 + (\nabla \times \mathbf{E})_{t_1}\hat{\mathbf{t}}_2\} \\ &= \nabla \times \{\hat{\boldsymbol{n}}(\nabla \times \mathbf{E})_n\} - k_0^2\mathbf{E}_t - \Delta\kappa\{(\nabla \times \mathbf{E})_{t_2}\hat{\mathbf{t}}_1 + (\nabla \times \mathbf{E})_{t_1}\hat{\mathbf{t}}_2\} \end{aligned} \tag{6.82}$$

where $\Delta\kappa = \kappa_1 - \kappa_2$.

Thus, we have shown that all three identities hold as long as the vector, **E** in this case, is divergenceless and satisfies the vector wave equation.

REFERENCES

[1] A. Chatterjee, J. M. Jin, and J. L. Volakis. Edge-based finite elements and vector ABCs applied to 3D scattering. *IEEE Trans. Antennas Propagat.*, 41(2):221–226, February 1993.

[2] X. Yuan. Three-dimensional electromagnetic scattering from inhomogeneous objects by the hybrid moment and finite element method. *IEEE Trans. Antennas Propagat.*, 38:1053–1058, 1990.

[3] J. M. Jin and J. L. Volakis. Electromagnetic scattering by and transmission through a three-dimensional slot in a thick conducting plane. *IEEE Trans. Antennas Propagat.*, 39(4):543–550, April 1991.

[4] J. Angelini, C. Soize, and P. Soudais. Hybrid numerical method for harmonic 3D Maxwell equations: Scattering by a mixed conducting and inhomogeneous anisotropic dielectric medium. *IEEE Trans. Antennas Propagat.*, 41(1):66–76, January 1993.

[5] J. D'Angelo and I. D. Mayergoyz. Three-dimensional RF scattering by the finite element method. *IEEE Trans. Magnetics*, 27(5):3827–3832, September 1991.

[6] I. D. Mayergoyz and J. D'Angelo. New finite element formulation for 3D scattering problem. *IEEE Trans. Magnetics*, 27(5):3967–3970, September 1991.

[7] L. C. Kempel and J. L. Volakis. Evaluation of new vector ABCs for conformal printed antennas. 1994 URSI Radio Science Meeting Digest, Seattle, WA.

[8] T. Özdemir and J. L. Volakis. A comparative study of an absorber boundary condition and an artificial absorber for truncating finite element meshes. *Radio Science*, 29:1255–1263, September–October 1994.

[9] W. C. Chew and H. W. Weedon. A 3D perfectly matched medium from modified Maxwell's equations with stretched coordinates. *Microwave Opt. Tech. Lett.*, 7(13), September 1994.

[10] C. H. Wilcox. An expansion theorem for electromagnetic fields. *Comm. Pure Appl. Math.*, 9:115–134, May 1956.

[11] C. T. Tai. *Generalized Vector and Dyadic Analysis*. IEEE Press, New York, 1992.

[12] D. S. Jones. An improved surface radiation condition. *IMA J. Appl. Math.*, 48:163–193, 1992.

[13] J. van Bladel. *Electromagnetic Fields*. Hemisphere Publishing Corp., New York, 1985.

[14] S. M. Rytov. Computation of the skin effect by the perturbation method. *J. Exp. Theor. Phys.*, 10:180–189, 1940. Translation by V. Kerdemelidis and K. M. Mitzner, Northrop Navair, Hawthorne, CA 90250.

[15] A. F. Peterson. Absorbing boundary conditions for the vector wave equation. *Microwave and Opt. Techn. Letters*, 1:62–64, April 1988.

[16] J. P. Webb and V. N. Kanellopoulos. Absorbing boundary conditions for finite element solution of the vector wave equation. *Microwave and Opt. Techn. Letters*, 2(10):370–372, October 1989.

[17] J. P. Berenger. A perfectly matched layer for the absorption of electromagnetic waves. *J. Comp. Phys.*, 114(2):185–200, October 1994.

[18] D. S. Katz, E. T. Thiele, and A. Taflove. Validation and extension to three dimensions of the Berenger PML absorbing boundary conditions for FD-TD meshes. *IEEE Microwave and Guided Wave Letters*, 4(8):268–270, August 1994.

[19] R. Mittra and U. Pekel. A new look at the Perfectly Matched Layer (PML) concept for the reflectionless absorption of electromagnetic waves. *IEEE Microwave and Guided Wave Lett.*, 5(3):84–86, March 1995.

[20] Z. J. Sacks, D. M. Kingsland, R. Lee, and J.-F. Lee. A perfectly matched anisotropic-absorber for use as an absorbing boundary condition. *IEEE Trans. Antennas Propagat.*, 43:1460–1463, 1995.

[21] D. M. Kingsland, J. Gong, J. L. Volakis, and J.-F. Lee. Performance of an anisotropic artificial absorber for truncating finite element meshes. *IEEE Trans. Antennas Propagat.*, 44:975–982, July 1996.

[22] S. Legault, T. B. A. Senior, and J. L. Volakis. Design of planar absorbing layers for domain truncation in FEM application. *Electromagnetics,* 16(4):451–464, July–August 1996.

[23] J. Gong and J. L. Volakis. Optimal selection of a uniaxial artificial absorber layer for truncating finite element meshes. *IEE Electronics Lett.*, 31(18):1559–1561, August 1995.

[24] W. C. Chew and J. M. Jin. Perfectly matched layers in the discretized space: An analysis and optimization. *Electromagnetics,* 16:325–340, 1996.

[25] A. Chatterjee, M. Nurnberger, J. Volakis, and M. Casciato. FEMATS: A general purpose scattering code using the finite element method. In *IEEE National Radar Conference Proceedings*, pp. 339–344, Ann Arbor, MI, May 1996.

[26] A. D. Yaghjian and R. V. McGahan. Broadside radar cross-section of a perfectly conducting cube. *IEEE Trans. Antennas Propagat.*, 33(3):321–329, March 1985.

[27] D. C. Ross, J. L. Volakis, and H. T. Anastassiu. A hybrid finite element-modal analysis of jet engine scattering. *IEEE Trans. Antennas Propagat.*, 45:277–285, March 1995.

[28] A. Woo, M. Schuh, M. Simon, H. T. G. Wang, and M. L. Sanders. Radar cross-section measurement data of a simple rectangular cavity. Technical Report NWC TM 7132, Naval Weapons Center, China Lake, CA, December 1991.

[29] V. Shankar, W. F. Hall, A Mohammedian, and C. Rowell. Development of a finite-volume, time-domain solver for Maxwell's equations. Technical report, Rockwell International, May 1993. Prepared for NASA/NDC under contract N62269-90-C-0257.

[30] M. Schuh, A. Woo, M. Sanders, and H. T. G. Wang. Radar cross-section measurement data of four small cavities. Technical Report 108782, NASA Ames, IA, November 1993.

[31] A. Glisson and D. R. Wilton. Simple and efficient numerical techniques for treating bodies of revolution. Technical Report 105, University of Mississippi, Oxford, 1982.

[32] J. M. Putnam and L. N. Medgyesi-Mitschang. Combined field integral equation formulation for axially inhomogeneous bodies of revolution. Technical Report MDC QA003, McDonnell-Douglas Research Labs, December 1987.

[33] L. C. Kempel and J. L. Volakis. Scattering by cavity-backed antennas on a circular cylinder. *IEEE Trans. Antennas Propagat.*, 42:1268–1279, September 1994.

[34] T. Özdemir and J. L. Volakis. Triangular prisms for edge-based vector finite element antenna analysis. *IEEE Trans. Antennas Propagat.*, pp. 788–797, May 1997.

[35] R. J. Sliva and H. T. G. Wang. Personal communication.

[36] T. Özdemir, M. W. Nurnberger, J. L. Volakis, R. Kipp, and J. Berrie. A hybridization of finite element and high frequency methods for pattern prediction of antennas on aircraft structures. *IEEE Antennas Propagat. Mag.*, 38(3):28–38, June 1996.

[37] J.-S. Wang and R. Mittra. Finite element analysis of MMIC structures and electronic packages using absorbing boundary conditions. *IEEE Trans. Microwave Theory Tech.*, 42(3):441–449, March 1994.

[38] V. N. Kanellopoulos and J. P. Webb, The importance of the surface divergence term in the finite element-vector absorbing boundary condition method. *IEEE Trans. Microwave Theory Tech.*, 43(9):2168–2170, September 1995.

7

Three-Dimensional FE-BI Method

7.1 INTRODUCTION

In this chapter, the finite element-boundary integral (FE-BI) method for full three-dimensional geometries is presented. This is one of the most powerful computational electromagnetics (CEM) techniques in use today and represents a hybridization of the traditional method of moments with the finite element method. Interest in FE-BI stems from the fact that volume integral equations have difficulty modeling combined metal and dielectric structures and they lead to more computationally intensive programs as compared to the finite element method. In the FE-BI method, the boundary integral (or integral equation) is used to satisfy the following requirements:

1. Bound or terminate the computational domain in which the finite element method is used.
2. Relate the electric and magnetic fields on the boundary.

The manner in which the FE-BI method satisfies these requirements will be presented.

The most general formulation will be given followed by several particular examples including: scattering and radiation by cavity-backed antennas recessed in either an infinite metallic plane or a metallic cylinder, scattering bodies or antennas placed within an axisymmetric boundary, and infinitely periodic structures such as array antennas. It is in these special cases that the FE-BI method has proven most valuable since each case individually combines the flexibility of the finite element method with the efficiency of a special boundary integral mesh closure. For both the general and special cases, comments regarding computational cost, flexibility, and accuracy will be addressed. For the case of cavity-backed antennas recessed in a ground plane, an extremely efficient solution technique that utilizes Fast Fourier Transforms (e.g., the CG-FFT method) will be presented in detail.

For the most part, the method of weighted residuals and Galerkin's technique (see Chapter 1) is used rather than the variational formulation (used in Chapters 5 and 6) unless otherwise noted. In addition, we assume isotropic media throughout the computational domain. However, the formulation presented herein can be extended to the more general anisotropic case without difficulty and this is done in the appendix for brick elements. To simplify the implementation of the method, the material within the computational domain is assumed to be homogeneous within each finite element, but can vary on an element-by-element basis. Also, most formulations presented in this chapter are in terms of the total electric field since this is most convenient in the case of antenna analysis. Magnetic field formulations can be obtained using duality; however, such formulations are not suitable when an infinite metallic surface is included in the geometry as will be explained later in this chapter.

We begin with the general three-dimensional formulation.

7.2 GENERAL FORMULATION

The most general situation to which the FE-BI method may be applied is a three-dimensional volume of arbitrary shape and composition. Such a region is shown in Fig. 7.1, and its boundary can be either physical or fictitious (a mathematical entity only). The finite element method permits a completely arbitrary material composition within the computational domain. For example, metallic and various dielectric structures can exist within the volume with no fundamental difference in the formulation. For metallic bodies, the appropriate field components are forced to zero

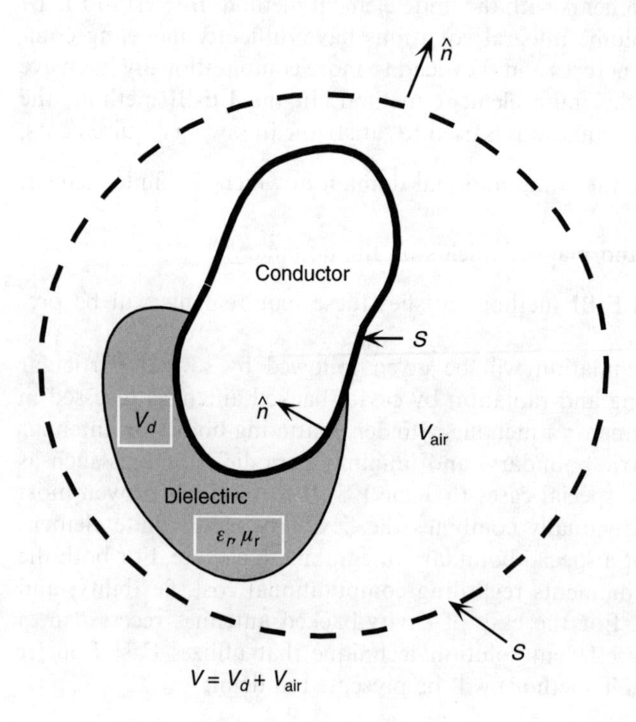

Figure 7.1 Computational volume in which the finite element method is to be applied.

(for the total electric field formulation considered herein). Various dielectric and magnetic materials are specified by appropriate permittivities and permeabilities on an element-by-element basis. This is in marked contrast to the surface integral equation (method of moments) since in that case the material must be homogeneous within each enclosed domain.

We begin with a derivation of the FE-BI equations using the physical equivalence principle.

7.2.1 Derivation of the FE-BI Equations

The derivation of the FE-BI equations begins with the vector wave equation. This second-order partial differential equation is solved by first taking the inner product of the vector wave equation and a vector sub-domain basis function, $\mathbf{W}_i$, thus forming the weighted residual (see Chapters 1 and 4). Our goal is to minimize this residual or equivalently to minimize the difference between the solution of the FE-BI discrete approximation and physical reality. This procedure generates N_e equations where N_e is the number of sub-domain basis functions associated with the electric field within and on the boundary of the volume. The resulting integro-differential equation is

$$\int_V \nabla \times \left[\frac{\nabla \times \mathbf{E}^{\text{int}}}{\mu_r}\right] \cdot \mathbf{W}_i \, dV - k_0^2 \int_V \epsilon_r \mathbf{E}^{\text{int}} \cdot \mathbf{W}_i \, dV =$$

$$-\int_V \nabla \times \left[\frac{\mathbf{M}^i}{\mu_r}\right] \cdot \mathbf{W}_i \, dV - jk_0 Z_0 \int_V \mathbf{J}^i \cdot \mathbf{W}_i \, dV \tag{7.1}$$

In this, the left hand side contains the unknown interior electric fields ($\mathbf{E}^{\text{int}}$) while the right hand side has the impressed sources ($\mathbf{J}^i$, $\mathbf{M}^i$). Since the excitation of the system is not relevant to the derivation of the FE-BI equations, the right hand side can be expressed as

$$f_i^{\text{int}} = -\int_V \left\{\nabla \times \left[\frac{\mathbf{M}^i}{\mu_r}\right] + jk_0 Z_0 \mathbf{J}^i\right\} \cdot \mathbf{W}_i \, dV \tag{7.2}$$

and its evaluation is left for specific applications. In practice, the electric current $\mathbf{J}^i$ in (7.2) is useful for modeling filamentary current sources such as the ones used to excite patch antennas, examples of which will be presented later in this chapter. The magnetic current $\mathbf{M}^i$ can be used to represent aperture feeds within the computational domain or on its boundary.

The FE-BI equation (7.1) contains second-order derivatives of the unknown electric field due to the use of the wave equation. It is desirable to transfer one of the derivatives from the unknown electric field onto the weight function so that linear weight and expansion functions may be used. This derivative transfer is accomplished by invoking the first vector Green's theorem [1] (see Chapter 5). Doing so, (7.1) becomes

$$\int_V \frac{\nabla \times \mathbf{E}^{\text{int}} \cdot \nabla \times \mathbf{W}_i}{\mu_r} \, dV - k_0^2 \int_V \epsilon_r \mathbf{E}^{\text{int}} \cdot \mathbf{W}_i \, dV$$

$$- jk_0 Z_0 \oint_S \hat{\mathbf{n}} \times \mathbf{H}^{\text{int}} \cdot \mathbf{W}_i dS = f_i^{\text{int}} \tag{7.3}$$

This is the weak form of the wave equation, and it possesses useful properties compared to (7.1). It has a symmetric volume contribution since an identical number of derivatives are required of both the unknown electric field and weight function and we will be using Galerkin's testing procedure. Hence, one may expect a symmetric linear system associated with the volume integral provided the material within the computational volume is reciprocal (i.e., not general anisotropic).

Recall that in the beginning of this chapter, we stated that there are two requirements which should be satisfied on the boundary of the finite element mesh: (1) mesh closure and (2) relating the tangential electric field to the tangential magnetic field. The latter requirement is clearly illustrated in (7.3) since the surface term includes the surface electric field through the testing function $\mathbf{W}_i$ and the tangential magnetic field $\hat{\mathbf{n}} \times \mathbf{H}^{\text{int}}$. In addition, with some foresight, we recognize that (7.3) represents an underdetermined system since the test functions are only associated with the electric field while both the electric field and surface magnetic field are unknown. We must therefore find a means of closing the mesh, relating the surface electric and magnetic fields, and providing additional equations.

The exterior excitation (e.g., a plane wave) can be introduced into (7.3) by considering the incident, reflected, and secondary (or scattered) fields as separable. Specifically, the total exterior magnetic field can be expressed as the sum

$$\mathbf{H}^{\text{ext}} = \mathbf{H}^{i} + \mathbf{H}^{\text{refl}} + \mathbf{H}^{\text{scat}} \tag{7.4}$$

Here, the incident field, $\mathbf{H}^{i}$, and the reflected field, $\mathbf{H}^{\text{refl}}$, are known while the secondary field ($\mathbf{H}^{\text{scat}}$) is obtained in terms of the interior fields using the surface equivalence principle. This field decomposition is useful in particular for analyzing structures that are infinite in extent such as a conformal antenna recessed in a metallic ground plane since the reflected field is already known and hence need not be computed. For finite structures, such as a scattering body encased within a surface of revolution domain (see the FE-SOR method discussed later in this chapter), the reflected field is omitted and only the impressed (incident) and secondary (scattered) fields are considered, where $\mathbf{H}^{\text{scat}}$ would have also to include any reflected fields that may be present. For radiation analysis, the impressed and reflected fields are omitted altogether and the total field is set equal to the secondary field.

A magnetic field integral equation (MFIE) [2], [3] can be formed once surface equivalent currents are used as illustrated in Fig. 7.2. These currents can then be used to express $\mathbf{H}^{\text{scat}}$ giving the MFIE [4][1]

$$\begin{aligned}-\hat{\mathbf{n}} \times [\mathbf{H}^{i}(\mathbf{r}) + \mathbf{H}^{\text{refl}}(\mathbf{r})] = -\frac{\mathbf{J}(\mathbf{r})}{2} - \oint_S \hat{\mathbf{n}} \times [\nabla \times \overline{\overline{G}}(\mathbf{r}, \mathbf{r}') \cdot \mathbf{J}(\mathbf{r}')]\, dS' \\ + jk_0 Y_0 \oint_S \hat{\mathbf{n}} \times \overline{\overline{G}}(\mathbf{r}, \mathbf{r}') \cdot \mathbf{M}(\mathbf{r}')\, dS'\end{aligned} \tag{7.5}$$

[1]The first right hand side term is due to the identity (valid just interior to the surface S)

$$\oint_S \hat{\mathbf{n}} \times \left[\nabla\left(\frac{e^{-jk_0R}}{4\pi R}\right) \times \mathbf{J}(\mathbf{r}')\right] dS' = \tfrac{1}{2}\mathbf{J}(\mathbf{r}) + \oint_S \hat{\mathbf{n}} \times \left[\nabla\left(\frac{e^{-jk_0R}}{4\pi R}\right) \times \mathbf{J}(\mathbf{r}')\right] dS'$$

where (see Chapter 4) the horizontal bar implies the principal value of the integral. However, as pointed out by Sancer [4], the principal value is not necessary since the numerical evaluation of the integral with $\mathbf{r}$ on S does not produce a singularity. The 1/2 factor is actually obtained without invoking the principal

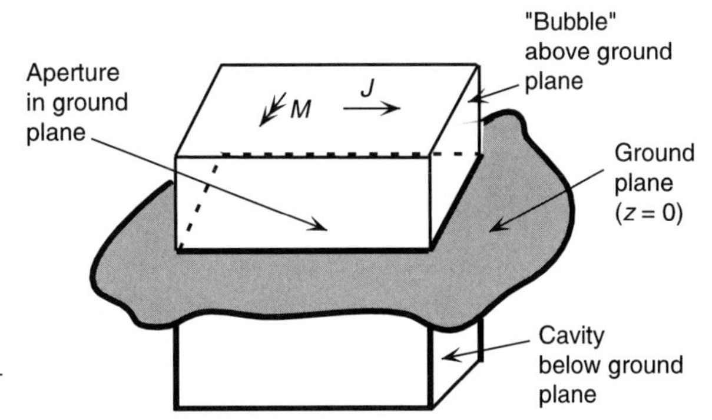

Figure 7.2 Computational volume and bounding surface involving a metallic ground plane.

As usual the electric and magnetic currents are associated with the tangential *external* fields, e.g., $\mathbf{J} = \hat{\mathbf{n}} \times \mathbf{H}^{\text{ext}}$ and $\mathbf{M} = \mathbf{E}^{\text{ext}} \times \hat{\mathbf{n}}$, respectively. Also, (7.5) enforces the identity $\mathbf{J} = \hat{\mathbf{n}} \times \mathbf{H}^{\text{ext}}$.

An alternative boundary integral equation can be derived by introducing the electric field integral equation (EFIE). to do so we decompose the electric fields as

$$\mathbf{E}^{\text{ext}} = \mathbf{E}^{\text{i}} + \mathbf{E}^{\text{refl}} + \mathbf{E}^{\text{scat}} \tag{7.6}$$

where the secondary field ($\mathbf{E}^{\text{scat}}$) can be written again in terms of $\mathbf{J}$ and $\mathbf{M}$ in Fig. 7.2 by invoking the equivalence principle (see Chapter 1). Doing so results in the EFIE [2], [5]

$$\begin{aligned} -\hat{\mathbf{n}} \times [\mathbf{E}^{i}(\mathbf{r}) + \mathbf{E}^{\text{refl}}(\mathbf{r})] = +\frac{\mathbf{M}(\mathbf{r})}{2} - \oint_S \hat{\mathbf{n}} \times [\nabla \times \overline{\overline{G}}(\mathbf{r}, \mathbf{r}') \cdot \mathbf{M}(\mathbf{r}')]\, dS' \\ + jk_0 Z_0 \oint_S \hat{\mathbf{n}} \times \overline{\overline{G}}(\mathbf{r}, \mathbf{r}') \cdot \mathbf{J}(\mathbf{r}')\, dS' \end{aligned} \tag{7.7}$$

A third alternative boundary integral equation is a linear combination of the MFIE and EFIE known as the combined field integral equation (CFIE) [6], [7]

$$(1 - \alpha)\,\text{MFIE} + \frac{\alpha}{Z_0}\,\text{EFIE} \tag{7.8}$$

where α is a parameter between 0 and 1 indicating the amount of emphasis given to EFIE or MFIE (e.g., 0.25 implies 25 percent EFIE and 75 percent MFIE). Both the MFIE and EFIE suffer from spurious resonances when S forms a closed surface, as in Fig. 7.1. That is, they break down at frequencies corresponding to the resonance frequencies of the enclosed volume. The CFIE avoids these resonances by mixing the MFIE and the EFIE [8]. We note that there are alternative forms of both the MFIE

value theorem by placing $\mathbf{r}$ slightly off the surface S and then taking the limit as $\mathbf{r}$ approaches S. Alternatively, the ∇ operator can be moved outside the integral and applied after completing the integration. Note that the identity

$$\nabla \times \overline{\overline{G}}_0(\mathbf{r}, \mathbf{r}') = -\nabla \times [\overline{\overline{I}} G_0(\mathbf{r}, \mathbf{r}')] = -\nabla G_0(\mathbf{r}, \mathbf{r}') \times \overline{\overline{I}}$$

implying $\nabla \times \overline{\overline{G}}_0(\mathbf{r}, \mathbf{r}') \cdot \mathbf{J}(\mathbf{r}') = -\nabla G_0(\mathbf{r}, \mathbf{r}') \times \mathbf{J}(\mathbf{r}')$ (given in Chapter 1), must be used to relate the curl of the free space dyadic Green's function with the scalar one.

and EFIE that may be more suitable for a particular application. However, for simplicity and illustration, we will only use (7.5), (7.7), or a variation of them. Neither of these boundary integral equations suffers from spurious resonances when used to simulate cavities or antennas that are recessed in a ground plane such as the case shown in Fig. 7.2.

Note that in both (7.5) and (7.7), the dyadic Green's function is left unspecified. For finite geometries (e.g., no infinite structures such as a metallic plane), the free-space dyadic Green's function $\overline{\overline{G}}_0$ should be used. In contrast, for structures involving an infinite metallic surface, such as a recessed conformal antenna, a practical choice is the second kind electric dyadic Green's function [9]. In this, the metallic ground plane condition is automatically included in the Green's function and need not be enforced via currents. Hence the mesh boundary S is restricted to the "bubble" above the ground plane, as shown in Fig. 7.2.

For cases where the volume is flush mounted with the ground plane, the "bubble" reduces to the aperture in the ground plane and for this limiting case the electric currents can be eliminated from the boundary integral equations [10], [11]. Also, the MFIE must be used for conformal antenna scattering calculations when the aperture lies in a ground plane since the incident electric field is exactly canceled in the aperture by the reflected electric field in the EFIE (7.7). Hence, no excitation for the system can be specified for the EFIE. In contrast, the EFIE must be used for open metal geometries since the MFIE is only valid for closed surfaces. A compromise is to use the CFIE since it has both the MFIE and EFIE in it. However, for the examples considered in this text, it is convenient to use the MFIE and hence for the rest of the chapter unless otherwise noted, we shall use the MFIE.

The weak form of the vector wave equation (7.3) involves the fields within the enclosing boundary while the fields in (7.5) are in the outer region. These fields must be coupled together to effect a hybridization of (7.3) and (7.5). This is done by enforcing tangential field continuity across the computational volume's boundary

$$\hat{\mathbf{n}} \times \mathbf{H}^{\text{int}} = \hat{\mathbf{n}} \times \mathbf{H}^{\text{ext}} \quad \text{on the surface S} \tag{7.9}$$

$$\hat{\mathbf{n}} \times \mathbf{E}^{\text{int}} = \hat{\mathbf{n}} \times \mathbf{E}^{\text{ext}} \quad \text{on the surface S} \tag{7.10}$$

The continuity condition associated with the magnetic fields (7.9) is often termed a "natural" condition and is enforced by setting $\mathbf{H}^{\text{ext}} = \mathbf{H}^{\text{int}}$ in (7.5). The electric field continuity condition (7.10) must be explicitly enforced in the formulation and is often termed an "essential" condition.

Combining (7.3), (7.5), and (7.9), we obtain the coupled FE and BI equations

$$\begin{aligned}
&\int_V \frac{\nabla \times \mathbf{E}^{\text{int}} \cdot \nabla \times \mathbf{W}_i}{\mu_r}\, dV - k_0^2 \int_V \epsilon_r \mathbf{E}^{\text{int}} \cdot \mathbf{W}_i\, dV - jk_0 Z_0 \oint_S \hat{\mathbf{n}} \times \mathbf{H}^{\text{int}} \cdot \mathbf{W}_i dS = f_i^{\text{int}} \\
&-\frac{1}{2}\oint_S [\mathbf{Q}_i \cdot (\hat{\mathbf{n}} \times \mathbf{H}^{\text{int}})]\, dS - \oint_S \oint_S \mathbf{Q}_i \cdot \left[\hat{\mathbf{n}} \times \nabla \times \overline{\overline{G}} \times \hat{\mathbf{n}}'\right] \cdot \mathbf{H}^{\text{int}}\, dS'\, dS \\
&\qquad - jk_0 Y_0 \oint_S \oint_S \mathbf{Q}_i \cdot \left[\hat{\mathbf{n}} \times \overline{\overline{G}} \times \hat{\mathbf{n}}'\right] \cdot \mathbf{E}^{\text{ext}}\, dS'\, dS = f_i^{\text{ext}}
\end{aligned} \tag{7.11}$$

where the exterior excitation term f_i^{ext} is given by

$$\tilde{f}_i^{\text{ext}} = -jk_0 Z_0 \oint_S \mathbf{Q}_i \cdot \hat{\mathbf{n}} \times [\mathbf{H}^{\text{i}} + \mathbf{H}^{\text{refl}}]\, dS = jk_0 Z_0 f_i^{\text{ext}} \tag{7.12}$$

and the testing functions associated with $\mathbf{H}^{\text{ext}}$ are indicated by $\mathbf{Q}_i$. Note that there are at present three classes of unknown fields in (7.11): $\mathbf{E}^{\text{int}}$, $\mathbf{E}^{\text{ext}}$, and $\mathbf{H}^{\text{int}}$. In (7.11), the magnetic field continuity condition (7.9) was explicitly enforced; however, the electric field continuity condition (7.10) must also be enforced to solve (7.11). This can be accomplished in one of two ways: (1) implicitly by using identical basis functions for $\mathbf{E}^{\text{int}}$ and $\mathbf{E}^{\text{ext}}$ on the surface S (hence, all occurrences of $\mathbf{E}^{\text{ext}}$ are replaced by $\mathbf{E}^{\text{int}}$ in (7.11)); or explicitly by enforcing the auxiliary relation

$$\oint_S [\mathbf{Q}_i \cdot \hat{\mathbf{n}} \times (\mathbf{E}^{\text{int}} - \mathbf{E}^{\text{ext}})]\, dS = 0 \tag{7.13}$$

which satisfies (7.10) in a weak or average sense. Also, the testing functions for the interior (FE) problem, $\mathbf{W}_i$, are not necessarily the same as the testing functions used for the exterior (IE) problem, $\mathbf{Q}_i$. In fact, the testing functions $\mathbf{W}_i$ are associated with the interior electric field while the testing functions $\mathbf{Q}_i$ are associated with the surface magnetic field. Hence, for the general case of a different expansion for the exterior and interior electric fields, we have the coupled equation

interior equation:

$$\int_V \frac{\nabla \times \mathbf{W}_i \cdot \nabla \times \mathbf{E}^{\text{int}}}{\mu_r}\, dV - k_0^2 \int_V \epsilon_r \mathbf{W}_i \cdot \mathbf{E}^{\text{int}}\, dV - jk_0 Z_0 \oint_S \mathbf{W}_i \cdot \hat{\mathbf{n}} \times \mathbf{H}^{\text{int}}\, dS = f_i^{\text{int}}$$

exterior magnetic field testing equation:

$$-\frac{1}{2}\oint_S [\mathbf{Q}_i \cdot (\hat{\mathbf{n}} \times \mathbf{H}^{\text{int}})]\, dS - \oint_S \oint_S \mathbf{Q}_i \cdot \left[\hat{\mathbf{n}} \times \nabla \times \overline{\overline{G}} \times \hat{\mathbf{n}}'\right] \cdot \mathbf{H}^{\text{int}}\, dS'\, dS$$
$$- jk_0 Y_0 \oint_S \oint_S \mathbf{Q}_i \cdot \left[\hat{\mathbf{n}} \times \overline{\overline{G}} \times \hat{\mathbf{n}}'\right] \cdot \mathbf{E}^{\text{ext}}\, dS'\, dS = f_i^{\text{ext}}$$

coupling equation:

$$\oint_S [\mathbf{Q}_i \cdot \hat{\mathbf{n}} \times (\mathbf{E}^{\text{int}} - \mathbf{E}^{\text{ext}})]\, dS = 0 \tag{7.14}$$

7.2.2 Solution of the FE-BI Equations

The solution of (7.14) proceeds by expanding the volume electric field and surface tangential electric and magnetic fields in terms of sub-domain basis functions

$$\mathbf{E}^{\text{int}} = \sum_{j=1}^{N_v} E_j \mathbf{W}_j \qquad \text{volume electric field}$$

$$\mathbf{E}^{\text{ext}} = \sum_{j=N_v+1}^{N_v+N_{es}} E_j \mathbf{V}_j \qquad \text{surface electric field}$$

$$\mathbf{H}^{\text{int}} = \sum_{j=N_v+N_{es}}^{N_v+N_{es}+N_{hs}} H_j \mathbf{Q}_j \qquad \text{surface magnetic field} \tag{7.15}$$

where N_v is the number of volume electric field unknowns, N_{es} is the number of surface (exterior) electric field unknowns, N_{hs} is the number of surface magnetic field unknowns, and the total number of unknowns is given by $N = N_v + N_{es} + N_{hs}$. The volume electric field is expanded in terms of volumetric basis functions that are based on the edges of the geometry (see Chapters 2 and 5). The surface electric and magnetic fields are expanded in terms of separate functions that have support only over sub-domain surface patches which are often triangular or rectangular in shape. In both cases, since we are using the Galerkin procedure for converting a continuous physical problem to a discrete approximation of that original problem, the same basis functions used for testing in (7.14) are used for field expansions, though $\mathbf{W}_i$ and $\mathbf{Q}_i$ are not necessarily identical. However, in (7.15) $\mathbf{V}_i$ is identical to $\mathbf{Q}_i$ though different symbols are used for clarity in this text.

As shown above, the enforcement of magnetic field continuity across the interface was performed by equating the exterior and interior magnetic fields across the volume boundary, S. Enforcement of the tangential electric field continuity must also be accomplished. One approach requires the use of the three equations in (7.14) including the coupling equation. When the three field expansions (7.15) are substituted into (7.14), we get

interior electric field testing:

$$\sum_{j=1}^{N_v} E_j \left\{ \int_V \frac{\nabla \times \mathbf{W}_i \cdot \nabla \times \mathbf{W}_j}{\mu_r} dV - k_0^2 \int_V \epsilon_r \mathbf{W}_i \cdot \mathbf{W}_j \, dV \right\}$$

$$- jk_0 Z_0 \sum_{j=N_e+1}^{N} H_j \left\{ \oint_S \mathbf{W}_i \cdot \hat{\mathbf{n}} \times \mathbf{Q}_j \, dS \right\} = f_i^{\text{int}}, \qquad i = 1, 2, \ldots, N_v$$

exterior surface magnetic field testing:

$$\sum_{j=N_e+1}^{N} H_j \left\{ -\frac{1}{2} \oint_S [\mathbf{Q}_i \cdot (\hat{\mathbf{n}} \times \mathbf{Q}_j)] \, dS - \oint_S \oint_S \mathbf{Q}_i \cdot \left[\hat{\mathbf{n}} \times \nabla \times \overline{\overline{G}} \times \hat{\mathbf{n}}' \right] \cdot \mathbf{Q}_j \, dS' \, dS \right\}$$

$$- jk_0 Y_0 \sum_{j=N_v+1}^{N_e} E_j \left\{ \oint_S \oint_S \mathbf{Q}_i \cdot \left[\hat{\mathbf{n}} \times \overline{\overline{G}} \times \hat{\mathbf{n}}' \right] \cdot \mathbf{V}_j \, dS' \, dS \right\} = f_i^{\text{ext}}$$

$$i = N_e + 1, N_e + 2, \ldots, N$$

interior/exterior coupling equation:

$$\sum_{j=1}^{N_v} E_j \left\{ \oint_S [\mathbf{Q}_i \cdot \hat{\mathbf{n}} \times \mathbf{W}_j] \, dS \right\} - \sum_{j=N_v+1}^{N_e} E_j \left\{ \oint_S [\mathbf{Q}_i \cdot \hat{\mathbf{n}} \times \mathbf{V}_j] \, dS \right\},$$

$$i = N_e + 1, N_e + 2, \ldots, N \qquad (7.16)$$

where $N_e = N_v + N_{es}$ is the total number of electric field unknowns and $N = N_v + N_{es} + N_{hs}$ is the total number of unknowns.

A different approach involves the implicit satisfaction of the electric field continuity requirement by employing identical basis functions. In this case, the surface basis functions $\mathbf{V}_j$ are chosen to be identical to a surface evaluation of the volume basis function $\mathbf{W}_j$, e.g., $\mathbf{V}_j = \mathbf{W}_j$ as $(x, y, z) \rightarrow$ surface. Accordingly, we rewrite the electric and magnetic field expansions as

$$\mathbf{E} = \sum_{j=1}^{N_e} E_j \mathbf{W}_j \qquad \text{volume and surface electric field}$$

$$\mathbf{H} = \sum_{j=N_e+1}^{N} H_j \mathbf{Q}_j \qquad \text{surface magnetic field} \tag{7.17}$$

Thus, by constraining the exterior field expansion to be identical to the interior field expansion, continuity is assured. This also results in a reduction of the matrix order since the interior surface electric field unknowns are now identical to the exterior surface electric field unknowns. We can omit the coupling equation in (7.14) to get

interior volume electric field:

$$\int_V \frac{\nabla \times \mathbf{W}_i \cdot \nabla \times \mathbf{E}^{\text{int}}}{\mu_r} dV - k_0^2 \int_V \epsilon_r \mathbf{W}_i \cdot \mathbf{E}^{\text{int}} dV - jk_0 Z_0 \oint_S \mathbf{W}_i \cdot \hat{\mathbf{n}} \times \mathbf{H}^{\text{int}} dS = f_i^{\text{int}}$$

exterior surface magnetic field:

$$-\frac{1}{2}\oint_S [\mathbf{Q}_i \cdot (\hat{\mathbf{n}} \times \mathbf{H}^{\text{int}})]\, dS - \oint_S \oint_S \mathbf{Q}_i \cdot \left[\hat{\mathbf{n}} \times \nabla \times \overline{\overline{G}} \times \hat{\mathbf{n}}'\right] \cdot \mathbf{H}^{\text{int}} dS'\, dS$$
$$- jk_0 Y_0 \oint_S \oint_S \mathbf{Q}_i \cdot \left[\hat{\mathbf{n}} \times \overline{\overline{G}} \times \hat{\mathbf{n}}'\right] \cdot \mathbf{E}^{\text{int}} dS'\, dS = f_i^{\text{ext}} \tag{7.18}$$

Using (7.17) in (7.18), we also get

interior volume electric field testing:

$$\sum_{j=1}^{N_e} E_j \left\{ \int_V \frac{\nabla \times \mathbf{W}_i \cdot \nabla \times \mathbf{W}_j}{\mu_r} dV - k_0^2 \int_V \epsilon_r \mathbf{W}_i \cdot \mathbf{W}_j\, dV \right\}$$
$$- jk_0 Z_0 \sum_{j=N_e+1}^{N} H_j \left\{ \oint_S \mathbf{W}_i \cdot \hat{\mathbf{n}} \times \mathbf{Q}_j\, dS \right\} = f_i^{\text{int}}, \qquad i = 1, 2, \ldots, N_e$$

exterior surface magnetic field testing:

$$\sum_{j=N_e+1}^{N} H_j \left\{ -\frac{1}{2}\oint_S [\mathbf{Q}_i \cdot (\hat{\mathbf{n}} \times \mathbf{Q}_j)]\, dS - \oint_S \oint_S \mathbf{Q}_i \cdot \left[\hat{\mathbf{n}} \times \nabla \times \overline{\overline{G}} \times \hat{\mathbf{n}}'\right] \cdot \mathbf{Q}_j\, dS'\, dS \right\}$$
$$- jk_0 Y_0 \sum_{j=1}^{N_e} E_j \left\{ \oint_S \oint_S \mathbf{Q}_i \cdot \left[\hat{\mathbf{n}} \times \overline{\overline{G}} \times \hat{\mathbf{n}}'\right] \cdot \mathbf{W}_j\, dS'\, dS \right\} = f_i^{\text{ext}}$$
$$i = N_e + 1, N_e + 2, \ldots, N \tag{7.19}$$

We observe that the number of unknowns is equal to the number of equations N and $N_e = N_v$. It is also understood that each contribution is nonzero only when both the test and expansion functions have support, e.g., although all electric field unknowns are shown in the second equation of (7.19), only those associated with the surface have support. The linear system represented by (7.19) is solved using either a direct or iterative matrix solver to determine the unknown electric and magnetic fields (see Chapter 9).

This latter set of coupled equations (7.19) yields the least number of unknowns and the simplest formulation to implement. However, this approach limits the flexibility of the method since the discretization required by the interior fields must also be used on the exterior electric field. Consequently, the boundary integral portion of the formulation may need to be oversampled to accommodate the geometrical requirements of the interior region. The converse is also true where many volume unknowns are essentially wasted to permit a detailed surface mesh. Regardless, the result is a potential waste of computational resources and flexibility.

7.2.3 Comments on the General FE-BI Formulation

The FE-BI equations, (7.11)–(7.19), are the most general forms and can be used under any circumstances. Such equations have been implemented by Antilla and Alexopoulos [2] while Eibert and Hansen [12] used a slightly different integral equation. An example of the kind of problems that can be modeled using such a flexible method is shown in Fig. 7.3 where the geometry is a coated finite height ogival cylinder. Figure 7.4 illustrates the RCS comparison between a FE-CFIE method [2] and measured data for $E_{\phi\phi}$-polarization.

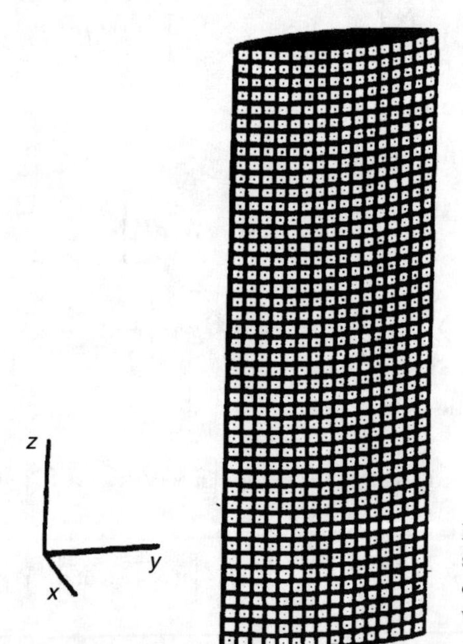

Figure 7.3 Coated cylinder with ogival cross section: 60.96 cm × 20.32 cm metallic cylinder having a 0.3175 cm dielectric coating with $\epsilon_r = 2.68 - j0.01$. [*After Antilla and Alexopoulos* [2], © J. Opt. Soc. Am., *1994*.]

The versatile FE-BI methods are still associated with large demands on computational resources. The FE portion of (7.11) will permit the specification of a complex inhomogeneous volume fill while imposing a low memory and compute cycle burden, principally due to the resulting sparse matrix. However, the two boundary terms are essentially identical to a surface method of moments formulation with the resulting fully populated matrix. Figure 7.5 illustrates the fill profile of a typical FE-BI matrix. The dark region indicates matrix entries that are nonzero while the white space denotes zeros and hence corresponds to portions of the matrix that

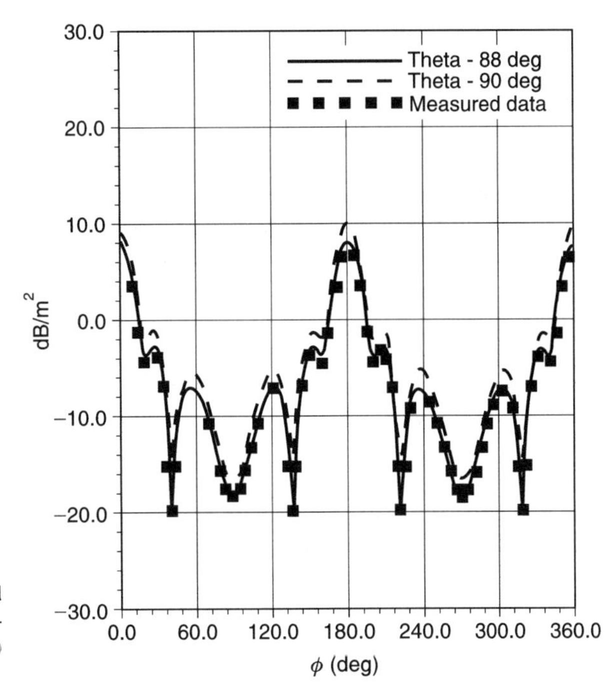

Figure 7.4 Radar cross section for the coated ogive cylinder shown in Fig. 7.3 for $E_{\phi\phi}$-polarization. [*After Antilla and Alexopoulos* [2], © J. Opt. Soc. Am., *1994*.]

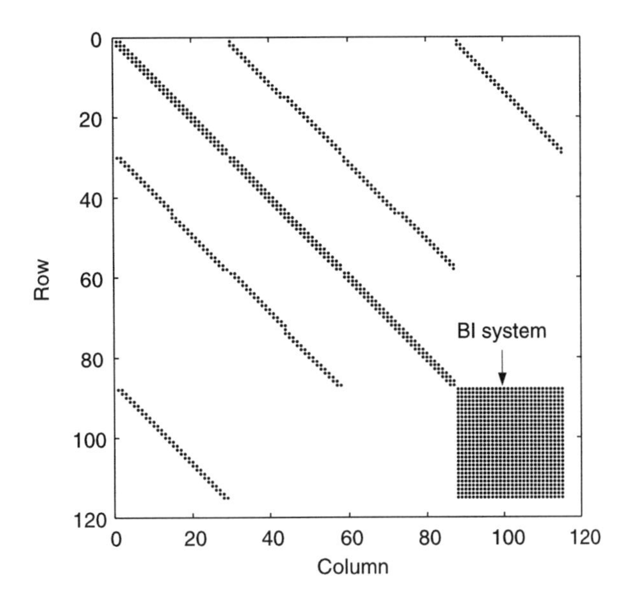

Figure 7.5 Fill profile for a typical finite element-boundary integral matrix. [*Courtesy of S. Bindiganavale.*]

need not be calculated or stored. Clearly, the volume finite element formulation results in a sparsely populated portion of the matrix while the boundary integral results in a fully populated portion of the matrix. The result is that for many circumstances, the boundary integral terms impose a higher demand on computational resources as compared to the FE portion. This is true even though the BI term is responsible for only a small fraction of the total number of unknowns, as shown in Fig. 7.5!

An example will illustrate this point. Consider a $\lambda \times \lambda \times \lambda$ volume that is discretized using edge-based brick elements with edge length a conservative $\lambda/10$. Each face of the volume will have 180 unknowns per field component. Including the edges at each face junction, the total number of surface unknowns is 1200 per field component or 2400 total surface unknowns. The boundary matrices associated with these unknowns would require approximately 12 MB of RAM if single precision complex number storage is used. This is obviously not a large burden even for many personal computers. However, consider the effect of doubling the sample rate as is often necessary for complex geometries or radiation problems. In this case, each edge of the mesh will be $\lambda/20$. The number of unknowns per field component is now 4800 or 9600 total surface unknowns, and these require over 184 MB of RAM! Although this amount of RAM can be found in high-end engineering workstations, it is clear that the boundary integral memory demand does not scale favorably.

Hence, the FE-BI method is usually implemented for certain special cases that reduce the BI's demand on resources. In the next several sections, we will present examples of these special cases beginning with the case of a cavity-backed volume recessed in a metallic ground plane. However, we first introduce the important topic of excitation and feed modeling.

7.3 EXCITATION AND FEED MODELING

Solution of the FE-BI system requires specification of the source(s). Several sources are typically used to excite the FE-BI system. These sources are broadly divided into two categories: exterior and interior. By far the most common exterior source is the plane wave. This is due to the fact that any incident field may be decomposed into a number of plane waves and superposition guarantees that the solution to a general excitation can be found by the sum of the solutions for each plane wave. However, an arbitrary wave source may be readily specified (for example, a Gaussian beam or a measured antenna pattern).

Interior sources are specified as either electric or magnetic currents. A number of these sources are used in practice to model both filamentary (probe) feeds as well as aperture feeds. The FE-BI method is particularly important for antenna modeling since sufficiently accurate and efficient artificial mesh truncation procedures are not available for accurate near-field calculations. Since the input impedance is required for antenna analysis and gain calculations, near-field accuracy is critical as well as a good feed model. In the following section, we present a brief discussion of feed modeling.

7.3.1 Plane Wave

One source, useful for determining the radar cross section (RCS), is the plane wave

$$\mathbf{E}^i = \hat{\mathbf{e}}^i e^{-jk_0(\hat{\mathbf{k}}^i \cdot \mathbf{r})}$$
$$\mathbf{H}^i = Y_0[\hat{\mathbf{k}}^i \times \mathbf{E}^i] \tag{7.20}$$

where $Y_0 = 1/Z_0$ is the free space admittance, the polarization of the incident field is indicated by $\hat{\mathbf{e}}^i$, and the incident field direction is denoted by $\hat{\mathbf{k}}^i = -\hat{r}$. That is, $\hat{\mathbf{k}}^i = -[\hat{x}\cos\phi^i \sin\theta^i + \hat{y}\sin\phi^i \sin\theta^i + \hat{z}\cos\theta^i]$ where (ϕ^i, θ^i) denote the angles of the incident plane wave field. If we assume an infinite, metallic ground plane, the reflected field is given by

$$\begin{aligned} \mathbf{E}^{\text{refl}} &= \hat{\mathbf{e}}^r e^{-jk_0(\hat{\mathbf{k}}^r \cdot \mathbf{r})} \\ \mathbf{H}^{\text{refl}} &= Y_0[\hat{\mathbf{k}}^r \times \mathbf{E}^{\text{refl}}] \end{aligned} \tag{7.21}$$

where $\hat{\mathbf{k}}^r = -\hat{x}\cos\phi^i \sin\theta^i - \hat{y}\sin\phi^i \sin\theta^i + \hat{z}\cos\theta^i$ and $\hat{\mathbf{e}}^r$ denotes the polarization of the reflected plane wave as dictated by the boundary conditions. Thus, in the ground plane aperture, the sum of the incident and the reflected magnetic field is twice the incident field while the sum of the incident and reflected electric fields is zero. Using these fields in (7.12) and assuming evaluation in the $z = 0$ plane, we get the generating functional for the system excitation vector

$$\tilde{f}_i^{\text{ext}} = -j2k_0 \oint_S \mathbf{Q}_i \cdot \hat{z} \times [\hat{\mathbf{k}}^i \times \hat{\mathbf{e}}^i e^{-jk_0(\hat{\mathbf{k}}^i \cdot \mathbf{r})}]\, dS \tag{7.22}$$

7.3.2 Probe Feed

Another useful source is an impressed electric current, $\mathbf{J}^i$, which can be used to simulate a probe feed. This feed has been used to model a patch antenna feed by the method of moments [13] as well as in the finite element method [10]. Assuming an infinitesimally thin current filament, the excitation term (7.2) for an arbitrarily oriented probe feed is given by

$$f_i^{\text{int}} = -jk_0 Z_0 \hat{l} \cdot \mathbf{J}^i = -jk_0 Z_0 I_0 \hat{l} \cdot \mathbf{W}_i(\mathbf{r}) \tag{7.23}$$

where $\hat{l}$ indicates the orientation of the probe, $\mathbf{W}_i$ is the weight or testing function associated with the ith unknown, and I_0 is the current flowing through the filament. For current probes lying along the x-, y-, or z-directions, we have

$$\begin{aligned} \hat{x} \cdot \mathbf{J}^i &= I_0 \delta(y - y_p)\delta(z - z_p) W_i^x(x, y, z) \\ \hat{y} \cdot \mathbf{J}^i &= I_0 \delta(x - x_p)\delta(z - z_p) W_i^y(x, y, z) \\ \hat{z} \cdot \mathbf{J}^i &= I_0 \delta(x - x_p)\delta(y - y_p) W_i^z(x, y, z) \end{aligned} \tag{7.24}$$

where the coordinates (x_p, y_p, z_p) indicate the location of the source within the computational domain. Note that the filament is not constrained other than lying within the computational volume. However, in practice a filament length of one finite element cell is usually assumed for simplicity (so that integration over that length will yield the moment $I_0 l$). With this assumption, the excitation vector generating function becomes

$$
\begin{aligned}
f_i^{\text{int}} &= -jk_0Z_0I_0\int_V [\delta(y-y_p)\delta(z-z_p)]\hat{\mathbf{x}}\cdot\mathbf{W}_i\,dV = -jk_0Z_0I_0\Delta_x W_i^x(\cdot, y_p, z_p)\\
&= -jk_0Z_0I_0\int_V [\delta(x-x_p)\delta(z-z_p)]\hat{\mathbf{y}}\cdot\mathbf{W}_i\,dV = -jk_0Z_0I_0\Delta_y W_i^y(x_p, \cdot, z_p)\\
&= -jk_0Z_0I_0\int_V [\delta(x-x_p)\delta(y-y_p)]\hat{\mathbf{z}}\cdot\mathbf{W}_i\,dV = -jk_0Z_0I_0\Delta_z W_i^z(x_p, y_p, \cdot)
\end{aligned}
\tag{7.25}
$$

where the integration volume in (7.25) includes all elements containing the source. Also, the three expressions in (7.25) correspond to x-, y-, and z-directed filaments, respectively. If more than one test edge is involved in the feed model (e.g., when the probe feed is in the interior of an element), the total contribution is given by evaluating (7.25) for all edges for which $\hat{l}\cdot\mathbf{W}_i \neq 0$.

A word of caution is in order. The probe feed presented herein has been criticized for being too simplistic in that all of the current is concentrated in an infinitesimally thin filament and that the assumption of a constant current is unrealistic. Admittedly, ignoring the diameter of the probe (and the coaxial aperture) leads to errors, particularly in the input reactance, and approximate formulae have been proposed to correct this error. However, these corrections can hardly be considered as satisfactory. Furthermore, the constant current assumption results in limiting the use of the probe feed to very thin regions where the fields are not expected to vary substantially (e.g., the thin substrate between a patch antenna and its underlying ground plane).

To correct these shortcomings, it is tempting to borrow solutions developed for the integral equation methods. In particular, Aberle and Pozar [14] use unknowns on the surface of the probe feed and enforce an EFIE on the probe to include the effects of its finite diameter and current variations along the probe. However, to use such a model requires use of the Green's function for the structure surrounding the probe. Typically, such a Green's function is not available for structures considered for finite element analysis. Use of a different Green's function, such as the half-space Green's function, is not correct and should not be used.

7.3.3 Voltage Gap Feed

Another commonly used excitation, often employed in the integral equation methods, is the voltage or gap generator. In this feed model, the voltage across a *small* gap is specified as fixed. Thus, Gauss' Law can be used to write

$$
\mathbf{E}^i\cdot\hat{l} = \frac{V}{\delta}
\tag{7.26}
$$

where $\mathbf{E}^i$ is the impressed electric field, V is the voltage assumed a priori across the gap, and δ is the length of the gap parallel to the incident field.

Assuming for simplicity that the gap coincides with the ith edge/unknown, the gap voltage feed is implemented by forcing the electric field (the unknown) to be equal to $\frac{V}{d}$ where d is the length of the ith edge. This can be done during each iteration of an iterative solver or via an appropriate auxiliary (additional equation) condition.

7.3.4 Coaxial Cable Feed

As noted above, the probe feed model is of acceptable accuracy for very thin substrates and for circumstances where the diameter of the probe may be safely ignored. For thicker substrates or for circumstances where the diameter of the probe and the size of the coaxial (or coax) aperture need be considered, an improved feed model is necessary. The coaxial feed geometry is shown in Fig. 7.6. The following derivation and model are based on the development given by Gong and Volakis [15].

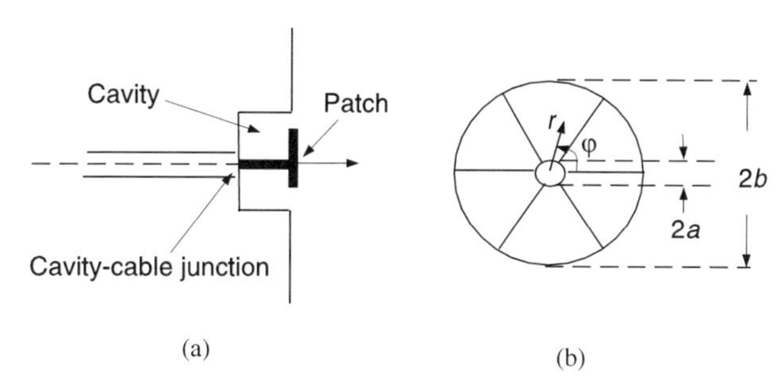

Figure 7.6 (a) Side view of a cavity-backed antenna with a coax cable feed; (b) Illustration of the FE mesh at the cavity-cable junction. [*After Gong and Volakis* [*15*], © *IEEE, 1995.*]

With the presence of the coax cable aperture, the boundary integral (not f_i^{int}) in (7.3) will include the term

$$f_i^C = -jk_0 Z_0 \iint_{S_f} (\mathbf{W}_i \times \mathbf{H}) \cdot \hat{\mathbf{z}}\, dS \tag{7.27}$$

where $\hat{\mathbf{n}} = \hat{\mathbf{z}}$ and S_f denotes the aperture of the coax cable. Assuming a TEM mode across S_f, the fields within the cable may be expressed as

$$\mathbf{E}^{\text{TEM}} = \frac{e_0}{\rho}\hat{\rho}, \qquad \mathbf{H}^{\text{TEM}} = \frac{h_0}{\rho}\hat{\phi} \tag{7.28}$$

where

$$e_0 = \frac{I_0 Z_0}{2\pi\sqrt{\epsilon_{rc}}}(1+\Gamma), \qquad h_0 = \frac{I_0}{2\pi}(1-\Gamma) \tag{7.29}$$

In these expressions, ρ is the radial variable from the center of the inner conductor, $\hat{\rho}$ and $\hat{\phi}$ are the usual cylindrical unit vectors in the coax aperture, I_0 is the current flowing through the coax cable, and ϵ_{rc} is the relative permittivity of the dielectric between the inner and outer conductors of the coax cable. Also, Γ is the reflection coefficient of the incoming coax TEM mode at the coax cable opening but can be eliminated from (7.29) to obtain the relation

$$h_0 = -\frac{\sqrt{\epsilon_{rc}}}{Z_0} e_0 + \frac{I_0}{\pi}. \tag{7.30}$$

We observe that (7.30) is the constraint at the cable junction in terms of the new quantities h_0 and e_0 which can be used as new unknowns in place of the fields $\mathbf{E}$ and $\mathbf{H}$. However, before introducing (7.27) into the system, it is necessary to relate e_0 and h_0 to the unknowns (edges) lying within the coaxial aperture. Since the actual field has a $1/\rho$ behavior in the cable, direct enforcement of field continuity on a point-by-point basis is not possible using edge elements having constant or linear variation. Instead, when using edge elements, one can only specify that the potential difference across the coax cable be equal to that across the bordering volume edges. Specifically, we find that

$$\Delta V = E_{\text{coax}}(b-a) = e_0 \ln \frac{b}{a}, \quad i = N_p \quad (p = 1, 2, \ldots, N_C) \tag{7.31}$$

where ΔV denotes the potential difference between the inner and outer surface of the cable, and E_{coax} denotes the field in the coax. Also, N_p denotes the global number for the edge across the coax cable and N_C is the number of edges in the coax aperture. When the condition (7.31) is used in the function (7.27), it introduces the excitation into the finite element system without a need to extend the mesh inside the cable or to employ a fictitious current probe. Specifically, we have

$$\begin{aligned} f_i^C &= -\frac{\pi}{3} jk_0 Z_0 (b-a) \left[\frac{I_0}{\pi} - \frac{\sqrt{\epsilon_{rc}}}{Z_0} \frac{b-a}{\ln(b/a)} E_{\text{coax}} \right] \\ &= C_i E_{\text{coax}} - f_i^{\text{coax}}, \quad i = N_p \end{aligned} \tag{7.32}$$

with

$$C_i = j \frac{\pi}{3} k_0 \sqrt{\epsilon_{rc}} \frac{(b-a)^2}{\ln(b/a)}$$

$$f_i^{\text{coax}} = j \frac{k_0 Z_0}{3} (b-a) I_0$$

Note that f_i^C is nonzero only for those edges which coincide with the aperture of the coaxial cable. It is also apparent that f_i^{coax} is a constant and becomes part of the excitation function f_i^{int} when moved to the right hand side of the matrix system. Basically, the excitation column entries will be zero or equal to f_i^{coax} for those edges coinciding with the coaxial cable aperture. Upon solution of the system, the input admittance at the coax aperture ($z = 0$) can be obtained using the expression

$$Y_{\text{in}} = \frac{1}{\Delta V} \oint_C \mathbf{H} \cdot \hat{\rho} \rho \, d\phi = \frac{2I_0}{e_0 \ln(b/a)} - \frac{1}{Z_c}$$

where Z_c is the characteristic impedance of the coax cable and the integration path C is around the center conductor.

7.3.5 Aperture-Coupled Microstrip Line

Figure 7.7 illustrates a microstrip antenna fed by a microstrip line network underneath the antenna via a coupling aperture. Care must be taken in incorporating such a feed into the FE method [16], [17]. This is because the microstrip line usually requires dramatically different discretization as compared to the cavity's geometries.

It is convenient to separate the computational domain to accommodate the smaller element dimensions required for the guided feed structure. Also, different finite element shapes may be favored for the antenna and the feed region. For example, consider a circular patch antenna. The cavity/antenna fields may be discretized using tetrahedral elements, whereas in the microstrip line region rectangular

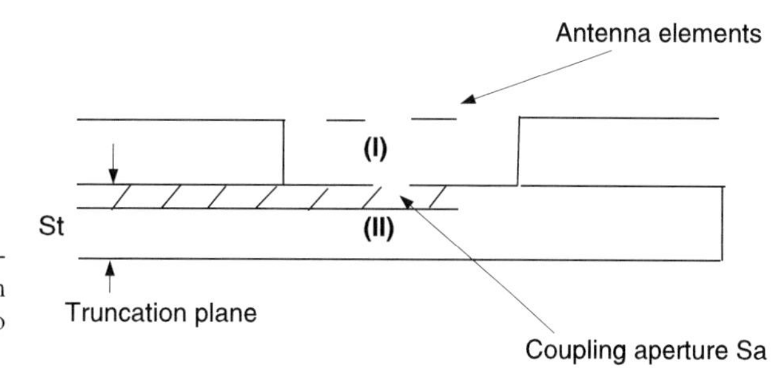

Figure 7.7 Cross section of an aperture coupled patch antenna, showing the cavity region I and the microstrip line region II for two different FEM computational domains.

bricks are the best candidates since the feed structure is rectangular in shape and the substrate has a constant thickness.

Although both types of elements employ edge-based field expansions, the meshes across the common area (coupling aperture) are different, and consequently some connectivity matrix must be introduced to relate the mesh edges across the aperture. This can be accomplished using the coupling equations (7.9), (7.10), and (7.13), introduced previously. However, since the aperture is very narrow, a 'static' field distribution may be assumed at any given frequency. Therefore, the *potential* concept may be applied to relate the fields on either side of the aperture. To do so, let us first classify the slot edges as follows [17]:

Cavity mesh (assumed to form right triangles across the aperture)

- E_j^{b1} $j = 1, 2, 3, \ldots$ parallel edges
- E_j^{b2} $j = 1, 2, 3, \ldots$ diagonal edges

Feed mesh (assumed to form square elements across the aperture)

- E_j^c $j = 1, 2, 3, \ldots$ parallel edges only

In these, parallel edges refer to edges parallel to the Cartesian axes. To ensure equal potential across the aperture requires that

$$\begin{aligned} E_j^{b1} &= \epsilon_j E_j^c \\ E_j^{b2} &= \frac{t}{2d}\left(\epsilon_j E_j^c + \epsilon_{j+1} E_{j+1}^e\right) \end{aligned} \tag{7.33}$$

in which

$$\epsilon_j = \begin{cases} +1 \\ -1 \end{cases} \tag{7.34}$$

and in these, t and d are the lengths of the parallel and diagonal edges, respectively. That is, t is simply the width of the *narrow* rectangular aperture between the two meshes. The coefficient ϵ_j is equal to ± 1 depending on the sign conventions associated with the meshes to either side of the coupling aperture. Essentially, (7.33) is a potential approximation to the electric field continuity conditions (7.10) and (7.13).

7.3.6 Mode Matched Feed

Alternatively, a mode matching procedure can be used at the aperture to rigorously include the higher order (evanescent) modes that are excited at the coax aperture. This amounts to including an additional boundary integral at the feed aperture and synthesizing a coax Green's function via a modal series. Modal series feed models are discussed by Reddy et al. [18]. In this, the authors present as an appendix the formulas for rectangular, circular, and coaxial feed apertures.

The electric field across the feed aperture can be expressed as a sum of incident and reflected fields

$$\mathbf{E} = \mathbf{E}^{\mathrm{i}} + \sum_m [a_m \mathbf{E}_m^{\mathrm{TE}}(x, y) + b_m \mathbf{E}_m^{\mathrm{TM}}(x, y)]\, e^{\gamma_m z} \tag{7.35}$$

with

$$\mathbf{E}^{\mathrm{i}}(x, y, z) = \mathbf{E}^{\mathrm{i}}(x, y)\, e^{-\gamma_{\mathrm{inc}} z} \tag{7.36}$$

where $\mathbf{E}^{\mathrm{i}}$ is the incident field, and $\mathbf{E}_m^{\mathrm{TE}}$ and $\mathbf{E}_m^{\mathrm{TM}}$ are the TE and TM modes, respectively. The reflection coefficients are given by a_m and b_m for TE and TM modes, respectively, while γ_m is the corresponding propagation constant within the waveguide. Note that in (7.36), the incident field is assumed to be propagating in the $+z$ direction; however, this is simply a convenient choice and other propagating fields can readily be implemented.

Since the waveguide modes are orthogonal, the reflection coefficients are given by

$$a_m = e^{-\gamma_m z} \int_{S_f} \mathbf{E}_m^{\mathrm{TE}} \cdot (\mathbf{E} - \mathbf{E}^{\mathrm{i}})\, dS \tag{7.37}$$

$$b_m = e^{-\gamma_m z} \int_{S_f} \mathbf{E}_m^{\mathrm{TM}} \cdot (\mathbf{E} - \mathbf{E}^{\mathrm{i}})\, dS \tag{7.38}$$

Across the feed aperture the magnetic field is obtained by Faraday's Law

$$\mathbf{H} = j\frac{Y_0}{k_0}\, \nabla \times \mathbf{E} \tag{7.39}$$

assuming a nonmagnetic material ($\mu_r = 1$). This is used across the feed aperture in (7.3) by equating it to the interior magnetic field ($\mathbf{H} = \mathbf{H}^{\mathrm{int}}$). Hence, the excitation for the system is given by

$$f_i^{\mathrm{mode}} = jk_0 Z_0 \int_{S_f} [\mathbf{W}_i \times \hat{\mathbf{z}}] \cdot \mathbf{H}\, dS \tag{7.40}$$

and upon applying (7.35)–(7.40) at $z = -L$ (e.g., within the waveguide, a distance L from the aperture), (7.40) becomes

$$\begin{aligned} f_i^{\mathrm{mode}} = {} & 2\gamma_m e^{\gamma_m L} \int_{S_f} \mathbf{W}_i \cdot \mathbf{E}^{\mathrm{i}}\, dS \\ & - \sum_m \gamma_m \left(\int_{S_f} \mathbf{W}_i \cdot \mathbf{E}_m^{\mathrm{TE}}\, dS \right) \left(\int_{S_f} \mathbf{E} \cdot \mathbf{E}_m^{\mathrm{TE}}\, dS \right) \\ & + \sum_m \frac{k_0^2}{\gamma_m} \left(\int_{S_f} \mathbf{W}_i \cdot \mathbf{E}_m^{\mathrm{TM}}\, dS \right) \left(\int_{S_f} \mathbf{E} \cdot \mathbf{E}_m^{\mathrm{TM}}\, dS \right) \end{aligned} \tag{7.41}$$

Note that this model assumes that the finite element mesh extends a distance L into the waveguide to reduce the number of modes that need be retained.

7.4 CAVITY RECESSED IN A GROUND PLANE

One of the most successful applications of the FE-BI method involves antennas situated in a cavity recessed in an infinite, metallic ground plane. The reason this application is so well-suited for the FE-BI method is that the costly boundary integral portion of the formulation is minimized while the material flexibility of the finite element method is retained. FE-BI was first applied to this problem by Jin and Volakis [10] where they utilized brick elements. The computer program developed under that effort, FEMA-BRICK, was exceptionally efficient and has been used by government, industry, and academia for the analysis and design of very large (hundreds of elements) patch and slot antenna arrays. The secret of this particular computer program's efficiency is the use of a biconjugate gradient-fast Fourier transform (BiCG-FFT) matrix solver [16], [19]. This approach will be discussed below as well as a description of the BiCG-FFT method. The FE-BI method has since been implemented using tetrahedral [11], [20] and prism [21] elements to enhance flexibility; however, use of such flexible finite elements generally preclude the use of a BiCG-FFT matrix solver. An additional enhancement of the basic recessed cavity formulation involves grafting geometrical theory of diffraction (GTD) coefficients to the FE-BI results to approximate a finite ground plane [22].

Figure 7.8 illustrates a cavity-backed aperture in a metallic ground plane. The aperture lies in the xy plane and can have arbitrary shape. The cavity can also have arbitrary composition, but we assume a metallic boundary on all its sides except for the aperture. The material within the volume is assumed to be inhomogeneous, and unless otherwise noted, the exterior region is assumed to be free space. The aperture can be either open or partially covered with an infinitesimally thin metallic patch. There can be more than one cavity/aperture, but for the sake of simplicity, we assume that all apertures lie in the $z = 0$ plane.

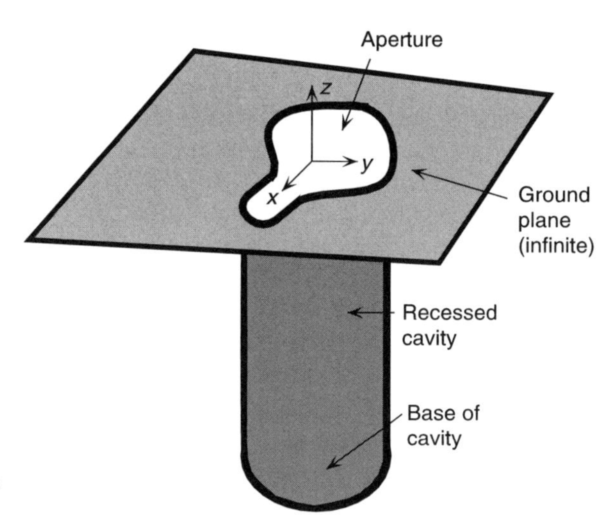

Figure 7.8 Illustration of a cavity recessed in a metallic ground plane.

7.4.1 Formulation

The efficiency of this method lies in the fact that the only exposed (nonmetallic) surface is the aperture. Since the cavity's other walls are metallic, for a total electric field formulation, the boundary conditions on those walls require a vanishing tangential electric field. The assumption of the aperture lying in an infinite metallic ground plane allows further simplification. The tangential surface electric fields ($\mathbf{E}^{\text{ext}}$) in (7.11) short out over the entire ground plane except for the aperture. Thus, the magnetic current ($-\hat{\mathbf{n}} \times \mathbf{E}^{\text{ext}}$) has support only over the aperture (it also vanishes over any surface patch lying within the aperture).

Normally, an electric current ($\hat{\mathbf{n}} \times \mathbf{H}^{\text{ext}}$) would be required over the entire, infinite ground plane in order to represent the metallic plane. However, a special dyadic Green's function can be used in place of the usual free-space Green's function, and since this Green's function satisfies the Neumann boundary condition, the electric currents are no longer required. Hence, the Green's function enforces the metallic plane boundary condition (without a need to introduce an equivalent electric current) thereby reducing the number of unknowns. This dyadic Green's function is an electric dyadic Green's function of the second kind [9], and for the case of an infinite metallic plane, it can be derived using image theory. Since this Green's function converts a surface tangential magnetic current to an exterior magnetic field, image theory states that the Green's function is simply twice the free-space Green's function

$$\overline{\overline{G}}_{e2} = -\left(\overline{\overline{I}} + \frac{\nabla\nabla}{k_0^2}\right)\left[\frac{e^{-jk_0 R}}{2\pi R}\right] \tag{7.42}$$

where both the source and test points lie in the xy plane. Note that (7.42) contains a minus sign, but alternative representations without the minus sign may be used accompanied with appropriate changes in the formulation. The evaluation of the exterior magnetic field requires only magnetic currents rather than both electric and magnetic currents, and the total magnetic field is given by

$$\hat{\mathbf{z}} \times \mathbf{H}^{\text{int}} = \hat{\mathbf{z}} \times \mathbf{H}^{\text{ext}} = \hat{\mathbf{z}} \times (\mathbf{H}^{\text{i}} + \mathbf{H}^{\text{refl}}) - jk_0 Y_0 \oint_S \left[\hat{\mathbf{z}} \times \overline{\overline{G}}_{e2} \times \hat{\mathbf{z}}\right] \cdot \mathbf{E}^{\text{int}}\, dS' \tag{7.43}$$

An important difference between (7.43) and (7.5) is that since the magnetic field in the aperture can be completely represented by the tangential electric field in the aperture, e.g., through the use of (7.43), we can *substitute* the surface magnetic field integral into (7.3). That is, the FE-BI equation, assuming surface field expansion terms identical to the volume expansion terms, is given by

$$\int_V \left[\frac{\nabla \times \mathbf{E}^{\text{int}} \cdot \nabla \times \mathbf{W}_i}{\mu_r} - k_0^2 \epsilon_r \mathbf{E}^{\text{int}} \cdot \mathbf{W}_i\right] dV - k_0^2 \oint_S \oint_S \left[\mathbf{W}_i \cdot \hat{\mathbf{n}} \times \overline{\overline{G}}_{e2} \times \hat{\mathbf{n}}' \cdot \mathbf{E}^{\text{int}}\right] dS'\, dS = f_i^{\text{int}} + \tilde{f}_i^{\text{ext}} \tag{7.44}$$

Hence, rather than having separate but coupled equations (7.19), we have a single equation (7.44).

The FE-BI system is then determined by substituting in suitable edge-based expansion functions, as was done in the previous section. Doing so, we get

$$\sum_{j=1}^{N} E_j \left\{ \int_V \left[\frac{\nabla \times \mathbf{W}_i \cdot \nabla \times \mathbf{W}_j}{\mu_r} - k_0^2 \epsilon_r \mathbf{W}_i \cdot \mathbf{W}_j \right] dV - k_0^2 \oint_S \oint_S \left[\mathbf{W}_i \cdot \hat{\mathbf{n}} \times \overline{\overline{G}}_{e2} \times \hat{\mathbf{n}}' \cdot \mathbf{W}_j \right] dS'\, dS \right\} = f_i^{\text{int}} + \tilde{f}_i^{\text{ext}}, \qquad i = 1, 2, 3, \ldots, N \tag{7.45}$$

where we have assumed that the volume expansion functions reduce to the surface expansion functions in the aperture, thus ensuring the enforcement of (7.10). Note that in (7.45), the surface term is nonzero only for test and source edges that lie in the aperture.

7.4.2 Solution Using Brick Elements

Although the formulation presented above can be and has been implemented using various different element volumes (e.g., bricks, prisms, or tetrahedrals) [2], [3], [11], [12], the use of brick elements with uniform discretization allows for a particularly efficient matrix solver: an iterative, FFT-based method. Bricks are attractive for discretizing rectangular volumes since they are easy to implement and readily conform to the cavity's dimensions. Bricks are not suitable for drum- or odd-shaped cavities since these volumes introduce stair-casing and thereby reduce the accuracy of the solution.

As stated above, if brick elements are used, then a particularly efficient implementation can be achieved. Experience has shown that the boundary integral portion of the FE-BI equation (7.44) dominates the computational cost for many applications both in terms of memory and compute cycles. This is because the sub-matrix associated with the boundary integral is fully populated and hence requires $\mathcal{O}(N_{\text{bi}}^2)$ storage and compute cycles *per iteration*, where N_{bi} is the number of unknowns associated with the boundary integral.

However, in the case of brick elements, a more efficient implementation can be used due to additional symmetry in the BI sub-matrix. Brick elements utilize a uniform surface discretization, and in this case, the boundary integral terms depend on the *physical* distance between unknowns in terms of rows and columns of the surface grid. Hence, the boundary integral sub-matrix entries can be written as

$$Y_{ij} = -k_0^2 \oint_S \oint_S \left\{ \mathbf{W}_i \cdot \hat{\mathbf{n}} \times \overline{\overline{G}}_{e2}[m - m', n - n'] \times \hat{\mathbf{n}}' \cdot \mathbf{W}_j \right\} dS'\, dS \tag{7.46}$$

where $[m, n]$ is the row and column (in the surface grid) of the test edge and $[m', n']$ is corresponding position of the source edge. In these terms, the arguments to the Green's function is the difference between the row $(m - m')$ and column $(n - n')$ of the source and test points. Hence, the sub-matrix that results from (7.46) is Block Toeplitz. This particular matrix structure lends itself well to efficient solution. Specific formulae for the boundary integral matrix entries are presented in the appendix.

Iterative solution algorithms refine an initial guess at the solution until a preset error threshold is satisfied. The major computational cost driver in any iterative algorithm is the matrix-vector multiply. All iterative solvers require at least one matrix-vector multiply per iteration (some require two) [23]. The FE-BI equations (7.44) can be solved using a symmetric matrix iterative solver such as the biconjugate gradient (BiCG) method. To apply this algorithm, it is useful to divide the matrix into individual operations for the FE and BI portions, viz.

$$[\mathscr{A}] \begin{matrix} \{E_j^{\text{bi}}\} \\ \{E_j^{\text{fe}}\} \end{matrix} + \begin{bmatrix} [\mathscr{G}] & [0] \\ [0] & [0] \end{bmatrix} \begin{matrix} \{E_j^{\text{bi}}\} \\ \{E_j^{\text{fe}}\} \end{matrix} = \begin{matrix} \{\tilde{f}_i^{\text{ext}}\} \\ \{f_i^{\text{int}}\} \end{matrix} \tag{7.47}$$

where $[\mathscr{A}]$ represents the FE matrix and $[\mathscr{G}]$ denotes the integral sub-matrix. As suggested by (7.47), the matrix-vector multiply required by the BiCG method can be performed as two separate operations. The FE matrix can be multiplied by the search vector (see Chapter 9 for the BiCG algorithm) and the result retained in a temporary vector. The BI sub-matrix can be independently multiplied by the portion of the search vector associated with boundary integral unknowns (the remainder of the search vector would be multiplied by null rows and is therefore omitted). This result is then added to the FE product, and it is identical to the product vector obtained if the entire matrix were multiplied by the search vector at one time. However, since the costly matrix-vector operation has been partitioned into a FE and BI portion, it can be optimized to exploit the sparseness of the FE matrix and the Block Toeplitz structure of the BI sub-matrix, respectively.

Sparse matrix multiplies are very efficient because only nonzero entries are considered and stored. In fact, since the FE matrix entries for a brick element are regular, there is no need to store more than a handful of entries. Each brick has 12 edges, four aligned along three orthogonal directions (x, y, z). Since each matrix entry represents an interaction, all possible interactions can be represented by 288 interactions: 144 (12 squared) for the first volume integral in (7.44) and another 144 for the second volume integral. Many of these potential interactions yield zeros, and thus significantly fewer interactions need be computed or stored. However, in practice, the logic required to determine significant and insignificant (null) interactions is extensive and usually 288 complex memory locations per layer of the mesh are allocated. Each layer is stored separately since the layer thickness need not be constant from layer to layer and therefore the matrix entries will be different. However, if the layers are of constant thickness (e.g., all bricks throughout the mesh are identical), only a single set of 288 interactions need be computed and retained. This is still insignificant compared to the storage cost of the actual FE interaction matrix! Specific formulae for the brick FE entries are provided in the appendix.

The Block Toeplitz structure of the BI sub-matrix, $[\mathscr{G}]$, can be exploited to yield impressive memory consumption and run-time efficiency. Careful inspection of (7.46) indicates that all interactions can be represented by computing and storing a single row of the interaction matrix and a portion of another row. For example, if the unknowns were numbered with all x-directed edges first followed by all y-directed edges (assuming the aperture lies in the $z = 0$ plane), then all interactions between two x-directed aperture edges are represented by the first N_x entries of the matrix's first row. The next N_y entries of the first row represent all interactions between x-directed test edges and y-directed source edges. Finally, the last N_y entries of the first row associated with the first y-directed edge represent all possible interactions between y-directed edges. This is the minimal set of interactions that need be computed, although in practice two full rows of the matrix may be computed and stored to simplify the required logic. (The additional partial row represents all interactions where the test edge is y-directed and the source edges are x-directed. For the situation considered herein, these are identical to the interactions involving an x-directed test edge and a y-directed source edge due to symmetry.)

These minimal interactions may be used in an iterative solver by utilizing the matrix partitioning scheme given by (7.47). That is, the FE entries may be represented by a sparse FE matrix, $[\mathscr{A}]$, whether stored or computed on-the-fly, and the BI interactions may be represented by $[\mathscr{G}]$. During the crucial matrix-vector multiply (per iteration), a sparse matrix operation is used for the FE portion and a FFT-based matrix-vector multiply is used for the BI portion. The FE matrix-vector multiply requires only $\mathcal{O}(N)$ while the BI matrix-vector multiply requires only $\mathcal{O}\{[M_{\max}\log_2(M_{\max})]^2\}$ operations, where N denotes the number of unknowns for the entire system and $M_{\max}$ is the maximum number of edges in either the x- or y-direction. The following section presents a detailed description concerning the implementation of a FFT-based matrix-vector multiply scheme.

7.4.3 FFT-Based Matrix-Vector Multiply Scheme

The BI matrix-vector multiply can be computed on the basis of the discrete, linear convolution theory and the Fast Fourier Transform (FFT) [16], [19]. In the following, the general coordinates (u, v) may be considered as $(u = x, v = y)$ for the planar case. As seen later, planar topologies are not the only cases that yield a Block Toeplitz matrix structure, and therefore use of general coordinates simplifies the extension of this presentation to other situations (e.g., apertures in an infinite, metallic cylinder).

The BI integral equation (7.46) can be represented by the following four sets of equations:

$$
\begin{aligned}
g_{uu}[t-t', s-s'] &= +k_0^2 \int_{S_e}\int_{S_{e'}} W_u(u',v')W_u(u,v)G^{uv}(u-u',v-v')\,du'\,dv'\,du\,dv \\
g_{uv}[t-t', s-s'] &= -k_0^2 \int_{S_e}\int_{S_{e'}} W_v(u',v')W_u(u,v)G^{uu}(u-u',v-v')\,du'\,dv'\,du\,dv \\
g_{vu}[t-t', s-s'] &= -k_0^2 \int_{S_e}\int_{S_{e'}} W_u(u',v')W_v(u,v)G^{vv}(u-u',v-v')\,du'\,dv'\,du\,dv \\
g_{vv}[t-t', s-s'] &= +k_0^2 \int_{S_e}\int_{S_{e'}} W_v(u',v')W_v(u,v)G^{vu}(u-u',v-v')\,du'\,dv'\,du\,dv
\end{aligned}
\tag{7.48}
$$

where g_{uu} represents the u-u interactions (first N_u entries of the first row of $[\mathscr{G}]$), g_{vv} represents the v-v interactions (last N_v interactions of the (N_u+1)th row), and so on. $W_{u,v}(u, v)$ are the edge-based testing/expansion functions, and e refers to the test element while e' denotes the source element. N_u is the number of grid points in the u-direction while N_v is the number of grid points in the v-direction, as displayed in Fig. 7.9.

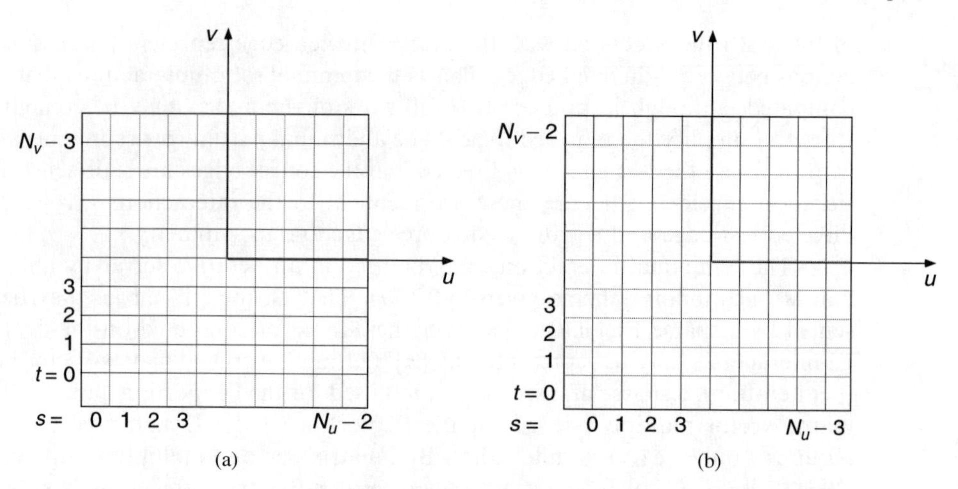

Figure 7.9 Illustration of collocated surface meshes. N_u is the number of nodes in the u-direction while N_v is the number of nodes in the v-direction.

In (7.48) the row of the *physical surface mesh* is denoted by t while the column of the physical surface mesh is indicated by s. For a mesh involving both u-directed and v-directed edges, there are two different collocated meshes: one for u-directed edges and one for v-directed edges, as shown in Fig. 7.9. Figure 7.9(a) corresponds to the u-directed edges while Fig. 7.9(b) refers to the v-directed edges. Notice that there are $N_v - 2$ rows and $N_u - 1$ columns for the unknowns on the u-directed edges. Also, there are $N_v - 1$ rows and $N_u - 2$ columns for the unknowns on the v-directed edges. This example illustrates our comment that there are two collocated mesh schemes with different numbering conventions.

These two separate meshes are important in implementing a FFT-based matrix-vector product. Since such a matrix-vector product relies upon the physical distance between the test and source functions rather than their matrix position, understanding the physical layout of the meshes permits proper filling of the data arrays and the correct calculations of the matrix-vector product. All of these comments are based upon the fact that each of the sub-matrices represented in (7.48) is used in a matrix-vector product and those products are truncated, discrete, linear convolutions, and hence amenable to the BiCG-FFT method.

To proceed, we define the two-dimensional discrete Fourier transform (DFT) pair

$$\tilde{F}[q,p] = \mathcal{F}_{2D}\{\tilde{f}[t,s]\} = \sum_{t=0}^{M_1-1}\sum_{s=0}^{M_2-1} \tilde{f}[t,s]e^{-j(2\pi/M_1M_2)(sp+tq)}$$

$$\tilde{f}[t,s] = \mathcal{F}_{2D}^{-1}\{\tilde{F}[q,p]\} = \frac{1}{M_1M_2}\sum_{t=0}^{M_1-1}\sum_{s=0}^{M_2-1} \tilde{F}[q,p]e^{j(2\pi/M_1M_2)(sp+tq)} \tag{7.49}$$

Using (7.49), the convolutions in (7.48) can then be rewritten as

$$\sum_{t'=0}^{M_1-1}\sum_{s'=0}^{M_2-1} E[t',s']g[t-t',s-s'] = \mathcal{F}_{2D}^{-1}\{\mathcal{F}_{2D}\{E[t,s]\} \bullet \mathcal{F}_{2D}\{g[t,s]\}\} \tag{7.50}$$

where "•" indicates a Hadamard (e.g., term-by-term) product. The order of the relevant DFTs must be $M_1 \geq 2$ (number of rows)−1 and $M_2 \geq 2$ (number of columns) − 1, where the number of rows and columns of the discretization may vary with each convolution. For example, the first convolution in (7.48) is associated with u-testing and u-source edges and hence the number of rows and columns is $(N_v - 2)$ and $(N_u - 1)$ respectively. The field sequences are loaded into an $M_1 \times M_2$ array in row/column order of the field discretization, and the remaining entries form a zero pad.

The Green's function sequence must be loaded into a similar array (in the same manner), and periodic replication must be performed to provide the necessary "negative lags." The data array (matrix) entries represented by (7.48) using the first u-directed edge for testing and all of the u-directed edges for sources (e.g., part of the first row of the matrix) represent all interactions where the source edge is to the right and above the test edge, as shown in Fig. 7.9. The "negative lags" are situations where the source edges are to the left and/or below the test edge.

If the sequence has the property, $g[t - t', s - s'] = g[t' - t, s' - s]$, then this replication process takes the form

$$\begin{aligned}
g[t, s] &= g[t, s] & 0 \leq t \leq \frac{M_1}{2} - 1 \quad & 0 \leq s \leq \frac{M_2}{2} - 1 \\
&= g[M_1 + 2 - t, s] & \frac{M_1}{2} \leq t \leq M_1 - 1 \quad & 0 \leq s \leq \frac{M_2}{2} - 1 \\
&= g[t, M_2 + 2 - s] & 0 \leq t \leq \frac{M_1}{2} - 1 \quad & \frac{M_2}{2} \leq s \leq M_2 - 1 \\
&= g[M_1 + 2 - t, M_2 + 2 - s] & \frac{M_1}{2} \leq t \leq M_1 - 1 \quad & \frac{M_2}{2} \leq s \leq M_2 - 1
\end{aligned} \tag{7.51}$$

The first group in (7.51) consists of cases where the source edge is to the right and above the test edge. The second group represents the cases where the source edge is to the right and below the test edge. The third group is for interactions where the source edge is to the left and above the test edge. Finally, the last group represents interactions when the source edge is to the left and below the test edge.

If such symmetry is not present, all possible lags must be computed requiring longer matrix build time since more than the first $\hat{u}$-directed and $\hat{v}$-directed edges need be used as sources. Whether even symmetry exists or not depends on the specific expansion functions and thus is implementation specific.

Once the periodic arrays are loaded, the required matrix-vector product for the $\hat{u}\hat{u}$-interactions may be performed in $\mathcal{O}((M_1 \log M_1)(M_2 \log M_2))$ operations rather than the $\mathcal{O}(((N_u - 2)(N_v - 3))^2)$ operations required for a standard matrix-vector product. The comparison is shown in Fig. 7.10 with $M_1 = 2(N_v - 3)$, $M_2 = 2(N_u - 2)$ and $N_v = N_u = N$. Clearly, when the number of nodes per side exceeds 10–15, the FFT-based matrix-vector product is more efficient than a conventional matrix-vector product. In practice, the FFT-based product is more efficient than a standard product in terms of wall clock time for $N < 10$ since in order to exploit the memory savings afforded by uniform zoning of a convolutional kernel without using FFTs, additional overhead is incurred to match the appropriate matrix entry with the correct search vector entries. Similar results are obtained for the other convolutions in (7.48).

The interested reader is referred to [16], [19], and [24] for additional details.

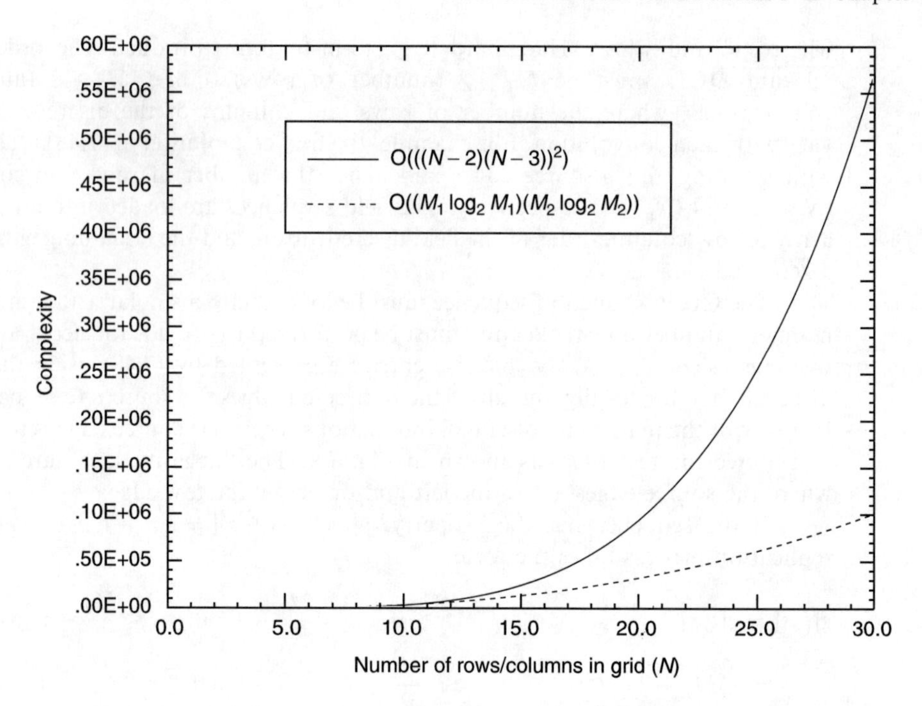

Figure 7.10 Comparison of the operation count (complexity) for a traditional matrix-vector product versus an FFT-based matrix-vector product. N is the number of nodes in each direction of the surface mesh.

7.4.4 Examples

For the convenience and educational development of the reader, information on obtaining a fully functional FE-BI brick program is provided at the end of this chapter. This program, *LMBRICK* (a.k.a. *Low Memory Brick*) utilizes the optimization techniques described previously for brick element implementations of the FE-BI method. These features include:

1. Automatic mesh generator for rectangular cavity-backed patch and slot antennas.
2. Precomputation of only necessary FE interactions and a custom sparse matrix-vector product that utilizes this minimal data set.
3. FFT-based matrix-vector product for the BI sub-matrix.

This program can compute the Radar Cross Section (RCS) of a cavity-backed patch or slot antenna, the radiation and gain pattern of a probe-fed conformal patch or slot antenna, and the input impedance of such antennas. Due to the efficient implementation, a large, finite array of similar or dissimilar elements may be modeled. Also, since the FE method is used in the cavity volume, the dielectric fill may be inhomogeneous on a brick-by-brick basis.

For example, consider calculating the RCS attributed to a 4 cm × 3 cm patch antenna recessed in an 8 cm × 6 cm × 0.1 cm cavity that is homogeneously filled with dielectric ($\epsilon_r = 2.0$). This antenna is shown in Fig. 7.11. Several different discretiza-

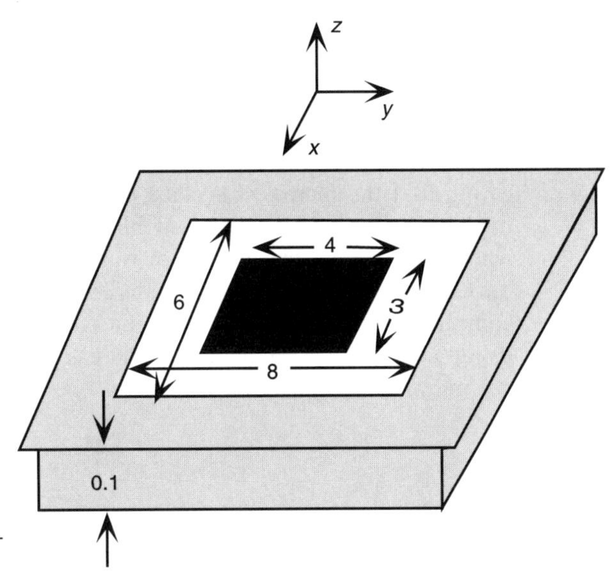

Figure 7.11 Cavity-backed patch antenna geometry used in the examples.

tions are used to illustrate the computational scaling associated with the FE-BI method using a BiCG-FFT solver. All calculations are made on a Pentium 60-MHz personal computer running Linux, and the iterative solver tolerance was set at 0.01. The first case involves 0.5 cm × 0.5 cm × 0.1 cm bricks which resulted in 411 unknowns. A total run time of 0.38 hours was required to compute the RCS for this problem at 201 different frequencies. For the same geometry, using 0.5 cm × 0.5 cm × 0.05 cm bricks, the number of unknowns was 932 and the run time was 1.63 hours. When the grid cell size was 0.5 cm × 0.5 cm × 0.025 cm, the number of unknowns was 1974 and the corresponding run time was 5.86 hours. Hence, as the number of *volume* unknowns increases, the memory consumption increases linearly, since the volume unknowns are associated with a sparse matrix whereas the solve time increases super-linearly (e.g., between linear and quadratic).

In the previous paragraph, the xy grid was kept constant (i.e., 0.5 cm × 0.5 cm) as the brick height was varied. Let us now consider the situation as the brick cell height is kept constant at 0.1 cm and the xy grid which is relevant to the boundary integral is varied from 0.5 cm × 0.5 cm to 0.25 cm × 0.25 cm and then down to 0.125 cm × 0.125 cm. As noted above, the solve time for the 0.5 cm × 0.5 cm × 0.1 cm grid cell size was 0.38 hours for 201 frequencies. In the case of a 0.25 × 0.25 cm × 0.1 cm grid cell size, the corresponding solve time is 2.22 hours (with 1781 unknowns). Finally, for the smaller cell size of 0.125 cm × 0.125 *cm* × 0.1 *cm* the number of unknowns grows to 7401 with a corresponding CPU time of 14.5 hours. Examining the ratio of solve time to number of unknowns, it is clear that boundary integral unknowns scale more favorably than volume unknowns. For example, in each set of runs, the cell grids of 0.5 cm × 0.5 cm × 0.025 cm and 0.25 cm × 0.25 cm × 0.1 cm have roughly the same number of unknowns ; however, when the increase in unknowns occurred within the volume rather than the surface, the solve time was roughly twice as long! The efficiency associated with the boundary unknowns, even though they lead to a fully populated matrix, is due to the use of a FFT-based solver and the improved convergence of the solver for dense systems. If a more traditional matrix-vector product is used for the boundary integral portion, it would result in a dramatically less favorable scaling. Figure 7.12 illustrates the computed RCS for these latter three discretizations. Note the RCS resonant frequency changes indicating that the increased surface unknowns are refining the solution, e.g., improving the estimate of the fields within the volume and on the aperture. In the cases where the aperture discretization was held constant but the volume was subdivided into thinner layers, no appreciable change was observed in the radar cross-section computations. Hence, the thin thickness of the cavity was sufficiently sampled using a single layer of elements whereas increased aperture discretization improved accuracy.

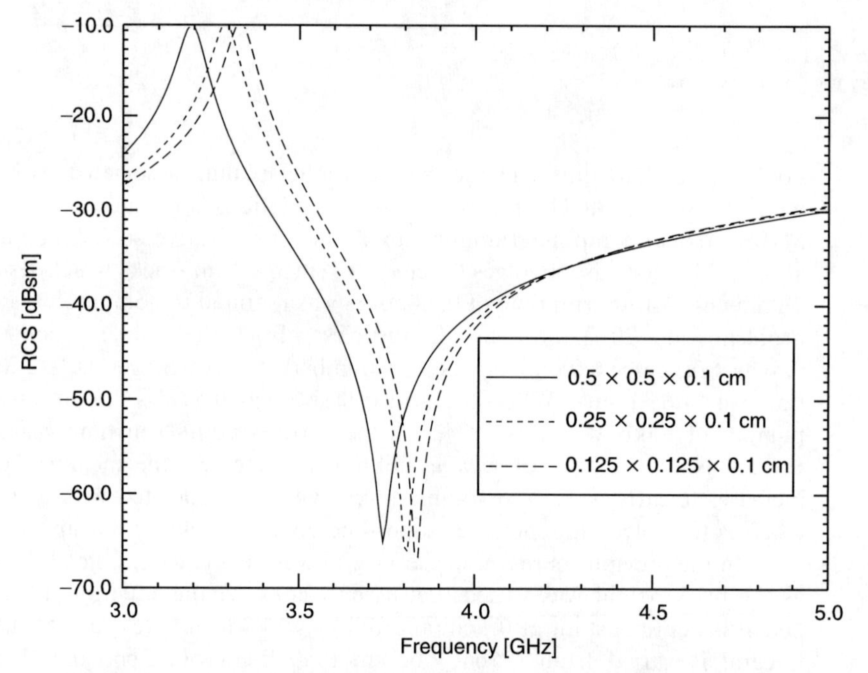

Figure 7.12 RCS as a function of frequency for a 4 cm × 3 cm patch antenna printed on a 8 cm × 6 cm × 0.1 cm cavity filled with a dielectric having $\epsilon_r = 2$; three different volume discretizations are shown.

To illustrate the accuracy of the FE-BI method, the radar cross section of a 6 cm × 5 cm × 2 cm cavity with a 3 cm × 2 cm aperture was computed using the FE-BI program cited above and a method of moments (MoM) program for the same geometry. The cavity was assumed to be filled with a material having a dielectric constant of $\epsilon_r = 2.17$. Figure 7.13 illustrates the comparison for normal incidence as a function of frequency and polarization (the $\theta\theta$-pol denotes the case where the electric field has no ϕ component).

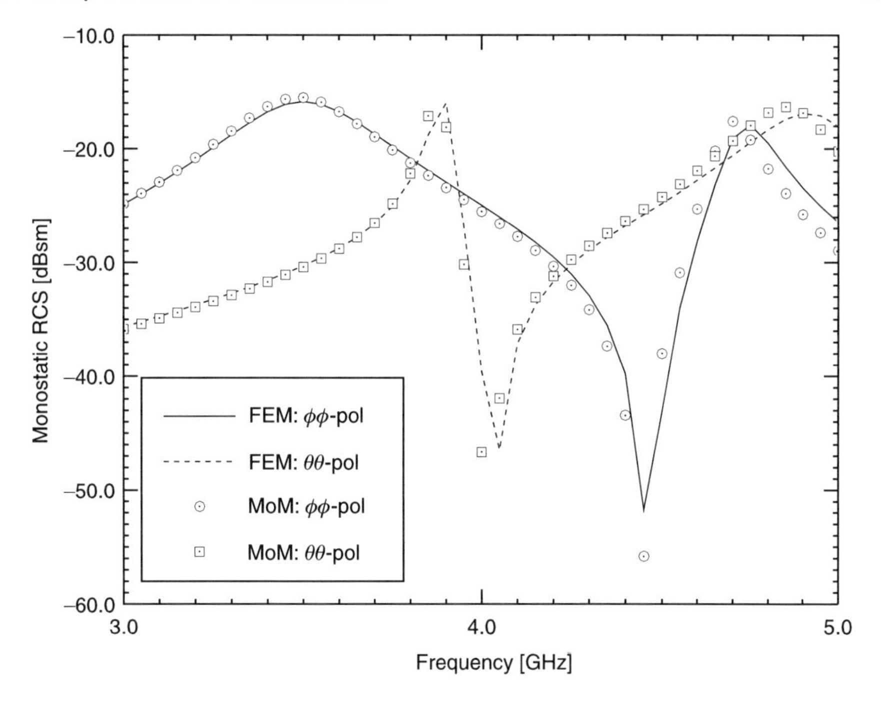

Figure 7.13 Comparison between the FE-BI method and the method of moments for computing the radar cross section of a 6 cm × 5 cm × 2 cm cavity with a 3 cm × 2 cm slot aperture. [*MoM data are Courtesy of James T. Aberle, 1997.*]

Another example of the use of a planar FE-BI computer program involves the design and analysis of finite conformal antenna arrays. The FE-BI method described above, where brick elements are used for subdividing the cavity volume and the FFT is used to handle the Block Toeplitz matrices, allows for the simulation of rather large antenna arrays on a modest computer. Figure 7.14 illustrates the gain pattern of a 5 × 5 patch antenna array. Each antenna element was similar to the one shown in Fig. 7.11 except that the dielectric constant of the substrate was $\epsilon_r = 13.9$ and the center-to-center spacing was 10 cm in each direction. The pattern at $\epsilon_r = 2.45$ GHz was allowed to radiate broadside and steered 30 degrees from broadside using standard beam steering techniques. Note that when the FE-BI method is used to simulate this array, all mutual coupling is included in the solution, thus increasing the fidelity of the model. This example was run on a Silicon Graphics workstation using approximately 14 MB of RAM. It involved 10,275 unknowns and took approximately seven minutes to compute each pattern!

7.4.5 Aperture in a Thick Metallic Plane

A variation of the planar cavity-backed geometry considered above is the case where the lower metallic surface is removed. The resulting geometry has two apertures: one in the upper metallic plane and one in the lower metallic plane. This geometry, which resembles an aperture in a thick metallic plane, is shown in Fig.

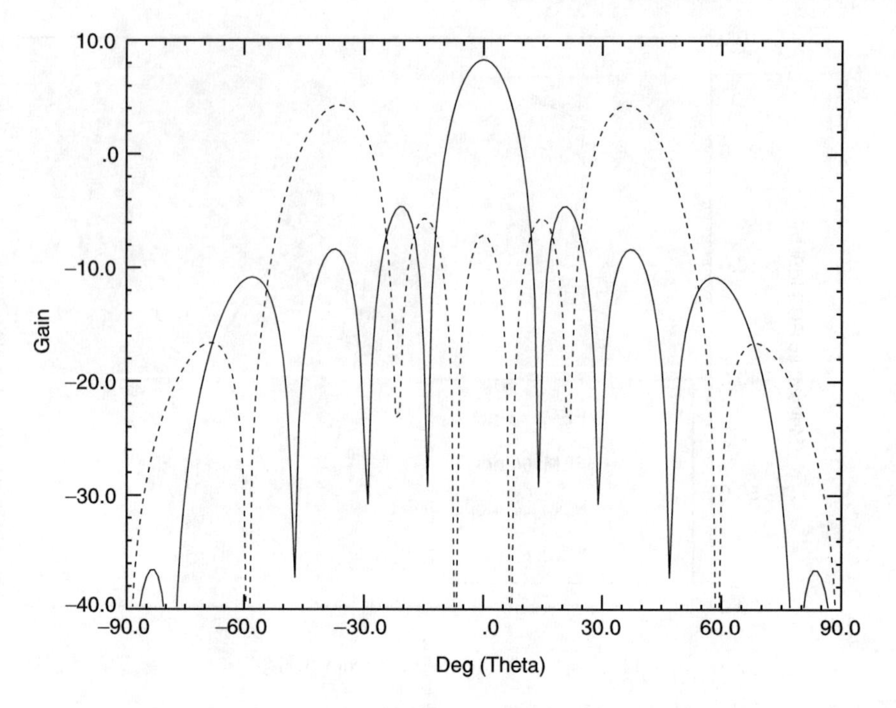

Figure 7.14 Radiation pattern of a 5 × 5 patch antenna array for broadside and 30 degrees off broadside. [*Courtesy of Jeffrey Tackett, 1997.*]

7.15. For this case, the lower aperture results in an additional BI integral resulting in the FE-BI equation

$$\int_V \left[\frac{\nabla \times \mathbf{E}^{\text{int}} \cdot \nabla \times \mathbf{W}_i}{\mu_r} - k_0^2 \epsilon_r \mathbf{E}^{\text{int}} \cdot \mathbf{W}_i \right] dV - k_0^2 \int_{S^+} \int_{S'^+} \left[\mathbf{W}_i \times \hat{\mathbf{z}} \cdot \overline{\overline{G}}_2 \cdot \hat{\mathbf{z}} \times \mathbf{E}^{\text{ext}} \right] dS' \, dS$$
$$- k_0^2 \int_{S^-} \int_{S'^-} \left[\mathbf{W}_i \times \hat{\mathbf{z}} \cdot \overline{\overline{G}}_2 \cdot \hat{\mathbf{z}} \times \mathbf{E}^{\text{ext}} \right] dS' \, dS = f_i^{\text{int}} + \tilde{f}_i^{\text{ext}} \qquad (7.52)$$

where S^+ and S^- indicate the upper and lower apertures, respectively.

Solution of (7.52) parallels that of (7.44) except that two boundary integrals are involved. As is the case with (7.44), (7.52) can be discretized using brick elements and

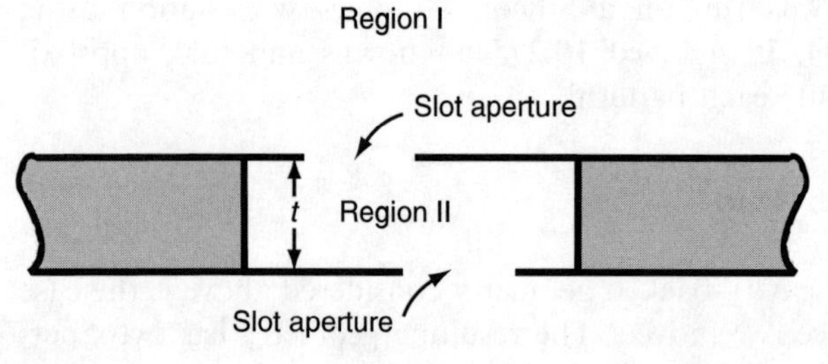

Figure 7.15 A thick metallic plane with finite apertures in the $z = 0$ and $z = -z_p$ planes.

a FFT-based matrix-vector product scheme to yield maximum efficiency. Note that the intra-aperture interactions possess the discrete convolutional property required for FFT solution. There are no inter-aperture interactions since the boundary integrals for each interaction are decoupled. Also, since brick elements are used, the same boundary matrix can be used for both apertures except for different geometrical parameters as appropriate.

Figure 7.16 illustrates the transmission coefficient as a function of aperture length (A/λ) while the other dimension of the aperture is fixed at $B = 0.1\lambda$. This example illustrates the capability of the FE-BI method to compute the transmission coefficient for an aperture in an increasingly thick plane; a difficult problem for the more traditional Method of Moments since such a solution approach would require the computationally expensive cavity Green's function [27].

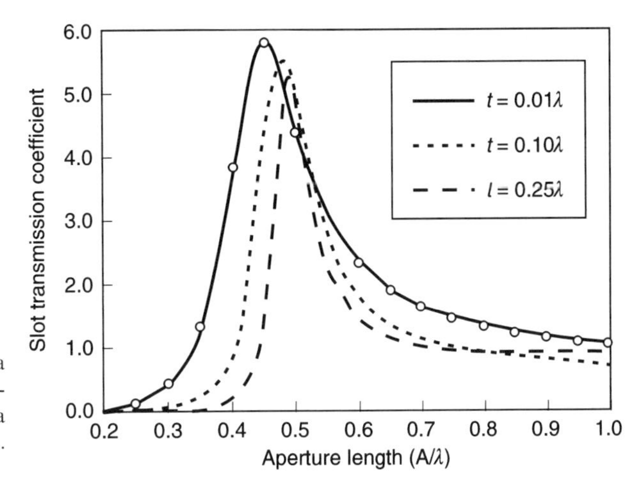

Figure 7.16 Transmission coefficient as a function of aperture length for normal incidence. The circles correspond to data for a thin conducting plane presented in [25]. [*After Jin and Volakis* [*26*], © *IEEE 1991.*]

The interested reader is referred to [25] and [26] for further details concerning the application of the FE-BI method to transmission problems.

7.5 CAVITY-BACKED ANTENNAS ON A CIRCULAR CYLINDER

A conformal antenna geometry, analogous to the case of a metallic ground plane, involves an infinite metallic circular cylinder. An example of one such antenna is shown in Fig. 7.17 where the patch elements are printed on a cavity-backed dielectric substrate that is recessed in the cylinder. In this case, (7.44) may be discretized using cylindrical shell elements similar to the brick. Cylindrical shell elements possess both geometrical fidelity and simplicity for cylindrical-rectangular cavities. Figure 7.18 illustrates a typical shell element which has eight nodes connected by 12 edges—four edges aligned along each of the three orthogonal directions of the cylindrical coordinate system. Each element is associated with 12 vector shape functions given by

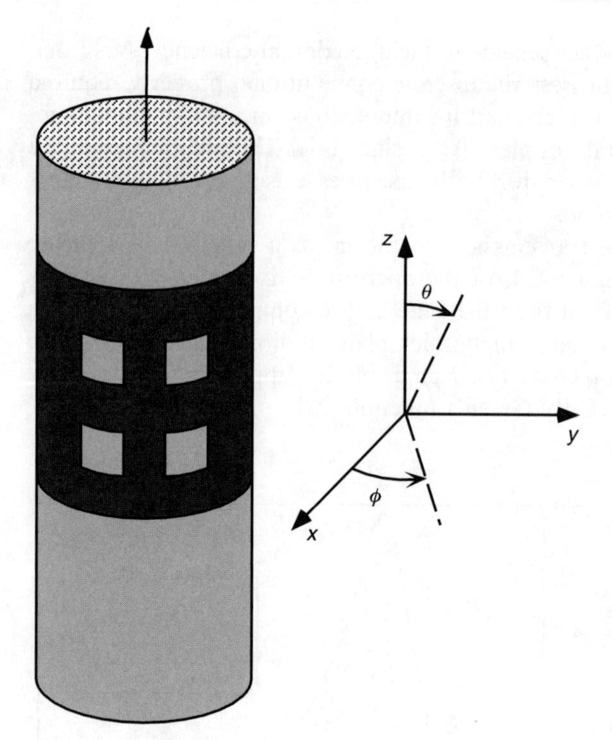

Figure 7.17 Illustration of four conformal patch antennas mounted flush with the surface of a metallic cylinder.

$$
\begin{aligned}
&\mathbf{W}_{12}(\rho,\phi,z)=\mathbf{W}_{\rho}(\rho,\phi,z;\cdot,\phi_r,z_t,+), && \mathbf{W}_{43}(\rho,\phi,z)=\mathbf{W}_{\rho}(\rho,\phi,z;\cdot,\phi_l,z_t,-)\\
&\mathbf{W}_{56}(\rho,\phi,z)=\mathbf{W}_{\rho}(\rho,\phi,z;\cdot,\phi_r,z_b,-), && \mathbf{W}_{87}(\rho,\phi,z)=\mathbf{W}_{\rho}(\rho,\phi,z;\cdot,\phi_l,z_b,+)\\
\\
&\mathbf{W}_{14}(\rho,\phi,z)=\mathbf{W}_{\phi}(\rho,\phi,z;\rho_b,\cdot,z_t,+), && \mathbf{W}_{23}(\rho,\phi,z)=\mathbf{W}_{\phi}(\rho,\phi,z;\rho_a,\cdot,z_t,-)\\
&\mathbf{W}_{58}(\rho,\phi,z)=\mathbf{W}_{\phi}(\rho,\phi,z;\rho_b,\cdot,z_b,-), && \mathbf{W}_{67}(\rho,\phi,z)=\mathbf{W}_{\phi}(\rho,\phi,z;\rho_a,\cdot,z_b,+)\\
\\
&\mathbf{W}_{15}(\rho,\phi,z)=\mathbf{W}_{z}(\rho,\phi,z;\rho_b,\phi_r,\cdot,+), && \mathbf{W}_{26}(\rho,\phi,z)=\mathbf{W}_{z}(\rho,\phi,z;\rho_a,\phi_r,\cdot,-)\\
&\mathbf{W}_{48}(\rho,\phi,z)=\mathbf{W}_{z}(\rho,\phi,z;\rho_b,\phi_l,\cdot,-), && \mathbf{W}_{37}(\rho,\phi,z)=\mathbf{W}_{z}(\rho,\phi,z;\rho_a,\phi_l,\cdot,+)
\end{aligned}
\tag{7.53}
$$

where $\mathbf{W}_{lk}$ is associated with the edge which is delimited by local nodes (l, k) as shown in Fig. 7.18 and (ρ, ϕ, z) denote the cylindrical coordinates. As can be inferred from (7.53), three fundamental vector weight functions are required for the complete representation of the shell element. They are

$$
\begin{aligned}
\mathbf{W}_{\rho}(\rho,\phi,z;\tilde{\rho},\tilde{\phi},\tilde{z},\tilde{s}) &= \frac{\tilde{s}\rho_b}{\alpha h}\,\frac{(\phi-\tilde{\phi})(z-\tilde{z})}{\rho}\,\hat{\rho}\\
\mathbf{W}_{\phi}(\rho,\phi,z;\tilde{\rho},\tilde{\phi},\tilde{z},\tilde{s}) &= \frac{\tilde{s}}{th}\,(\rho-\tilde{\rho})(z-\tilde{z})\,\hat{\phi}\\
\mathbf{W}_{z}(\rho,\phi,z;\tilde{\rho},\tilde{\phi},\tilde{z},\tilde{s}) &= \frac{\tilde{s}}{t\alpha}\,(\rho-\tilde{\rho})(\phi-\tilde{\phi})\,\hat{z}
\end{aligned}
\tag{7.54}
$$

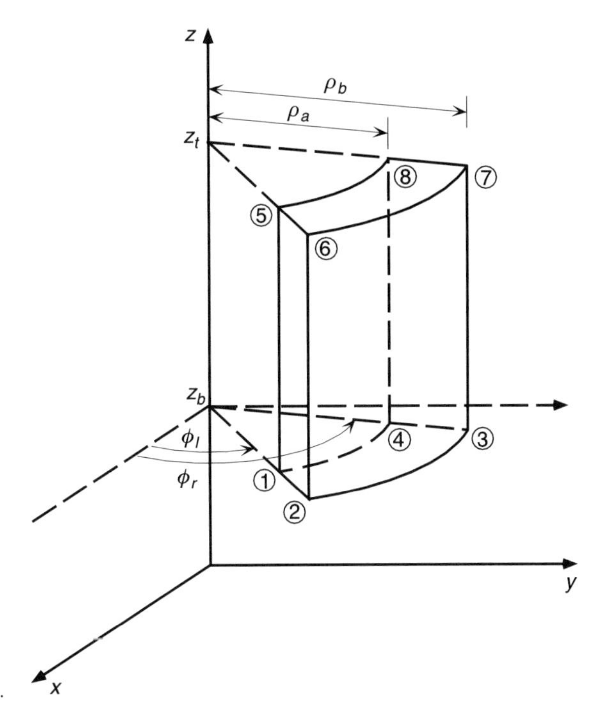

Figure 7.18 Cylindrical shell element.

where the element parameters $(\rho_a, \rho_b, \phi_l, \phi_r, z_b, z_t)$ are defined in Fig. 7.18, $t = \rho_b - \rho_a$, $\alpha = \phi_r - \phi_l$ and $h = z_t - z_b$. Each local edge is distinguished by $\tilde{\rho}, \tilde{\phi}, \tilde{z}$, and $\tilde{s}$ as given in (7.53). The $\frac{1}{\rho}$- term which appears in the definition of the $\hat{\rho}$-directed weight (7.54) is essential in satisfying the divergence-free requirement, i.e., so that $\nabla \cdot \mathbf{W}_j = 0$. Note that as the radius of the cylinder becomes large, the curvature of these elements decreases, resulting in weight functions which are functionally similar to the bricks presented by Jin and Volakis [26].

In addition to the use of cylindrical shell elements, the dyadic Green's function in (7.44) must also be changed to account for the metallic boundary condition on the cylindrical metallic surface. Specifically, the dyadic Green's function $(\overline{\overline{G}}_{e2})$ must now satisfy both the radiation condition and the Neumann boundary condition at $\rho = a$

$$\hat{\rho} \times \nabla \times \overline{\overline{G}}_{e2}(a, \phi, z|a, \phi', z') = 0 \tag{7.55}$$

This dyadic Green's function may be expressed exactly as an eigenfunction series [9]

$$\begin{aligned}
G^{zz}(a, \bar{\phi}, \bar{z}) &= -\frac{1}{(2\pi)^2}\sum_{n=-\infty}^{\infty}\int_{-\infty}^{\infty}\left(\frac{k_\rho}{k_0}\right)^2 \frac{1}{\gamma}\frac{H_n^{(2)}(\gamma)}{H_n^{'(2)}(\gamma)}\, e^{j(n\bar{\phi}-k_z\bar{z})}\, dk_z \\
G^{\phi z}(a, \bar{\phi}, \bar{z}) &= -\frac{1}{(2\pi)^2}\sum_{n=-\infty}^{\infty}\int_{-\infty}^{\infty}\left(\frac{nk_z}{k_0^2 a\gamma}\right)\frac{H_n^{(2)}(\gamma)}{H_n^{'(2)}(\gamma)}\, e^{j(n\bar{\phi}-k_z\bar{z})}\, dk_z \\
G^{\phi\phi}(a, \bar{\phi}, \bar{z}) &= \frac{1}{(2\pi)^2}\sum_{n=-\infty}^{\infty}\int_{-\infty}^{\infty}\frac{1}{\gamma}\left[\frac{H_n^{'(2)}(\gamma)}{H_n^{(2)}(\gamma)} - \left(\frac{nk_z}{k_0 a k_\rho}\right)^2 \frac{H_n^{(2)}(\gamma)}{H_n^{'(2)}(\gamma)}\right] e^{j(n\bar{\phi}-k_z\bar{z})}\, dk_z
\end{aligned} \tag{7.56}$$

where $G^{z\phi}(a, \bar{\phi}, \bar{z}) = G^{\phi z}(a, \bar{\phi}, \bar{z})$, $\gamma = k_\rho a$ and $k_\rho = \sqrt{k_0^2 - k_z^2}$.

A different representation of this Green's function was used in [28]. In this, (7.56) is converted into a creeping wave series using Watson's transformation. The resulting series converges in two terms or less for large radius cylinders, and hence it is more efficient than (7.56) as the radius increases. For reference, (7.56) typically requires approximately $2k_0a$ series terms where k_0 is the wavenumber in free space and a is the radius of the cylinder. For large radius cylinders (greater than 5λ or so), 60 series terms or more for each component in (7.56) are required for each combination of test and source edges! The creeping wave series expansion for the cylinder Green's function is presented in [29].

Replacing the Green's function components in (7.44) with (7.56), the fields associated with a conformal cavity recessed in a metallic cylinder can be solved in the same manner as the planar metallic ground plane example given previously. This includes the use of an efficient FFT-based solver since the boundary integral submatrix is Block Toeplitz provided the surface is discretized using uniform patches (e.g., cylindrical shell elements).

7.5.1 Examples

The FE-BI method has been applied to scattering and radiation by cavities recessed in an infinite metallic cylinder. Consider the scattering by a cavity-backed patch antenna recessed in a circular cylinder. The patch is 3 cm × 2 cm and is placed

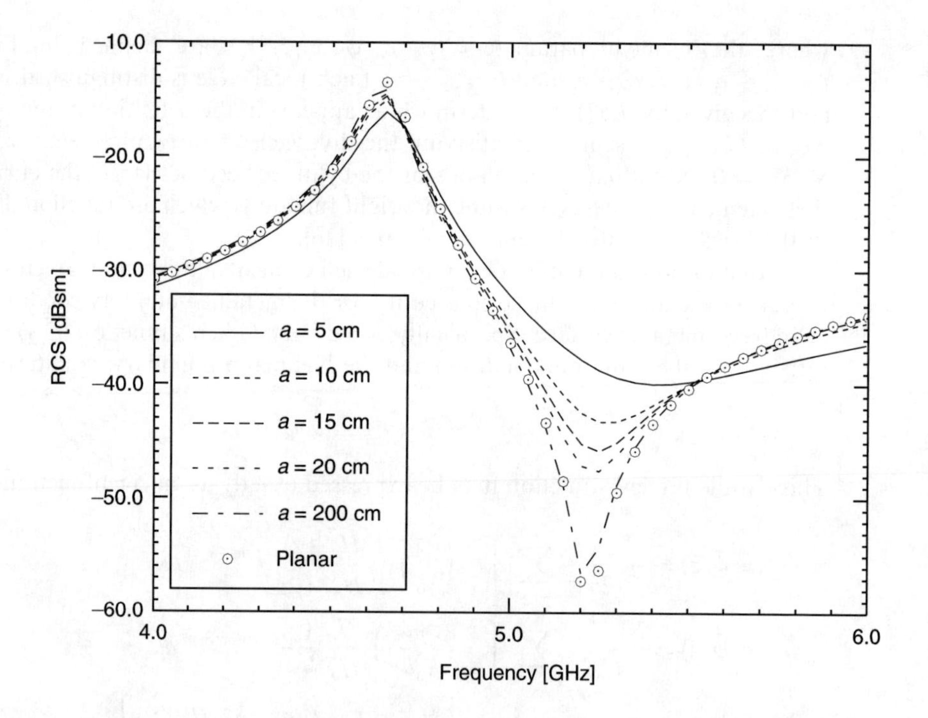

Figure 7.19 Radar cross section of a conformal patch antenna for transverse magnetic (TM) polarization. The various curves correspond to different cylinder radii, and the antenna is flush mounted to the surface of the cylinder.

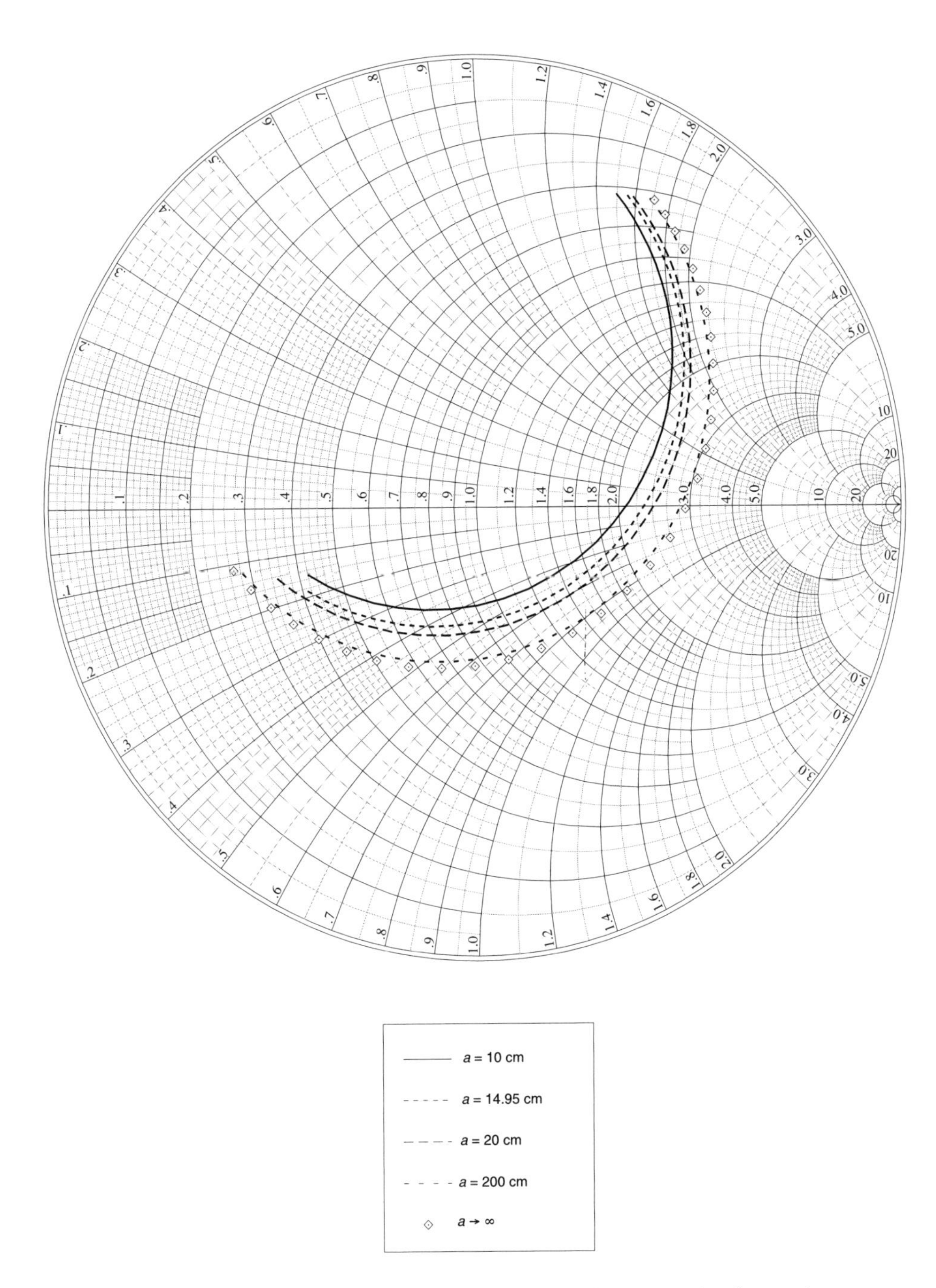

Figure 7.20 Smith chart for a 3.5 cm × 3.5 cm patch antenna for frequencies between 2.4 and 2.7 GHz. The cylinder radius varied between 10 cm and quasi-planar (200 cm).

on top of a dielectric filled cavity that is 6 cm × 5 cm × 0.07874 cm. The dielectric constant of the substrate is $\epsilon_r = 2.17$, and the antenna is flush mounted with the cylinder surface. One of the strengths of the FE-BI method for singly curved conformal antennas is its ability to investigate the effects of curvature on the RCS of a patch such as the one shown in Fig. 7.17. The RCS for various cylinder radii (denoted as "a") is shown in Fig. 7.19.

One antenna parameter expected to be sensitive to curvature is the input impedance. This is because the input impedance is strongly influenced by the modal fields within the cavity, and for certain excitations (polarizations), these fields are strongly dependent on curvature. One such polarization is obtained by feeding the patch antenna parallel to the cylinder axis (e.g., if the patch is centered on $(\phi, z) = (0, 0)$, this feed is along the z-axis). Figure 7.20 is a Smith Chart representation of the input impedance for a 3.5 cm × 3.5 cm patch antenna on different cylinder curvatures. The input impedance is strongly influenced by the curvature of the cavity/antenna.

7.6 RECENT ADVANCES IN THE FE-BI METHOD

The material presented thus far in this chapter was developed during the late 1980s and the early 1990s. Recent published results have extended the flexibility of the FE-BI method as more research groups have applied the method to a greater variety of problems. The examples presented previously in this chapter involved a cavity-backed inhomogeneous region recessed in either a ground plane or an infinite metallic cylinder. The principle reason for these restrictions involves the computational cost of the boundary integral. For the cases considered so far, boundary integral solvers are available which are efficient in terms of both memory consumption and compute cycle requirements.

Below we mention other extensions of the FE-BI method. Among them is the finite element-periodic moment method (FE-PMM) approach for the simulation of infinitely periodic structures with unprecedented flexibility. Also, recent papers have introduced nonplanar, noncylindrical boundaries in the implementation of the FE-BI. Further, techniques such as the surface of revolution (SOR) and fast multipole method (FMM) offer increased flexibility with moderate cost. An overview of recent publications follows. The interested reader should consult the cited papers for implementation details.

7.6.1 Finite Element–Periodic Method of Moments

An extension of the FE-BI method was proposed by McGrath and Pyati [30] and Lucas and Fontana [31] to investigate phase antenna arrays. In this, the finite element method is used to model an antenna element that forms the unit cell of an infinitely periodic array. Traditionally, periodic moment method (PMM) formulations have been used to investigate the performance of infinitely periodic antennas. However, the use of pure integral equation formulations limit the utility of the PMM approach to layered substrates and simpler antenna elements which can be easily modeled using the method of moments.

The flexibility of the finite element method permits the antenna element to be constructed on arbitrary materials and can have arbitrary shape. Rather than using one of the integral equations introduced at the beginning of this chapter, the finite element–periodic moment method (FE-PMM) utilizes a periodic integral equation. Specifically, Floquet modes are used to periodically replicate the boundary and radiation conditions imposed on the unit cell. Note that in this application, periodic boundary conditions are not only applied to the aperture of the antenna, but also on all nonmetallic sides of the unit cell.

An example of the use of the FE-PMM method is illustrated in Fig. 7.21 where a notch radiator is immersed within the unit cell mesh. This is an example of the type of detail readily modeled via the finite element method. Figure 7.22 illustrates the E-

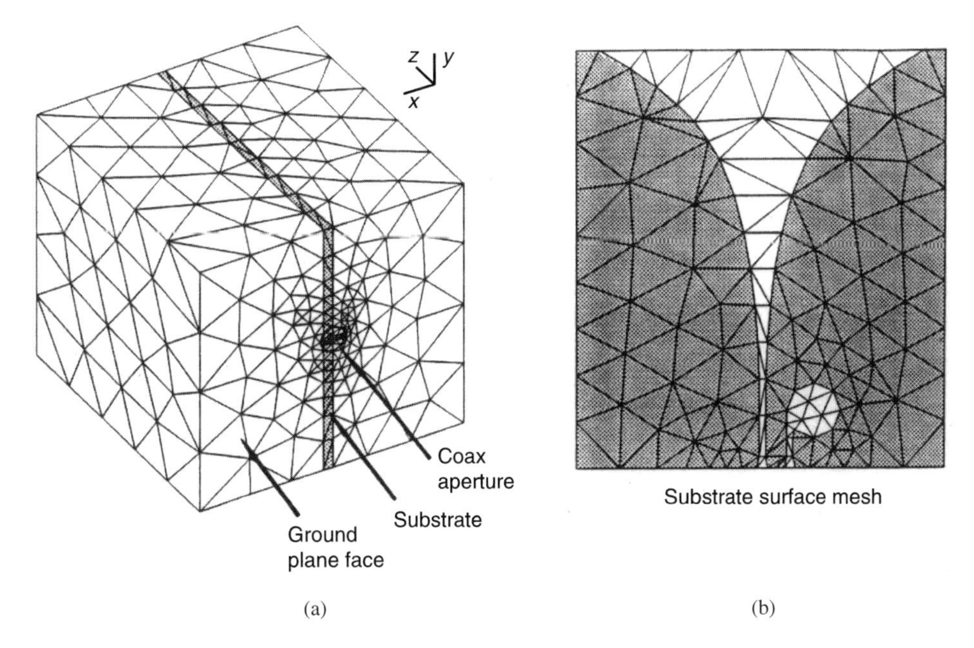

Figure 7.21 Finite element mesh for flared notch antenna: (a) unit cell illustrating coax aperture; (b) substrate surface mesh. [*After McGrath* [32].]

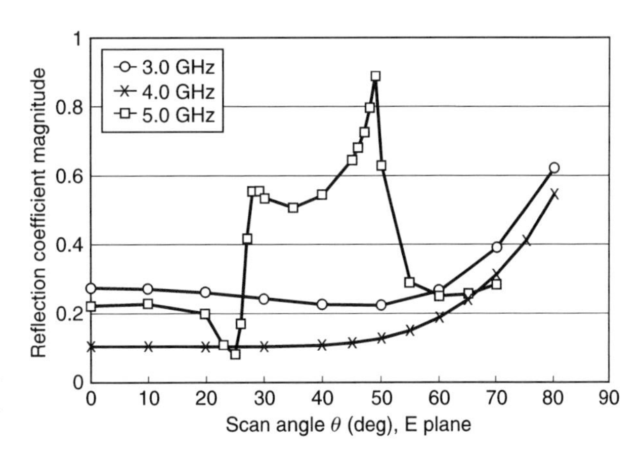

Figure 7.22 E-plane active reflection coefficient versus scan angle for the notch antenna shown in Fig. 7.21. [*After McGrath* [32].]

plane active reflection coefficient for an infinite array of such radiators. The H-plane active reflection coefficient for that same array is given in Fig. 7.23.

Recently, the FE-PMM method has been applied to a variety of additional applications such as reinforced concrete, artificial dielectrics, and pyramidal cone foam absorber [33]. The FE-PMM analysis has also been used to study bandgap materials [34].

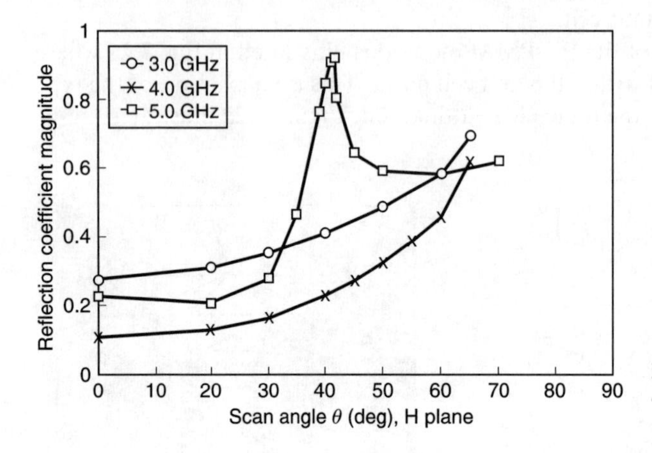

Figure 7.23 H-plane active reflection coefficient versus scan angle for the notch antenna shown in Fig. 7.21. [*After McGrath* [32].]

7.6.2 Finite Element–Surface of Revolution Method

At the beginning of this chapter, the most general form of the finite element-boundary integral formulation was presented. As noted, the BI consumes much of the CPU and memory requirements unless special techniques are used such as the BiCG-FFT. This is, of course, a specialized implementation of the BI since it is only applicable to planar or cylindrical apertures. Other specialized finite element-boundary integral formulations have been presented in the literature, and these offer a practical solution methodology for an important class of problems. One approach couples the interior finite element solution with an external region that is a surface of revolution (SOR). This special boundary permits the efficient enforcement of the boundary and radiation conditions and can be used to simulate several cubic wavelength geometries. The major uses of this finite element–surface of revolution (FE-SOR) method are to compute the scattering by complex objects and the radiation by axisymmetric antennas.

The FE-SOR method has been considered by Boyse and Seidl [35] for scattering by nearly axisymmetric bodies. Their formulation involved the use of a finite element expansion of the fields within the SOR and an eigenfunction series expansion in the azimuthal direction for the boundary integral. They used node-based tetrahedral elements within the mesh and a Fourier modal-azimuthal expansion utilizing Hermite polynomials.

A group at the Jet Propulsion Laboratory (JPL) has also published a series of articles [36], [37], [38], [39] detailing their hybrid FE-SOR formulation. In their implementation vector edge finite element expansion functions are used rather than the node-based elements in [35]. Also, dissimilar boundary integral basis functions are used and the finite element and boundary integral regions are coupled using the method presented in [36]. Recently Zuffada, Cwik, and Jamnejad presented an analysis of a circular waveguide antenna with a choke collar [39]. In this, they used a magnetic field finite element formulation

$$\frac{Z_0}{jk_0}\int_V\left[\frac{1}{\epsilon_r}(\nabla\times\mathbf{T}^*\cdot\nabla\times\mathbf{H})-k_0^2\mu_r\mathbf{T}^*\cdot\mathbf{H}\right]dV-\oint_S\mathbf{T}^*\cdot\mathbf{M}\,dS=0 \tag{7.57}$$

and a combined field integral equation

$$Z_m\left[\frac{\mathbf{M}}{Z_0}\right]+Z_J[\mathbf{J}]=0 \tag{7.58}$$

where the integro-differential operators Z_M and Z_J are those used by the body of revolution (BOR) integral equation formulation [40]. Also in [39], the essential boundary conditions are enforced using

$$\oint_S[\mathbf{T}^*\cdot(\mathbf{E}\times\hat{\mathbf{n}}-\mathbf{M})]\,dS=0 \tag{7.59}$$

and tetrahedral finite elements are used to discretize (7.57). Finally, the SOR basis functions are given by

$$\begin{aligned}\mathbf{U}^t&=\hat{t}\,\frac{T_k(t)}{\rho(t)}\,e^{jn\phi}\\ \mathbf{U}^\phi&=\hat{\phi}\,\frac{T_k(t)}{\rho(t)}\,e^{jn\phi}\end{aligned} \tag{7.60}$$

where $T_k(t)$ is a triangle function spanning the kth annulus on the SOR. The variables t and ϕ refer to the local SOR coordinates, and n refers to the mode in the Fourier series expansion.

As an example of the implementation in [39], Fig. 7.24 illustrates a circular waveguide antenna with a choke collar. The H-plane pattern for this antenna (see Fig. 7.25) was computed using the FE-SOR method to determine the fields in the shaded region shown in Fig. 7.24.

We note that in [39], a mode matching technique was used to model the feed. This is in effect a separate integral equation applied across the feed aperture and

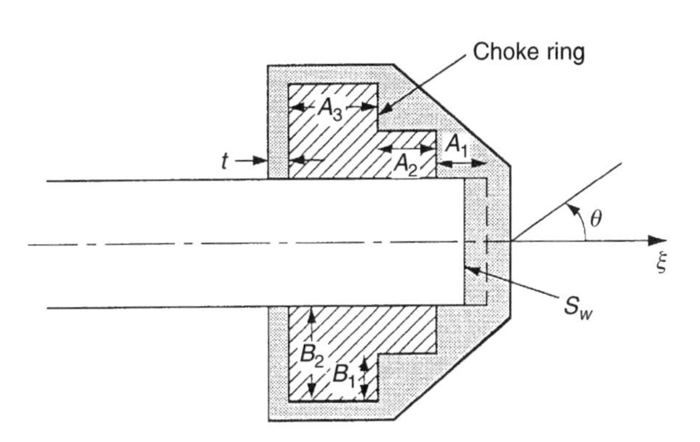

Figure 7.24 JPL circular waveguide with choke ring. The FE-SOR method was used to model the shaded area. [*After Zuffada, Cwik, and Jamnejad* [39], © *IEEE, 1997.*]

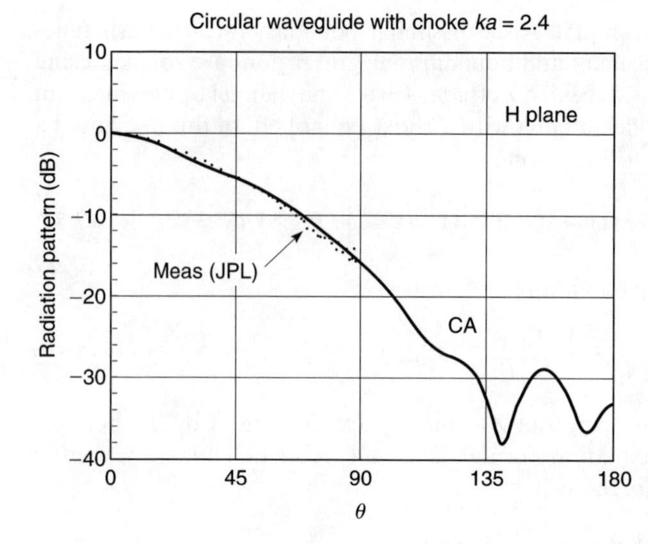

Figure 7.25 H-plane pattern for the choked circular waveguide shown in Fig. 7.24. [*After Zuffada, Cwik, and Jamnejad* [39], © *IEEE, 1997.*]

illustrates an additional application for the FE-BI method. When using the mode matching technique, the sum of the modes synthesizes the required Green's function used to represent the feed structure as discussed previously.

7.6.3 Fast Integral Solution Methods

Recent advances of the FE-BI technique for three-dimensional geometries involve use of fast integral equation solution methods to speed up the computation of the BI matrix-vector products in the iterative solver. Among them, the fast multipole method (FMM) to implement the boundary integral for two-dimensional [41] and three-dimensional FE-BI formulations [42] has been considered. This particular formulation is suitable for planar cavity-backed structures (it is based on the same approach as was presented earlier in this chapter), and the authors demonstrated a speed-up in solve time for large aperture problems. The requirement for a large aperture, hence a large separation between boundary groups, is inherent in the FMM. Details on the theory and implementation of the FMM are given in Chapter 8.

In the FMM, the boundary integral unknowns are grouped together and the interaction between groups is computed rather than the interaction between individual unknowns. These group interactions are then disaggregated to provide the necessary interactions. The resulting order of operations and memory are $\mathcal{O}(N^{\alpha})$ rather than $\mathcal{O}(N^2)$, where $\alpha \leq 1.5$. Hence, an improvement in the solution efficiency provided the distance between groups is large. If this condition is not met, accuracy issues arise since the interactions of elements can no longer be approximated via FMM grouping. Also, the overhead associated with grouping the boundary unknowns and then disaggregating them can be detrimental to the method's efficiency, but this is problem dependent.

Another fast integral method, referred to as the *adaptive integral method* (AIM), was recently introduced by Bleszynski et al. [43]. AIM is also an $\mathcal{O}(N^{1.5})$ method, and its improved speed is based on mapping the boundary integral mesh on a uniform grid so that the $\mathcal{O}(N \log N)$ FFT can be used for carrying out the matrix-vector products. Research on fast integral methods is currently very active, and the reader should consult future publications on the development and application of these techniques.

APPENDIX 1: EXPLICIT FORMULAS FOR BRICK ELEMENTS

The FE-BI equation for a cavity recessed in a metallic plane is given by (7.45) and is reproduced here for convenience

$$\sum_{j=1}^{N} E_j \left\{ \int_V \left[\frac{\nabla \times \mathbf{W}_i \cdot \nabla \times \mathbf{W}_j}{\mu_r} - k_0^2 \epsilon_r \mathbf{W}_i \cdot \mathbf{W}_j \right] dV - k_0^2 \int_S \int_{S'} [\mathbf{W}_i \cdot \hat{\mathbf{z}} \times \overline{\overline{G}}_{e2} \times \hat{\mathbf{z}} \cdot \mathbf{W}_j] \, dS' \, dS \right\} = f_i^{\text{int}} + \tilde{f}_i^{\text{ext}}, \qquad i = 1, 2, 3, \ldots, N \tag{7.61}$$

As suggested in (7.47), the linear system (7.61) can be written in matrix form as

$$[\mathcal{A}] \begin{matrix} \{E_j^{\text{bi}}\} \\ \{E_j^{\text{fe}}\} \end{matrix} + \begin{bmatrix} [\mathcal{G}] & [0] \\ [0] & [0] \end{bmatrix} \begin{matrix} \{E_j^{\text{bi}}\} \\ \{E_j^{\text{fe}}\} \end{matrix} = \begin{matrix} \{\tilde{f}_i^{\text{ext}}\} \\ \{f_i^{\text{int}}\} \end{matrix} \tag{7.62}$$

where $[\mathcal{A}]$ represents the FE matrix and $[\mathcal{G}]$ denotes the boundary integral submatrix. In this appendix, we give explicit formulas for the $[\mathcal{A}]$ matrix entries and formulas that permit numerical evaluation of $[\mathcal{G}]$. We begin with $[\mathcal{A}]$ for anisotropic media. The corresponding formulae for isotropic media are presented in Section 5.4.

Element Matrices for Brick Elements: Anisotropic Media

The unknown field, $\mathbf{E}$, can be written in terms of three components corresponding to edges lying parallel to the x-, y-, and z-axes, e.g.,

$$\begin{aligned} E_x &= \sum_{j=1}^{4} E_{xj}^e W_{xj}^e(x, y, z) \\ E_y &= \sum_{j=5}^{8} E_{yj}^e W_{yj}^e(x, y, z) \\ E_z &= \sum_{j=9}^{12} E_{zj}^e W_{zj}^e(x, y, z) \end{aligned} \tag{7.63}$$

where E_{dj} are the unknown field expansion coefficients associated with the jth local edge which is parallel to the $\hat{d}$-axis where $d = \{x, y, z\}$ and W_{dj} are the expansion functions associated with each local edge. In (7.63), there are four field expansions per brick for each component corresponding to the 12 edges of the brick. The superscript e denotes the global element number, and the index j denotes the local edge numbers. The local edges are defined as follows:

$$\text{edge}_1 = \text{node}_1 \rightarrow \text{node}_2 \qquad \text{edge}_2 = \text{node}_4 \rightarrow \text{node}_3$$
$$\text{edge}_3 = \text{node}_5 \rightarrow \text{node}_6 \qquad \text{edge}_4 = \text{node}_8 \rightarrow \text{node}_7$$

$$\text{edge}_5 = \text{node}_1 \rightarrow \text{node}_4 \qquad \text{edge}_6 = \text{node}_5 \rightarrow \text{node}_8$$
$$\text{edge}_7 = \text{node}_2 \rightarrow \text{node}_3 \qquad \text{edge}_8 = \text{node}_6 \rightarrow \text{node}_7$$

$$\text{edge}_9 = \text{node}_1 \rightarrow \text{node}_5 \qquad \text{edge}_{10} = \text{node}_2 \rightarrow \text{node}_6$$
$$\text{edge}_{11} = \text{node}_4 \rightarrow \text{node}_8 \qquad \text{edge}_{12} = \text{node}_3 \rightarrow \text{node}_7 \tag{7.64}$$

Figure 7.26 illustrates this brick element, the local node numbers used in (7.64), and the edge lengths, $\{h_x^e, h_y^e, h_z^e\}$.

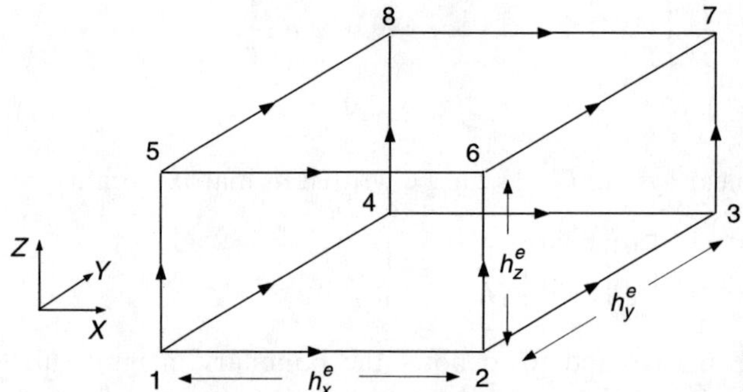

Figure 7.26 Brick element. The numbers indicate the local node numbers.

As mentioned in Chapter 7, for brick elements it is sufficient to compute and store a 12 × 12 matrix for each layer of the mesh that corresponds to the first term of (7.61), which for anisotropic media is written as

$$A^{(1)} = \int_V [\nabla \times \mathbf{W}_t \cdot (\overline{\overline{\mu}}_r^{-1} \cdot \nabla \times \mathbf{W}_s)] \, dV_e \tag{7.65}$$

A second set of 12 × 12 data arrays for each layer of the mesh is required for the second integral in (7.61), which for anisotropic media becomes

$$A^{(2)} = \int_V [\mathbf{W}_t \cdot (\overline{\overline{\epsilon}}_r \cdot \mathbf{W}_s)] \, dV_e \tag{7.66}$$

The material parameters are now tensors and are given by

$$\overline{\overline{\mu}}_r^{-1} = \begin{pmatrix} \tilde{\mu}_{xx} & \tilde{\mu}_{xy} & \tilde{\mu}_{xz} \\ \tilde{\mu}_{yx} & \tilde{\mu}_{yy} & \tilde{\mu}_{yz} \\ \tilde{\mu}_{zx} & \tilde{\mu}_{zy} & \tilde{\mu}_{zz} \end{pmatrix} \tag{7.67}$$

$$\overline{\overline{\epsilon}}_r = \begin{pmatrix} \epsilon_{xx} & \epsilon_{xy} & \epsilon_{xz} \\ \epsilon_{yx} & \epsilon_{yy} & \epsilon_{yz} \\ \epsilon_{zx} & \epsilon_{zy} & \epsilon_{zz} \end{pmatrix} \tag{7.68}$$

The interactions associated with (7.65) can be written as

$$
\begin{aligned}
A_{xx}^{(1)} &= \frac{h_x^e h_z^e \tilde{\mu}_{zz}}{6h_y^e}[K]_1 + \frac{h_x^e h_y^e \tilde{\mu}_{yy}}{6h_z^e}[K]_2 + \frac{h_x^e \tilde{\mu}_{zy}}{4}[K]_3 + \frac{h_x^e \tilde{\mu}_{yz}}{4}[K]_3^T \\
A_{xy}^{(1)} &= \frac{-h_z^e \tilde{\mu}_{zz}}{6}[K]_5 + \frac{h_x^e \tilde{\mu}_{zx}}{4}[K]_4 + \frac{h_y^e \tilde{\mu}_{yz}}{4}[K]_6 + \frac{h_x^e h_y^e \tilde{\mu}_{yx}}{4h_z^e}[K]_3^T \\
A_{xz}^{(1)} &= \frac{-h_y^e \tilde{\mu}_{yy}}{6}[K]_5^T + \frac{h_x^e h_z^e \tilde{\mu}_{zx}}{4h_y^e}[K]_3 + \frac{h_z^e \tilde{\mu}_{zy}}{4}[K]_4 + \frac{h_x^e \tilde{\mu}_{yx}}{4}[K]_6 \\
A_{yx}^{(1)} &= \frac{-h_z^e \tilde{\mu}_{zz}}{6}[K]_5 + \frac{h_y^e \tilde{\mu}_{zy}}{4}[K]_4 + \frac{h_x^e \tilde{\mu}_{xz}}{4}[K]_6 + \frac{h_y^e h_x^e \tilde{\mu}_{xy}}{4h_z^e}[K]_3^T \\
A_{yy}^{(1)} &= \frac{h_x^e h_y^e \tilde{\mu}_{xx}}{6h_z^e}[K]_1 + \frac{h_y^e h_z^e \tilde{\mu}_{zz}}{6h_x^e}[K]_2 + \frac{h_y^e \tilde{\mu}_{xz}}{4}[K]_3 + \frac{h_y^e \tilde{\mu}_{zx}}{4}[K]_3^T \\
A_{yz}^{(1)} &= \frac{-h_x^e \tilde{\mu}_{xx}}{6}[K]_5 + \frac{h_y^e h_z^e \tilde{\mu}_{zy}}{4h_x^e}[K]_3 + \frac{h_y^e \tilde{\mu}_{xy}}{4}[K]_4 + \frac{h_z^e \tilde{\mu}_{zx}}{4}[K]_6 \\
A_{zx}^{(1)} &= \frac{-h_y^e \tilde{\mu}_{yy}}{6}[K]_5^T + \frac{h_z^e h_x^e \tilde{\mu}_{xz}}{4h_y^e}[K]_3 + \frac{h_x^e \tilde{\mu}_{xy}}{4}[K]_4 + \frac{h_z^e \tilde{\mu}_{yz}}{4}[K]_6 \\
A_{zy}^{(1)} &= \frac{-h_x^e \tilde{\mu}_{xx}}{6}[K]_5 + \frac{h_z^e h_y^e \tilde{\mu}_{yz}}{4h_x^e}[K]_3 + \frac{h_z^e \tilde{\mu}_{xz}}{4}[K]_4 + \frac{h_y^e \ddot{\mu}_{zy}}{4}[K]_6 \\
A_{zz}^{(1)} &= \frac{h_y^e h_z^e \tilde{\mu}_{yy}}{6h_x^e}[K]_1 + \frac{h_x^e h_z^e \tilde{\mu}_{xx}}{6h_y^e}[K]_2 + \frac{h_z^e \tilde{\mu}_{yx}}{4}[K]_3 + \frac{h_z^e \tilde{\mu}_{xy}}{4}[K]_3^T
\end{aligned}
\tag{7.69}
$$

For (7.69), the values for the brick element matrices are given by

$$
\begin{aligned}
[K]_1 &= \begin{bmatrix} 2 & -2 & 1 & -1 \\ -2 & 2 & -1 & 1 \\ 1 & -1 & 2 & -2 \\ -1 & 1 & -2 & 2 \end{bmatrix} &
[K]_2 &= \begin{bmatrix} 2 & 1 & -2 & -1 \\ 1 & 2 & -1 & -2 \\ -2 & -1 & 2 & 1 \\ -1 & -2 & 1 & 2 \end{bmatrix} \\
[K]_3 &= \begin{bmatrix} -1 & -1 & 1 & 1 \\ 1 & 1 & -1 & -1 \\ -1 & -1 & 1 & 1 \\ 1 & 1 & -1 & -1 \end{bmatrix} &
[K]_4 &= \begin{bmatrix} 1 & -1 & 1 & -1 \\ -1 & 1 & -1 & 1 \\ 1 & -1 & 1 & -1 \\ -1 & 1 & -1 & 1 \end{bmatrix} \\
[K]_5 &= \begin{bmatrix} 2 & 1 & -2 & -1 \\ -2 & -1 & 2 & 1 \\ 1 & 2 & -1 & -2 \\ -1 & -2 & 1 & 2 \end{bmatrix} &
[K]_6 &= \begin{bmatrix} 1 & 1 & -1 & -1 \\ 1 & 1 & -1 & -1 \\ -1 & -1 & 1 & 1 \\ -1 & -1 & 1 & 1 \end{bmatrix}
\end{aligned}
\tag{7.70}
$$

The reader is cautioned that these matrices are not the same as the ones used for isotropic media (see Chapter 5).

The element matrix terms for anisotropic dielectric media are given by

$$\begin{aligned}
A_{xx}^{(2)} &= \frac{h_x^e h_y^e h_z^e \epsilon_{xx}}{36}[L]_1 & A_{xy}^{(2)} &= \frac{h_x^e h_y^e h_z^e \epsilon_{xy}}{24}[L]_2 & A_{xz}^{(2)} &= \frac{h_x^e h_y^e h_z^e \epsilon_{xz}}{24}[L]_2^T \\
A_{yx}^{(2)} &= \frac{h_x^e h_y^e h_z^e \epsilon_{yx}}{24}[L]_2^T & A_{yy}^{(2)} &= \frac{h_x^e h_y^e h_z^e \epsilon_{yy}}{36}[L]_1 & A_{yz}^{(2)} &= \frac{h_x^e h_y^e h_z^e \epsilon_{yz}}{24}[L]_2 \\
A_{zx}^{(2)} &= \frac{h_x^e h_y^e h_z^e \epsilon_{zx}}{24}[L]_2 & A_{zy}^{(2)} &= \frac{h_x^e h_y^e h_z^e \epsilon_{zy}}{24}[L]_2^T & A_{zz}^{(2)} &= \frac{h_x^e h_y^e h_z^e \epsilon_{zz}}{36}[L]_1
\end{aligned} \tag{7.71}$$

where $[L]_1$ is given in Chapter 5 and $[L]_2$ is given by

$$[L]_2 = \begin{bmatrix} 2 & 1 & 2 & 1 \\ 2 & 1 & 2 & 1 \\ 1 & 2 & 1 & 2 \\ 1 & 2 & 1 & 2 \end{bmatrix} \tag{7.72}$$

Boundary Integral Contribution

The boundary integral present in (7.61) can be written as

$$\begin{aligned}
I^{\mathrm{BI}} &= -\int_S \int_S [(\hat{\mathbf{z}} \times \mathbf{W}_t) \cdot (2\overline{\overline{G}}_0) \cdot (\hat{\mathbf{z}} \times \mathbf{W}_s)]\, dS'\, dS \\
I^{\mathrm{BI}} &= -2\int_S \int_S [\mathbf{W}_t \cdot (\hat{\mathbf{z}} \times \overline{\overline{G}}_0 \times \hat{\mathbf{z}}) \cdot \mathbf{W}_s]\, dS'\, dS \\
I^{\mathrm{BI}} &= I^{\mathrm{BI}(1)} + I^{\mathrm{BI}(2)}
\end{aligned} \tag{7.73}$$

where $\overline{\overline{G}}_0$ is the free-space dyadic Green's function defined in Chapter 1. From this Green's function, we recognize that

$$I^{\mathrm{BI}(1)} = \frac{1}{2\pi} \int_S \int_S [\mathbf{W}_t \cdot (\hat{\mathbf{z}} \times \overline{\overline{I}} \times \hat{\mathbf{z}}) \cdot \mathbf{W}_s] \frac{e^{-jk_0 R}}{R}\, dS'\, dS \tag{7.74}$$

and

$$I^{\mathrm{BI}(2)} = \frac{1}{2\pi k_0^2} \int_S \int_S \left[\mathbf{W}_t \cdot \left(\hat{\mathbf{z}} \times \nabla\nabla \frac{e^{-jk_0 R}}{R} \times \hat{\mathbf{z}}\right) \cdot \mathbf{W}_s\right] dS'\, dS \tag{7.75}$$

where $R = \sqrt{(x - x')^2 + (y - y')^2}$.

For convenience, we can express the surface basis functions as

$$\begin{aligned}
W_x(x, y; \tilde{x}, \tilde{y}, \tilde{s}) &= \frac{\tilde{s}}{h_y^e}(y - \tilde{y}) \\
W_y(x, y; \tilde{x}, \tilde{y}, \tilde{s}) &= \frac{\tilde{s}}{h_x^e}(x - \tilde{x})
\end{aligned} \tag{7.76}$$

where the parameters $\tilde{s}$, $\tilde{x}$, and $\tilde{y}$ can be adjusted to match the basis functions presented in Chapter 2. Using (7.76) in (7.74), we find

$$I^{\mathrm{BI}(1)} = I_{xx}^{\mathrm{BI}(1)} + I_{yy}^{\mathrm{BI}(1)} \tag{7.77}$$

where

$$
\begin{aligned}
I_{xx}^{\mathrm{BI}(1)} &= -\frac{\tilde{s}_t\tilde{s}_s}{2\pi(h_y^e)^2}\int_{y_l^t}^{y_u^t}\int_{x_l^t}^{x_u^t}\int_{y_l^s}^{y_u^s}\int_{x_l^s}^{x_u^s}\left[(y-\tilde{y}_t)(y'-\tilde{y}_s)\frac{e^{-jk_0R}}{R}\right]dx'\,dy'\,dx\,dy \\
I_{yy}^{BI(1)} &= -\frac{\tilde{s}_t\tilde{s}_s}{2\pi(h_x^e)^2}\int_{y_l^t}^{y_u^t}\int_{x_l^t}^{x_u^t}\int_{y_l^s}^{y_u^s}\int_{x_l^s}^{x_u^s}\left[(x-\tilde{x}_t)(x'-\tilde{x}_s)\frac{e^{-jk_0R}}{R}\right]dx'\,dy'\,dx\,dy
\end{aligned}
\tag{7.78}
$$

in which the limit subscripts "l" and "u" denote the lower and upper limits, respectively. The integrals in (7.78) cannot be evaluated in closed form; however, they may be evaluated numerically without much difficulty. For the self-cell case, the integration should be broken into subregions over which the terms within $(\cdot)$ are assumed constant. The pertinent integral is then of the form

$$
\begin{aligned}
I^{\mathrm{BI(c)}} &= \int_{y_l^t}^{y_u^t}\int_{x_l^t}^{x_u^t}\int_{y_l^s}^{y_u^s}\int_{x_l^s}^{x_u^s}\left[\frac{e^{-jk_0R}}{R}\right]dx'\,dy'\,dx\,dy \\
&= \int_{y_l^t}^{y_u^t}\int_{x_l^t}^{x_u^t}\int_{y_l^s}^{y_u^s}\int_{x_l^s}^{x_u^s}\left[\frac{e^{-jk_0R}}{R}-\frac{1}{R}\right]dx'\,dy'\,dx\,dy + \int_{y_l^t}^{y_u^t}\int_{x_l^t}^{x_u^t}\int_{y_l^s}^{y_u^s}\int_{x_l^s}^{x_u^s}\frac{1}{R}\,dx'\,dy'\,dx\,dy
\end{aligned}
\tag{7.79}
$$

where the latter integral can now be evaluated analytically. Specifically we have [27]

$$
\begin{aligned}
\iiiint\frac{1}{R}\,dx'\,dy'\,dx\,dy &= \iiiint\frac{dx'\,dy'\,dx\,dy}{\sqrt{(x-x')^2+(y-y')^2}} \\
&= \iint[x\ln(y+R)+y\ln(x+R)]\,dx\,dy \\
&= \Bigg[\frac{(x-x')(y-y')}{2}\{(x-x')\ln[(y-y')+R] \\
&\quad + (y-y')\ln[(x-x')+R]\} \\
&\quad - \frac{(x-x')(y-y')[(x-x')+(y-y')]}{4} - \frac{R^3}{6}\Bigg]
\end{aligned}
\tag{7.80}
$$

The second boundary term, (7.75), requires further manipulation before evaluation. Specifically, applying the divergence theorem twice and converting one of the gradient operators from ∇ to ∇' (e.g., $\nabla' G_0 = -\nabla G_0$), we have

$$
I^{\mathrm{BI}(2)} = -\frac{1}{2\pi k_0^2}\int_S\int_S\left[\nabla\cdot(\hat{\mathbf{z}}\times\mathbf{W}_t)\,\nabla'\cdot(\hat{\mathbf{z}}\times\mathbf{W}_s)\frac{e^{-jk_0R}}{R}\right]dS'\,dS
\tag{7.81}
$$

Evaluating (7.81) using (7.76), we get

$$
I^{\mathrm{BI}(2)} = I_{xx}^{\mathrm{BI}(2)} + I_{xy}^{\mathrm{BI}(2)} + I_{yx}^{\mathrm{BI}(2)} + I_{yy}^{\mathrm{BI}(2)}
\tag{7.82}
$$

where

$$
\begin{aligned}
I_{xx}^{\mathrm{BI}(2)} &= +\frac{\tilde{S}_t\tilde{S}_s}{2\pi k_0^2(h_y^e)^2} I^{\mathrm{BI(c)}} \\
I_{xy}^{\mathrm{BI}(2)} &= -\frac{\tilde{S}_t\tilde{S}_s}{2\pi k_0^2(h_x^e h_y^e)} I^{\mathrm{BI(c)}} \\
I_{yx}^{\mathrm{BI}(2)} &= -\frac{\tilde{S}_t\tilde{S}_s}{2\pi k_0^2(h_x^e h_y^e)} I^{\mathrm{BI(c)}} \\
I_{yy}^{\mathrm{BI}(2)} &= +\frac{\tilde{S}_t\tilde{S}_s}{2\pi k_0^2(h_x^e)^2} I^{\mathrm{BI(c)}}
\end{aligned}
\tag{7.83}
$$

Once again, $I^{\mathrm{BI(c)}}$ is evaluated using analytical formulas for the self-cell and standard numerical integral techniques for all other interactions.

For triangular elements, the interested reader is referred to [5], [44], [45], and [46]. These papers provide various coordinate-free evaluations of the self-cell integrals.

APPENDIX 2: BRICK FINITE ELEMENT–BOUNDARY INTEGRAL COMPUTER PROGRAM

To assist the reader in understanding some of the difficult concepts in Chapter 7, Dr. Leo C. Kempel of Mission Research Corporation has made available a fully functional finite element–boundary integral computer program (source code) and user's guide via the World Wide Web:

http://www-personal.engin.umich.edu/~volakis/

Some features of this computer program, *LMBRICK* (a.k.a. *Low Memory Brick*), are as follows:

1. Automatic mesh generator for rectangular cavity-backed patch and slot antennas.
2. Precomputation of only necessary FE interactions and a custom sparse matrix-vector product that utilizes this minimal data set.
3. FFT-based matrix-vector product for the BI sub-matrix, hence the ability to model large arrays.
4. Probe and plane wave sources. Probes are parallel to the x-, y-, and z-axes, though they may be arbitrarily placed within the antenna cavity.
5. Capability of computing monostatic and bistatic radar cross section (RCS) and antenna gain.
6. Capability of computing input impedance at each feed probe.
7. Ability to model layers, blocks, or fully inhomogeneous (brick-by-brick) isotropic dielectric and magnetic materials.
8. Carefully documented to aid the user in understanding the method.

Readers are encouraged to report problems and suggestions for further capabilities to Dr. Kempel via his e-mail address: l.kempel@ieee.org.

REFERENCES

[1] C. T. Tai. *Generalized Vector and Dyadic Analysis*. IEEE Press, New York, 1992.

[2] G. E. Antilla and N. G. Alexopoulos. Scattering from complex three-dimensional geometries by a curvilinear hybrid finite element-integral equation approach. *J. Opt. Soc. Am. A*, 11(4):1445–1457, April 1994.

[3] J. M. Jin, J. L. Volakis, and J. D. Collins. A finite element-boundary integral method for scattering and radiation by two- and three-dimensional structures. *IEEE Antennas Propagat. Soc. Mag.*, 33(3):22–32, June 1991.

[4] R. E. Collin. *Field Theory of Guided Waves*. IEEE Press, New York, 1991.

[5] S. M. Rao, D. R. Wilton, and A. W. Glisson. Electromagnetic scattering by surfaces of arbitrary shape. *IEEE Trans. Antennas Propagat.*, 30:409–418, May 1982.

[6] J. R. Mautz and R. F. Harrington. A combined-source formulation for radiation and scattering from a perfectly conducting body. *IEEE Trans. Antennas Propagat.*, 27:445–454, July 1979.

[7] E. Arvas, A. Rahhal-Arabi, A. Sadigh, and S. M. Rao. Scattering from multiple conducting and dielectric bodies of arbitrary shape. *IEEE Antennas Propagat. Soc. Mag.*, 33:29–36, April 1991.

[8] A. F. Peterson. The *interior resonance* problem associated with surface integral equations of electromagnetics: numerical consequences and a survey of remedies. *Electromagnetics*, 10(3): 293–312, July–September 1990.

[9] C. T. Tai. *Dyadic Green's Functions in Electromagnetic Theory*. IEEE Press, New York, 1994.

[10] J. M. Jin and J. L. Volakis. A hybrid finite element method for scattering and radiation by microstrip patch antennas and arrays residing in a cavity. *IEEE Trans. Antennas Propagat.*, 39(11):1598–1604, November 1991.

[11] J. Gong, J. L. Volakis, A. Woo, and H. Wang. A hybrid finite element boundary integral method for analysis of cavity-backed antennas of arbitrary shape. *IEEE Trans. Antennas Propagat.*, 42(9):1233–1242, September 1994.

[12] T. Eibert and V. Hansen. Calculation of unbounded field problems in free space by a 3D FEM/BEM-hybrid approach. *J. Electromagn. Waves Appl.*, 10(1):61–78, 1996.

[13] D. M. Pozar. Input impedance and mutual coupling of rectangular microstrip antennas. *IEEE Trans. Antennas Propagat.*, 30(6):1191–1196, November 1982.

[14] J. T. Aberle and D. M. Pozar. Accurate and versatile solutions for probe-fed microstrip patch antennas and arrays. *Electromagnetics*, 11(1–2):1–19, 1991.

[15] J. Gong and J. L. Volakis. An efficient and accurate model of the coax cable feeding structure for FEM formulations. *IEEE Trans. Antennas Propagat.*, 43(12):1474–1478, December 1995.

[16] J. L. Volakis and K. Barkeshli. Applications of the conjugate gradient FFT method to radiation and scattering. In T. K. Sarkar, editor, *Application of the*

Conjugate Gradient Method to Electromagnetics and Signal Analysis, Chapter 6, Elsevier, New York, 1991.

[17] J. L. Volakis, T. Özdemir, and J. Gong. Hybrid finite element methodologies for antennas and scattering. *IEEE Trans. Antennas Propagat.*, 45(3):493–507, March 1997.

[18] C. J. Reddy, M. D. Deshpande, C. R. Cockrell, and F. B. Beck. Analysis of three-dimensional cavity-backed aperture antennas using a combined finite element method/method of moments/geometrical theory of diffraction technique. Technical Report 3548, NASA Langley Research Center, Hampton, VA, November 1995.

[19] T. J. Peters and J. L. Volakis. Application of the conjugate gradient FFT method to scattering from thin planar material plates. *IEEE Trans. Antennas Propagat.*, 36:518–526, April 1988.

[20] A. C. Polycarpou, J. T. Aberle, and C. A. Balanis. Analysis of arbitrary shaped cavity-backed patch antennas using a hybridization of the finite element and spectral domain methods. In *IEEE Int. Symp. on Antennas and Propagation Digest*, pp. 130–133, Baltimore, MD, July 1996.

[21] T. Özdemir, 1997. Personal communication.

[22] C. J. Reddy, M. D. Deshpande, C. R. Cockrell, and F. B. Beck. Radiation characteristics of cavity backed aperture antennas in finite ground plane using the hybrid FEM/MoM technique and geometrical theory of diffraction. *IEEE Trans. Antennas Propagat.*, 44(10):1327–1333, October 1996.

[23] J. L. Volakis. Iterative solvers. *IEEE Antennas Propagat. Soc. Mag.*, 37:94–96, December 1995.

[24] J. M. Jin and J. L. Volakis. Biconjugate gradient FFT solution for scattering by planar plates. *Electromagnetics*, 12(1):105–119, January–March 1992.

[25] J. R. Mautz and R. F. Harrington. Electromagnetic transmission through a rectangular aperture in a perfectly conducting plane. Sci. Rpt. 10, Air Force Cambridge Res. Labs., Hanscom AFB, MA, February 1976. Contract F19628-73-C-0047.

[26] J. M. Jin and J. L. Volakis. Electromagnetic scattering by and transmission through a three-dimensional slot in a thick conducting plane. *IEEE Trans. Antennas Propagat.*, 39(4):543–550, April 1991.

[27] K. Barkeshli and J. L. Volakis. Electromagnetic scattering from an aperture formed by a rectangular cavity recessed in a ground plane. *J. Electromagn. Waves Appl.*, 5:715–734, 1991.

[28] L. C. Kempel and J. L. Volakis. Scattering by cavity-backed antennas on a circular cylinder. *IEEE Trans. Antennas Propagat.*, 42:1268–1279, September 1994.

[29] P. H. Pathak and N. N. Wang. An analysis of the mutual coupling between antennas on a smooth convex surface. Technical Report 784583-7, Ohio State University ElectroScience Laboratory, Columbus, OH, October 1978.

[30] D. T. McGrath and V. P. Pyati. Phased array antenna analysis with the hybrid finite element method. *IEEE Trans. Antennas Propagat.*, 42(12):1625–1630, December 1994.

[31] E. W. Lucas and T. P. Fontana. A 3-D hybrid finite element/boundary element method for the unified radiation and scattering analysis of general infinite periodic arrays. *IEEE Trans. Antennas Propagat.*, 43(2):145–153, February 1995.

[32] D. T. McGrath. *Phase array antenna analysis using hybrid finite element methods*. PhD thesis, Air Force Institute of Technology, Dayton, OH, 1993. AFIT/DS/ENG/93–4.

[33] D. T. McGrath. Prediction of high power and wideband transmissivity of periodic structures. In *AMEREM Conf.*, Albuquerque, NM, May 1996.

[34] D. T. McGrath and V. P. Pyati. Periodic structure reflection and transmission calculation using the hybrid finite element method. In *IEEE Int. Symp. on Antennas and Propagation Digest*, pp. 142–145, Baltimore, MD, July 1996.

[35] W. Boyse and A. Seidl. A hybrid finite element method for near bodies of revolution. *IEEE Trans. Mag.*, 27:3833–3836, September 1991.

[36] T. Cwik. Coupling finite element and integral equation solutions using decoupled boundary meshes. *IEEE Trans. Antennas Propagat.*, 40:1496–1504, December 1992.

[37] T. Cwik, C. Zuffada, and V. Jamnejad. Efficient coupling of finite element and integral equation representations for three-dimensional modeling. In T. Itoh, G. Pelosi, and P. Silvester, editors, *Finite Element Software for Microwave Engineering*. John Wiley and Sons, New York, 1996.

[38] T. Cwik, C. Zuffada, and V. Jamnejad. Modeling three-dimensional scatterers using a coupled finite element–integral equation formulation. *IEEE Trans. Antennas Propagat.*, 44(4):453–459, April 1996.

[39] T. Cwik, C. Zuffada, and V. Jamnejad. Modeling radiation with an efficient hybrid finite element–integral equation–waveguide mode matching technique. *IEEE Trans. Antennas Propagat.*, 45(1):34–39, January 1997.

[40] L. N. Medgeysi-Mitschang and J. M. Putnam. Electromagnetic scattering from axially inhomogeneous bodies of revolution. *IEEE Trans. Antennas Propagat.*, 32:797–806, 1984.

[41] S. S. Bindiganavale and J. L. Volakis. A hybrid FE-FMM technique for electromagnetic scattering. *IEEE Trans. Antennas Propagat.*, 45(1):180–181, January 1997.

[42] N. Lu and J.-M. Jin. Application of the fast multipole method to finite-element boundary-integral solution of scattering problems. *IEEE Trans. Antennas Propagat.*, 44(6):781–786, June 1996.

[43] E. Bleszynski, M. Bleszynski, and T. Jaroszewicz. AIM: Adaptive integral method for solving large-scale electromagnetic scattering and radiation problems. *Radio Sci.*, 31(5):1225–1251, 1996.

[44] D. R. Wilton, S. M. Rao, A. W. Glisson, D. H. Shaubert, O. M. Al-Bundak, and C. M. Butler. Potential integrals for uniform and linear source distributions on polygonal and polyhedral domains. *IEEE Trans. Antennas Propagat.*, 32(3):276–281, March 1984.

[45] R. D. Graglia. On the numerical integration of the linear shape functions times the 3D Green's function or its gradient on a plane triangle. *IEEE Trans. Antennas Propagat.*, 41(10):1448–1455, October 1993.

[46] T. F. Eibert and V. Hansen. On the calculation of potential integrals for linear source distributions on triangular domains. *IEEE Trans. Antennas Propagat.*, 43(12):1499–1502, December 1995.

8

Fast Integral Methods

S. Bindiganavale and J. L. Volakis

8.1 THE ADAPTIVE INTEGRAL METHOD

When iterative methods are used for the solution of hybrid finite element-boundary integral (FE-BI) systems, such as that in (4.133), most of the CPU time is typically spent in computing the matrix-vector product appearing in

$$[\tilde{G}_\nabla]\{H_z^{\text{boundary}}\} + [\tilde{G}]\{\Psi\} = \{V\} \tag{8.1}$$

which is repeated here from (4.132). The greater CPU time is due to the fully populated matrices $[\tilde{G}_\nabla]$ and $[\tilde{G}]$. Consequently, the CPU time for carrying out the matrix-vector products is $O(N_b^2)$, whereas the corresponding CPU time for sparse matrices approaches $O(N)$. As usual, N denotes the total number of unknowns in the domain and N_b refers to the mesh boundary unknowns. It was noted in Chapter 7 that the FE-BI method is a robust solution approach which combines the best features of partial differential and integral equation methods. The BI reduces the computational volume to a minimum without compromising accuracy. However, the $O(N_b^2)$ CPU and memory growth of the BI compromises the method's utility for large-scale simulations. Efforts have therefore been ongoing to reduce the computing resources consumed for the solution of the BI subsystem. Use of the FFT in the case of Toeplitz subsystems reduces the CPU and memory requirements down to $O(N_b \log N_b)$. This approach was discussed in Chapter 7 and has been generalized to triangular grids by Gong et al. [1]. More recently, a procedure was introduced to cast arbitrary surface grids onto overlaying equivalent uniform grids, as illustrated in Fig. 8.1. As a result, the resulting boundary matrix is again Toeplitz and the FFT can be used to carry out the matrix vector products. One such procedure was introduced by Bleszynski et al. [2] and employs equivalent delta sources to represent the fields exterior to the radiator or scatterer. These delta sources are placed on an equi-spaced grid and are evaluated by matching moments of the fields generated by the original

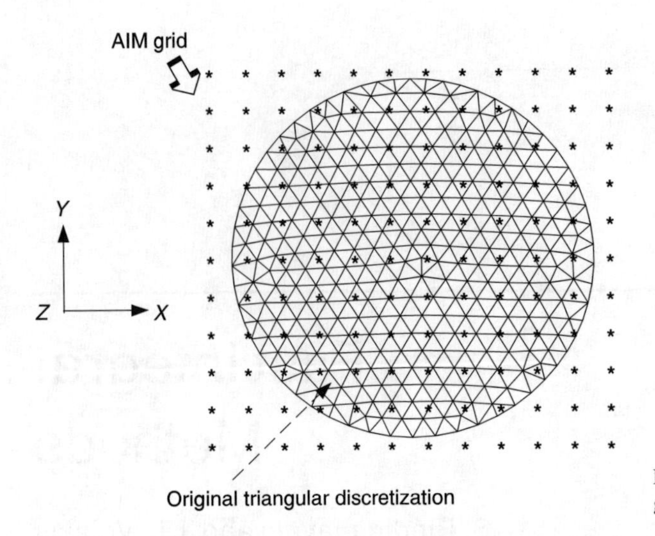

Figure 8.1 Mapping of the original triangular grid to a uniform AIM grid.

surface currents/fields and those due to the delta sources on the new equi-spaced grid. For planar BI surfaces, the delta sources are placed on a rectangular grid, whereas in three dimensions the equi-spaced grid is cubical. Therefore, three-dimensional FFTs must be used in the same manner as done with k-space methods [3], [4].

The method introduced by Bleszynski et al. [2] is referred to as the Adaptive Integral Method (AIM) and has been implemented for scattering and radiation [2], [5]. In all these applications of AIM, the FFT is only used to compute the matrix-vector products associated with the far zone fields, whereas the near zone interactions are computed using the original fields/currents on the BI surface. That is, $[\tilde{G}]$ in (8.1) is decomposed as

$$[\tilde{G}] = [\tilde{G}_{\text{near}}] + [\tilde{G}_{\text{far}}]$$

where $[\tilde{G}_{\text{near}}]$ is a banded or sparse matrix and $[\tilde{G}_{\text{far}}]$ is Toeplitz in form. With this decomposition, the overall CPU and memory requirements of the BI subsystem are reduced down to $O(N_b^{1.5})$ or less. The constant in front of $N_b^{1.5}$ though varies with the bandwidth of $[\tilde{G}_{\text{near}}]$. Typically, $[\tilde{G}_{\text{near}}]$ includes those elements which are a distance of 0.3λ to 0.5λ from the testing point to maintain the accuracy of the solution. Clearly, this implies that AIM is more efficient for large-scale simulations involving bodies which span many wavelengths. However, it has been observed [5] that AIM is particularly attractive even for small bodies which include fine details as is the case with antennas. In some situations with only 1150 BI unknowns, as much as tenfold reduction in CPU and memory has been observed.

AIM belongs to the category of matrix compression methods. At this time other techniques are also being investigated to speed up the matrix vector products and reduce memory requirements of large-scale systems which may involve hundreds of thousands of volume and boundary integral unknowns. Among these, the fast multipole method (FMM) is being considered by several research groups and is discussed below.

8.2 FAST MULTIPOLE METHOD

The fast multipole method (FMM) is an efficient approach for calculating the matrix-vector products associated with dense subsystems as that in (8.1). One of the first applications of FMM was given by Barnes and Hut [6] for calculating interstellar body interactions. More recently, the FMM was used quite successfully to handle very large-scale interactions [7], [8]. The reader is referred to [9] for an early overview of the FMM.

The first application of FMM to acoustics and electromagnetics appeared in [10] and [11]. These articles demonstrate the $O(N_b^{1.5})$ CPU and memory requirement of FMM. However, even lower CPU requirements are possible by using the multilevel FMM discussed in [12] or the windowed FMM [13].

Below we describe the FMM method at a tutorial level for two-dimensional applications. The reader is cautioned that the speed-up achieved by the various compression schemes can compromise the accuracy of the solution [14].

8.2.1 Boundary Integral Equation

For simplicity let us consider the solution of the boundary integral

$$H_z(\mathbf{r}) = 2H_z^{\text{inc}} + \frac{j}{2}\int_C \Psi(\mathbf{r}')\, H_0^{(2)}(k_0|\mathbf{r}-\mathbf{r}'|)\, dl', \qquad \mathbf{r}, \mathbf{r}' \in C \tag{8.2}$$

where H_z^{inc} denotes the TE incident/excitation field and $H_0^{(2)}(\cdot)$ is the zeroth-order Hankel function of the second kind. This integral is a specialization of (4.119) and can be combined with the finite element system (4.133) for the solution of H_z and Ψ. Physically, (8.2) describes the field relation at the aperture of a dielectrically filled groove, as illustrated in Fig. 8.2. It is constructed by enforcing the condition

$$H_z = H_z^{\text{inc}} + H_z^{\text{scat}}$$

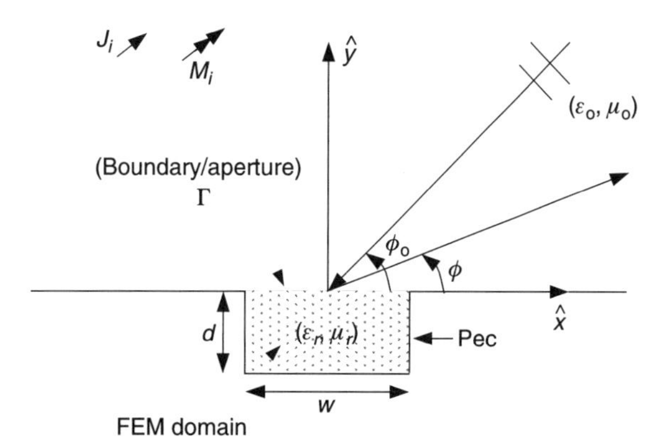

Figure 8.2 Geometry of the groove recessed in a ground plane.

on the aperture, where

$$H_z^{\text{scat}}(\mathbf{r}) = +\frac{j}{2}\int_C \frac{\partial H_z(\mathbf{r}')}{\partial y} H_0^{(2)}(k_0|\mathbf{r}-\mathbf{r}'|)\,dx', \qquad \mathbf{r},\mathbf{r}' \in C$$

$$= +\frac{j}{2}\int_C \Psi(\mathbf{r}')\,H_0^{(2)}(k_0|\mathbf{r}-\mathbf{r}'|)\,dx' \tag{8.3}$$

From (4.76) we can identify that

$$\frac{\partial H_z}{\partial y} = -\frac{jk_0}{Z_0}E_t = \frac{+jk_0}{Z_0}E_x$$

and from (4.114)

$$\frac{\partial H_z}{\partial y} = \Psi = jk_0 Y_0 M_z$$

where M_z denotes the equivalent magnetic current over the aperture. As discussed in Section 4.4.5, (8.3) accounts for image theory which resulted in the introduction of the factor of 2 in the right-hand side of (8.2). In the next few subsections we examine the discretization and evaluation of the integral (8.3) using various versions of the FMM. This exposition provides a close look at the characteristics of FMM for electromagnetic applications and demonstrate the features which are responsible for the CPU speed-up and memory reduction.

8.2.2 Exact FMM

In accordance with the FMM (see Figs. 8.3 and 8.4), the N_b boundary unknowns introduced for the discretization of (8.2) or (8.3) are subdivided into

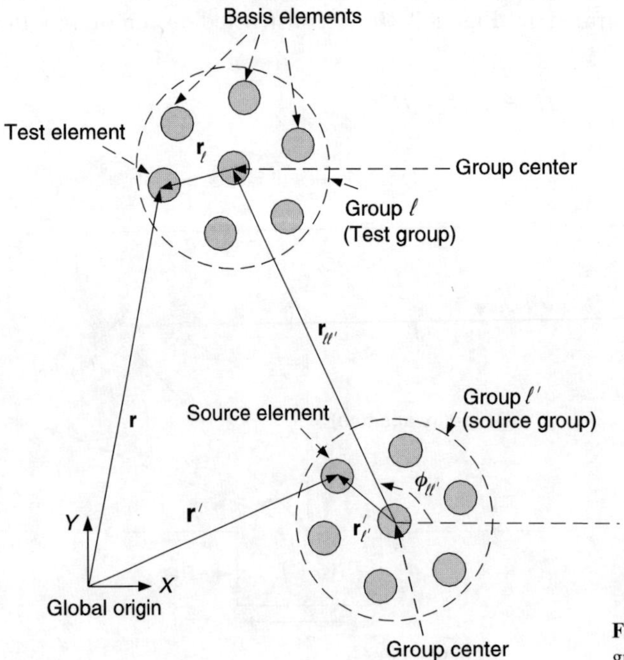

Figure 8.3 Computation of the boundary integral matrix vector product using exact FMM.

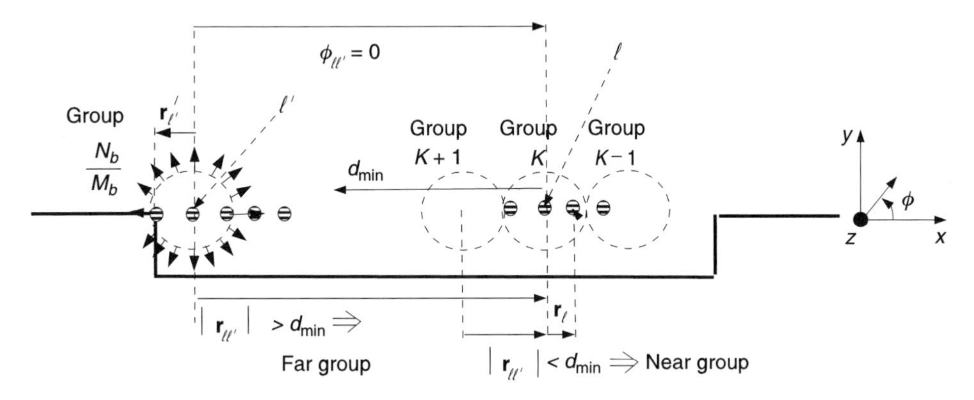

Figure 8.4 Computation of the boundary integral matrix vector product using exact FMM.

groups with each group assigned M_b unknowns. Thus, a total of $L_b \approx N_b/M_b$ groups are constructed. The key step in all FMM procedures is to rewrite the integral (8.3) as a product of terms each being a function of $\mathbf{r}$ (observation point) or $\mathbf{r}'$ (integration point) but not both. In this manner, the evaluation of the integral is carried out by considering the group-to-group interactions separately from the intergroup interactions. Beyond the math, this breakdown of interactions/operations can be viewed in the context of the manager–worker model. Basically, we can view each group as managed by the center element with the workers comprising the elements of the group. Communication/interaction among the groups takes place through the managers who in turn interact with the group elements. The decompositions reduce the direct interdependence of each group member with the other elements belonging to different groups, and this is at the heart of the CPU speed-up afforded by FMM. As stated earlier, though, there are inherent approximations as part of the group decomposition process which must be understood in order to assess the accuracy of each FMM algorithm.

To achieve the decomposition of (8.3) into a product of functions in $\mathbf{r}$ and $\mathbf{r}'$, we first invoke the addition theorem to rewrite the Hankel function as

$$H_0^{(2)}(k_0|\mathbf{r}_{ll'} + \mathbf{r}_l - \mathbf{r}'_{l'}|) \approx \sum_{n=-Q/2}^{Q/2} J_n(k_0|\mathbf{r}'_{l'} - \mathbf{r}_l|)H_n^{(2)}(k_0 r_{ll'})e^{\mathrm{j}n(\phi_{ll'} - \phi_{r'r})}, \; r_{ll'} > |\mathbf{r}_l - \mathbf{r}'_{l'}| \tag{8.4}$$

where $r_{ll'}$ denotes the distance between the centers of the l and l' groups, as illustrated in Figs. 8.3 and 8.4. Also, $\phi_{ll'}$ and $\phi_{r'r}$ are the angles between the vectors $\mathbf{r}_{ll'}$ and $\mathbf{r}'_{l'} - \mathbf{r}_l$ with the x-axis, respectively. The source and observation points $\mathbf{r}'_{l'}$ and $\mathbf{r}_l$ have their origin at the center of the l' and l groups, respectively, while $\mathbf{r}'$ and $\mathbf{r}$ are measured from the origin. Typically, the semi-empirical formula

$$Q/2 = k_0 D + 5\ln(k_0 D + \pi) \tag{8.5}$$

is used to truncate the sum (8.4), where D is the diameter of the circle enclosing the groups. This is consistent with the radius of convergence associated with the Hankel function. In general, $Q/2 = M_b$, ensures convergence. (It will be shown that Q is the number of directions in which the radiation of the group is sampled. With M_b being the number of basis elements in the group, $Q = 2M_b$ satisfies the Nyquist criterion for faithful replication of the source group radiation.)

Next we introduce the Fourier integral of the Bessel function

$$J_n(k_0|\mathbf{r}'_{l'} - \mathbf{r}_l|) = \frac{1}{2\pi}\int_{2\pi} e^{\mathbf{jk}\cdot(\mathbf{r}'_{l'}-\mathbf{r}_l)-\mathrm{j}n(\phi-\phi_{r'_{l'}r}+\pi/2)}\, d\phi \tag{8.6}$$

and in conjunction with (8.4) we can now rewrite (8.3) as

$$H_z^{\mathrm{scat}}(\mathbf{r}) = -\frac{k_0 Y_0}{4\pi}\int_{2\pi} V_{l'}(\phi)T_{ll'}(\phi)e^{-\mathbf{jk}\cdot\mathbf{r}_l}\, d\phi \tag{8.7}$$

where $\mathbf{k} = k_0(\hat{x}\cos\phi + \hat{y}\sin\phi)$ is measured from the x-axis. In this,

$$V_{l'}(\phi) = \int_\Gamma M_z(\mathbf{r}')\, e^{\mathbf{jk}\cdot\mathbf{r}'_{l'}}\, dl' \tag{8.8}$$

is identified as the far-field pattern of the source group and

$$T_{ll'}(\phi) = \sum_{n=-Q/2}^{Q/2} H_n^{(2)}(k_0 r_{ll'})\, e^{-\mathrm{j}n(\phi-\phi_{ll'}+\pi/2)} \tag{8.9}$$

is referred to as the translation operator providing the group-to-group (l to l') interactions. From (8.7)–(8.9), we observe that the integral (8.3) has now been decomposed into terms which separate out the dependence on $\mathbf{r}$ and $\mathbf{r}'$. The final evaluation of H_z^{scat} proceeds by discretizing the integral over ϕ to yield the expression

$$H_z^{\mathrm{scat}}(\mathbf{r}) = -\frac{k_0 Y_0}{4\pi}\Delta\phi\sum_{q=1}^{Q} T_{ll'}(\phi_q)\, V_{l'}(Q_q)\, e^{-\mathbf{jk}_q\cdot\mathbf{r}_l} \tag{8.10}$$

which is the radiated fields from some location in the source group l' to a point within the receiving group l. Note that $\Delta\phi = 2\pi/Q$ indicates the angular spacing between the propagation vectors of plane waves emanating from a group. Thus $\phi_q = q\Delta\phi$, $q = 1\ldots Q$, whereas $\mathbf{k}_q = k_0(\hat{x}\cos\phi_q + \hat{y}\sin\phi_q)$. As mentioned earlier, the number of plane wave directions is set equal to twice the number of elements in the group ($Q = 2M_b$), thus satisfying the Nyquist sampling theorem with respect to the integration over ϕ. Given the above steps, the exact FMM procedure for carrying out the matrix-vector product can be summarized as follows:

1. Compute pattern of the source group (aggregation). Mathematically, this corresponds to evaluating $V_{l'}(\phi_q)$ given in (8.8). The evaluation of $V_{l'}(\phi_q)$ for a single source group and at a single direction requires M_b operations, corresponding to the number of elements in the group (the integration over the line segment is performed as a summation). Consequently for L_b groups and Q directions for each group, the operation count is QM_bL_b.
2. The next step is to employ the translation operator to evaluate the pattern of a source group at the center of the test group. Mathematically, this operation amounts to computing the coefficient $A_l(\phi_q) = V_{l'}(\phi_q)T_{ll'}(\phi_q)$. The evaluation of $A_l(\phi_q)$ involves an operation count of QL_b^2, where again L_b denotes the number of groups and Q is the number of directions.

3. Finally, at the receiving group, the fields are redistributed (disaggregation). Mathematically, this amounts to computing the expression $H_z^{\text{scat}}(\mathbf{r})$, as given in (8.10). Evaluating the sum at a single point requires Q operations. Thus for L_b groups, each containing M_b unknowns, the operation count is QL_bM_b.

From the above we conclude that the total operation count of the above three steps is $C_1QM_bL_b + C_2QL_b^2$. Also, the near field (by this it is meant that groups in the near vicinity of each other are treated using the standard moment method procedure) operation count is of $O(N_bM_b)$. On choosing $Q \sim O(M_b)$, the operation count of the three steps reduces to $C_1M_bN_b + C_2(N_b^2/M_b)$. On setting $M_b \sim \sqrt{N_b}$, implying $L_b = \sqrt{N_b}$ (an optimal choice), the final operation count is $N_b^{1.5}$ and this should be compared to the usual N_b^2 operation count of direct solvers. The reduction of the operation count from $O(N_b^2)$ down to $O(N_b^{1.5})$ is indeed dramatic. An appreciation of the CPU reduction can be acquired by setting, for example, $N_b = 2000$ which is a relatively small number of elements. However, further improvements can still be achieved by nesting groups leading to multilevel FMM [12].

8.2.3 Windowed FMM

In the exact FMM, the translation operation between groups assumed isotropic radiation. However, it is suggestive that the groups would interact strongly along the line joining them and less so in other directions. Indeed, it was shown in [13] that the translation operator could be contemplated as composed of a geometrical optics (GO) term (along the line joining the source and test group) and two diffraction terms associated with the shadow boundaries of the GO term. To illustrate the validity of this concept, we plot in Fig. 8.5 the translation operator

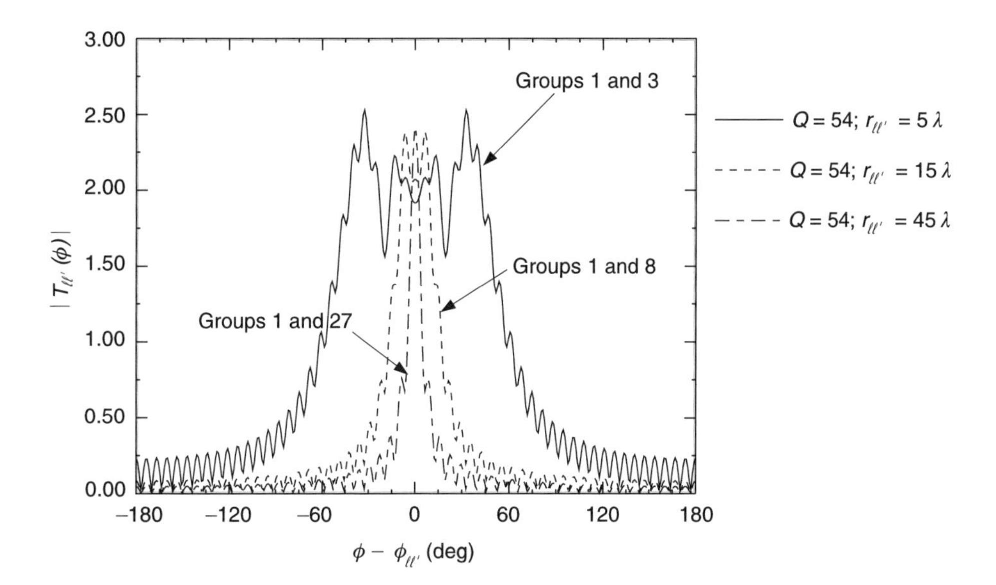

Figure 8.5 The Translation operator for different groups on the boundary of a 50λ wide groove; 750 BI unknowns; 27 groups.

for different group separation distances along the groove of width 50λ. For this example, the number of unknowns on the boundary was 750, resulting in 27 groups. As seen, the "lit" region of the translation operator narrows as the group separation distance is increased, eventually displaying the predictable sinc function behavior for large group separation distances. The tapering off of the translation operator from a value oscillating around 2 down to zero for larger $\phi - \phi_{ll'}$ values is characteristic of the geometrical optics plus diffraction terms in the context of traditional high frequency methods. We may also comment that this high frequency model enables the identification of a lit region even for groups which are not widely separated (for example, see Fig. 8.5 for the translation operator between groups 1 and 3).

The key characteristic of the windowed FMM is the exploitation of the diminished value of $T_{ll'}(\phi)$ for large $\phi - \phi_{ll'}$. Basically, in the windowed FMM, the computation of $T_{ll'}(\phi)$ for these angles is avoided altogether. This can be accomplished by multiplying $T_{ll'}(\phi)$ with the filter (windowing) function

$$W_{ll'}(\phi_q) = \begin{cases} 1 & (|\phi_q - \phi_{ll'}| < \beta_s) \\ e^{-\alpha(|\phi_q - \phi_{ll'}| - \beta_s)^2} & (|\phi_q - \phi_{ll'}| > \beta_s) \end{cases} \tag{8.11}$$

where

$$\beta_s = \sin^{-1}\left(\frac{Q+1}{2k_0 r_{ll'}}\right) \tag{8.12}$$

and α is a taper factor to be specified. Note also that β_s was selected to provide a larger bandpass window when $r_{ll'}$ is smaller as dictated from high frequency analysis.

The discretized plane wave expansion can now be written as

$$H_z^{\text{scat}} = -\frac{k_0 Y_0}{4\pi} \Delta\phi \sum_{q=1}^{Q} W_{ll'}(\phi_q) T_{ll'}(\phi_q) V_{l'}(\phi_q) e^{-\mathbf{j}\mathbf{k}_q \cdot \mathbf{r}_l} \tag{8.13}$$

By taking into account only the nonzero sector of $W_{ll'}(\phi)$, the operation count of the translation process is now reduced to $C_3 L_b^2 \sim N_b^2/M_b^2$ with the corresponding total operation count given by $C_1 M_b N_b + C_4(N_b^2/M_b^2)$. Grouping the unknowns into $N_b^{1/3}$ elements per group results in a total operation count of $O(N_b^{4/3})$. This should be compared with the $O(N_b^{3/2})$ operation count of the exact FMM.

The computation of the boundary integral matrix vector product by employing the windowed FMM is depicted pictorially in Fig. 8.6 illustrating that the filter function has the effect of eliminating plane wave interactions at directions away from the line joining the interacting groups.

8.2.4 Fast Far Field Algorithm (FAFFA)

This is an approximate version of the FMM since the algorithm is based on introducing the large argument approximation of the Hankel function. That is, the approximation

$$H_0^{(2)}(k_0|\mathbf{r} - \mathbf{r}'|) \sim e^{-\mathbf{j}k_0 \hat{r}_{l'l} \cdot \mathbf{r}_{lm}} \sqrt{\frac{2\mathbf{j}}{\pi}} \frac{e^{-\mathbf{j}k_0 \mathbf{r}_{l'l}}}{\sqrt{k_0 r_{l'l}}} e^{-\mathbf{j}k_0 \hat{r}_{l'l} \cdot \mathbf{r}_{nl'}} \tag{8.14}$$

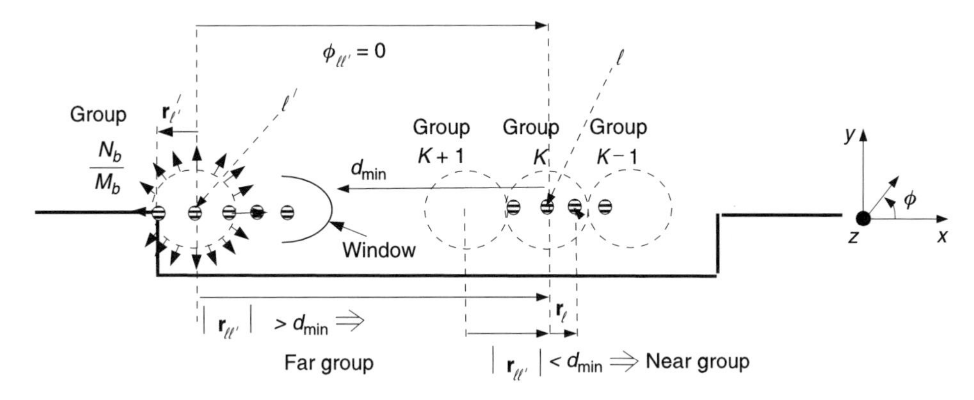

Figure 8.6 Computation of the boundary integral matrix vector product using windowed FMM.

is used. As shown in Fig. 8.7, $r_{l'l}$ is the distance between the center of the test group l and the center of the source group l'; $r_{nl'}$ is the distance between the nth source element and its group center; and r_{lm} is the distance between the mth test element and its group center.

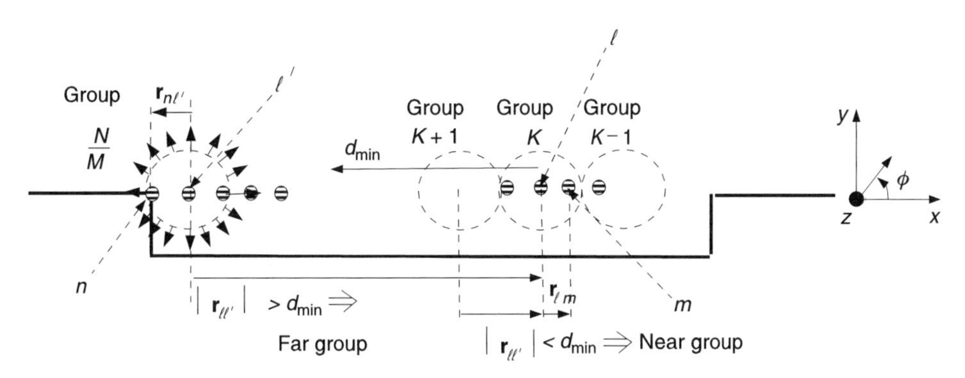

Figure 8.7 Computation of the boundary integral matrix vector product using the FAFFA.

The introduction of the large argument expansion necessitates that the FMM procedure be used only for groups which are very well separated. However, (8.14) allows for the immediate decoupling of the test-source element interactions, thus enabling the computation of the matrix-vector product for far-field groups with a reduced operation count. This is illustrated below by going through the steps of the FAFFA corresponding to the three steps of the FMM:

1. The aggregation of source elements in a single source group now involves M_b operations, corresponding to the number of elements in the source group. Specifically,

$$V_{l'l} = \sum_{j=1}^{M_b} M_j e^{-jk_0 \hat{r}_{l'l} \cdot \mathbf{r}_{nl'}} \tag{8.15}$$

and since the above aggregation needs to be done for all source groups, the operation count becomes $O[(N_b/M_b)M_b] \sim O(N_b)$, where N_b/M_b represents the total number of groups. Also, this operation, being dependent only on the test *group* rather than the test *element*, needs to be repeated for all $Q = N_b/M_b$ test groups leading to a total operation count of $O(N_b^2/M_b)$ for aggregation. It should be noted that use of the large argument expansion, rather than the addition theorem for the Hankel function, results in the aggregation sum being a function of the test group also ($V_{l'l}$) unlike the exact FMM where the aggregation sum is a function of source group only ($V_l'(\phi)$). Thus, the technique by which the Exact FMM and the FAFFA reduce the operation count differ in the fact that while in the exact FMM the aggregation sum is characterized by a source group (l') and a direction (ϕ) which is not interwined with the test group direction, the aggregation sum in the FAFFA is characterized by the source group (l') and the test group (l).

2. The main advantage of FAFFA is due to the faster computation of the translation operator. We have

$$A_{l'l} = T_{l'l} V_{l'l} \tag{8.16}$$

where in the FAFFA the translation operator simplifies to

$$T_{l'l} = \sqrt{\frac{2\mathbf{j}}{\pi}} \frac{e^{-\mathrm{j}k_0 r_{l'l}}}{\sqrt{k_0 r_{l'l}}} \tag{8.17}$$

This should be compared to the sum (8.9) for the exact FMM. Clearly, (8.17) needs to be done only at the group level and involves $O(N_b^2/M_b^2)$ operations for all possible test and source group combinations, making it the least computationally intensive step.

3. The disaggregation or redistribution process is again the operation

$$H_z^{\mathrm{scat}}(r) = -\frac{k_0 Y_0}{2} \sum_{l'=1}^{N_b/M_b} A_{l'l} e^{-\mathrm{j}k_0 \hat{r}_{l'l} \cdot \mathbf{r}_{lm}} \tag{8.18}$$

Since this operation involves only the source *group* instead of the source *element*, it needs to be done for each source group, implying $O(N_b/M_b)$ operations to generate a single row of the matrix-vector product. To generate M_b rows, corresponding to a test group, the operation count would be $O(N_b)$. With N_b/M_b test groups, the operation count is $O(N_b^2/M_b)$.

Consolidating the above three steps for the FAFFA algorithm, we have

$$Op.count \sim C_1 N_b M_b + C_2 \frac{N_b^2}{M_b} \tag{8.19}$$

where the first term refers to the operations associated with the near-field terms. As before, $M_b = \sqrt{N_b}$ and the total operation count is $O(N_b^{1.5})$. While the operation count for this algorithm could be further reduced down to $O(N_b^{1.33})$ by performing the process of "interpolation" and "anterpolation" as described in [15] for very large objects, we found that the accuracy deteriorated for the considered applications. Hence, only the $O(N_b^{1.5})$ version was used.

8.3 LOGIC FLOW

The operation counts described in the previous section for the various algorithms are illustrated with the help of flow diagrams and sections of code from the computation of the matrix vector products for the far groups. Figures 8.8 and 8.11 depict the flow diagram and code for computing the matrix-vector product in the exact FMM. It is seen in Fig. 8.11 that each of the aggregation, translation, and disaggregation operations consists of a single multiplication which is described below.

- The aggregation operation consists of the product of an entry of the trial vector (represented as `Dum(J)` in Fig. 8.11) with an aggregation factor, represented in Fig. 8.11 for the Jth element and Kth direction as `SrcGc(J,K)`. This is given by `SrcGc(J,K)` $= \Delta_J e^{jk_0(\hat{x}\cos\phi_K + \hat{y}\sin\phi_K)\cdot \mathbf{r}'_{JGr}}$ where Δ_J is the length of the Jth discretization element, ϕ_K is the Kth radiation direction and $\mathbf{r}'_{JGr}$ is the direction vector of the Jth element, measured from the center of the group `(JGr)` it belongs to. The result of the aggregation operation yields a term characterized by only the source group and radiation direction (represented in Fig. 8.11 as `V(JGr,K)`).
- The translation operation involves the multiplication of the aggregation sum, `V(JGr,K)`, with a translation factor, represented in Fig. 8.11 for the IGrth test group, JGrth source group, and Kth radiation direction by `Trans(IGr,JGr,K)`. This is given by

$$\texttt{Trans(IGr,JGr,K)} = \sum_{n=-K/2}^{K/2} e^{-jn(\phi_K - \phi_{IGr,JGr} + \pi/2)} H_n^{(2)}(k_0 r_{IGr,JGr})$$

  where ϕ_K is the Kth radiation direction, $r_{IGr,JGr}$ and $\phi_{IGr,JGr}$ are the distance and angle between the IGrth and JGrth groups. The result of the translation operation yields a term dependent only on the test group and radiation direction (represented in Fig. 8.11 as `GrGr(IGr,K)`).
- The disaggregation operation involves the multiplication of the translation sum, `GrGr(IGr,K)`, with a disaggregation factor, represented in Fig. 8.11 for the Ith test element and the Kth radiation direction by `TestGc(I,K)`. This is given by

$$\texttt{TestGc(I,K)} = e^{-jk_0(\hat{x}\cos\phi_K + \hat{y}\sin\phi_K)\cdot \mathbf{r}_{IGr}}$$

  where $\mathbf{r}_{IGr}$ is the direction vector of the Ith element measured from the center of the group `(IGr)` it belongs to. The result of the disaggregation operation yields a term dependent only on the test element alone and is the contribution to the Ith entry of the product vector.

The windowed FMM differs from the exact FMM in the translation phase, and this is illustrated in Figs. 8.9 and 8.12. These figures illustrate that the windowed FMM achieves its reduced operation count by eliminating some of the directions in which plane wave interaction takes place. The innermost loop in the translation phase has an operation count which is a constant (15–25 in our simulations, 20 in

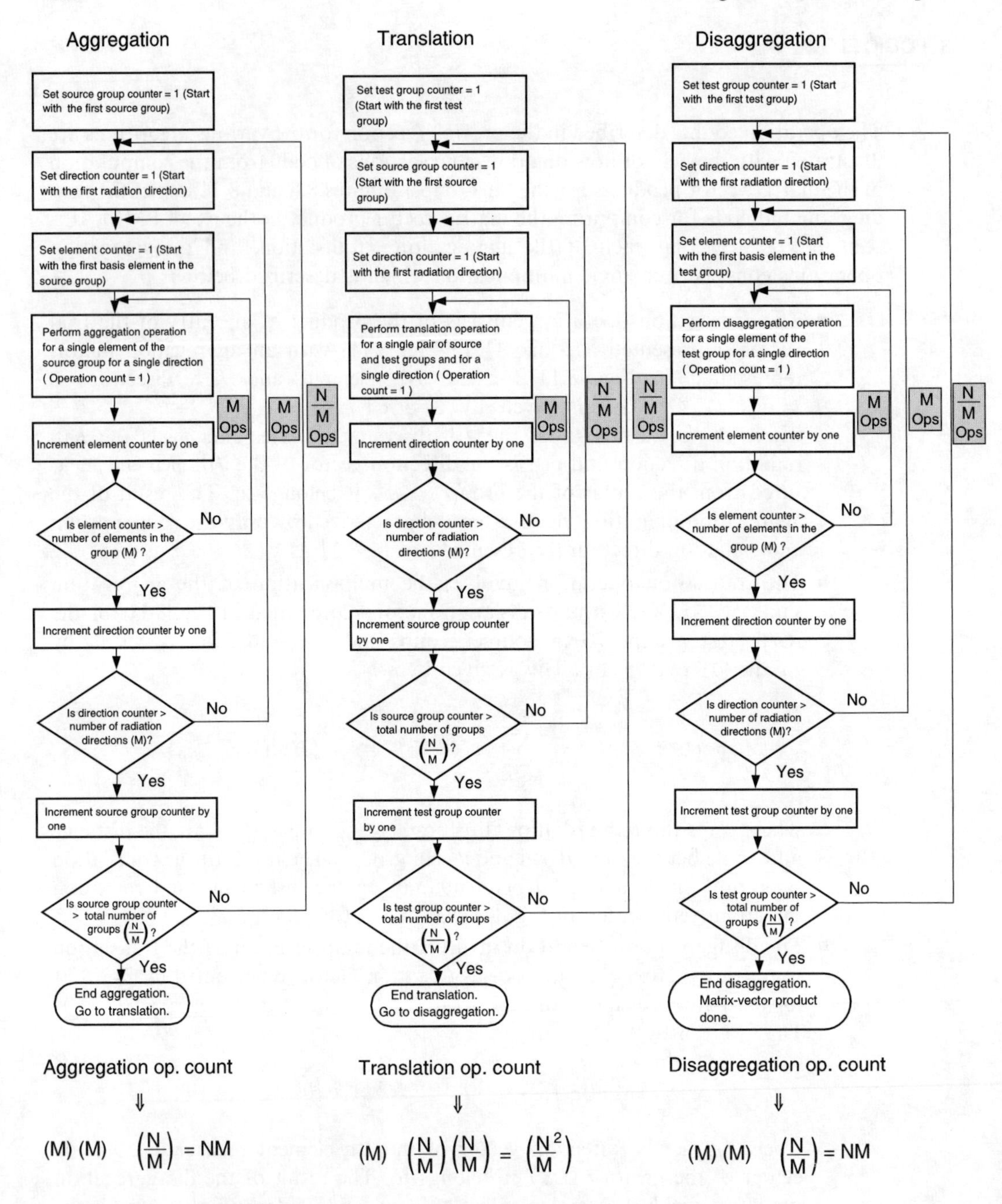

Figure 8.8 Sequence of operations to be performed in the Exact FMM.

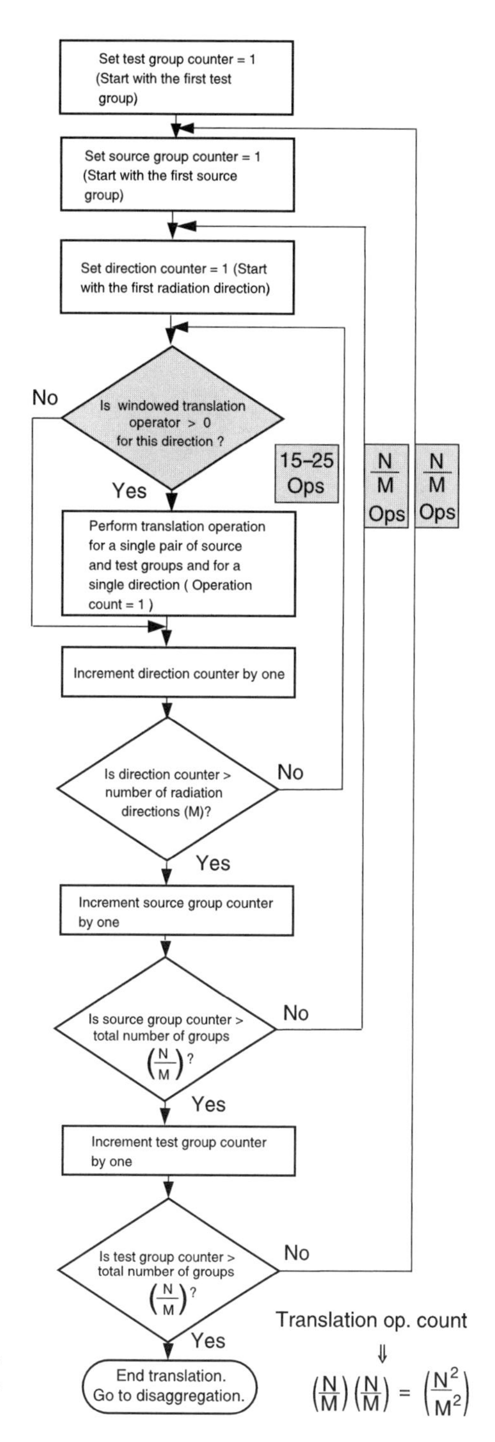

Figure 8.9 Sequence of operations to be performed in the translation process of the windowed FMM.

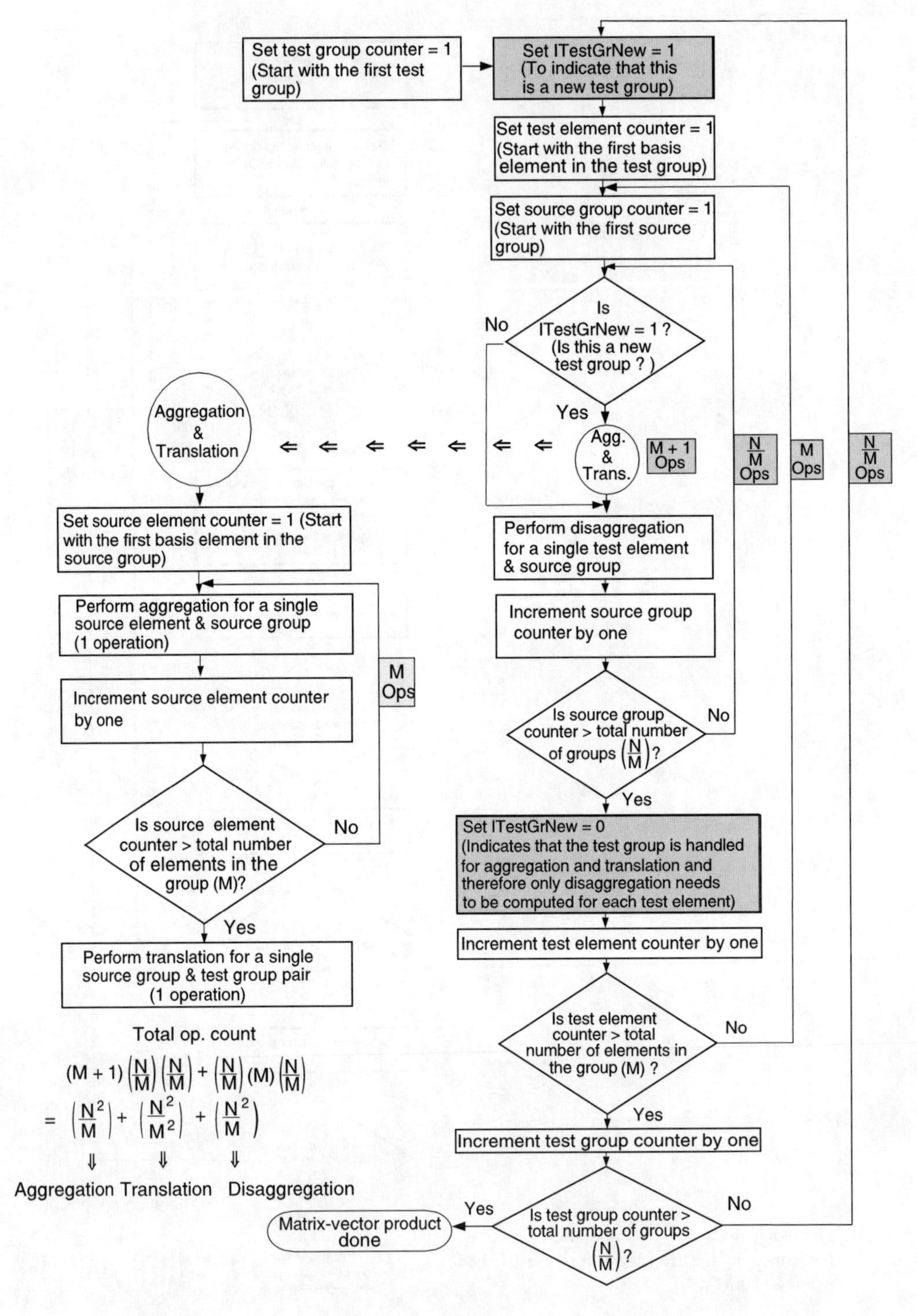

Figure 8.10 Sequence of operations to be performed in the FAFFA.

```
c    NGr - Number of groups; IQAngles - Number of radiation directions
c    DPhi - Angular spacing between radiation directions
c    NElGr(NGr) - Array containing number of elements in each group
c    GetGlobal - function which gets the global element # given a group
c              number and local element number
c    DMin - Minimum distance beyond which two group are treated as far
c        groups
c    Distance - Function which returns the distance between groups
c    Dum - Vector multiplying the matrix - in the CG algorithm
c        will be the search and residual vectors
c    AX - Array which is the matrix-vector product
c    SrcGc,Trans & TestGc - aggregation, translation & disaggregation factors
c        respectively described in the text

     IQAngles = 2*NGr
     DPhi = 2*Pi/IQAngles

c    Aggregation - Operation count O(NM)
     do JGr =1,NGr
       do K = 1,IQAngles
         do JEl =1,NElGr(JGr)
           J = GetGlobal(JGr,JEl)
           if (S.eq.'*') then
             V(JGr,K) = V(JGr,K) + Dum(J)*conjg(SrcGc(J,K))
           else
             V(JGr,K) = V(JGr,K) + Dum(J)*SrcGc(J,K)
           endif
         enddo
       enddo
     enddo

c    Translation - Operation count O(N^2/M)
     do IGr = 1,NGr
       do JGr = 1,NGr
         if (Distance(IGr,JGr).gt.DMin) then
           do K = 1,IQAngles
             if (S.eq.'*') then
               GrGr(IGr,K) = GrGr(IGr,K) + conjg(Trans(IGr,JGr,K))*V(JGr,K)
             else
               GrGr(IGr,K) = GrGr(IGr,K) + Trans(IGr,JGr,K) * V(JGr,K)
             endif
           enddo
         endif
       enddo
     enddo

c    Disaggregation - Operation count O(NM)
     do IGr = 1,NGr
       do K = 1,IQAngles
         do IEl = 1,NElGr(IGr)
           I = GetGlobal(IGr,IEl)
           if (S.eq.'*') then
             AX(I) = AX(I) + GrGr(IGr,K)*conjg(TestGc(I,K))
           else
             AX(I) = AX(I) + GrGr(IGr,K)*TestGc(I,K)
           endif
         enddo
       enddo
     enddo
```

Loop counts indicated by braces: Aggregation — N/M, M, M; Translation — N/M, N/M, M; Disaggregation — N/M, M, M.

Figure 8.11 Code indicating the computation of the matrix-vector product in the Exact FMM.

```
c   The symbols in this section of code represent the
c      same quantities as in the code  for the exact FMM

c   Translation - Operation count O(N^2/M^2)

      do IGr = 1,NGr
        do JGr = 1,NGr
          if (Distance(IGr,JGr).gt.DMin) then
            do K = 1,IQAngles
              if (S.eq.'*') then
                if (abs(Trans(IGr,JGr,K)).eq.0) then
                  continue
                else
                  GrGr(IGr,K) = GrGr(IGr,K) + conjg(Trans(IGr,JGr,K)) * V(JGr,K)
                endif
              else
                if (abs(Trans(IGr,JGr,K)).eq.0) then
                  continue
                else
                  GrGr(IGr,K) = GrGr(IGr,K) + Trans(IGr,JGr,K) * V(JGr,K)
                endif
              endif
            enddo
          endif
        enddo
      enddo
```

(Braces at left labelled $\frac{N}{M}$ and $\frac{N}{M}$ for the two outer loops.)

Figure 8.12 Code indicating the computation of the matrix-vector product in the translation phase of the windowed FMM.

[13]) and is a significant reduction from the corresponding operation count in the exact FMM.

The technique by which the FAFFA achieves its speed-up is depicted in Figs. 8.10 and 8.13. It is seen that the FAFFA "recycles" the plane wave spectra of the source group. For a given test group, the aggregation and translation operations are performed only once for each source group, necessitating that only the disaggregation operation needs to be performed for each individual element of the test group. Similar to the exact FMM, the aggregation, translation, and disaggregation processes consist of a single multiplication. However, the factors used in the three processes and the method by which the reduced operation count is achieved are different.

- The aggregation operation again consists of the product of an entry of the trial vector with an aggregation factor, represented in Fig. 8.13 for the Jth element and IGrth test group as `SrcGc(IGr,J)`. This is given by `SrcGc(IGr,J)` $= \Delta_J e^{-\mathrm{j}k_0 \hat{r}_{JGr,IGr}\cdot \mathbf{r}_{J,JGr}}$ where Δ_J is the length of the Jth discretization element, $\hat{r}_{JGr,IGr}$ is the unit vector along the line joining the source and test groups while $\mathbf{r}_{J,JGr}$ is the vector along the line joining the source element with its group center. Thus, an aggregation sum is formed for each combination of source and test groups.
- The translation operation involves the multiplication of the aggregation sum with a translation factor, represented in Fig. 8.13 for the IGrth test group

```
c     The symbols in this section of code represent the
c        same quantities as in the code  for the exact FMM
c     ITestGrNew is a counter which is set to 1 if a test group
c        is "new". This means that since the aggregation and
c        translation operations involve only the test group
c        rather than the test element, they need to be done
c        only once (for the first element in the test group).
c        For the rest of the elements in the test group only
c        the disaggregation operation needs to be done (corresponds
c        to ITestGrNew = 0).
      do IGr = 1,NGr
        ITestGrNew = 1
        do IEl = 1,NElGr(IGr)
          I = GetGlobal(IGr,IEl)
          do JGr = 1,NGr
            if (Distance(IGr,JGr).gt.Dmin) then
              if (ITestGrNew.eq.1) then
                V = (0.,0.)
                do JEl = 1,NElGr(JGr)
                  J = GetGlobal(JGr,JEl)
                  if (S.eq.'*') then
                    V = Dum(J)*conjg(SrcGc(IGr,J)) + V
                  else
                    V = Dum(J)*SrcGc(IGr,J) + V
                  endif
                enddo
                if (S.eq.'*') then
                  GrGr(JGr) = conjg(Trans(IGr,JGr))*V
                else
                  GrGr(JGr) = Trans(IGr,JGr)*V
                endif
              endif
              if (S.eq.'*') then
                AX(I) = AX(I) + GrGr(JGr)*conjg(TestGc(I,JGr))
              else
                AX(I) = AX(I) + GrGr(JGr)*TestGc(I,JGr)
              endif
            endif
          enddo
          ITestGrNew = 0
        enddo
      enddo
```


Figure 8.13 Code indicating the computation of the matrix-vector product in the FAFFA.

and JGrth source group by `Trans(IGr,JGr)` and given by `Trans(IGr,JGr)` $= e^{-jk_0 r_{JGr,IGr}}/\sqrt{k_0 r_{JGr,IGr}}$. Again, the translation operation needs to be done for each pair of test and source groups.

- The disaggregation operation involves the multiplication of the translation sum, `GrGr(JGr)`, with a disaggregation factor, represented in Fig. 8.13 for the Ith test element and the JGrth source group by `TestGc(I,JGr)`. This is given by `TestGc(I,JGr)` $= e^{-jk_0 \hat{r}_{JGr,IGr}\cdot \mathbf{r}_{IGr,I}}$, where $\mathbf{r}_{IGr,I}$ is the vector along the line joining the test element with its group center. It should be noted that to compute the interactions between a pair of groups, the aggregation and translation need to be done only once, and thus the crux of the FAFFA is indicated with highlighted sections of the code in Fig. 8.13.

8.4 RESULTS

The results presented in this section [16], [17] are based on an FMM computer code, incorporating a conjugate gradient solver, and executed on an HP 9000/750 workstation with a peak flop rate of 23.7 MFLOPS. The geometry considered was the rectangular groove shown in Fig. 8.2. Table 8.1 compares the execution time and RMS error [14] of the standard FE-BI to the FE-Exact FMM, FE-FAFFA and the FE-Windowed FMM (FE-WFMM) for grooves of widths 25λ, 35λ and 50λ. The depth of the groove was 0.35λ with a material filling of $\epsilon_r = 4$ and $\mu_r = 1$ and was illuminated at normal incidence. The data reveal that the FE-FMM^{Exact} offers almost a 50 percent savings in execution time with almost no compromise in accuracy. While the FE-FAFFA is the fastest of the three algorithms, the RMS error was substantially higher (> 1 dB). If the maximum tolerable RMS error is set at 1 dB [14], the FE-Windowed FMM is the most attractive option since it meets the error criterion and is only slightly slower than the FE-FAFFA.

Table 8.2 gives the exact BI operation count rather than merely stating its order. The knowledge of the constants associated with each exponent of N_b enables us to compare the requirements of two algorithms which might have the same order of operation count. In Table 8.2, N_{NG} is the number of near groups (groups which are treated with the exact moment method procedure owing to their electrical proximity) which depends on the algorithm and the problem geometry. N_{NG} is smallest for the FE-FMM^{Exact} and largest for the FE-FAFFA, due to the use of the far-zone Green's function in the latter. Table 8.2 also gives the number of multiplications in a single BI matrix-vector product for the 50λ groove. For the FE-FMM^{Exact} the number of multiplications is reduced by a factor of three over the FE-BI. However, the actual CPU time is reduced by a smaller factor due to computational overhead for the various function calls. Of interest is the comparison of the residual error as a function of the number of iterations in the conjugate gradient solver. Such a comparison is shown in Fig. 8.14 and it is seen that the curves for the FE-BI and the FE-FMM^{Exact} overlap to graphical accuracy whereas the FE-WFMM shows a very small deviation from the exact result. Thus, the

TABLE 8.1 CPU Times and RMS Error of the Hybrid Algorithms

Groove width	Total Unknowns	BI Unknowns	CPU time for BI (minutes, seconds) FE-BI (CG)	FE-FAFFA	FE-FMM^{Exact}	FE-WFMM
25λ	2631	375	(8,48)	(3,26)	(5,25)	(4,13)
35λ	3681	525	(16,34)	(5,55)	(10,31)	(7,22)
50λ	5256	750	(45,1)	(14,31)	(26,18)	(16,10)

Groove Width	RMS error (dB) FE-FAFFA	FE-FMM^{Exact}	FE-WFMM
25λ	1.12	0.0752	0.6218
35λ	1.2	0.1058	0.721
50λ	1.36	0.1123	0.843

TABLE 8.2 Exact BI Operation Count of the Hybrid Algorithms. N_b is the Number of BI Unknowns. N_{NG} is the Number of near Groups (groups which are treated with the exact moment method). Q_{WIN} is the Width of the Window in the Windowed FMM.

Algorithm	Operation count for BI Computation (as implemented)	Operations (multiplications) Required for BI Computation for the 50λ Groove
FE-BI	N_b^2	562500
FE-FAFFA	$(N_{\mathrm{NG}}+2)N_b^{1.5}+(1-2N_{\mathrm{NG}})N_b-N_{\mathrm{NG}}N_b^{0.5}$	136890
FE-FMM^{Exact}	$(N_{\mathrm{NG}}+6)N_b^{1.5}-2N_{\mathrm{NG}}N_b$	180356
FE-WFMM	$(4+N_{\mathrm{NG}}+Q_{\mathrm{WIN}})N_b^{1.33}-N_{\mathrm{NG}}Q_{\mathrm{WIN}}N_b^{0.667}$	153780

hybridization of the FMM does not have any adverse effect on the condition of the FE-BI system. The time for *each iteration* is reduced and the total number of iterations remains approximately the same, resulting in reduced overall solution time for the Fast BI algorithms.

The performance of the hybrid algorithms at a more stressing angle of incidence is depicted in Fig. 8.15. For this example calculation, the width of the groove was 10λ and it is seen that the RMS error follows the same trend as for normal incidence illumination. However, even for this smaller size aperture, the scalability of the speed-up is maintained. The employed near-group radius was 1λ implying that the matrix-vector products for groups separated by a distance less than a wavelength was computed using the exact method of moments procedure. Smaller near-group distances can be employed to reduce the CPU time even further, and near-group distances down to 0.3λ have been found to yield sufficiently accurate results.

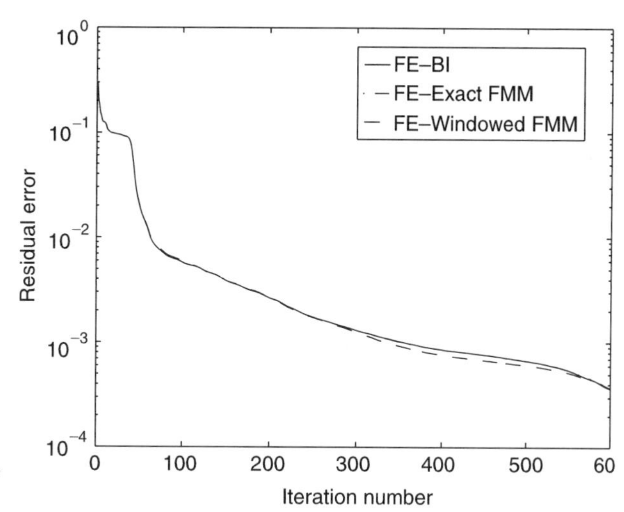

Figure 8.14 Convergence curves for the hybrid algorithms for the groove of width 25λ.

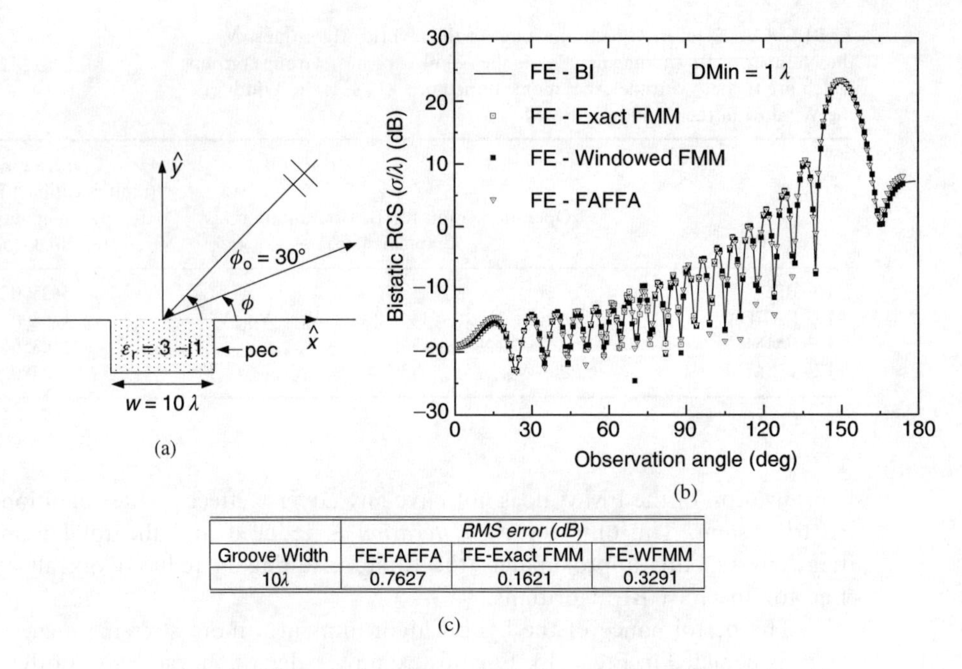

	RMS error (dB)		
Groove Width	FE-FAFFA	FE-Exact FMM	FE-WFMM
10λ	0.7627	0.1621	0.3291

(c)

Figure 8.15 Scalability of the hybrid techniques to smaller problems: (a) Problem geometry, (b) Bistatic patterns, (c) Error table.

REFERENCES

[1] J. Gong, J. L. Volakis, A. Woo, and H. Wang. A hybrid finite element boundary integral method for analysis of cavity-backed antennas of arbitrary shape. *IEEE Trans. Antennas Propagat.*, 42(9):1233–1242, September 1994.

[2] E. Bleszynski, M. Bleszynski, and T. Jaroszewicz. AIM: Adaptive integral method for solving large-scale electromagnetic scattering and radiation problems. *Radio Sci.*, 31(5):1225–1251, 1996.

[3] N. N. Bojarski. *k*-space formulation of the electromagnetic scattering problem. Technical Report AFAL-TR-71-5, U.S. Air Force, March 1971.

[4] C. Y. Shen, K. J. Glover, M. I. Sancer, and A. D. Varvatsis. The discrete Fourier Transform method of solving differential integral equations in scattering theory. *IEEE Trans. Antennas Propagat.*, AP-37:1032–1049, August 1989.

[5] S. Bindiganavale, T. Özdemir, J. L. Volakis, and J. Berrie. Broadband antenna analysis and scattering from planar structures using a fast integral method. Int. Radio Science Meeting, Montreal, CA, 1997.

[6] J. Barnes and P. Hut. A hierarchical $O(N \log N)$ force-calculation algorithm. *Nature*, 324(4):446–449, 1986.

[7] M. S. Warren and J. K. Salmon. Astrophysical *n*-body simulations using hierarchical tree data structures. In *Proceedings Supercomputing 92*, pp. 570–576, 1992.

[8] H-. Q. Ding, N. Karasawa, and W. A. Goddard III. Atomic level simulations on a million particles: The cell multipole method for Coulomb and London nonbond interactions. *J. Chem. Phys.*, 97(6):4309–4315, 1992.

[9] L. F. Greengard. *The Rapid Evaluation of Potential Fields in Particle Systems.* The MIT Press, 1988.

[10] V. Rokhlin. Rapid solution of integral equations for scattering theory in two dimensions. *Journal of Computational Physics*, 86(2):414–439, 1990.

[11] R. Coifman, V. Rokhlin, and S. Wandzura. The fast multipole method for the wave equation: A pedestrian prescription. *IEEE Antennas and Propagation Magazine*, 35(3):7–12, 1993.

[12] J. M. Song and W. C. Chew. Multilevel fast multipole algorithm for solving combined field integral equation of electromagnetic scattering. *Microwave and Optical Technology Letters*, 10:14–19, 1995.

[13] R. J. Burkholder and D. H. Kwon. High-frequency asymptotic acceleration of the fast multipole method. *Radio Science*, 31(5):1199–1206, 1996.

[14] S. S. Bindiganavale and J. L. Volakis. Guidelines for using the fast multipole method to calculate the RCS of large objects. *Microwave and Optical Technology Letters*, 11(4):190–194, 1996.

[15] C. C. Lu and W. C. Chew. Fast far field approximation for calculating the RCS of large objects. *Microwave and Optical Technology Letters*, 8(5):238–241, 1995.

[16] S. S. Bindiganavale and J. L. Volakis. A hybrid FEM-FMM technique for electromagnetic scattering. *IEEE Transactions on Antennas and Propagation*, 45(1):180–181, 1997. (Also in *Proc. 12th Ann. Rev. Progress Appl. Computat. Electromagn. (ACES)*, Monterey, CA, March 1996, pp. 563–570.)

[17] S. S. Bindiganavale and J. L. Volakis. Comparison of three FMM techniques for solving hybrid FE-BI systems. *IEEE Antennas and Propagation Magazine*, **39**(4), 47–60, 1997.

9

Numerical Issues

9.1 INTRODUCTION

In the previous chapter, we outlined the formulation of the finite element method as applied to problems in electromagnetics. In three dimensions, FEM is primarily a volume formulation and the number of unknowns escalates rapidly as the size of the problem increases. Therefore, the limiting factor in dealing with three-dimensional problems is the unknown count and the associated demands on storage and solution time. Techniques which have $O(N)$ storage and solution times are thus necessary to tackle three-dimensional problems. This is one of the principal reasons for the popularity of partial differential equation techniques over integral equation (IE) approaches, as the latter lead to dense matrices with $O(N^2)$ storage. As the problem size increases, the IE and hybrid methods, both of which need $O(N^l)$, $1 < l \leq 2$, storage, quickly become unmanageable in terms of storage and solution time. Another concern while solving problems having more than 100,000 unknowns—a scenario that can be envisioned for most practical problems—is to avoid software bottlenecks. The algorithmic complexity of any part of the program should increase at most linearly with the number of unknowns. This is not possible in many cases but as a rule of thumb, it is generally true that schemes can be devised to manipulate sparse matrices using $O(N)$ storage and operation count.

In this chapter, we will present some numerical considerations for writing efficient sparse matrix codes, of which FEM is an example. The trade-offs associated with the various data structures used to represent sparse matrices and their impact on vectorization and parallelization are first discussed. Next, we review some techniques to solve linear systems of equations. Direct solvers like Cholesky decomposition and Gaussian elimination are discussed along with ordering algorithms for optimizing memory and CPU resources. We then focus on iterative solvers, point and block preconditioning strategies and their corresponding trade-offs. A modified incomplete LU (ILU) preconditioner is presented, which seems to work better than the original ILU preconditioner for the weakly positive-definite matrix systems we have encountered. Iterative solvers for unsymmetric matrix systems are also mentioned to handle anisotropic geometries and situations where the boundary conditions make the system unsymmetric. We devote an entire section to sparse eigenanalysis where we focus mainly on solving the generalized eigenvalue problem using sparse and full matrix methods. To solve large problems, the computationally intensive portions of the finite element code need to be parallelized on massively parallel architectures. A parallelization paradigm is discussed in connection with a distributed memory multiprocessor such as the KSR1 (Kendall Square Research) machine.

9.2 SPARSE STORAGE SCHEMES

The matrix systems in finite elements and related PDE methods are very sparse and the percentage of sparsity increases with the number of unknowns. In an average three-dimensional tetrahedral mesh with edge basis functions, the minimum number of nonzero elements per row can be 9 and the maximum number of nonzeros per row is about 30. The total number of nonzeros varies between $15N$ and $16N$, where N is the number of unknowns. Assuming a square matrix, the matrix is 99.84 percent sparse for a 10,000-unknown problem whereas for 100,000 unknowns, 99.984 percent of the matrix entries are zero. Clearly, it makes little sense to store these zero entries which motivates us to find the *best* possible scheme for storing such matrices. As we shall see in the subsequent paragraphs and indeed throughout this chapter, the definition of *best* is not unique and is governed by computer architecture.

There are various storage schemes for sparse matrices. In this chapter, we will discuss the more viable ones: Compressed Sparse Row (CSR) format, ITPACK format [1], and the jagged diagonal format. Knowledge of the storage formats is important since the speed of computation on vector or parallel processors is directly linked to the data structure used for matrix storage.

The most commonly used format for storing sparse matrices is referred to as the Compressed Sparse Row (CSR) format. In this scheme, the values of the nonzero elements of the sparse matrix $\mathcal{A}$ are stored by rows along with their corresponding column indices, in two long vectors $\mathcal{VAL}$ and $\mathcal{ROW}$, respectively.[1] The dimension of $\mathcal{VAL}$ and $\mathcal{COL}$ equals the number of nonzero elements in $\mathcal{A}$. Another pointer array—$\mathcal{ROWPNTR}$—of dimension N is used to store the number of nonzero elements per row. Thus the position of each element in the sparse matrix is uniquely defined. For example, if we have the 5×5 unsymmetric matrix $\mathcal{A}$

[1]For convenience the matrices and column vectors will be denoted by bold letters in this chapter only and the columns will be treated as vectors in the dot/inner product definitions.

$$\mathcal{A} = \begin{bmatrix} 3 & 0 & 0 & 4 & 5 \\ 7 & 0 & 4 & 0 & 2 \\ 4 & 0 & 7 & 0 & 0 \\ 0 & 0 & 8 & 0 & 0 \\ 9 & 7 & 0 & 0 & 0 \end{bmatrix}$$

Then according to the CSR scheme, $\mathcal{VAL}$ and $\mathcal{ROW}$ will take the form

$$\mathcal{VAL} = [3 \quad 4 \quad 5 \quad 7 \quad 4 \quad 2 \quad 4 \quad 7 \quad 8 \quad 9 \quad 7]$$
$$\mathcal{COL} = [1 \quad 4 \quad 5 \quad 1 \quad 3 \quad 5 \quad 1 \quad 3 \quad 3 \quad 1 \quad 2]$$

and the row pointers will be stored in the array $\mathcal{ROWPNTR}$

$$\mathcal{ROWPNTR} = [3 \quad 6 \quad 8 \quad 9 \quad 11]$$

Note the first value of $\mathcal{ROWPNTR}$ implies that after reading three entries of $\mathcal{VAL}$, we will then start reading entries that belong to the second row of $\mathcal{A}$. After reading the sixth entry of $\mathcal{VAL}$ we will then begin reading entries of $\mathcal{VAL}$ that belong to the third row of $\mathcal{A}$ and so on. The last entry of $\mathcal{ROWPNTR}$ is always equal to the length of the vector $\mathcal{VAL}$.

In the above example, the matrix entries for each row were stored in ordered fashion, i.e., in increasing order of column indices, but this is not necessary for commutative operations like addition and multiplication. A similar data structure which stores the row indices instead of the column indices is called the *Compressed Sparse Column* (*CSC*) format. The CSC format is sometimes used when the matrix is to be accessed along the rows and not the columns, e.g., in the multiplication of the transpose of a sparse matrix with a vector. The CSR/CSC scheme is very convenient for addition, multiplication, permutation, and transposition of sparse matrices. However, if the matrix is extremely sparse, then row-wise traversal can lead to short vector lengths and a significant hit in performance on vector supercomputers. Thus alternative storage schemes are necessary for sparse matrix codes to run efficiently on vector machines.

Of course, storage can be further reduced if the matrix is symmetric since only the upper or lower triangle of the matrix can be stored without loss of information. However, there is a significant performance hit since the increased indirect addressing causes tremendous memory contention. For this reason, it is advisable to store the entire matrix since storage requirements are usually very low for PDE methods and the resulting increase in runtime is not worth the storage trade-off.

In the ITPACK storage scheme, a sparse matrix of order N is stored using two arrays $\mathcal{VAL}$ and $\mathcal{COL}$. Then, according to the ITPACK scheme, the rows of the array $\mathcal{VAL}$ will contain the nonzero elements of the corresponding rows of the original matrix. The number of columns of $\mathcal{VAL}$ will be equal to the maximum number of nonzeros in a row; rows containing fewer nonzero elements will be zero padded. Again, considering the sparse matrix $\mathcal{A}$, the corresponding ITPACK array $\mathcal{VAL}$ can be represented as

$$\mathcal{VAL} = \begin{bmatrix} 3 & 4 & 5 \\ 7 & 4 & 2 \\ 4 & 7 & 0 \\ 8 & 0 & 0 \\ 9 & 7 & 0 \end{bmatrix}$$

The column indices of the elements in $\mathcal{VAL}$ are stored in an integer array $\mathcal{COL}$ defined as

$$\mathcal{COL} = \begin{bmatrix} 1 & 4 & 5 \\ 1 & 3 & 5 \\ 1 & 3 & * \\ 3 & * & * \\ 1 & 2 & * \end{bmatrix}$$

The asterisk denotes that the corresponding elements of $\mathcal{COL}$ are zeros. The ITPACK storage scheme is attractive for generating finite element matrices since the number of comparisons required while augmenting the matrix depends only on the locality of the corresponding variable and not on the number of unknowns. *This feature can also be used for implementing fast searches and comparisons, whenever the matrix is extremely sparse.* Moreover, the sparse matrix-vector multiplication process can be highly vectorized because of large vector lengths when the number of nonzeros in all rows is nearly equal. This is because the multiplication operation is carried out by traversing the columns of $\mathcal{VAL}$ and $\mathcal{COL}$ whose dimensions are $O(N)$. However, for our application, almost half the space is lost in storing zeros. As a result, a lot of storage as well as computational effort is wasted in storing and operating on zeros, respectively.

The modified ITPACK scheme [2] does alleviate this problem to a certain degree by sorting the rows of the matrix by decreasing number of nonzero elements. However, 30 percent of the allotted space is still lost in zero padding.

The other storage format that has been found to be useful for sparse matrices is the jagged diagonal storage scheme [3]. On a vector machine, this format gives better performance in terms of vectorizability. In this scheme, the vector lengths are approximately equal to the order of the system being solved. The rows are first sorted by increasing degree of sparsity. The first jagged diagonal is constructed by taking the first element from each row of the CSR data structure of the ordered matrix. The rest of the jagged diagonals can be obtained in a similar fashion. The matrix is thus stored as a collection of subvectors of decreasing length. The number of jagged diagonals equals the number of nonzeros in the first row of the sorted matrix. An additional vector is required as before to store the corresponding column numbers from the original sparse matrix. The inner loop of the matrix-vector multiplication routine traverses the entire length of a jagged diagonal, the maximum dimension of which is the same as that of the sparse matrix. This feature enhances vectorization massively. The storage requirement of the above format can be made to be the same as the previously mentioned CSR format through careful programming. Again, taking the earlier sparse matrix example, we see in Fig. 9.1 how the matrix is stored in the jagged diagonal format. The arrays VAL and COL store the matrix values and the corresponding column numbers of the sparse matrix, respec-

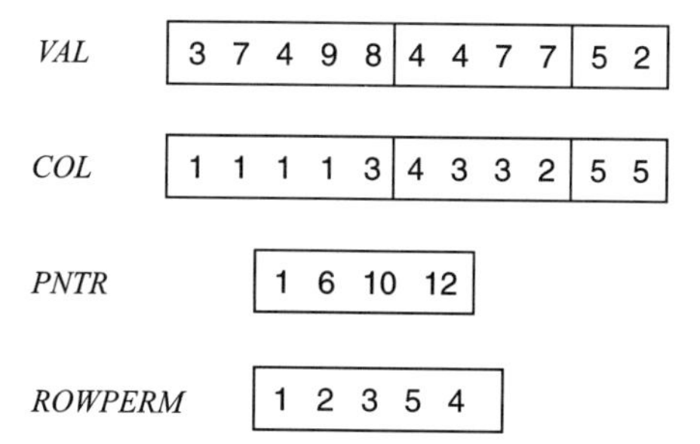

Figure 9.1 Jagged diagonal storage format.

tively. The *PNTR* vector has a length of 4 in this case which equals one plus the most populated row of the sparse matrix. This vector therefore stores the starting location of each jagged diagonal in the arrays *VAL* and *COL*. The array *ROWPERM* indicates the locations of the permuted rows of the original sparse matrix. The altered code then runs at around 275 Mflops on a Cray C-90. The dot product reaches speeds of 550 Mflops and the vector updates execute at 600 Mflops. It must be mentioned that the CRAY C-90 is a substantially faster machine than the Cray YMP but the CSR formatted matrix-vector multiplication routine runs about four times slower on the C-90. Moreover, the speed of operation is a direct consequence of the amount of sparsity in the matrix system. In spite of these caveats, the *jagged diagonal* method is much better than the CSR or ITPACK storage methods for implementation on vector processors.

9.3 DIRECT EQUATION SOLVER

In this and the two subsequent sections, we will concentrate on the various techniques for solving the linear equation system

$$\mathcal{A}x = b \tag{9.1}$$

When $\mathcal{A}$ is dense, i.e., most of the elements of $\mathcal{A}$ are nonzero, the decision is somewhat straightforward. The inversion of the matrix can be carried out in $O(N^3)$ operations using popular methods like LU decomposition or Cholesky factorization. However, in the case of sparse matrices, a simple application of the traditional methods can prove catastrophic, as storage and processor demands will far exceed acceptable levels. Our focus in this section will, therefore, be on sparse factorization techniques.

9.3.1 Factorization Schemes

LU decomposition is a method of reducing a general matrix into two triangular matrices: L (lower triangular) and U (upper triangular with unit diagonals). Thus, in mathematical notation, this can be written as

$$\mathcal{A} = \mathcal{L}\mathcal{U} \tag{9.2}$$

The advantage of this representation is that the subsequent systems

$$\mathcal{L}w = b \tag{9.3}$$

$$\mathcal{U}x = w \tag{9.4}$$

can be solved in $O(N^2)$ operations using forward and backward substitutions, respectively. Moreover, the decomposition is unique if $\mathcal{A}$ is nonsingular. For symmetric positive definite matrices, the procedure is called *Cholesky factorization* and the decomposition of $\mathcal{A}$ can be written as

$$\mathcal{A} = \mathcal{L}\mathcal{L}^T \tag{9.5}$$

The Cholesky factorization thus preserves symmetry of the factored matrix and is also a unique factorization of $\mathcal{A}$. Public domain codes for both Cholesky factorization and LU decomposition can be found in [4].

9.3.2 Error Control

Due to finite precision arithmetic, floating point errors can creep into matrix factorization schemes. For a full matrix order n, the error bound can be expressed as [5]

$$|e| < 3.01 n \epsilon_M a_M \tag{9.6}$$

where ϵ_M is the machine precision and a_M is the largest element of the original or factorized matrix.

In sparse matrix factorizations, the error bound is actually lower since only a few operations are performed on each nonzero element. The error matrix for the LU decomposition is expressed as

$$\mathcal{L}\mathcal{U} = \mathcal{A} + \mathcal{E} \tag{9.7}$$

where $\mathcal{E}_{ij}$ is bounded by the expression

$$|\mathcal{E}_{ij}| < 3.01 n \epsilon_M a_{ij} n_{ij} \tag{9.8}$$

and n_{ij} is a matrix of integers given by

$$\begin{aligned} n_{ij} &= \sum_{k=1}^{\min(i,j)} n_{ij}^{(k)} \\ n_{ij}^{(k)} &= 1, \text{ if both } L_{ik} \text{ and } U_{kj} \text{ are nonzero} \\ n_{ij}^{(k)} &= 0, \text{ otherwise} \end{aligned} \tag{9.9}$$

As can be observed from the error bounds, the growth of the error is directly proportional to the maximum value of any element that occurs in the original matrix or due to factorization. Thus, a strategy for monitoring element growth and then reducing it points the way for error control. One of the most popular strategies is pivoting. Scaling with the largest element in the corresponding row of the submatrix (*partial pivoting*) or with the largest element in the entire submatrix (*complete pivoting*) usually stabilizes the factorization process and provides accurate answers. Complete pivoting comes with a severe price tag in computational expense; partial pivoting is, therefore, a method of choice for most factorization schemes. It should be mentioned here that if a matrix is positive definite, diagonal elements are chosen as pivots since diagonal dominance is a natural consequence of positive definiteness.

For sparse factorizations, even partial pivoting can be too expensive and too rigid. In such cases, *threshold pivoting* is employed which strives to maintain sparsity and employs a user-defined threshold parameter to determine the choice of the pivot [5]. Threshold pivoting is quite popular and is used in production level codes. Zlatev's strategy [6] is a variation of threshold pivoting; in addition to maintaining sparsity, it reduces the number of search rows for the pivot to a user-defined value.

9.3.3 Matrix Ordering Strategies

Sparse matrix techniques can be used to solve very large problems since the storage required increases as $O(N)$, where N denotes the degrees of freedom for the problem. However, when a sparse matrix is factorized, the upper or lower triangular factors may not reflect the sparsity pattern of the respective upper or lower triangle of the original sparse matrix. This phenomenon is called *fill-in*, which corresponds to the additional nonzero elements generated during factorization. If matrix fill is left uncontrolled, serious storage and performance penalties ensue. Fill is undesirable for three compelling reasons:

- Additional storage must be allocated for the extra nonzeros. An extreme example is a matrix with full first row, full first column and main diagonal and zeros elsewhere would be completely filled on factorization.
- Number of operations needed for factorization increases with increasing fill.
- The error bounds defined earlier increase as the matrix becomes filled with more and more nonzero entries.

Strategies for the reduction of fill-in have their origins in graph theory. Since the amount of fill depends on the row/column permutation selected, a convenient ordering of the matrix will drastically reduce the computation time and storage requirements of the factorization. However, it is extremely difficult to find an optimum ordering which will *guarantee* the smallest possible fill-in or operation count. In fact, no general algorithm exists to generate an optimal ordering for an arbitrary graph. Existing strategies attempt to find an ordering for which the fill-in and operation count are low, without guaranteeing a true minimum. In finite elements, all matrices are *structurally symmetric*, i.e., the positions of the nonzeros form a symmetric pattern, even though the corresponding values may break the matrix symmetry. Thus we will mention ordering strategies for symmetric matrices only.

A graph consists of a set of vertices together with a set of edges. Thus, a finite element mesh can be considered to be an undirected graph since the edge pair between nodes (u, v) and (v, u) are indistinguishable due to symmetry. Figure 9.2 shows a symmetric matrix and its corresponding labeled undirected graph. A graph with n vertices is *labeled* when there exists a one-to-one correspondence between the vertices and the integers $1, 2, \ldots, n$. Ordering strategies for symmetric matrices hinge on the fact that the graph of a symmetric matrix remains invariant under a symmetric permutation of its rows and columns; what changes is merely the vertex labeling.

Before discussing the various algorithms involved with matrix ordering, we need to be familiar with a few basic terms in graph theory. Any square matrix $\mathcal{A}$ of order N can be considered to be an undirected graph with N labeled *vertices*, $v_1, v_2, \ldots, v_n$. The pair (v_i, v_j) is an *edge* of the graph if and only if $\mathcal{A}_{ij} \neq 0$. The

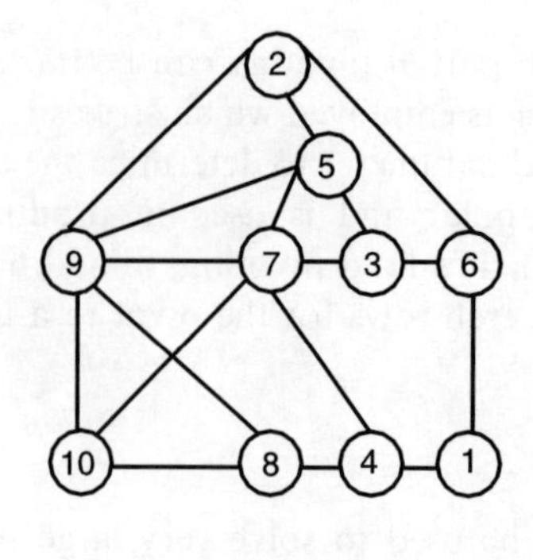

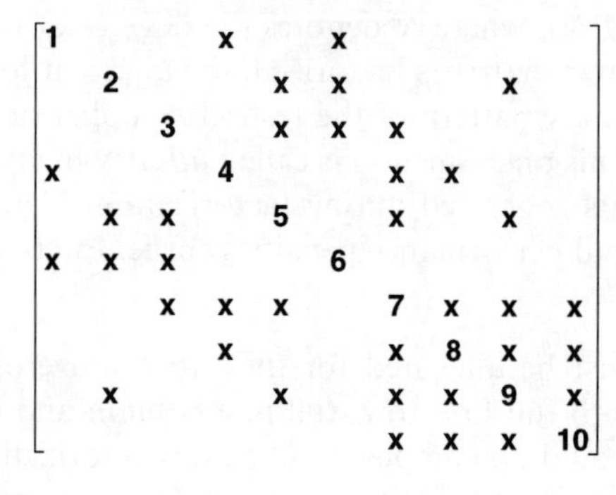

Figure 9.2 (a) Corresponding graph; (b) symmetric sparse matrix structure.

diagonal element $\mathcal{A}_{ii}$ corresponds to a *loop* or *selfedge* and is always present for a nonsingular matrix. If (v_i, v_j) forms a valid edge, vertices v_i, v_j are said to be *adjacent* to each other and the corresponding edge is *incident* on each of the vertices. The number of edges incident on a vertex denotes the *degree* of the vertex. The *distance* between two vertices $d(v_i, v_j)$ is the length of the shortest connected path between them. The largest distance between v_i and any other vertex of the graph is called the *eccentricity* of the vertex $e(v_i)$. The vertex with the largest eccentricity is called the *peripheral vertex*. Since no efficient algorithms are available for determining a peripheral vertex, a *pseudo-peripheral* vertex is used. v_i is a pseudo-peripheral vertex if $d(v_i, v_j) = e(v_i)$ implies that $e(v_i) = e(v_j)$, thereby guaranteeing that the eccentricity of the selected vertex is large.

Most matrix reordering algorithms start with a vertex of minimum degree or a pseudo-peripheral vertex. The *bandwidth reduction* algorithm mentioned here is due to Cuthill and McKee [7]. Starting with a pseudo-peripheral vertex, all unlabeled vertices adjacent to it are labeled successively in order of increasing degree. The *reverse Cuthill-McKee algorithm* is used when the matrix profile needs to be minimized. In this case, the orderings of the Cuthill-McKee algorithm are merely reversed to arrive at the minimized profile. Figure 9.3 shows a typical profile reduction algorithm at work. Notice the bandedness of the final system compared with the arbitrary sparsity pattern of the original matrix. King [8] also proposed a profile reduction algorithm with similar performance characteristics as the reverse Cuthill-McKee algorithm. The profile reduction and bandwidth reduction algorithms are useful since they save both storage and operation count in the triangular factorization process. However, none of them explicitly minimize the fill-in of the factors.

The algorithm commonly used for reducing fill-in during factorization of a sparse matrix is called the *minimum degree algorithm*. The idea behind the algorithm

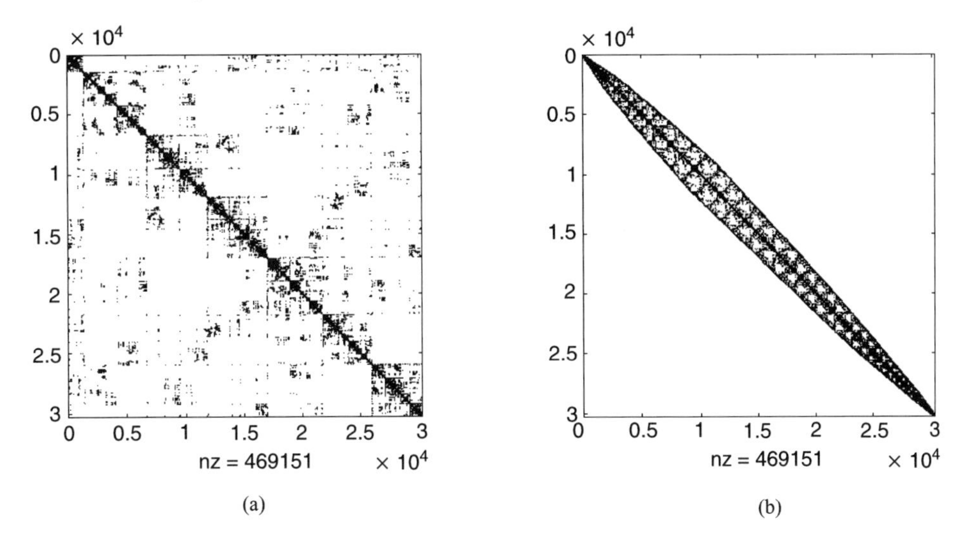

Figure 9.3 (a) Original matrix structure; (b) matrix structure after re-ordering using a profile reduction algorithm.

is simple intuitively and one of the cheapest and the most effective computationally. Fill-in and operation count is minimized locally by selecting, at each stage of the elimination process and among all possible diagonal entries, that row and column which introduces the least number of nonzeros in the resulting factor. It is quite amazing that such a simple idea works so effectively. One of the problems with the original implementation of the algorithm was that the total storage could not be predicted beforehand. To alleviate this problem, George and Liu [9] introduced the concept of *indistinguishable vertices*. Two vertices are said to be indistinguishable from each other if they have the same degree. Also, once they are indistinguishable at an intermediate factorization step, they continue to be so till one of them is eliminated. If one of them is of minimum degree, then elimination of one is directly followed by the elimination of its partner in the next step. A detailed description of the minimum degree algorithm is well beyond the scope of this book. It can be found in most texts on graph theory as well as in Chapters 4 and 5 of [5].

9.4 ITERATIVE EQUATION SOLVERS

In PDE techniques like finite elements or finite differences, the order N of the system of linear equations may be very large. Three-dimensional problems lead to even bigger equation systems. As shown in the earlier section, direct solvers usually suffer from fill-in to an extent that these large problems cannot be solved at a reasonable cost even on state-of-the-art parallel machines. It is, therefore, essential to employ solvers whose memory requirements are a small fraction of the storage demand of the coefficient matrix. This necessitates the use of iterative algorithms instead of direct solvers to preserve the sparsity pattern of the finite element matrix. Especially attractive are iterative methods that involve the coefficient matrices only in terms of matrix-vector products with $\mathcal{A}$ or $\mathcal{A}^T$. The most powerful iterative algorithm of this type is the conjugate gradient algorithm for solving positive definite linear systems [10]. In this section, we will discuss some algorithms which have been found effective in solving the sparse matrices that occur in our application. These are the biconjugate gradient (BiCG) and the quasi-minimal residual (QMR) [11] algorithms. A version of the generalized minimal residual (GMRES) method is also presented. These algorithms can also be used for solving unsymmetric matrix systems as is the case with anisotropic materials.

The convergence pattern of the CG method for self-adjoint positive definite (SPD) systems can be described by

$$0 < \frac{E_n}{E_0} \leq 4\left[\frac{\sqrt{K_a}-1}{\sqrt{K_a}+1}\right]^{2n} \tag{9.10}$$

where $E_n = \boldsymbol{r}_n^T \mathcal{A}^{-1} \boldsymbol{r}_n$, $K_a = \lambda_{\max}/\lambda_{\min}$ is the spectral condition number and n is the number of iterations. Axelsson [12] and Van der Vorst [13] examine the convergence of the CG algorithm in detail. Its convergence is shown to be superlinear, with a convergence rate that depends on the distribution of the (mainly smallest) eigenvalues of $\mathcal{A}$, rather than on the spectral condition number. However, if the matrix $\mathcal{A}$ is not too far from being positive definite, which is the case with the matrix systems emerging from edge element implementations, the BiCG and CG algorithms should still converge. Some implementations espouse premultiplication of $\mathcal{A}$ by $\mathcal{A}^T$ to ensure positive definiteness of the system. Unless the condition number is known a priori to be small or the matrix is unitary, this is a *very bad idea*, since the convergence is going to be drastically slow as is evident from (9.10).

The *biconjugate gradient* (*BiCG*) method is a variation of the CG algorithm. This scheme is useful for solving unsymmetric systems; however, it performs equally well when applied to symmetric systems of linear equations. For symmetric matrices, BiCG differs from CG in the way the inner product of the vectors are taken. BiCG usually converges much faster than CG; however, the convergence is highly erratic. The BiCG algorithm for unsymmetric matrix problems is given in Fig. 9.4. For symmetric positive definite matrix systems, the BiCG algorithm needs approximately half the computational work and only one matrix-vector multiply operation. The complete algorithm is presented later in the chapter (Fig. 9.18).

The *conjugate gradient squared* (*CGS*) algorithm [14] performs best when applied to unsymmetric systems of linear equations. A big advantage of CGS over BiCG when solving unsymmetric equation systems is that the matrix-vector product only involves the matrix $\mathcal{A}$ and not $\mathcal{A}^T$. Figure 9.5 shows the CGS algorithm for positive definite matrices, as it appears in [14]. It should be noted that $\boldsymbol{s}$ requires a second, arbitrary nonzero starting vector. Usually, $\boldsymbol{s}$ is set to $\boldsymbol{r}_0$ or $\boldsymbol{r}_0^*$ if $\boldsymbol{r}_0$ is full or to random entries when $\boldsymbol{r}_0$ is sparse. CGS usually converges faster than BiCG but is more unstable since the residual polynomials are the squared BiCG polynomials and hence exhibit even more erratic behavior than the BiCG residuals. Moreover, there are cases where CGS diverges, while BiCG still converges. The erratic convergence pattern of BiCG and CGS has to do with the nonpositive definiteness of the inner products used [14]. The BiCGSTAB algorithm [15] alleviates this problem by damping out the wild fluctuations.

Initialization

$\boldsymbol{x}$ given

$\boldsymbol{r}_0 = \boldsymbol{s}_0 = \boldsymbol{b} - \mathcal{A}\boldsymbol{x}_0; \boldsymbol{p}_{-1} = \boldsymbol{q}_{-1} = 0; \rho_{-1} = 1;$

$n = 0;$

Repeat until $(resd \leq tol)$

$\rho_n = \boldsymbol{s}_n \cdot \boldsymbol{r}_n$ (1)

$\beta_n = \rho_n / \rho_{n-1}$ (2)

$\boldsymbol{p}_n = \boldsymbol{r}_n + \beta_n \boldsymbol{p}_{n-1}$ (3)

$\boldsymbol{q}_n = \boldsymbol{s}_n + \beta_n \boldsymbol{q}_{n-1}$ (4)

$\sigma_n = \boldsymbol{q}_n \cdot \mathcal{A}\boldsymbol{p}_n; \alpha_n = \rho_n / \sigma_n$ (5)

$\boldsymbol{r}_{n+1} = \boldsymbol{r}_n - \alpha_n \mathcal{A}\boldsymbol{p}_n$ (6)

$\boldsymbol{s}_{n+1} = \boldsymbol{s}_n - \alpha_n \mathcal{A}^T \boldsymbol{q}_n$ (7)

$\boldsymbol{x}_{n+1} = \boldsymbol{x}_n + \alpha_n \boldsymbol{p}_n$ (8)

$resd = \sqrt{|\boldsymbol{r} \cdot \boldsymbol{r}^*|}$

$n = n + 1$

EndRepeat

$\mathcal{A}$ is a sparse complex unsymmetric matrix.

Figure 9.4 Unsymmetric biconjugate gradient algorithm.

Initialization

$\boldsymbol{x}$ and $\boldsymbol{s}$ given

s:Arbitrary

$\boldsymbol{r}_0 = \boldsymbol{b} - \mathcal{A}\boldsymbol{x}_0;\ resd = \sqrt{\boldsymbol{r}_0^* \cdot \boldsymbol{r}_0};$

$\boldsymbol{q}_0 = \boldsymbol{p}_{-1} = 0;\ \rho_{-1} = 0;\ n = 0;$

Repeat until $(resd \leq tol)$

$\rho_n = \boldsymbol{s} \cdot \boldsymbol{r}_n$ (1)

$\beta_n = \rho_n / \rho_{n-1}$ (2)

$\boldsymbol{u}_n = \boldsymbol{r}_n + \beta_n \boldsymbol{q}_n$ (3)

$\boldsymbol{p}_n = \boldsymbol{u}_n + \beta_n(\boldsymbol{q}_n + \beta_n \boldsymbol{p}_{n-1})$ (4)

$\boldsymbol{v}_n = \mathcal{A}\boldsymbol{p}_n$ (5)

$\sigma_n = \boldsymbol{s} \cdot \boldsymbol{v}_n; \alpha_n = \rho_n / \sigma_n$ (6)

$\boldsymbol{q}_{n+1} = \boldsymbol{u}_n - \alpha_n \boldsymbol{v}_n$ (7)

$\boldsymbol{r}_{n+1} = \boldsymbol{r}_n - \alpha_n \mathcal{A}(\boldsymbol{u}_n + \boldsymbol{q}_{n+1})$ (8)

$\boldsymbol{x}_{n+1} = \boldsymbol{x}_n + \alpha_n(\boldsymbol{u}_n + \boldsymbol{q}_{n+1})$ (9)

$resd = \sqrt{\boldsymbol{r}_{n+1}^* \cdot \boldsymbol{r}_{n+1}}$

$n = n + 1$

EndRepeat

$\mathcal{A}$ is a sparse complex unsymmetric matrix.

Figure 9.5 Conjugate gradient squared algorithm.

Besides the problem of erratic convergence, BiCG suffers from the problem of breakdowns. Although the breakdowns occur only in exact arithmetic, the robustness of the iterative solver is compromised and near-breakdowns can adversely affect convergence accuracy. The BiCG algorithm is said to have broken down if

$$\boldsymbol{q}_n \cdot \boldsymbol{p}_n = 0, \text{ when } \boldsymbol{s}_n \neq 0, \boldsymbol{r}_n \neq 0 \tag{9.11}$$

or if

$$\boldsymbol{s}_n \cdot \boldsymbol{r}_n = 0, \text{ when } \boldsymbol{s}_n \neq 0, \boldsymbol{r}_n \neq 0 \tag{9.12}$$

According to Freund et al. [16], the above breakdown can be attributed to the Galerkin condition that a BiCG residual must satisfy. The second type of breakdown parallels the breakdown in the unsymmetric Lanczos process.

Freund [11] has proposed the *quasi-minimal residual* (*QMR*) algorithm with look-ahead for solving linear equation systems. QMR eliminates the oscillations in the BiCG residual norm and generates smooth, near monotonically converging iterates. QMR also avoids breakdowns of the first and second kind: the latter is corrected by using a look-ahead technique. Moreover, transpose-free QMR algorithms exist for solving unsymmetric matrix systems. The reader is referred to [11] for algorithms pertaining to unsymmetric matrices with look-ahead. It should be mentioned in passing that in most cases, breakdowns do not occur; however, for the sake of robustness, the look-ahead feature should be included in commercial codes. The QMR algorithm for complex symmetric systems without look-ahead is presented in Fig. 9.6.

In most cases, QMR converges in about the same number of iterations as BiCG with one significant difference. Since the algorithm minimizes the residual at

Initialization

$\boldsymbol{x}$ given

$\boldsymbol{r}_0 = \boldsymbol{b} - \mathcal{A}\boldsymbol{x}_0$;

$resd = \sqrt{\boldsymbol{r}_0^* \cdot \boldsymbol{r}_0}$; $\rho_1 = \sqrt{\boldsymbol{r}_0^* \cdot \boldsymbol{r}_0}$; $\boldsymbol{v}_1 = \boldsymbol{r}_0/\rho_1$;

$\boldsymbol{p}_0 = \boldsymbol{d}_0 = 0$; $c_0 = \epsilon_0 = 1, s_0 = \theta_0 = 1, \eta_0 = -1$;

$\boldsymbol{v}_1 = \boldsymbol{r}_0/resd$;

$n = 1$;

Repeat for $n = 1, 2, \ldots$

if $\epsilon_{n-1} = 0$ *then stop* $\rightarrow$ *Code Breakdown*

$\delta_n = \boldsymbol{v}_n \cdot \boldsymbol{v}_n$ (1)

if $\delta_n = 0$ *then stop*

$\boldsymbol{p}_n = \boldsymbol{v}_n - (\rho_n \delta_n/\epsilon_{n-1})\boldsymbol{p}_{n-1}$ (2)

$\epsilon_n = \boldsymbol{p}_n \cdot \mathcal{A}\boldsymbol{p}_n$; $\beta_n = \epsilon_n/\delta_n$ (3)

$\boldsymbol{v}_{n+1} = \mathcal{A}\boldsymbol{p}_n - \beta_n \boldsymbol{v}_n$ (4)

$\rho_{n+1} = \sqrt{\boldsymbol{v}_{n+1}^* \cdot \boldsymbol{v}_{n+1}}$; $\theta_n = \rho_{n-1}/c_{n-1}|\beta_n|$ (5)

$c_n = 1/\sqrt{1+\theta_n^2}$; $s_n = (\beta_n/|\beta_n|)(\rho_{n+1}c_n/c_{n-1}|\beta_n|)$; $\eta_n = \eta_{n-1}(\rho_n c_n^2/\beta_n c_{n-1}^2)$ (6)

$\boldsymbol{d}_n = \eta_n \boldsymbol{p}_n + (\theta_{n-1}c_n)^2 \boldsymbol{d}_{n-1}$ (7)

$\boldsymbol{x}_n = \boldsymbol{x}_{n-1} + \boldsymbol{d}_n$; $\boldsymbol{v}_{n+1} = \boldsymbol{v}_{n+1}/\rho_{n+1}$ (8)

$n = n + 1$

End

$\mathcal{A}$ is a sparse complex symmetric matrix.

Figure 9.6 Quasi-minimal residual (QMR) algorithm for complex symmetric matrices without look-ahead.

each iteration, convergence criteria can be set at a maximum number of iterations rather than at the residual norm for large problems. It has also been observed that QMR stagnates for those iterations when BiCG yields wildly oscillating residual norms.

Another mildly bothersome aspect of QMR is that the residual vector cannot be recovered without going through a matrix-vector multiplication. However, there is an excellent upper bound for the residual norm defined as

$$\|\boldsymbol{r}_n\| \leq \|\boldsymbol{r}_0\|\sqrt{n+1}|s_1 s_2 \ldots s_n| \tag{9.13}$$

which is approximately five times larger than the true residual norm. Usually, the upper bound is computed until we are very close to the tolerance and then the residual is computed exactly by carrying out an additional matrix-vector multiply for about five to ten more iterations.

The *GMRES* (*Generalized Minimal Residual*) algorithm (see Fig. 9.7), first proposed by Saad and Schulz [17] is another iterative solver for sparse systems. It leads to the smallest residual for a fixed number of iteration steps. However, each of these iteration steps becomes increasingly more expensive because GMRES stores the iteration vectors. To limit the increasing storage requirements and workload per iteration step, it is necessary to restart the algorithm. The proper choice of spanning

Initialization
    $\boldsymbol{x}$ is given
Define the $(m+1)\times m$ matrix $\mathcal{H}_m = \{h_{ij},\ 1 \leq i \leq m+1 \text{ and } 1 \leq j \leq m\}$
    Specify the number of spanning vectors m
START: $\mathbf{r}_0 = \mathbf{b} - \boldsymbol{\mathcal{A}}\mathbf{x}$
    $\beta = \sqrt{\mathbf{r}_0^* \cdot \mathbf{r}_0}$
    $\mathbf{v}_1 = \mathbf{r}_0/\beta$
Repeat2 for $j = 1, \ldots, m$
    $\mathbf{w}_j = \boldsymbol{\mathcal{A}}\mathbf{v}_j$
Repeat1 for $i = 1, \ldots, j$
    $h_{i,j} = \mathbf{w}_j^* \cdot \mathbf{v}_i$
    $\mathbf{w}_j = \mathbf{w}_j - h_{i,j}\mathbf{v}_i$
End Repeat1
    $h_{j+1,j} = \sqrt{\mathbf{w}_j^* \cdot \mathbf{w}_j}$
    $\mathbf{v}_{j+1} = \mathbf{w}_j/h_{j+1,j}$
End Repeat2

Compute $\mathbf{y}_m$ to minimize $||\beta\mathbf{e}_1 - \mathcal{H}_m\mathbf{y}||_2$
    $\boldsymbol{\mathcal{V}}_m = \{\mathbf{v}_1, \mathbf{v}_2, \ldots, \mathbf{v}_m\}$
    $\mathbf{x} = \mathbf{x} + \boldsymbol{\mathcal{V}}_m\mathbf{y}_m$
If convergence is achieved, then stop, else go to START

Figure 9.7 Restarted GMRES algorithm ($\mathbf{e}_j$ refers to the jth column of the identity matrix).

vectors m requires a priori experience, and is dependent on the system parameters and sampling rate. With regard to the number of operations, GMRES requires only one matrix-vector product but the number of inner products increases linearly with the iteration steps. In addition to the matrix storage requirements, $(m+1)N$ complex words are needed to store the m search or spanning vectors. The only drawback of GMRES is that CPU time and storage per iteration increases linearly with the iteration count. A way to overcome this is to choose the initial spanning vectors m in an optimal way. If m is too small, GMRES may exhibit slow convergence or it may not converge at all. On the other hand, if m is large, an excessive amount of work is needed since the CPU time is $O(m^2N)$. More details about this solver are given in [18] and [17].

Figure 9.8 shows a comparison of the performance among the three different solvers (BiCG, QMR, and GMRES). This comparison is done for a sparse symmetric three-dimensional FEM system of approximately 6700 unknowns simulating an enclosed transmission line. The GMRES solver was applied with only $m = 12$ and converged after about 75 restarts. Its error history was monotonic, a typical characteristic of GMRES.

Finally, we make mention of the BiCGSTAB algorithm, originated by Van der Vorst [15]. BiCGSTAB combines monotonic convergence with the superior storage properties of BiCG.

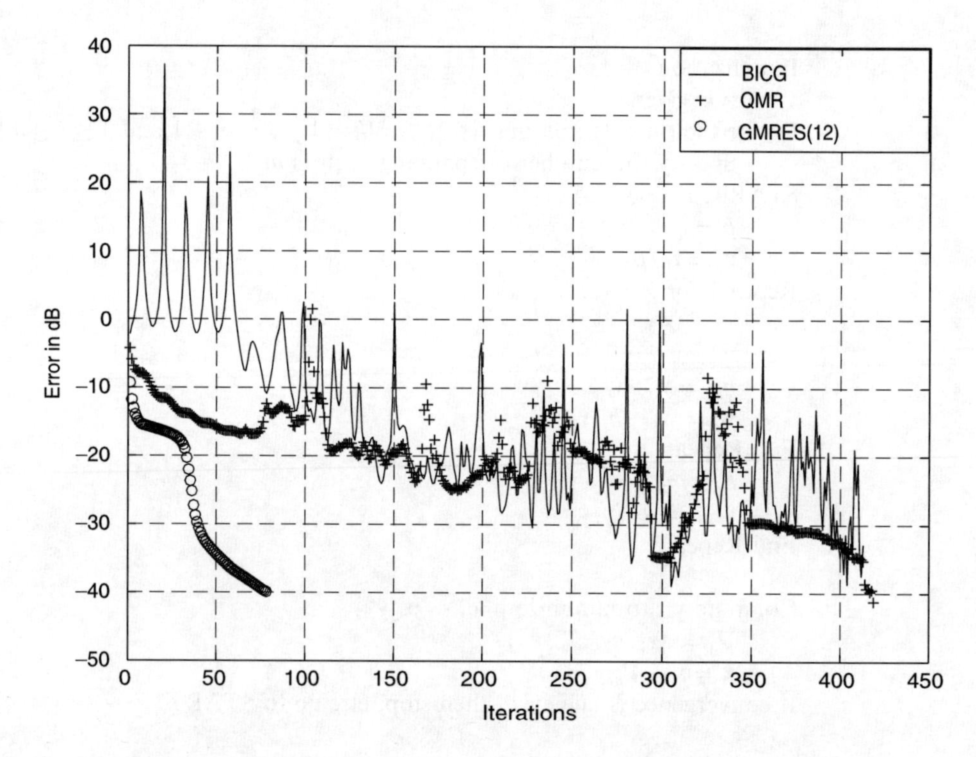

Figure 9.8 Example convergence history of the BiCG, QMR and GMRES algorithms. For GMRES, iterations refer to restarts. [*Courtesy of Y. Botros.*]

9.5 PRECONDITIONING

The condition number of a system of equations usually increases with the number of unknowns. It is then desirable to precondition the coefficient matrix such that the modified system is well conditioned and converges in significantly fewer iterations than the original system. The equivalent preconditioned system is of the form

$$\mathcal{C}^{-1}\mathcal{A}\mathbf{x} = \mathcal{C}^{-1}\mathbf{b} \tag{9.14}$$

The nonsingular preconditioning matrix $\mathcal{C}$ must satisfy the following conditions. It

1. should be a good approximation to $\mathcal{A}$.
2. should be easy to compute.
3. should be invertible in $O(N)$ operations.

The preconditioners mentioned in the following section are the diagonal and the ILU point and block preconditioners. Block preconditioners are usually preferable due to reduced data movement between memory level hierarchies as well as a decreased number of iterations required for convergence. Block algorithms are also suited for high-performance computers with multiple processors since all scalar, vector, and matrix operations can be performed with a high degree of parallelism.

9.5.1 Diagonal Preconditioner

The simplest preconditioner that can be used in iterative solvers is the point diagonal preconditioner. The preconditioning matrix $\mathbf{C}$ is a diagonal matrix which is easy to invert and has a storage requirement of N complex words, where N is the number of unknowns. The entries of $\mathbf{C}$ are given by

$$C_{ij} = \delta_{ij}\mathcal{A}_{ij}, \quad i = 1, \ldots, N; \quad j = 1, \ldots, N \tag{9.15}$$

where δ_{ij} is the Kronecker delta. The matrix $\mathcal{C}^{-1}$ contains the reciprocal of the diagonal elements of $\mathcal{A}$. For a positive definite matrix, the diagonal preconditioner is very effective both in terms of cost and performance. In the matrix systems we dealt with, the diagonally preconditioned systems converged in about 35 percent of the number of iterations required for the unpreconditioned cases. The diagonal preconditioner is also easily vectorizable and consumes 4.1 microseconds per iteration per unknown on a Cray YMP, a marginal slowdown over the unpreconditioned system.

A more general diagonal preconditioner is the block diagonal preconditioner. The point diagonal preconditioner is a block diagonal preconditioner with block size 1. The block diagonal preconditioning matrix consists of $m \times m$ symmetric blocks, as shown in Fig. 9.9. The inverse of the whole matrix is simply the inverse of each individual block put together. If the preconditioning matrix $\mathcal{C}$ is broken up into n blocks of size m, the storage requirement for the preconditioner is at most $m^2 \times N$. However, this method suffers a bit from fill-in since the inverted $m \times m$ blocks are dense even though the original blocks may have been sparse. Due to this reason, large blocks cannot be created since the inverted blocks would lead to full matrices and take a significant fraction of the total CPU time for inversion. However, since

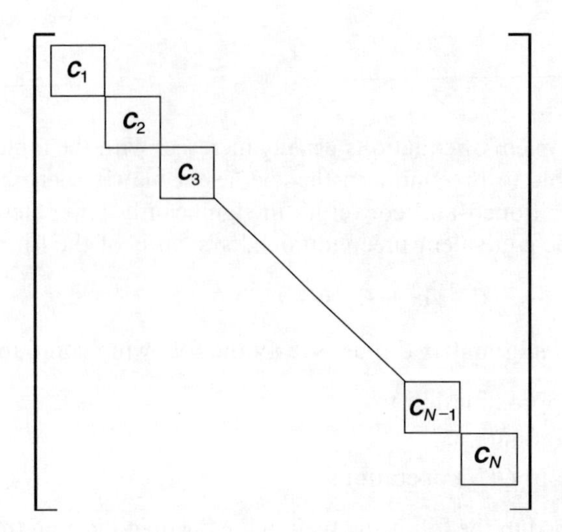

Figure 9.9 Structure of block preconditioning matrix.

the structure of the preconditioning matrix is known a priori, this preconditioner vectorizes well and has been observed to run at 194 MFLOPS on the Cray-YMP for a block size of 8. As an example, for a system of 20,033 unknowns, a block size of 2 caused the maximum reduction in the number of iterations (14 percent) and ran at 197 MFLOPS.

The simple diagonal preconditioner was implemented and tested for solving the problem (with approximately 6700 unknowns) of a shielded microstrip line. The significance of the preconditioner is illustrated in Fig. 9.10, where the number of iterations dropped from 174 to 78 when the diagonal preconditioner is employed. In general, the diagonal preconditioner saves from 30 percent to 60 percent of the total number of iterations and CPU time for all of the iterative solvers. This percentage, of course, is highly dependent on many factors and parameters. The element shape

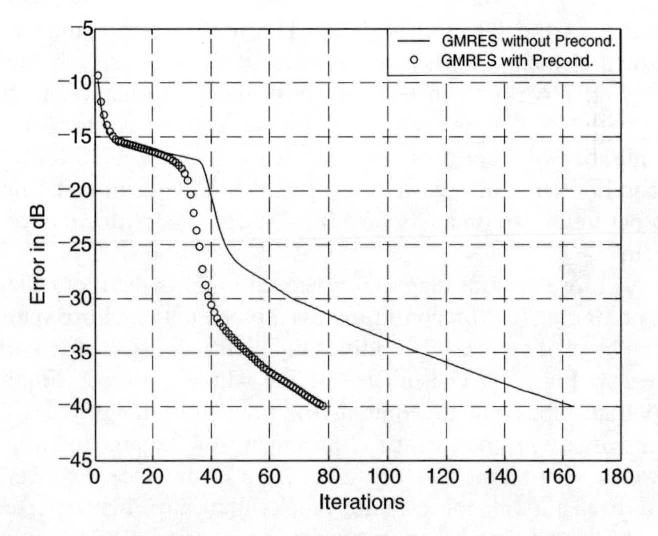

Figure 9.10 Convergence pattern for GMRES with and without diagonal preconditioning for a standard problem. [*Courtesy of Y. Botros.*]

employed to discretize the computational domain, the sampling or discretization rate, the layer parameters, etc., all affect the performance.

9.5.2 Incomplete LU (ILU) Preconditioner

The ILU preconditioner is considered an improvement over the diagonal preconditioner. As mentioned earlier, the convergence of conjugate-gradient methods depends on the clustering of eigenvalues in the spectrum of the preconditioned system. The ILU preconditioner enhances the eigenvalue clustering leading to improved convergence. In our test cases, the traditional ILU preconditioner [19] was employed with zero fill-in. Higher values of fill-in usually improve the convergence even further until a trade-off is reached between storage, matrix-vector multiplication time, and iteration count for convergence [12]. There is another flavor of ILU called ILUT [20] which stores matrix elements only if they exceed a certain threshold value relative to the diagonal. In the cases mentioned below, no attempt was made to employ higher values of fill-in since the preconditioner already occupied storage space equal to that of the coefficient matrix.

Algorithm 1: Modified ILU Preconditioner with Zero Fill-In

It is assumed that the data is stored in CSR format and that the column numbers for each row are sorted in increasing order. The sparse matrix is stored in the vector $\mathcal{D}$ and the column numbers in $\mathcal{PC}$. $\mathcal{SIG}(i)$ contains the total number of nonzeros till the ith row. The locations of the diagonal entries for each row are stored in the vector $\mathcal{DIAG}$. The preconditioner is stored in a complex vector, $\mathcal{LU}$.

```
for i=1 step 1 until n-1 do
begin
  lbeg=diag(i)
  lend=sig(i)
  d=lu(lbeg)
  for j=lbeg+1 step 1 until lend do
  begin
    jj=pc(j)
    ij=srch(jj,i)
    if (ij.ne.0) then
    begin
      lu(ij)=e=lu(ij)/d
      for k=lbeg+1 step 1 until lend do
      begin
        kk=pc(k)
        ik=srch(kk,i)
        if (ik.ne.0) lu(ik)=lu(ik)-e*lu(k)
      end
    end
  end
end
```

In comparison with the traditional ILU preconditioner given above, the modified ILU preconditioner eliminates the inner loop over the integer variable k. The modified algorithm basically scales the off-diagonal elements in the lower triangular portion of the matrix by the column diagonal. Since the matrix is symmetric, it retains the LDL^T form and is also positive definite if the coefficient matrix is positive definite. For our test cases, the modified ILU was especially helpful since the traditional ILU preconditioned system may not have been positive definite, as documented in [21]. The modified ILU preconditioner is also less expensive to generate and converges in about 1/3 the number of iterations taken by the point diagonal preconditioner. However, on vector architectures, the time taken by the two preconditioning strategies is approximately the same since each iteration of the ILU preconditioned system is about three times more expensive [22]. The forward and backward substitutions are very difficult to parallelize and prove to be the bottleneck since they are inherently sequential processes with vector lengths approximately half that of the sparse matrix-vector multiplication process. The triangular solver is also extremely difficult to parallelize [23].

One way to improve the parallelization of the ILU preconditioner is to use level scheduling and self-scheduling [23]. In particular, level scheduling can be used to increase parallelizability by taking advantage of matrix structure and sparsity. For solving any lower triangular system $Lx = b$, the ith unknown in the forward solution is given by

$$x_i = \frac{1}{l_{ii}}\left(b_i - \sum_{j<i} l_{ij}x_j\right) \tag{9.16}$$

If L is dense, all the components $x_1, \ldots, x_{i-1}$ need to be computed before x_i can be obtained. However, when L is sparse, most of the l_{ij}s are zero; hence, we may not need to compute all of the unknowns $x_1, \ldots, x_{i-1}$ before solving for x_i. Level scheduling is based on this simple observation. The dependencies between the unknowns can be modeled using a graph in which node i corresponds to the unknown x_i and an edge from node j to node i indicates that $l_{ij} \neq 0$ implying that the value of x_j is needed for solving x_i. The operation shown in (9.16) can then be rewritten as

$$x_i = \frac{1}{l_{ii}}\left(b_i - \sum_{j<i:l_{i,j}\neq 0} l_{ij}x_j\right) \tag{9.17}$$

Thus x_i can be solved at the kth step if all the components x_j in (9.17) have been computed in the earlier steps.

To implement the level scheduling algorithm, it is first necessary to define the *depth* of a node and the *level* of the graph. The depth of a node is defined as the maximum distance from the root [3]. Therefore, let us place an imaginary root node with links to the nodes having no predecessors so that the depth of each node will be defined from the same point. The depth of each node can now be computed with one pass through the structure of the coefficient matrix L by

$$depth(i) = \begin{Bmatrix} 1, & \text{if } l_{ij} = 0 \text{ for all } j < i \\ 1 + \max_{j<i}\{depth(j) : l_{ij} \neq 0\}, & \text{otherwise} \end{Bmatrix} \tag{9.18}$$

The level of the graph can then be defined as the set of nodes with the same depth. The level scheduling algorithm can now be implemented without physically ordering the matrix, but solving the system in increasing order of node depth and distributing the nodes at each depth among the available processors.

Algorithm 2: Forward Elimination Step with Level Scheduling

The number of levels of the graph, *nlev*, can be easily determined from the depth information. To do so, let us define two other integer vectors: *ORDER*(i) stores the ordering of the rows of L in terms of increasing node depth and *LEVEL*(i) which stores the index to the start of each level in *ORDER*(i).

```
do k=1,...,nlev
  do j=ilevel(k),...,ilevel(k+1)-1 (parallel loop)
    i=iorder(j)
    execute Equation (9.17)
  enddo
enddo
```

However, in our experience, parallelizing the ILU preconditioned system with level scheduling did not lead to significant speedup mainly due to the enormous amount of memory traffic that was generated. This observation was also noticed in [23], where the authors estimated that the parallel algorithm generated as much as ten times more traffic than the sequential code. The block ILU preconditioner considered next reduces memory traffic and is thus more effective to parallelize.

In implementing the block ILU preconditioner, one block is distributed to each processor in a multiprocessor architecture, thus achieving load balancing as well as minimizing fill-in. The modified ILU decomposition outlined earlier is then carried out on each of these individual blocks. Further, since the blocks are much larger than the block diagonal version, the preconditioner is a closer approximation to the coefficient matrix. Moreover, the triangular solver is fully parallelized since each processor solves an independent system of equations through forward and backward substitution. For example, in an equation system with 20,033 unknowns, the number of iterations was reduced by approximately half the number required by the diagonal preconditioner. However, since the work done is less than twice that for the diagonal preconditioner, only marginal savings of CPU time was achieved in this case. Also, the number of iterations required for convergence is highly sensitive to block size, as shown in Table 9.1 for the 20,033 unknown system. Table 9.1 clearly shows that a larger block size (smaller number of blocks) does not guarantee faster convergence. Nevertheless, there is an approximately 50 percent decrease in the number of iterations over the point diagonal preconditioner, regardless of block size. The optimum block size is dependent on the sparsity pattern of the matrix and can only be determined empirically. The savings in the number of iterations over the point diagonal preconditioner for 28 blocks is given in Table 9.2 for a system having 224,476 unknowns. From the table, it is clearly observed that the block ILU preconditioner is very effective in reducing the iteration count; however, the CPU time required is about 10 percent less than that required by the point diagonal preconditioner for the best case.

TABLE 9.1 No. of Iterations versus Number of Blocks for a Block ILU Preconditioned Biconjugate Gradient Solution Method

No. of Blocks	No. of Iterations
1	127
2	176
4	185
8	172
12	162
16	174
24	223
28	177

TABLE 9.2 Number of Iterations Required for Convergence of a 224,476 Unknown System Using the Point Diagonal and Block ILU Preconditioning Strategies

	Iterations		
Angle of Incidence	Point Diagonal (I)	Block ILU (II)	Ratio (II/I)
0	2943	2758	.937
10	5985	3834	.641
20	5464	3984	.729
30	6048	3651	.604
40	5770	3256	.564
50	5107	3720	.728
60	6517	4162	.639
70	5076	4108	.809
80	5305	3551	.669
90	2898	2832	.977

In a nutshell, the simplest and the most effective preconditioner was found to be the diagonal preconditioner. It is also amenable to vectorization and parallelization. The ILU preconditioner should be employed when the matrix system is ill-conditioned and vectorizability is not an issue. On most high performance PCs and scalar workstations, the ILU preconditioner performs better than the diagonal one. For parallel architectures, block ILU is clearly the method of choice. Matrix ordering strategies that minimize matrix profile can further enhance the block ILU performance since only a small fraction of resultant nonzeros will lie outside the blocks.

9.5.3 Approximate Inverse Preconditioner

When the matrix $\mathcal{A}$ is indefinite, standard preconditioning techniques may fail to achieve quick convergence. Most of the previous preconditioners (such as the diagonal and ILU) will perform poorly in this case.

The Approximate Inverse Preconditioner (*AIPC*) is typically more robust and operates by minimizing the Frobenius norm of the matrix **R**,

$$\mathbf{R} = \bar{\bar{I}} - \mathcal{A}\mathcal{M} \tag{9.19}$$

where $\mathcal{M}$ is the preconditioning matrix and $\bar{\bar{I}}$ refers to the identity matrix.

The Frobenius norm of any matrix S of dimension $(m \times n)$ is given by

$$\|S\|_F = \sqrt{\sum_{j=1}^{m} \sum_{i=1}^{n} |s_{ij}|^2} \tag{9.20}$$

Referring to [17], this minimization can be achieved in two different ways. The first is via the *Global Iteration* approach which treats the matrix $\mathcal{M}$ as an unknown sparse matrix and minimizes the objective function in (9.19). One of the well-known techniques that employs this approach is the *Global Steepest Descent Method.* Its implementation is as follows:

```
Initialize M
Repeat for i=1 till convergence
     R = I - AM
     G = A^T R
     α = ||G||_F^2 ||AG||_F^2
     M = M + αG
     Update M
EndRepeat
```

The drawback of this technique is its high CPU time and memory cost (both of order n^2). This is because the entire matrix is used during the minimization process. However, the *Column-Oriented Algorithm* minimizes the individual functions

$$f_j(\mathbf{m}) = \|\mathbf{e}_j - \mathcal{A}\mathbf{m}_j\|_F^2, \qquad j = 1, 2, \ldots, n \tag{9.21}$$

where $\mathbf{e}_j$ and $\mathbf{m}_j$ are the jth columns of the identity and preconditioner matrices, respectively. The subscript F implies the Frobenius norm defined in (9.20).

The algorithm for the Approximate Inverse Preconditioner is given in Fig. 9.11. Note that n_i in the 'Repeat2' loop of this algorithm can be set as small as

Initialize M
Repeat1 for each column $j = 1, 2, \ldots, n$
  $\mathbf{m}_j = \mathcal{M}\mathbf{e}_j$
Repeat2 for $i = 1, 2, \ldots, n_i$
  $\mathbf{r}_j = \mathbf{e}_j - \mathcal{A}\mathbf{m}_j$
  $\alpha_j = \dfrac{\mathbf{r}_j \cdot \mathcal{A}\mathbf{r}_j^*}{\mathcal{A}\mathbf{r}_j \cdot \mathcal{A}\mathbf{r}_j^*}$
  $\mathbf{m}_j = \mathbf{m}_j + \alpha_j \mathbf{r}_j$
  Apply numerical dropping to $\mathbf{m}_j$
EndRepeat2
EndRepeat1

n_i = # of MR iterations used to minimize $\mathcal{R}$. Large values lead to a better preconditioner.

Figure 9.11 Approximate inverse preconditioner using MR (minimum residual) iteration.

unity. However, higher values of n_i will lead to a better preconditioner and thus reduce the number of iterations needed for convergence. The trade-off is that the preconditioner will become more dense.

9.5.4 Flexible GMRES with Preconditioning

In all the preconditioners discussed in the previous sections, it was implicitly assumed that the preconditioning matrix $\mathcal{M}$ is fixed. However, in many cases, $\mathcal{M}$ may not be a constant operator and therefore, iterative solvers preconditioned with constant operators may not converge. *Flexible* iterative solvers are those that include variations or changes of the preconditioner from one iteration to another. One of these solvers is the Flexible GMRES (FGMRES) algorithm. A pseudocode for the FGMRES algorithm is given in Fig. 9.12 [17].

Initialization
  $\mathbf{x}$ is given
  Specify the number of spanning vectors m
Define the $(m+1)\times m$ matrix $\mathcal{H}_m = \{h_{ij}, 1 \le i \le m+1 \text{ and } 1 \le j \le m\}$
START: $\mathbf{r}_0 = \mathbf{b} - \mathcal{A}\mathbf{x}$
  $\beta = \sqrt{\mathbf{r}_0^* \cdot \mathbf{r}_0}$
  $\mathbf{v}_1 = \mathbf{r}_0/\beta$
Repeat1 for $j = 1, \ldots, m$
  $\mathbf{z}_j = \mathcal{M}^{-1}\mathbf{v}_j,\ \mathbf{w} = \mathcal{A}\mathbf{z}_j$
Repeat2 for $i = 1, \ldots, j$
  $h_{i,j} = \mathbf{w} \cdot \mathbf{v}_i^*$
  $\mathbf{w} = \mathbf{w} - h_{i,j}\mathbf{v}_i$
EndRepeat2
  $h_{j+1,j} = \sqrt{\mathbf{w}^* \cdot \mathbf{w}}$
  $\mathbf{v}_{j+1} = \mathbf{w}/h_{j+1,j}$
EndRepeat1
Compute $\mathbf{y}_m$ to minimize $\|\beta\mathbf{e}_1 - \mathcal{H}_m\mathbf{y}\|_2$
  $\mathcal{V}_m = \{\mathbf{v}_1, \mathbf{v}_2, \ldots, \mathbf{v}_m\}$
  $\mathcal{Z}_m = \{\mathbf{z}_1, \mathbf{z}_2, \ldots, \mathbf{z}_m\}$
  $\mathbf{x} = \mathbf{x} + \mathcal{Z}_m\mathbf{y}_m$
  If convergence is achieved, then stop, else go to START

Figure 9.12 Flexible GMRES algorithm.

9.6 EIGENANALYSIS

In the area of microwave circuits, accurate determination of dominant eigenvalues or propagation constant of a system is essential for a viable design. Since dominant mode operation is usually desired, determination of those eigenvalues influences the operating bandwidth of the design. In finite elements, eigenanalysis is complicated by three factors: the matrices are sparse, the method gives rise to a generalized eigenproblem, and only a few selected eigenvalues are desired. The dense eigenvalue problem is essentially solved, and EISPACK has excellent black box routines to do the job. However, the sparse eigenproblem is still an area of active research and will be the main focus of this section.

It should be pointed out that the eigenvalue problem for a general matrix is usually more difficult to solve than a set of linear equations. Since determination of eigenvalues requires finding the roots of an nth-order polynomial, it is an essentially iterative process as polynomial equations cannot be solved algebraically for fourth- and higher order polynomials. However, before the onset of iterations, the system is usually reduced to a convenient form for fast calculation of eigenvalues. The reduction does not come cheaply and usually takes longer than the actual eigenvalue calculation process. It is also in this reduction process that dense and sparse problems are treated differently.

The standard eigenproblem is defined as

$$\mathcal{A}\mathbf{x} = \lambda\mathbf{x} \tag{9.22}$$

where λ denotes the eigenvalue of the matrix $\mathcal{A}$ and x represents the corresponding eigenvector. The generalized eigenproblem found most commonly in finite element analysis is given as

$$\mathcal{A}\mathbf{x} = \lambda\mathcal{B}\mathbf{x} \tag{9.23}$$

where λ is the eigenvalue of the $\mathcal{A}$, $\mathcal{B}$ pencil. Usually, $\mathcal{A}$ is symmetric and $\mathcal{B}$ is symmetric positive definite for finite element systems. However, as shown in Chapter 5, $\mathcal{B}$ can be symmetric indefinite which increases the computational rigor significantly. The eigenproblem is usually reduced to a simpler form before starting the iterative solution. The reduction is achieved in dense matrices by means of a congruence transformation. The resulting eigenproblem amounts to solving for the eigenvalues and eigenvectors of a symmetric tridiagonal matrix for symmetric problems or an upper Hessenberg matrix when the original matrix is unsymmetric. A tridiagonal matrix has nonzeros only in its diagonal and in its first upper and lower codiagonals. An upper Hessenberg matrix has nonzeros only in its upper triangle, diagonal and first lower codiagonal. The reduction to tridiagonal or upper Hessenberg form is achieved by a series of rotations, called *Givens* rotations, or Householder reflections. For detailed information on these algorithms, the reader is referred to [24]. Givens rotations are used when the matrix is sparse *and* banded since it is possible to carry out these rotations such that no nonzeros are introduced outside the band. Once the tridiagonal or the upper Hessenberg form has been obtained, there exist powerful techniques like the QR and QL algorithms to determine all the eigenvalues of the system. Such algorithms are readily available in source code from the EISPACK library through *netlib* [4]. However, there are two restrictions in the above approach: (1) the matrix needs to be banded and (2) the entire spectrum of eigenvalues is computed. The necessity of a banded matrix is a prohibitive requirement for finite element meshes with arbitrary sparsity patterns. The computation of the entire spectrum can be avoided using the bisection method based on Sturm sequences for calculating eigenvalues within a specified interval [24].

9.6.1 Direct and Inverse Iteration

For a truly sparse eigensolver, the original matrix $\mathcal{A}$ should be used in the iterative process only as a matrix-vector product. This avoids the need for an explicit inversion and resulting loss of sparsity. One of the earliest methods to accomplish this is known as the *power method*, also known as the method of direct iteration. It is best for computing a few select eigenvalues of large sparse matrices. A bonus is the automatic availability of the desired eigenvector. The algorithm is simple and has been given in Fig. 9.13 for solving (9.22). Thus, $\lambda = \langle \boldsymbol{x}, \mathcal{A}\boldsymbol{x}\rangle / \langle \boldsymbol{x}, \boldsymbol{x}\rangle$ is an eigenvalue of $\mathcal{A}$ and $\boldsymbol{x}$ is the corresponding eigenvector. The starting vector $\boldsymbol{e}_1$ is a zero vector with unit value at the first location. The more dominant the eigenvalue, the faster the convergence to the desired eigenvalue. Interior eigenvalues can be found by using a starting vector which is orthogonal to the existing eigenvectors. For example, if $\lambda_1, \lambda_2, \ldots, \lambda_n$ have been found along with the eigenvectors $\boldsymbol{x}^1, \boldsymbol{x}^2, \ldots, \boldsymbol{x}^n$, then the starting vector should be orthogonal to each of the preceding eigenvectors

$$\boldsymbol{q} = \boldsymbol{p} - (\boldsymbol{p}^T\boldsymbol{x}^1)\boldsymbol{x}^1 - (\boldsymbol{p}^T\boldsymbol{x}^2)\boldsymbol{x}^2 - \cdots - (\boldsymbol{p}^T\boldsymbol{x}^n)\boldsymbol{x}^n$$

where $\boldsymbol{p}$ is the starting vector. Thus dominant eigenvalues with corresponding eigenvectors can be obtained without modifying the nonzero structure of the matrix $\mathcal{A}$. However, the outlook is not so rosy in practice: round-off errors rapidly lead to loss of orthogonality, and re-orthogonalization is necessary from time to time. The criterion for re-orthogonalization is usually when $|\lambda_1/h|$ is larger than a prespecified tolerance, where h is the current estimate of the eigenvalue and λ_1 is the dominant eigenmode. As we will see later, loss of orthogonality is the bane of sparse iterative eigensolvers and cannot be avoided in finite precision arithmetic, leading to storage and time constraints if interior eigenvalues are required in addition to the extremal ones.

Initialization
    Choose any column vector, say $\boldsymbol{e}_1$
    $\boldsymbol{x}_0 = \mathcal{A}\boldsymbol{e}_1/|\mathcal{A}\boldsymbol{e}_1|$
Repeat until
    $\mathcal{A}\boldsymbol{x}_k/|\mathcal{A}\boldsymbol{x}_k| = \boldsymbol{x}_{k+1} \approx \boldsymbol{x}_k$

Figure 9.13 Power method.

In some cases, eigenvalues immediately adjacent to a desired cutoff value are sought. The determination of the dominant eigenmode of a two-dimensional microstrip line is a case in point. The dominant mode is usually the one closest to the maximum wavenumber supported by the dielectric medium. However, predicting an eigenvalue close to the desired one is often not an easy task. Algorithms like the determinant search method use preliminary guesses and the properties of Sturm sequences to predict eigenvalues near the desired ones.

Once an educated guess can be made regarding the desired eigenvalue, the eigenproblem becomes somewhat easier to solve. Shifts of origin can be carried out to improve the performance of the algorithm. This works on the principle that if λ_i, $i = 1, \ldots, n$, denotes the spectrum of $\mathcal{A}$, then $\mathcal{A} - \sigma \boldsymbol{I}$ has eigenvalues $\lambda_i - \sigma$ and the same eigenvectors as $\mathcal{A}$. In this way, interior eigenvalues can be calculated once the neighborhood of the eigenvalue can be determined. The method converges linearly with the factor λ_p/λ_q, where λ_p is the dominant eigenvalue and λ_q is the next dominant one. Since convergence rate depends heavily on the separation between eigenvalues, closely spaced eigenvalues can cause the algorithm to stagnate.

The explanation as to why direct iterations converge can be found in [5]. Basically, if the eigenvectors are taken to form an orthonormal basis in n-dimensional space, then pre-multiplication of an arbitrary vector u with $\mathcal{A}$ produces a tilt toward the first dominant eigenvector by the factor λ_p/λ_q, where λ_p, λ_q are successive eigenvalues. Successive premultiplication followed by normalization will converge to the dominant eigenvector. Once an eigenvector is found, we choose another arbitrary starting vector for computing the next dominant eigenvalue. The starting vector must be orthogonal to the dominant eigenvector. This process is called *deflation* by which the iteration vector is restricted to the invariant subspace which is the complement of the known eigenvectors.

The method of inverse iteration is used for solving the eigenvalues of the inverse problem

$$\mathcal{A}^{-1}\boldsymbol{x} = \frac{1}{\lambda}\boldsymbol{x} \tag{9.24}$$

where $\mathcal{A}$ is symmetric and nonsingular. The method of inverse iteration with shifts of origin can be efficiently applied to the computation of eigenvectors when a set of eigenvalues are known from methods such as bisection.

Both methods, direct and inverse iteration, can be used for solving the generalized eigenproblem. Thus, the power method with shift σ can be extended as given in Fig. 9.14. The shift factor is usually taken close to the desired eigenvalue. The method requires the solution of a linear system of equations at each step. It can be accelerated by factorizing the $\mathcal{B}$ matrix using sparse techniques at the beginning of the iterative procedure. In the inverse method with shifts of origin, the system to be solved at every iteration is

$$\mathcal{B}\mathbf{y}_{\mathbf{k+1}} = (\mathcal{A} - \sigma\mathcal{B})\mathbf{x}_{\mathbf{k}}$$

Initialization

Choose any column vector, say $\boldsymbol{e}_1$

$\boldsymbol{x}_0 = \boldsymbol{e}_1$

$k = 0$

Repeat

Solve $\mathcal{B}\boldsymbol{y}_{k+1} = (\mathcal{A} - \sigma\mathcal{B})\boldsymbol{x}_k$

$\boldsymbol{x}_{k+1} = \boldsymbol{y}_{k+1}/|\boldsymbol{y}_{k+1}|$

If $\boldsymbol{x}_{k+1} \approx \boldsymbol{x}_k$ then stop

Figure 9.14 Shifted power method for the generalized eigenproblem.

9.6.2 Simultaneous Iteration

The method of *simultaneous iteration* (also known as *subspace iteration*) is a generalization of the power method described in the previous section. This method as well as the subsequent Lanczos algorithm rests on the concept of Rayleigh matrices, Ritz values, and Ritz vectors. Let us consider $\mathcal{Q} = (\boldsymbol{q}_1, \boldsymbol{q}_2, \ldots, \boldsymbol{q}_m)$ as an orthonormal basis of a subspace $\boldsymbol{\mathcal{S}}$, invariant under $\boldsymbol{\mathcal{A}}$, arranged in the form of an $n \times m$ matrix $\mathcal{Q}$. Then $\boldsymbol{C} = \mathcal{Q}^T \boldsymbol{\mathcal{A}} \mathcal{Q}$ is a square symmetric *Rayleigh matrix* of order m. It can be shown that the eigenvalues of $\boldsymbol{C}$ are the same as that of $\boldsymbol{\mathcal{A}}$ and the eigenvectors of $\boldsymbol{C}$ equals $\mathcal{Q}\boldsymbol{y}$, where $\boldsymbol{y}$ is the eigenvector of $\boldsymbol{\mathcal{A}}$. The advantage is that a much smaller matrix $\boldsymbol{C}$ of order $m \ll n$ yields the desired extremal eigenvalues of $\boldsymbol{\mathcal{A}}$. The eigenvalues of $\boldsymbol{C}$ are the best approximations to the eigenvalues of $\boldsymbol{\mathcal{A}}$ and are known as *Ritz values* and the corresponding eigenvectors are known as *Ritz vectors*. Thus, if $\boldsymbol{r}_i = \boldsymbol{\mathcal{A}}\boldsymbol{x}_i - \mu_i \boldsymbol{x}_i$ is the residual vector for the Ritz pair $(\mu_i, \boldsymbol{x}_i)$, then there is an eigenvalue of $\boldsymbol{\mathcal{A}}$ in the interval $[\mu_i - \|\boldsymbol{r}_i\|, \mu_i + \|\boldsymbol{r}_i\|]$.

In the method of simultaneous iteration, the orthonormal matrix $\mathcal{Q}$ is made up of Ritz vectors at each step. The subspace $\boldsymbol{\mathcal{S}} = span(\boldsymbol{Q})$ is not initially invariant under $\mathcal{A}$ but becomes nearly so as the iteration progresses. The algorithm by Reinsch [25] is given in Fig. 9.15. (For proof that the columns of $\boldsymbol{U}_m$ are Ritz vectors, see [5].) Simultaneous iteration in conjunction with shifts can be a powerful tool for solving large eigenvalue problems with minimal usage of computational resources. Thus the k leftmost or k rightmost eigenvalues can be found along with their corresponding eigenvectors, though the speed of convergence depends on the size k of the subspace.

Initialization
- Choose any orthonormal basis $\boldsymbol{U}_0$ of dimension $n \times k$, where $k \ll n$.
- $m = 1$

Repeat for $m = 1, 2, \ldots$
- $\boldsymbol{C} = \mathcal{A}\boldsymbol{U}_{m-1}$
- Orthonormalize $\boldsymbol{C}$ such that
  $\boldsymbol{C} = \boldsymbol{Q}\boldsymbol{R}$, with $\boldsymbol{Q}$ unitary and $\boldsymbol{R}$ upper triangular
- Find eigenvalues of $\boldsymbol{R}\boldsymbol{R}^T$
- Spectral decomposition of $\boldsymbol{R}\boldsymbol{R}^T = \boldsymbol{P}\boldsymbol{D}\boldsymbol{P}^T$,
  where $\boldsymbol{D}$ is diagonal with $D_{ii} = \mu_i^2$
  and $\boldsymbol{P}$ is unitary
- $\boldsymbol{U}_m = \boldsymbol{Q}\boldsymbol{P}$
- If residual vector $<$ tolerance, convergence achieved

EndRepeat

Figure 9.15 Simultaneous iteration for the standard eigenproblem.

In finite elements, where the generalized eigenproblem is to be solved in most cases, simultaneous iteration provides a powerful tool for extracting the extremal eigenvalues. If an educated guess can be made regarding the eigenvalue, the inverse iteration with shift is used for finding the eigenvalues of large, sparse systems. As shown earlier, the shifted generalized eigenproblem is defined as

$$(\mathcal{A} - \sigma\mathcal{B})\,X = \Lambda\mathcal{B}X \tag{9.25}$$

where σ is the shift and Λ is the vector of eigenvalues. The algorithm is given in Fig. 9.16 and is taken from [5]. This algorithm works even if the $\mathcal{B}$ matrix in the $\mathcal{A}$, B pencil is symmetric indefinite. Note that the sparse system $\mathcal{A} - \sigma\mathcal{B}$ needs to be solved at each step for multiple right-hand sides. It is convenient to do this using one of the sparse direct solving strategies outlined in the earlier section. The solution of the generalized order of the H_A, H_B pencil, is much smaller than n, the order of the $\mathcal{A}$, B pencil.

Initialization
  Choose any $\boldsymbol{U}_0$ of dimension $n \times k$, where $k \ll n$. $\boldsymbol{\Lambda}_0$ is also available.
  $m = 1$
Repeat for $m = 1, 2, \ldots$
  $\boldsymbol{R} = \mathcal{B}\boldsymbol{U}_{m-1}\Lambda_{m-1}$
  Solve $(\mathcal{A} - \sigma\mathcal{B})\,\boldsymbol{C} = \boldsymbol{R}$ for $\boldsymbol{C}$
  Construct two Rayleigh matrices:
$$\boldsymbol{H}_A = \boldsymbol{C}^T\boldsymbol{R}$$
$$\boldsymbol{H}_B = \boldsymbol{C}^T\mathcal{B}\boldsymbol{C}$$
  Solve the generalized eigenproblem:
$$(\boldsymbol{H}_A - \sigma\boldsymbol{H}_B)\,\boldsymbol{P} = \Lambda_m\boldsymbol{H}_B\boldsymbol{P},$$
  where $\boldsymbol{P}$ is $\boldsymbol{H}_B$-orthogonal
  $\boldsymbol{U}_m = \boldsymbol{C}\boldsymbol{P}$
  If residual vector < tolerance, convergence achieved
EndRepeat

Figure 9.16 Simultaneous iteration with shift for the generalized eigenproblem.

One of the problems with this method is that k, the size of the desired subspace, is not known a priori. However, k can be modified within the iteration process by adding new columns to the basis or by deflating the basis from the converged eigenvectors.

9.6.3 Lanczos Algorithm

The Lanczos algorithm results when the initial guess for the orthonormal basis is drawn from the Krylov subspace. Therefore, if $\boldsymbol{b}$ is an arbitrary nonzero vector, a Krylov subspace is defined as

$$\mathcal{K}^m = span\,(\boldsymbol{b}, \mathcal{A}\boldsymbol{b}, \ldots, \mathcal{A}^{m-1}\boldsymbol{b}) \tag{9.26}$$

It can be reasoned from the convergence of the power method that $\mathcal{K}^m$ will be nearly invariant under $\mathcal{A}$, when m is sufficiently large. Choosing the vectors from a Krylov subspace lends the Lanczos algorithm some remarkable properties:

- The Rayleigh matrix $\boldsymbol{C}$ is tridiagonal which simplifies the computation of Ritz pairs.

- The computation of the orthonormal Lanczos basis can be done through a three-term recurrence relation.
- Convergence to the eigenvalues is very rapid.

However, the chief flaw in the algorithm lies in the loss of orthogonality of the Lanczos bases due to round-off errors. Selective, and sometimes, complete re-orthogonalization is needed to correct this problem. However, this method can be a bit expensive when interior eigenvalues are required. For extremal eigenvalues, the Lanczos method is one of the most efficient for large sparse matrices as it has far superior convergence properties than the power method. Lanczos converges approximately as $(2k)^{2(n-1)}$ whereas the power method converges as $k^{2(n-1)}$, where n is the iteration number and $k = (\lambda_1/\lambda_2) < 1$. The convergence rate is especially critical if the separation between the two adjacent eigenvalues is small. The basic algorithm without re-orthogonalization is given in Fig. 9.17. For a definitive account of the Lanczos algorithm and its features, the reader is referred to [26]. The source code is also available via *netlib* [4].

Initialization
  Choose any $\boldsymbol{b}$ of dimension n.
  Set $\boldsymbol{q}_0 = 0$; $\boldsymbol{r}_0 = \boldsymbol{b}$; $\beta_0 = \|\boldsymbol{b}\|_2$
  $m = 1$
Repeat for $m = 1, 2, \ldots$
  $\boldsymbol{q}_m = \boldsymbol{r}_{m-1}/\beta_{m-1}$
  $\boldsymbol{r}_m = \mathcal{A}\boldsymbol{q}_m - \beta_{m-1}\boldsymbol{q}_{m-1}$
  $\alpha_m = \boldsymbol{q}_m \cdot \boldsymbol{r}_m$
  $\boldsymbol{r}_m = \boldsymbol{r}_m - \alpha_m\boldsymbol{q}_m$; $\beta_m = \|\boldsymbol{r}_m\|$
  Solve the eigenproblem:
    $\boldsymbol{T}_m\boldsymbol{h}_i = \mu_i\boldsymbol{h}_i$, $\boldsymbol{T}_m$ is tridiagonal
  If $\beta_m|h_{im}| <$ tolerance, convergence achieved
EndRepeat

Figure 9.17 Lanczos algorithm for the standard eigenproblem.

The generalized eigenproblem gets considerably tougher to solve if the matrix $\mathcal{B}$ in the $\mathcal{A}$, $\mathcal{B}$ pencil is not positive definite or unsymmetric. If $\mathcal{A}$ is symmetric and $\mathcal{B}$ is symmetric indefinite, its Cholesky factorization does not exist and consequently, the product $\mathcal{B}^{-1}\mathcal{A}$ will usually be unsymmetric. The QZ algorithm is the method of choice for solving full unsymmetric generalized eigenproblems. The Lanczos tri-diagonalization is particularly effective for solving generalized eigenproblems with symmetric, indefinite matrices since the inversion and hence pre-multiplication is not carried out explicitly. A procedure for solving the generalized symmetric eigenproblem where $\mathcal{A}$, $\mathcal{B}$ are both indefinite is given in [27].

In conclusion, it should be remarked that the speed of convergence of the sparse eigensolution techniques is quite amazing compared to full matrix methods. This is only to be expected, but a 500-fold speedup for the extremal eigenvectors of a generalized eigenvalue problem of order 800 caught the author by surprise. When this fact is coupled with the meager storage requirements, usage of sparse methods for eigenvalue problems is clearly a win–win situation. The trade-off comes in the increased programming labor and the inability to use public domain black-box routines.

9.7 PARALLELIZATION

As mentioned earlier, there are two problems which limit the vectorizability of a sparse matrix code: short vector lengths and indirect addressing. There is not much to be done about the second problem since sparse matrices must have indirect addressing to exploit the $O(N)$ storage feature. However, the first problem can be removed by storing the matrix in an optimizable machine-dependent format. The jagged diagonal method of matrix storage is a case in point. The still slower execution speeds of the matrix-vector multiply compared with the vector update is due to the indirect addressing in the inner loop which causes memory contention. On a distributed memory architecture, the second problem can also be partly removed by keeping local copies of the desired vector in each processor. The subsequent gather and scatter operations of the updated vectors then consume the majority of the processor communication time.

Below we discuss the implementation of a finite element code on two different types of massively parallel architectures:[2] the KSR1 and the Intel iPSC/860. The KSR1 is a parallel machine which implements a shared virtual memory, although the memory is physically distributed for the sake of scalability. The Intel iPSC/860, on the other hand, is a distributed memory, Multiple Instruction, Multiple Data (MIMD) system in which the nodes process information independently of one another and communicate by passing messages to each other. The conversion of sequential or vectorized code to parallel code involves two primary tasks:

- *parallelization of DO loops*
  Parallelism is introduced by allowing each processor to execute a portion of the DO loop.
- *distribution of arrays among the processor set*
  Since each processor only has a limited amount of memory, each array is divided into smaller units that reside on each node. This also allows array accesses from each processor to be serviced by different nodes, thus reducing contention for resources on any single node.

On a cache-only memory machine such as the KSR1, only the first step is necessary since the hardware cache system automatically takes care of data distribution among the processors. This makes porting codes to the KSR1 quite easy. However, the increased control of data distribution and communication on the iPSC/860 can translate into improved performance for some applications. The data distribution on message-passing architectures is controlled by a preprocessing step in which the computation domain is partitioned *efficiently* among the processors. This efficient partitioning step is more commonly referred to as *Domain Decomposition* which aims to minimize interprocessor communication and to achieve load balancing. A detailed discussion of domain decomposition is outside the scope of this book; for further information, the reader is referred to an excellent review paper by Hamandi et al. [28].

[2]The KSR1 is no longer available, and the Intel iPSC/860 is being phased out. Nevertheless, these parallel platforms represent examples of distributed memory architectures.

1. KSR1 Port

The most important aspect of parallelization involves optimizing the iterative solver. For the sake of simplicity, we consider the complex symmetric BiG solver.

Figure 9.18 shows the symmetric BiCG algorithm; the unsymmetric method given in Fig. 9.4 contains an additional matrix-vector multiply and a few additional vector updates. For a system of equations containing N unknowns, all vectors in the algorithm are of size N and the sparse matrix $\mathcal{A}$ is of order N. Table 9.3 shows the operation counts per iteration for each type of vector operation, where *nze* denotes the number of nonzero elements in the sparse matrix. In the finite element code, each vector operation is implemented as a loop and parallelization is achieved by tiling these loops. For P processors, the vectors are divided into P sections of N/P consecutive elements. Each processor is assigned the same section of each vector. This partitioning attempts to reduce communication while balancing load. To guarantee correctness, synchronization points are added after lines 2, 7, and 9. Lines 2 and 7 require synchronization to guarantee that the dot products are computed correctly.

Initialization:

$\boldsymbol{x}$ given

$\boldsymbol{r} = \boldsymbol{b} - \mathcal{A}\boldsymbol{x};\ \boldsymbol{p} = \boldsymbol{r};\ tmp = \boldsymbol{r} \cdot \boldsymbol{r}$

Repeat until ($resd \leq tol$)

$$\left.\begin{array}{ll} \boldsymbol{q} = \mathcal{A}\boldsymbol{p} & (1) \\ \alpha = tmp/(\boldsymbol{q} \cdot \boldsymbol{p}) & (2) \end{array}\right\} \textit{Step } 1$$

$$\left.\begin{array}{ll} \boldsymbol{x} = \boldsymbol{x} + \alpha\boldsymbol{p} & (3) \\ \boldsymbol{r} = \boldsymbol{r} - \alpha\boldsymbol{q} & (4) \\ \boldsymbol{q} = \boldsymbol{C}^{-1} * \boldsymbol{r} & (5) \\ resd = \sqrt{|\boldsymbol{r} \cdot \boldsymbol{r}^*|} & (6) \\ \beta = (\boldsymbol{r} \cdot \boldsymbol{q})/tmp & (7) \end{array}\right\} \textit{Step } 2$$

$$\left.\begin{array}{ll} tmp = \beta \times tmp & (8) \\ \boldsymbol{p} = \boldsymbol{q} + \beta\boldsymbol{p} & (9) \end{array}\right\} \textit{Step } 3$$

EndRepeat

$\mathcal{A}$ is a sparse complex symmetric matrix.
$\boldsymbol{C}$ is the preconditioning matrix.
$\boldsymbol{q}, \boldsymbol{p}, \boldsymbol{x}, \boldsymbol{r}$ are complex vectors
α, β, *tmp* are complex scalars; *resd*, *tol* are real scalars.

Figure 9.18 Symmetric biconjugate gradient method with preconditioning.

TABLE 9.3 Floating Point Operations per Iteration

	Complex		Real	
Operation	*	+	*	+
Matrix multiply	*nze*	*nze* − N	4*nze*	4*nze* − 2N
Vector updates	4N	3N	16N	12N
Dot products	3N	3N	12N	12N

Note that the dot products in lines 6 and 7 require only one synchronization. The line 9 synchronization guarantees that $\boldsymbol{p}$ is completely updated before the matrix multiply for the next iteration begins.

In the sparse matrix vector multiplication, each processor computes a block of the result vector by multiplying the corresponding block of rows of the sparse matrix with the operand vector. Since the operand vector is distributed among the processors, data communication is required. The communication pattern is determined by the sparsity structure of the matrix, which is derived from the unstructured mesh. Therefore, the communication pattern is unstructured and irregular. Vector updates and dot products are easily parallelized using the same block distribution as in the sparse matrix vector multiply.

Sparse computations are known to be hard to implement efficiently on a distributed memory machine, mainly because of the unstructured and irregular communication pattern. However, the previous scheme was easily and efficiently implemented on the KSR1 Massively Parallel Processing machine thanks to the global address space [29]. Table 9.4 shows the execution time of one iteration (in seconds) and the speedup for different numbers of processors and for two problem sizes.

TABLE 9.4 Execution Time and Speedup for the Iterative Solver

	$N = 20{,}033$		$N = 224{,}476$	
Procs	Execution Time (secs per iter)	Speedup	Execution Time (secs per iter)	Speedup
1[a]	.515	1	10.8	1
8	.071	7.3	1.4	7.7
16	.040	12.9	.671	16.1
29	.027	19.1	.304	35.6
60[b]			.149	76.2

[a]For 1, 8, and 16 processors, only the first 100 iterations were run.
[b]Code run on a 64-node KSR at Cornell University.

For both problems, the performance scales surprisingly well up to a large number of processors. For the 20,033-unknown problem, the speedup for the parallelized sparse solver varies from 1 to 38 as the number of processors is increased from 1 to 56 (Fig. 9.19). The overall performance of the solver on 28 processors is more than three times that of a single processor on the Cray-YMP. The large problem (224,476 unknowns) exhibits superlinear speedup which can be attributed to a memory effect. As a matter of fact, the large data set does not entirely fit in the local cache of a single node in the KSR which results in a large number of page faults. However, as the number of processors increases, the large data set is more evenly distributed over the different processors' memories.

The global matrix assembly is the second largest computation in terms of execution time. Typically, the elemental matrices are computed for each element in the 3D mesh and assembled in a global sparse matrix. A natural way of parallelizing the global matrix assembly is to distribute the elements over the processors, have

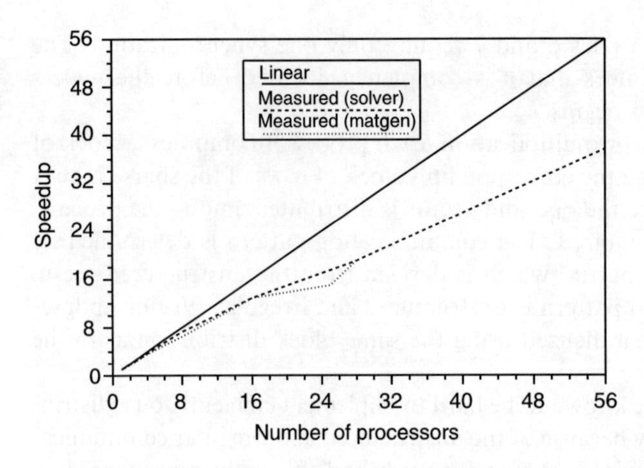

Figure 9.19 Speedup curve for the linear equation solver on the KSR1.

each processor compute the elemental matrix of the elements it owns, and update the global sparse matrix. Since the global sparse matrix is shared by all processors, the update needs to be done automatically. On the KSR1 this can be done by using the hardware lock mechanism.

The performance for the matrix assembly is given in Table 9.5 and also in Fig. 9.19.

TABLE 9.5 Execution Time and Speedup for the Matrix Generation and Assembly (20,033 unknowns)

Procs	Execution time in seconds	Speedup
1	24.355	1
2	13.376	1.8
4	6.811	3.6
8	3.744	6.5
16	1.89	12.9
25	1.625	15.0
28	1.276	19.1

9.7.1 Analysis of Communication

In the main loop (Fig. 9.18), significant communication between processors takes place only during the sparse matrix vector multiply (line 1) and the vector update of p (line 9). The rest of the vector operations incurs little or no communication at all. The distribution of the nonzero entries in the matrix affects the amount and nature of communication. In this section, an analysis is presented of the communication pattern incurred by the sparse matrix vector multiplication as derived from analysis of the sparsity structure of the matrix.

Line 1. In the matrix-vector multiply, each processor computes an N/P-sized subsection of the product $\boldsymbol{q}$. The processor needs the elements of $\boldsymbol{p}$ that correspond to the nonzero elements found in the N/P rows of $\boldsymbol{A}$ that are aligned with its subsection. Because the matrix $\boldsymbol{A}$ remains constant throughout the program, the set of elements of $\boldsymbol{p}$ that a given processor needs is the same for all iterations in the loop. However, since $\boldsymbol{p}$ is updated at the end of each iteration, all copies of its element set are invalidated in each processor's local cache except for the ones that the processor itself updates. As a result, in each iteration, processors must obtain updated copies of the required elements of $\boldsymbol{p}$ that they do not own.

These elements can be updated by a read miss to the corresponding subpage, by an automatic update, or by an explicit prefetch or poststore instruction. Figure 9.20 lists the number of subpages that each of the 28 processors needs to acquire from other processors. Automatic update of an invalid copy of a subpage becomes more likely as the number of processors sharing this subpage grows. The number of processors that need a given subpage (excluding the processor that updates the subpage) is referred to as the *degree of sharing* of that subpage. Figure 9.21 shows the degree of sharing histogram for the example problem. Since the only subpage misses occurring in Step 1 of the sparse solver are coherence misses due to the vector $\boldsymbol{p}$, the use of the poststore instruction to broadcast the updated sections of the vector $\boldsymbol{p}$ from Step 3 should eliminate the subpage misses in Step 1. However, the overhead

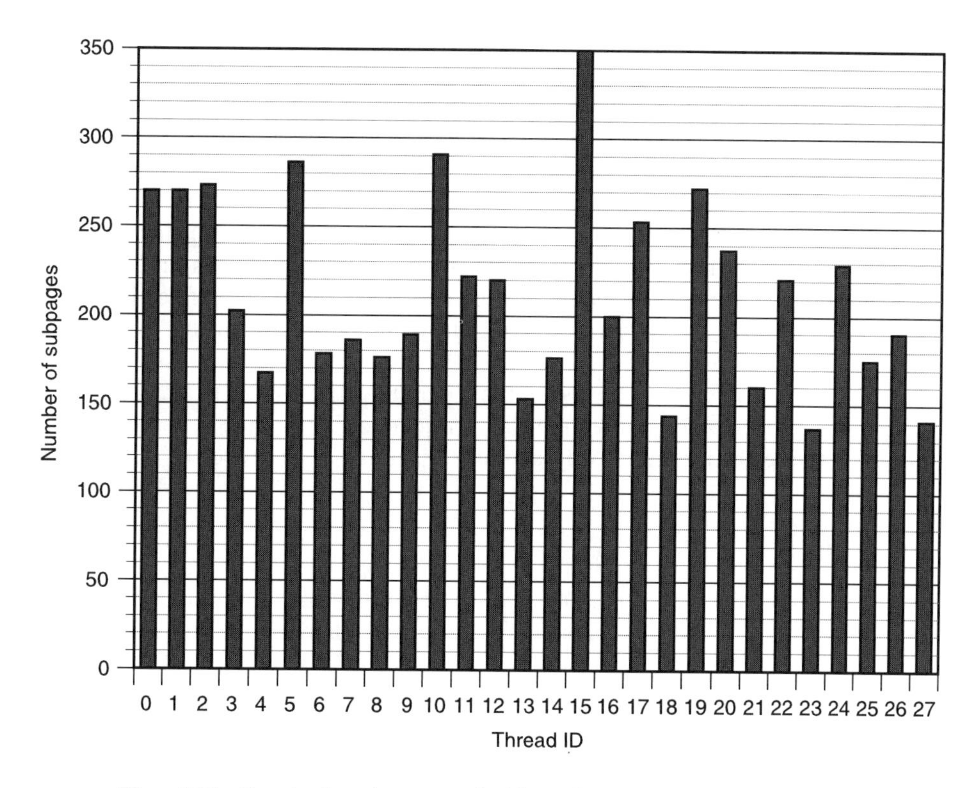

Figure 9.20 Counts of **p** subpages required by each processor sparse matrix-vector multiply (total copies = 5968).

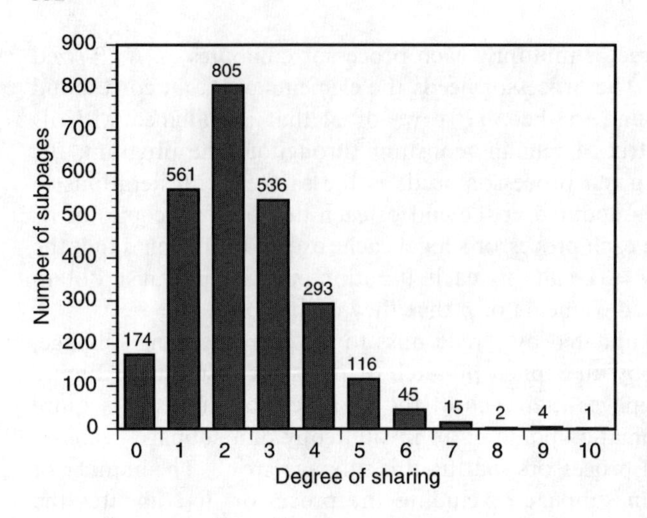

Figure 9.21 Degree of sharing histogram of **p** subpages during sparse matrix-vector multiply (28 processors).

of executing the poststore instruction in Step 3 offsets the reduction in execution time of Step 1. On a poststore, the processor typically stalls for 32 cycles while the local cache is busy for 48 cycles. As a result, the net reduction in execution time is only 3 percent.

Line 9. Before proceeding with the updates of the N/P elements of $\boldsymbol{p}$ for which it is responsible, each processor must acquire exclusive ownership for those elements. Because a cache line holds eight consecutive elements, each processor will generate $N/8P$ requests for ownership (assuming all subpages are shared). In order to hide access latencies, the request for ownership can be issued in the form of a prefetch instruction after step 1. This could lead to an eightfold decrease in the number of subpage misses. However, as with the poststore instruction, the benefit of prefetching is offset by the overhead of processing the prefetch instructions in Step 2. This is because the processor stalls for at least two cycles on a prefetch and the local cache cannot satisfy any processor request until the prefetch is put on the ring. The overall execution time is reduced by only 4 percent in this case.

Lines 2, 6, 7. The rest of the communication is due to the three dot products. Each processor computes the dot product for the vector subsection that it owns. These are then gathered and summed up on a single processor.

2. Intel iPSC/860 Port

The parallelization of the DO loops is one of the main tasks since the majority of the computer time is spent on solving the linear system of equations. The basic strategy for parallelizing the DO loops on the iPSC/860 is similar to the KSR1 with each processor executing a portion of the DO loop. This scheme works fine as long as there are no dependencies in the body of the loop, as is the case for the vector updates and the sparse matrix-vector multiply of the linear solver. However, the main loop in the matrix generation/assembly phase contains a dependency between loop iterations. As on the KSR1, this problem is solved by using a mechanism in which each processor locks a row of the matrix while performing an update. Since the locking of each row is maintained by the processor whose memory holds the particular row, processors lock and unlock rows by sending messages to the appropriate row owner.

Even though the parallelization of loops enables programs to run faster on multiprocessors, the distribution of arrays must be done for the code to run at all. Arrays are distributed in the code by partitioning data along one array dimension among the processors. Thus for a 1000-element array, processor 1 holds the first 100 elements, processor 2 the next 100 and so on. The straightforward method for accessing this distributed array involves the translation of array references into subroutine calls. Thus an expression $x = a(i)$ is translated into the call *call fetcha*(i, x). The subroutine *fetcha* then sends a message to the processor that holds element $a(i)$, which in turn sends a reply message with the value of $a(i)$. Although this scheme requires the implementation of a new subroutine for each distributed array and the replacement of each array access with a subroutine call, the process is easy and mechanical.

The scheme mentioned above does not, however, result in good performance. The primary reason for this is that the overhead for sending a message is much higher than that of sending a single byte. The cost for sending ten or even 100 bytes is usually not much higher than that of sending 1 byte. Thus, messages need to be 'bundled' for fast and efficient operation. However, the simple strategy mentioned above is in direct contrast to message bundling. One way of overcoming this conflict is to implement the simple scheme for parts of the code that do not take up a significant portion of computation time like the matrix generation/assembly phase and a better scheme for accessing the distributed arrays in the equation solver phase.

The primary operation in the solver that generates communication is the sparse matrix-vector product. Since the matrix-vector product involves performing a dot product of each row with the distributed vector, each processor must obtain the values for the entire vector from the other processors. The dot product operation must be carried out in several phases as each processor may not be able to hold the entire vector in memory. Thus, each processor P begins the matrix-vector multiply by sending its portion of the vector to other processors, then performs the following tasks for every other processor P':

- Reads the portion of the vector owned by P'.
- Updates the partial dot product for each row by adding the product of the appropriate matrix element with the elements of the partial vector.

After performing the above operations for all the processors, the dot product is complete. Unfortunately, each phase requires a pass over all the sparse matrix rows owned by the processor. For better parallel performance, each row of the matrix must be sorted to allow the phases to pass over the rows in order. It was found that the problem scaled reasonably well for a small number of processors. However, as the number of processors increased, much of the time was spent on communication and book-keeping than on true computation.

REFERENCES

[1] D. R. Kincaid and T. C. Oppe. ITPACK on supercomputers. *Numerical Methods: Lecture Notes in Mathematics*, 1005:151–161, 1982.

[2] G. V. Paolini and G. Radicati di Brozolo. Data structures to vectorize CG algorithms for general sparsity patterns. *BIT*, 29:703–718, 1989.

[3] E. Anderson and Y. Saad. Solving sparse triangular linear systems on parallel computers. *International Journal of High Speed Computing*, 1(1):73–95, 1989.

[4] netlib. Available through the World Wide Web at http://www.netlib.org.

[5] S. Pissanetzky. *Sparse Matrix Techniques*. Academic Press, New York, 1984.

[6] Z. Zlatev. On some pivotal strategies in Gaussian elimination by sparse technique. *SIAM J. Numer. Anal.*, 17:18–30, 1980.

[7] E. Cuthill and J. McKee. Reducing the bandwidth of sparse symmetric matrices. In *Proceedings of the 24th National Conference of the ACM*. Brandon Systems Press, NJ, 1969.

[8] I. P. King. An automatic reordering scheme for simultaneous equations derived from network systems. *Int. J. Numer. Meth. Eng.*, 2:523–533, 1970.

[9] A. George and J. W. H. Liu. *Computer Solution of Large Sparse Positive Definite Systems*. Prentice-Hall, Englewood Cliffs, NJ, 1981.

[10] M. R. Hestenes and E. Stiefel. Methods of conjugate gradients for solving linear systems. *J. Res. Natl. Bur. Stand.*, 49:409–436, 1952.

[11] R. Freund. Conjugate-gradient type methods for linear systems with complex symmetric coefficient matrices. *SIAM J. Sci. Stat. Comput.*, 13:425–448, 1992.

[12] O. Axelsson. *Solution of Linear Systems of Equations: Iterative Methods*. Number 572 in Lecture Notes in Mathematics. Springer-Verlag, Germany, 1977.

[13] H. A. Van der Vorst. *Preconditioning by Incomplete Decompositions*. PhD thesis, University of Utrecht, Holland, 1982. ACCU-series 32.

[14] P. Sonneveld. CGS, a fast solver for nonsymmetric linear systems. *SIAM J. Sci. Stat. Comput.*, 10:35–52, 1989.

[15] H. A. Van der Vorst. Bi-CGSTAB: A fast and smoothly converging variant of Bi-CG for the solution of nonsymmetric linear systems. *SIAM J. Sci. Stat. Comput.*, 13:631–644, 1992.

[16] R. W. Freund, G. H. Golub, and N. M. Nachtigal. Iterative solution of linear systems. *Acta Numerica*, pp. 57–100, 1991.

[17] Y. Saad. *Iterative Methods for Sparse Linear Systems*, PWS Pub. Co., Boston, 1996.

[18] R. Barret et al. *Templates for the Solution of Linear Systems: Building Blocks for Iterative Solvers*. SIAM, 1994.

[19] H. P. Langtangen. Conjugate gradient methods and ILU preconditioning of nonsymmetric matrix systems with arbitrary sparsity patterns. *Int. J. Numer. Meth. Fluids*, 9:213–233, 1989.

[20] Y. Saad. ILUT: A dual threshold incomplete LU factorization. Technical Report UMSI 92/38, University of Minnesota Supercomputer Institute, Minneapolis, MN, March 1992.

[21] J. R. Lovell. Hierarchical basis functions for 3D finite element methods. *ACES Digest*, pp. 657–663, 1993.

[22] A. Chatterjee, J. L. Volakis, and D. Windheiser. Parallel computation of 3D electromagnetic scattering using finite elements. *Int. J. Num. Modeling*, 7:329–342, 1994.

[23] E. Rothberg and A. Gupta. Parallel ICCG on a hierarchical memory multiprocessor—Addressing the triangular solve bottleneck. *Parallel Computing*, 18:719–741, 1992.

[24] G. H. Golub and C. F. Van Loan. *Matrix Computations*. Johns Hopkins Univ. Press, Baltimore, MD, 1983.

[25] C. H. Reinsch. A stable rational QR algorithm for the computation of eigenvalues of a hermitian, tridiagonal matrix. *Numer. Math.*, 25:591–597, 1971.

[26] J. Cullum and R. A. Willoughby. *Lanczos algorithms for large symmetric eigenvalue computations*. Progress in Scientific Computing Series. Birkhauser Boston Inc., 1983.

[27] J. F. Lee, D. K. Sun, and Z. J. Cendes. Full-wave analysis of dielectric waveguides using tangential vector finite elements. *IEEE Trans. Microwave Theory Tech.*, MTT-39(8):1262–1271, August 1991.

[28] Hamandi, Lee, and Ozguner. Survey of domain decomposition methods. In *ACES Conference Digest*, March 1995.

[29] D. Windheiser, E. Boyd, E. Hao, S. G. Abraham, and E. S. Davidson. KSR1 multiprocessor: Analysis of latency hiding techiques in a sparse solver. In *Proc. of the 7th International Parallel Processing Symposium*, Newport Beach, April 1993.

Index

About the Authors

John L. Volakis is a professor in the Department of Electrical Engineering and Computer Science at the University of Michigan. He received his Ph.D. from Ohio State University in 1982 and spent two years at Rockwell International working on the B-1B program before going to Michigan. He has 20 years of experience in numerical and analytical methods and pioneered the application and development of hybrid finite element methods to high-frequency electromagnetics. His publications include over 140 refereed journal articles, 12 book chapters on analytical and numerical methods, and numerous conference articles. He also coauthored the book, *Approximate Boundary Conditions in Electromagnetics* (Institute of Electrical Engineers, London, 1995). Dr. Volakis is a Fellow of the IEEE and has advised over 20 Ph.D. students. He has served as an associate editor to *IEEE Transactions on Antennas and Propagation* and *Radio Science*. Currently, he is an associate editor for the *IEEE Antennas and Propagation Society Magazine* and the *Journal of Electromagnetic Waves and Applications*.

Arindam Chatterjee obtained his Ph.D. from the University of Michigan in 1994. From 1989 to 1994, he served as a research assistant and later as a Research Fellow in the Radiation Laboratory, University of Michigan, Ann Arbor. His work there dealt with the development, implementation, and application of the finite element method and absorbing boundary conditions in modeling electromagnetic radiation and scattering from arbitrary three-dimensional structures. From 1995 to 1996, he worked at Compact Software. He is presently with the HP-EEsof division of Hewlett Packard and works on the development of the HP HFSS (High-Frequency Structure Simulator) finite element modeling package for CAD simulation.

Leo C. Kempel is a senior research engineer in Mission Research Corporation's Electromagnetic Observables Sector. He received his Ph.D. from the University of Michigan in 1994. In addition to conducting research in scattering reduction, Dr.

Kempel developed the finite element-boundary integral method for singly-curved structures and modeled conformal antennas with complex material loading using the FEM. His current interest has expanded to antennas on doubly-curved conformal platforms, modeling anisotropic substrate, and to developing novel hybridization strategies designed to marry the best properties of the finite element method with other computational electromagnetics methods such as integral equations or physical optics.